Modelling Congenital Heart Disease

Gianfranco Butera • Silvia Schievano
Giovanni Biglino • Doff B. McElhinney
Editors

Modelling Congenital Heart Disease

Engineering a Patient-specific Therapy

Springer

Editors
Gianfranco Butera
Department of Pediatric Cardiology,
Cardiac Surgery and Heart Lung
Transplantation - ERN GUARD HEART
Bambin Gesù Hospital and Research
Institute
Rome, Italy

Giovanni Biglino
University of Bristol
Bristol Medical School
Bristol, UK

Silvia Schievano
UCL Institute of Cardiovascular Science and
Great Ormond St Hospital for Children
London, UK

Doff B. McElhinney
Stanford University
Lucile Packard Children's Hospital
Palo Alto, CA
USA

ISBN 978-3-030-88891-6 ISBN 978-3-030-88892-3 (eBook)
https://doi.org/10.1007/978-3-030-88892-3

This Springer imprint is published by the registered company Springer Nature Switzerland AG
The registered company address is: Gewerbestrasse 11, 6330 Cham, Switzerland

Foreword

Innovation in medicine requires the identification of a clinical need and technical solutions to address the identified problem to assure that the technical novelty brings benefit to the patient population in need. Doctors working in clinical practice and biomedical engineers need to work in optimal coordination in order to be productive and innovative. In real life, this interaction is very often full of misunderstandings, competition and even rivalry. Each professional group thinks that the other does not understand the problem and fails to get the point. For instance, the engineer might not understand that some risks of device failure may be acceptable in a clinical setting and sets test parameters so severely that a project might fail. Clinicians are not trained to understand technology and also might not understand failure modes, markets and business.

No doubt there is huge benefit in bringing clinicians and engineers together. A better understanding of the clinical problem by the engineers and an understanding by the clinicians of the technical opportunities will lead to better and faster innovation. Computational models have the potential to simulate behaviour of new devices or a modification of a surgical procedure better than animal experiments.

The dream of having patient-specific approaches with custom-made devices developed with virtual device testing could well come true using the computational techniques and power which are available already today.

This is one of the concepts of the *in silico medicine* which is gaining popularity due to its effectiveness, as it aims to be used for diagnosis, treatment and/or prevention of a disease, with a reduction in animal experimentation as well.

The authors of this book, whom we have been associated in the past, show that the fusion of biomedical engineering and clinical medicine already exists with applications in congenital heart diseases since the early 1990s.

On an important note, this book can be easily approached by both communities of clinicians and engineers given the introductory part where basic concepts are made clear to a general audience. In our opinion the authors' approach will lead to a fruitful and vast area of a today underexplored academic field.

Francesco Migliavacca
Department of Chemistry, Material and
Chemical Engineering "Giulio Natta"
Politecnico di Milano, Milan, Italy
francesco.migliavacca@polimi.it

Philipp Bonhoeffer
Science and Music, Montecastelli Pisano, Italy
bohop1705@gogglemail.com; pb@scienceandmusic.com

Preface

From Leonardo to Quantum Physics and Back to Leonardo

Leonardo Da Vinci is considered a world genius of all times. He investigated every sector of human knowledge. He was a painter, draughtsman, musician, sculptor and architect, but also a scholar of hydraulics, optics, urban planning, engineering and anatomy.

He always tried to search for the science behind the events occurring in nature. Furthermore he tried to use this knowledge to develop new tools in many aspects of life.

Obviously it was a time when the amount of knowledge was not as huge as today and it was somehow possible for a single person to gain knowledge across different aspects of science and life. In particular, for our purpose, he implemented the understanding from science, engineering and nature to understand each aspect and back to develop new tools.

From then, engineering and natural sciences (including anatomy and physiology) have followed an independent path leading to a great divide between them.

Nowadays, it is probably the right time to go back in order to look for an integrated approach to science and in particular to medicine. This is called the translational approach.

This book goes exactly in that direction. In fact, the book aims to present applications of engineering modelling (computational, mainly, but also experimental, including 3D printing) in the context of facilitating a patient-centred approach in treating congenital heart disease (CHD). Considering the unique anatomical arrangement pre/post repair in CHD, as well as the possibility of opting for different surgical strategies or selecting different devices (e.g. stents) for interventions, a patient-specific approach is certainly warranted in this area of medicine. Engineering techniques (e.g. computational fluid dynamics, fluid-structure interaction, structural simulations, virtual surgery, advanced image analysis, 3D printing) are now sufficiently ripe to be explored from a translational angle.

The effort here is to integrate the knowledge coming from two different disciplines: medical and engineering. In fact, more frequently during clinical daily life, engineering is helping to improve the understanding of the single patient and his or her peculiar anatomy and physiology. In this way it is possible to tailor the treatment on the basis of each subject characteristics. Furthermore, the field necessitates a dialogue between engineers and clinicians.

In that respect, involving in many of the chapters of the book a cardiologist and a bioengineer incorporates both voices in the description of state-of-the-art models for different CHDs.

Finally, nowadays, it is not possible that a single person retains all the needed knowledge. However, Leonardo's approach of deeply integrating both aspects is still valid and it is going to be able to uncover new and unexpected scenarios!

Rome, Italy Gianfranco Butera
London, UK Silvia Schievano
Bristol, UK Giovanni Biglino
Palo Alto, CA, USA Doff B. McElhinney

Contents

Part I

Basic Concepts

Overview of Computational Methods

1

Giovanni Maria Formato, Silvia Schievano, and Giovanni Biglino

Congenital heart disease (CHD) is characterized by heterogeneous anatomical arrangements where small differences can dramatically impact functional outcomes. Therefore, in order to make a correct diagnosis or plan an intervention, it is essential for the clinician to accurately visualize the anatomical structures and obtain good insight into functional alterations. Advanced medical imaging, based on ultrasound, computed tomography and magnetic resonance, provides exhaustive functional and anatomical information of the pathology, and is an essential tool in the clinical management of CHD. However, medical imaging cannot be used as a *predictive* tool.

Computational modelling can provide physiological insight and predictive information. Models are an approximation of the reality; in engineering, the physics of a complex phenomenon can be modelled by partial differential equations (PDEs). For example, fluid motion is modelled by the Navier–Stokes equations, while electromagnetism is modelled by Maxwell's equations. Solving these equations provides a description of the model variables within the spatial domain; e.g., the solution of the Navier–Stokes equations provides the map of the velocity vectors and pressure of the fluid flow in all the points belonging to the domain. PDEs can be solved analytically only in simplified problems (e.g. steady flow in a perfectly cylindrical tube with rigid walls), providing a closed-form solution for the quantity of interest. However, in the cardiovascular system the physical domain is always characterized by a complex geometry (blood vessels are never straight), and often, the physics involved are so complex that several assumptions cannot be made without oversimplifying the problem. In these cases, PDEs must be solved numerically with the help of a computer.

Different computational techniques can be employed to model different physical phenomena. Structural simulations are used to model solid problems, such as the expansion of a stent

G. M. Formato
3D and Computer Simulation Laboratory, IRCCS Policlinico San Donato, San Donato Milanese, Italy

Bristol Heart Institute, Bristol Medical School, University of Bristol, Bristol, UK
e-mail: giovannimaria.formato@grupposandonato.it

S. Schievano
UCL Institute of Cardiovascular Science and Great Ormond St Hospital for Children, London, UK
e-mail: s.schievano@ucl.ac.uk

G. Biglino (✉)
University of Bristol, Bristol Medical School, Bristol, UK
e-mail: g.biglino@bristol.ac.uk

G. Butera et al. (eds.), *Modelling Congenital Heart Disease*,
https://doi.org/10.1007/978-3-030-88892-3_1

inside a vessel; fluid dynamic simulations are used to model flow problems, such as blood flowing through a stenosis; fluid–structure simulations are used to model coupled problems, e.g. the opening of a calcified aortic valve during ventricular ejection.

Computational modelling and simulation in CHD can both assist the clinicians preoperatively and be used as a research tool to investigate the pathophysiology in different congenital scenarios. For this reason, several clinical centres worldwide today work in connection with biomedical engineering laboratories to ensure a comprehensive and multidisciplinary approach to CHD.

1.1 General Framework

The transfer of information between the clinical and the engineering domain, the creation of computational models and then the translation from bench to bedside are a process requiring several fundamental steps, and each one may change depending on the anatomy being inspected, the clinical question being addressed and the methodological approaches preferred by each laboratory. Nonetheless, the general workflow is described here, highlighting possible differences stemming from different computational methodologies.

1.1.1 Geometry Generation

The first step consists of building a 3D geometric representation of the anatomical structure of interest, whose boundaries delineate the physical space where the PDEs are solved. Depending on the type of simulation, the computational domain of interest changes, and thus different elements are reconstructured from the available medical images. For instance, in a structural simulation of a vessel wall deformation after stent deployment, the boundaries of the model are the internal and external wall surfaces, while in the haemodynamic simulation the boundary would be the luminal surface.

Patient-specific models are reconstructed from medical images in two steps. Firstly, image segmentation identifies pixels of each image in the data set sharing a common characteristic (e.g. the colour) and delineates the contours. Secondly, the contours are used to build a 3D surface, which, depending on the technique, can be discrete (i.e. represented by a certain number of connected triangles in the space) or analytical (i.e. represented by 3D analytical functions such as non-uniform rational B splines—NURBS).

1.1.2 Computational Mesh

Once the geometry of the model is defined, the spatial domain must be discretized into a computational mesh. The volume is broken down into a collection of elementary cells and points, called elements and nodes, respectively. The equations are then solved inside the elements, and the solution is given as a collection of values at the nodes. The impact of the mesh on the quality of the final solution can be thought as that of pixel size on image resolution. Fine meshes are able to better capture the characteristics of a physical phenomenon, like a high-resolution image contains more details of the impressed scene. On the other hand, meshes with a large number of elements result in problems with a large number of unknowns, which require longer computational time or the use of supercomputers. Therefore, it is recommended to use fine meshes where the solution has a clinical interest (e.g. at the vessel walls and at the site of stenosis), and a coarser mesh elsewhere to reduce computational cost. Moreover, it is crucial to ensure that the numerical solution is mesh-independent. It is good practice to carry out a mesh independence study ("sensitivity analysis"), whereby solutions are iteratively obtained with increasingly finer meshes until no variation is observed, hence achieving the spatial convergence of the results.

1.1.3 Material Properties

The next step consists of assigning the material properties of the constituent parts of the model, namely the constitutive equations. This phase demands the development of tissue constitutive models, which can vary depending on the type of tissue and its location.

For structural simulations [1], the deformation of soft tissues is represented by reversible constitutive models to reflect the elastic nature of the material. Linear elasticity has simple implementation and computational speed, but can be used only with very small deformations. Hyperelastic models can be adopted for tissues undergoing large deformations; here, the stress–strain relationship is nonlinear, but irreversible processes are still not allowed. Finally, isotropic material models are simpler and faster, although the anisotropy of soft tissues should be considered by including one or more layers of fibres in the material.

For fluid dynamic simulations [2], the assumption that the blood in large arteries behaves as a Newtonian fluid is reliable only with high shear rates; otherwise, the aggregation of the erythrocytes can cause the blood to have a pseudoplastic behaviour, which requires the use of more complex constitutive models.

Remarkably, the adopted model depends on the scale of the problem. If macroscopically the biological materials are often considered homogeneous, at smaller scales the presence of microscopic entities should be taken into account.

1.1.4 Boundary Conditions

Finally, the model must be completed by imposing boundary conditions. Boundary conditions are the values of the solution (or of its spatial derivatives) of the PDEs prescribed at the domain boundaries. For instance, in structural simulations, boundary conditions are prescribed as displacements, while in fluid dynamic simulations they are imposed on the velocity or its spatial derivative, i.e. the pressure. There are multiple possibilities for the application of the boundary conditions, depending on the type of simulation, model complexity and availability of clinical data.

In fluid dynamic simulations, the typical boundary conditions are the velocity waveform at the inlet and the pressure waveform at the outlet. The blood velocity waveform can be estimated non-invasively and, thus, used as patient-specific boundary condition, whereas the pressure waveform is typically modelled using different approaches, such as a constant pressure value or a Windkessel lumped parameter model. However, the Navier–Stokes equations can also be solved by applying the velocity waveforms both at the inlet and at the outlets. In this case, particular attention must be reserved in order to satisfy the mass conservation law for the blood flow.

1.1.5 Computational Simulation

Having provided the correct boundary condition and material properties, and having discretized the spatial domain into a computational mesh, the PDEs can be solved to model the physical phenomenon of interest. The computational methods used to discretize the equations differ based on the mathematical formulation underpinning them. Traditionally, the finite element method is used for structural problems while the finite volume method is used for flow problems, although there are numerous examples of finite element techniques adopted in computational fluid dynamic codes. Nevertheless, the solution of the PDEs relies on iterative schemes, which seek for the solution until a fixed accuracy is reached. For a broader description of numerical methods used for engineering computations, the reader is referred to Ref. [3].

1.1.6 Post-Processing of Results

The final phase of the simulation workflow consists of the post-processing of the results. The solution of the equations provides the value of the variables of interest at all the points of the computational mesh. Although such quantities suffice to describe the physics of the phenomenon, they can further be combined to build metrics that, in turn, can be particularly meaningful in different clinical scenarios. In the following section, a review of such quantities and their clinical importance for cardiovascular applications is presented.

1.2 Clinical Significance of Computational Modelling for Cardiovascular Application

1.2.1 Computational Fluid Dynamic Simulations

The key role of haemodynamics on the development and progression of cardiovascular pathologies has been extensively demonstrated. Many cardiovascular disorders occur in the presence of "disturbed" flow in different geometrical configurations. The carotid and coronary bifurcations, the inner aortic arch and the insertion points of aortic branches are preferred sites for endothelial lining lesions [4], while the turbulence downstream an artificial valve can promote platelet adhesion to the valve surface [5]. In essence, a complex geometry leading to changes in flow can also lead to biological events.

Blood flow velocity and pressure fields from CFD simulation can be mapped inside the anatomic structure with a spatial and temporal resolution unachievable by any clinical methodology. The velocity field is used to draw the streamlines and visualize flow patterns, which is particularly useful to evaluate detailed haemodynamics or predict these during the design of cardiovascular devices [6] and interventional planning [7]. On the other hand, the blood pressure field on an aneurysm can provide significant information on risk of rupture [8], or be used to compute the power loss across a cardiovascular component, such as a stenotic artery [9], the right ventricle in a repaired tetralogy of Fallot [10] or the total cavopulmonary connection junction in the Fontan circulation [11]. In coronary artery disease, the computed pressure can also be used to derive the fractional flow reserve (FFR), overcoming the difficulties of the invasive procedure [12].

The velocity field can be used to derive other quantities also of clinical significance. The frictional force exerted by blood flow on the vessel wall is quantified by the wall shear stress (WSS), whose spatial distribution can be obtained by multiplying the gradient of the velocity and the blood viscosity. WSS is a vector, thus has a modulus, direction and verse that vary in time during the cardiac cycle. The importance of WSS on vascular remodelling and pathology has been extensively demonstrated: low and oscillatory WSS lead to endothelial dysfunction, which initiates the atherogenic cascade [13], and high WSS may damage the endothelium and lead to aneurysm formation [14] or dissection [15]. WSS-derived metrics have been adopted to synthesize the temporal evolution of WSS and describe a specific biological effect. For instance, time-averaged WSS (TAWSS) is the temporal average of the WSS modulus, Oscillatory Shear Index (OSI) describes the oscillation of the WSS and particle residence time (PRT) is linked to the stagnation of blood particles at the vessel wall.

Finally, CFD can be used to describe the bulk flow topology. Vorticity is the curl of the velocity field and measures the local rotation of fluid particles. Importantly, vorticity does not necessarily imply the presence of a vortex in the flow, rather being related to the spinning of each fluid particle. In viscous laminar flows, fluid particles move straight but have non-zero vorticity, while in the ocean's waves the fluid particles may move circularly but have zero vorticity. Nevertheless, the importance of such 4-dimensional structures has been shown in numerous cases, including drug transport from drug-eluting stent [16], or atherogenesis and thrombogenesis, being responsible for the transport of nutrients and oxygen from the blood to the wall [17, 18], the damping of WSS [19], platelet activation and adhesion [20], and the reduction in turbulence [21]. Therefore, a number of bulk flow descriptors (e.g. local normalized helicity, streamwise vorticity and spanwise vorticity) have been obtained by combing the velocity and vorticity fields, the first relating to the translational and the second to the rotational motion of the fluid particles [22].

1.2.2 Structural Simulations

While CFD provides a quantitative description of the haemodynamics and of the forces exerted by blood flow, for some applications it is more relevant to focus our analysis at the structural level, disregarding flow-related effects. Structural simulations allow to compute the displacement field occurring in a loaded structure, which in turn can be used to compute strain and stress distributions.

Structural simulations can support the design of cardiovascular devices such as stents. A specific strut design can be tested by simulating stent expansion and deriving, from the displacement field, geometric quantities like the radial and longitudinal recoil, the foreshortening and the dogboning of the stent [23]. The map of strain and stress can give information on the mechanical properties [24]. It has been shown that the mechanical forces exerted by the implanted stent on the vessel wall are linked to the risk of luminal injury, a factor leading to vascular restenosis [25]. Simulations of stent deployment can be used in the preoperative phase to virtually deploy the stent inside a specific vessel and estimate the stress distribution on the vessel wall, as well as the stent position, the lumen gain and the vessel straightening in the post-stenting configuration [26].

Structural simulations have also been extensively used as a tool to predict atherosclerotic plaque rupture. From a mechanical standpoint, plaque rupture represents a structural failure starting at regions of high stress due to morphological, material and geometrical discontinuities. It should be noted that histological features contribute to increasing stress concentration, including the lipid pool size [27], fibrous cap thickness [28] and the presence of microcalcifications [29, 30]. The type of stress to consider to build robust rupture criteria is still being debated. Maximum normal stresses and Von Mises stresses are inadequate to describe the complex stress state inside the plaque; thus, the maximum normal and shear stresses have been recommended [31].

1.2.3 Fluid–Structure Simulations

Fluid–structure simulations comprise both the fluid and solid domains: the blood flow exerts haemodynamic forces on the solid boundaries, which are able to deform and in turn exert mechanical forces on the fluid. The final solution provides information for both domains, and the same metrics described in the previous sections can be derived.

In this case, however, the measured quantity is affected by both the fluid flow and structural stress. It is increasingly recognized that considering both domains is important to ensure the clinical relevance of simulation results. For instance, including the compliance of the vascular walls affects the power loss computation in the Fontan circulation [32] or WSS distribution and magnitude in coronary arteries [33]. Over the past decade, there has been a push towards developing fluid–structure interaction models for a number of cardiovascular applications, including complex multiphysics models of heart contraction [34], Fontan procedure [35], cardiac valves [36], atherosclerotic plaque [37] and aneurysm rupture [38]. Such models can be integrated within more complex frameworks to provide additional insight. For instance, fluid–solid growth models include multiple time scales to predict vascular remodelling [39] or aneurysm growth [40], whereby, instead of fluid–structure simulations running at small time scale (e.g. order of seconds), remodelling stimuli metrics such as WSS or tensile stress are fed into mathematical models of vascular growth and remodelling operating at longer time scales (e.g. order of months). Although these models can give interesting insights into the complex mechanisms of tissue remodelling and growth, the time scale involved renders their clinical validation and applicability particularly challenging.

References

1. Owen B, Bojdo N, Jivkov A, Keavney B, Revell A. Structural modelling of the cardiovascular system. Biomech Model Mechanobiol. 2018;17(5):1217–12428.
2. Fedosov DA, Noguchi H, Gompper G. Multiscale modeling of blood flow: from single cells to blood rheology. Biomech Model Mechanobiol. 2014;13(2):239–58.

3. Chapra SC, Canale RP, et al. Numerical methods for engineers. Boston: McGraw-Hill Higher Education; 2010.
4. Morbiducci U, Kok AM, Kwak BR, Stone PH, Steinman DA, Wentzel JJ. Atherosclerosis at arterial bifurcations: evidence for the role of haemodynamics and geometry. Thromb Haemost. 2016;115(03):484–92.
5. Bluestein D, Yin W, Affeld K, Jesty J. Flow-induced platelet activation in a mechanical heart valve. J Heart Valve Dis. 2004;13(3):501–8.
6. Jiménez JM, Davies PF. Hemodynamically driven stent strut design. Ann Biomed Eng. 2009;37(8):1483–94.
7. Slesnick TC. Role of computational modelling in planning and executing interventional procedures for congenital heart disease. Can J Cardiol. 2017;33(9):1159–70.
8. Cebral JR, et al. Aneurysm rupture following treatment with flow-diverting stents: computational hemodynamics analysis of treatment. Am J Neuroradiol. 2011;32(1):27–33.
9. Liu X, et al. Functional assessment of the stenotic carotid artery by CFD-based pressure gradient evaluation. Am J Physiol-Heart Circ Physiol. 2016;311(3):H645–53.
10. Das A, Banerjee RK, Gottliebson WM. Right ventricular inefficiency in repaired tetralogy of Fallot: proof of concept for energy calculations from cardiac MRI data. Ann Biomed Eng. 2010;38(12):3674–87.
11. Dasi LP, et al. Fontan hemodynamics: importance of pulmonary artery diameter. J Thorac Cardiovasc Surg. 2009;137(3):560–4.
12. Morris PD, van de Vosse FN, Lawford PV, Hose DR, Gunn JP. "Virtual" (computed) fractional flow reserve. JACC Cardiovasc Interv. 2015;8(8):1009–17.
13. Malek AM, Alper SL, Izumo S. Hemodynamic shear stress and its role in atherosclerosis. JAMA. 1999;282(21):2035–42.
14. Dolan JM, Kolega J, Meng H. High wall shear stress and spatial gradients in vascular pathology: a review. Ann Biomed Eng. 2013;41(7):1411–27.
15. Tse KM, Chiu P, Lee HP, Ho P. Investigation of hemodynamics in the development of dissecting aneurysm within patient-specific dissecting aneurismal aortas using computational fluid dynamics (CFD) simulations. J Biomech. 2011;44(5):827–36.
16. Kolachalama VB, Tzafriri AR, Arifin DY, Edelman ER. Luminal flow patterns dictate arterial drug deposition in stent-based delivery. J Control Release. 2009;133(1):24–30.
17. Ma P, Li X, Ku DN. Convective mass transfer at the carotid bifurcation. J Biomech. 1997;30(6):565–71.
18. Sluimer JC, et al. Hypoxia, hypoxia-inducible transcription factor, and macrophages in human atherosclerotic plaques are correlated with intraplaque angiogenesis. J Am Coll Cardiol. 2008;51(13):1258–65.
19. Morbiducci U, Ponzini R, Grigioni M, Redaelli A. Helical flow as fluid dynamic signature for atherogenesis risk in aortocoronary bypass. A numeric study. J Biomech. 2007;40(3):519–34.
20. Zhan F, Fan Y, Deng X. Swirling flow created in a glass tube suppressed platelet adhesion to the surface of the tube: its implication in the design of small-caliber arterial grafts. Thromb Res. 2010;125(5):413–8.
21. Moffatt HK, Tsinober A. Helicity in laminar and turbulent flow. Annu Rev Fluid Mech. 1992; 24(1):281–312.
22. Morbiducci U, et al. Quantitative analysis of bulk flow in image-based hemodynamic models of the carotid bifurcation: the influence of outflow conditions as test case. Ann Biomed Eng. 2010;38(12):3688–705.
23. Migliavacca F, Petrini L, Colombo M, Auricchio F, Pietrabissa R. Mechanical behavior of coronary stents investigated through the finite element method. J Biomech. 2002;35(6):803–11.
24. Etave F, Finet G, Boivin M, Boyer J-C, Rioufol G, Thollet G. Mechanical properties of coronary stents determined by using finite element analysis. J Biomech. 2001;34(8):1065–75.
25. Welt FGP, Rogers C. Inflammation and restenosis in the stent era. Arterioscler Thromb Vasc Biol. 2002;22(11):1769–76.
26. Auricchio F, Conti M, De Beule M, De Santis G, Verhegghe B. Carotid artery stenting simulation: from patient-specific images to finite element analysis. Med Eng Phys. 2011;33(3):281–9.
27. Imoto K, et al. Longitudinal structural determinants of atherosclerotic plaque vulnerability. J Am Coll Cardiol. 2005;46(8):1507–15.
28. Virmani R, Kolodgie FD, Burke AP, Farb A, Schwartz SM. Lessons from sudden coronary death: a comprehensive morphological classification scheme for atherosclerotic lesions. Arterioscler Thromb Vasc Biol. 2000;20(5):1262–75.
29. Bluestein D, et al. Influence of microcalcifications on vulnerable plaque mechanics using FSI modeling. J Biomech. 2008;41(5):1111–8.
30. Maldonado N, et al. A mechanistic analysis of the role of microcalcifications in atherosclerotic plaque stability: potential implications for plaque rupture. Am J Physiol-Heart Circ Physiol. 2012;303(5):H619–28.
31. Holzapfel GA, Mulvihill JJ, Cunnane EM, Walsh MT. Computational approaches for analyzing the mechanics of atherosclerotic plaques: a review. J Biomech. 2014;47(4):859–69.
32. Orlando W, Shandas R, DeGroff C. Efficiency differences in computational simulations of the total cavopulmonary circulation with and without compliant vessel walls. Comput Methods Programs Biomed. 2006;81(3):220–7.
33. Malvè M, García A, Ohayon J, Martínez MA. Unsteady blood flow and mass transfer of a human left coronary artery bifurcation: FSI vs. CFD. Int Commun Heat Mass Transf. 2012;39(6):745–51.
34. Baillargeon B, Rebelo N, Fox DD, Taylor RL, Kuhl E. The Living Heart Project: a robust and integrative simulator for human heart function. Eur J Mech – A Solids. 2014;48:38–47.

35. Long CC, Hsu M-C, Bazilevs Y, Feinstein JA, Marsden AL. Fluid-structure interaction simulations of the Fontan procedure using variable wall properties. Int J Numer Methods Biomed Eng. 2012;28(5):513–27.
36. De Hart J, Peters GWM, Schreurs PJG, Baaijens FPT. A three-dimensional computational analysis of fluid–structure interaction in the aortic valve. J Biomech. 2003;36(1):103–12.
37. Dalin T, et al. Sites of rupture in human atherosclerotic carotid plaques are associated with high structural stresses. Stroke. 2009;40(10):3258–63.
38. Borghi A, Wood NB, Mohiaddin RH, Xu XY. Fluid–solid interaction simulation of flow and stress pattern in thoracoabdominal aneurysms: a patient-specific study. J Fluids Struct. 2008;24(2):270–80.
39. Figueroa CA, Baek S, Taylor CA, Humphrey JD. A computational framework for fluid-solid-growth modeling in cardiovascular simulations. Comput Methods Appl Mech Eng. 2009;198(45–46):3583–602.
40. Watton PN, Hill NA, Heil M. A mathematical model for the growth of the abdominal aortic aneurysm. Biomech Model Mechanobiol. 2004;3(2):98–113.

Congenital Heart Diseases: Basic Concepts from a Pediatric Cardiology Perspective

2

Mario Giordano and Gianfranco Butera

2.1 Anatomy and Segmental Analysis of Congenital Heart Diseases

A segmental analysis of the different heart structures is required to have a comprehensive approach to congenital heart disease. The needed steps for a complete segmental analysis are as follows: thoracoabdominal situs, cardiac position, atrial situs, ventricular loop, great artery position, atrioventricular (AV) and ventriculoarterial (VA) connections, infundibular anatomy, and associated anomalies (Fig. 2.1).

The thoracoabdominal situs describes the position of asymmetrical viscera inside the thoracic and abdominal cavities. Three different types of situs may be identified: solitus, inversus, and ambiguous. Situs solitus is the most common, and it is characterized by a right-sided position of liver with a left-sided stomach and spleen. The right lung is trilobed, while the left one is bilobed, and the right bronchus is short and epiarterial, whereas the left one is long and hypoarterial. Situs inversus is a complete inversion of situs solitus, characterized by a left-sided liver and a right-sided stomach and spleen. The left lung is trilobed, while the right one is bilobed, and the left bronchus is short and epi-arterial, whereas the right one is long and hypo-arterial. In some cases, neither a solitus nor an inversus deployment of viscera may be identified and the situs is defined ambiguous. The two most subsets of situs ambiguous are as follows: right isomerism (bilateral trilobed lungs, large symmetric median liver, and asplenia) and left isomerism (bilateral trilobed lungs, small symmetric median liver, and polysplenia). Right isomerism is usually characterized by bilateral superior vena cava, a total anomalous return of pulmonary veins, and severe intracardiac malformations (often with a functionally univentricular heart), whereas left isomerism shows a bilateral superior vena cava with an interruption of inferior vena cava with azygos/hemiazygos continuation and extracardiac malformations (severe intracardiac malformation is less frequent) [1, 2].

The cardiac orientation within the thorax is described as levocardia, dextrocardia, and mesocardia [3]. In levocardia, the most of heart is within the left hemithorax and the cardiac apex is left-sided, whereas in dextrocardia the heart is predominantly in the right hemithorax and the apex is right-sided. The prevalent position within the hemithorax and the apex direction are usually concordant; however, a discordance is possible. In mesocardia, the heart has a median position

M. Giordano (✉)
Department Pediatric Cardiology, Monaldi Hospital, University of Campania L. Vanvitelli, Naples, Italy

G. Butera
Department of Pediatric Cardiology, Cardiac Surgery and Heart Lung Transplantation - ERN GUARD HEART, Bambin Gesù Hospital and Research Institute, Rome, Italy

G. Butera et al. (eds.), *Modelling Congenital Heart Disease*,
https://doi.org/10.1007/978-3-030-88892-3_2

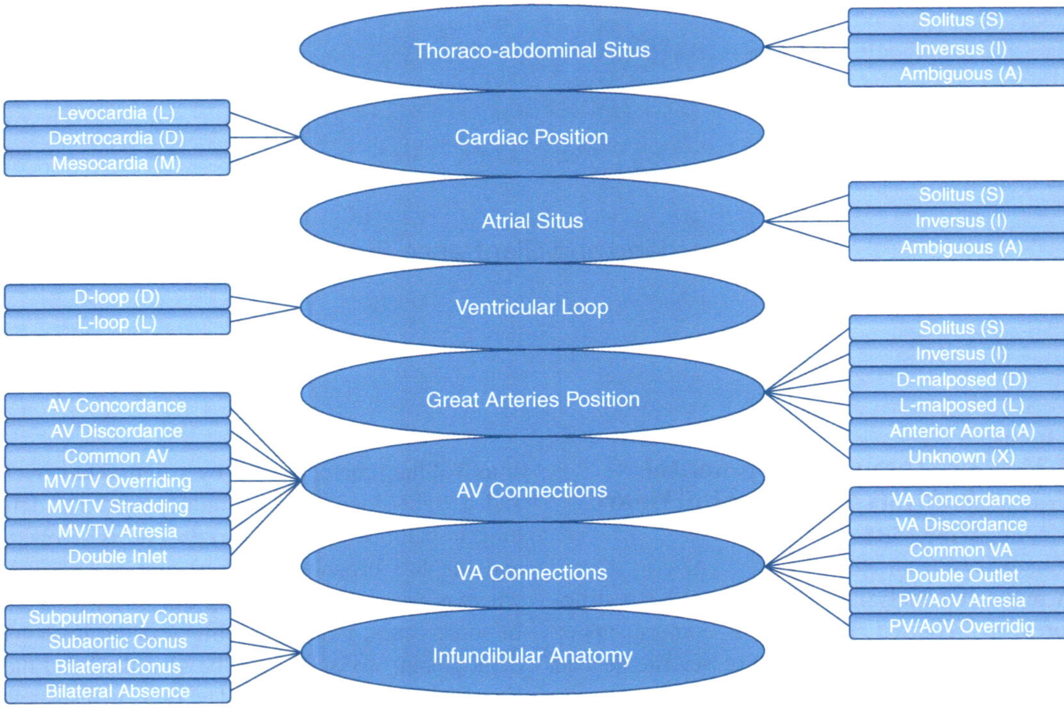

Fig. 2.1 Step-by-step anatomical segmental analysis to identify a CHD. *AV* atrioventricular, *MV* mitral valve, *TV* tricuspid valve, *VA* ventriculoarterial, *PV* pulmonary valve, *AoV* aortic valve

below the sternum and the apex is facing down. Rarely, the heart can be partially or completely exteriorized, and this condition is known as ectopia cordis.

Atrial situs is related to thoracoabdominal situs. Right atrium shows an appendage with broad base and an extension of pectinate muscles beyond the base, while the left atrium has a long narrow appendage with confined pectinate muscles.

The determination of ventricular loop requires to identify the morphologically right ventricle and the morphologically left one. Right ventricle is triangular in shape and characterized by septal attachments of the tricuspid valve, coarse trabeculations, and a distinct septal surface that includes the moderator bands. Left ventricle has a conic shape with fine inner apical trabeculations, papillary muscles attach to the left ventricular free wall without attachments to the interventricular septum, and a fibrous continuity in between mitral ring and the aortic annulus is present. The recognition of ventricular morphology allows to identify the ventricular loop. In ventricular D loop (right-handedness), the right ventricle wraps around the left ventricle simulating a right-hand orientation with the palmar surface placed over the right ventricle, the thumb in the tricuspid valve, and the fingers in the right ventricular outflow tract. In ventricular L loop (left-handedness), the right ventricle wraps around the left ventricle simulating a left-hand orientation with the palmar aspect around the right ventricle, the thumb in the tricuspid valve, and the fingers in the right ventricular outflow tract. Ventricular loop is related to the embryologic rotation of cardiac tube, and D loop is considered the normal orientation [4].

The usual spatial relationship between the great arteries is characterized by the aorta placed posteriorly and right-sided, while the pulmonary artery is anterior and left-sided. Anomalous relationships are as follows: inversus (posterior left-sided aorta and anterior right-sided pulmonary artery), D-malposed (anterior right-sided aorta and posterior left-sided pulmonary artery), L-malposed (anterior left-sided aorta and posterior right-sided pulmonary artery), anterior deployment (anterior aorta and posterior pulmonary artery), and side by side and unknown placement.

Normal atrioventricular valve alignment and connection are characterized by a mitral valve between the left atrium and the left ventricle, and a tricuspid valve between the right atrium and the right ventricle. Anomalous AV connections are as follows: atrioventricular discordance (with a tricuspid valve between a right atrium and a left ventricle and a mitral valve between a left atrium and a right atrium), a common AV valve, a double inlet ventricle (with both AV valves connected to a single ventricle), and a mitral/tricuspid valve atresia. Malalignments of AV valves are overriding and straddling. The AV "overriding" is characterized by a valve committed to two ventricular chambers, whereas the AV "straddling" is characterized by a valve with a chordal apparatus connected either to both sides of an interventricular septum or to papillary muscles of both ventricles [5].

A concordant ventriculoarterial connection is a left ventricle draining to the aorta and a right ventricle draining to the pulmonary artery. The switching of normal relationships between ventricles and great arteries determines a discordant VA connection (the aorta arising from a right ventricle and the pulmonary artery from the left one). Other anomalies of VA connections are as follows: a common VA valve (truncus arteriosus) or a double outlet arising from one ventricle (double outlet right ventricle and double outlet left ventricle). About VA alignment, the "overriding" is a possible phenomenon, whereas "straddling" is not possible as the absence of chordal attachment and subvalvular apparatus.

The assessment of infundibular anatomy is the last step of segmental analysis of CHDs. The usual anatomy is characterized by a muscular infundibulum (subpulmonary conus) in the outlet of right ventricle, whereas a direct fibrous continuity connects the aortic and the mitral valves. Anomalies of infundibular anatomy are as follows: a single subaortic conus, a bilateral conus (both subaortic and subpulmonary), and a bilateral absence of muscular conus with bilateral fibrous continuity between the AV valve and the homolateral VA valve.

The segmental analysis allowed to achieve a complete anatomical assessment of congenital heart diseases; however, the identification of associated cardiac anomalies is necessary for a comprehensive pathophysiological approach.

2.2 Pathophysiology and Hemodynamic of Congenital Heart Diseases

A pathophysiological approach to CHDs allowed to identify five most categories: left-to-right shunt lesions, right-to-left shunt lesions, right-sided obstructive lesions, left-sided obstructive lesion, and complex congenital lesions.

Shunt lesions. In humans, a normal cardiovascular circulation is characterized by a complete septation between oxygenated blood and deoxygenated blood with two in-series circulations. An anomalous connection between pulmonary and systemic circulation allows a blood passage from a kind of circulation to the other one by determining a "shunt." A simple pulmonary–systemic connection (atrial septal defects, ventricular septal defects, atrioventricular septal defects, patent arterial duct, etc.) develops a left-to-right shunt; however, the association of a right-sided obstruction determines the development of a right-to-left shunt (tetralogy of Fallot, pulmonary stenosis/atresia with ventricular septal defect). In left-to-right shunts, a portion of oxygenated blood flows from the left-sided circulation to the lungs. The augmented pulmonary output realizes a diastolic overload of right ventricle in the pre-tricuspid shunt and of left ventricle in the post-tricuspid shunt; moreover, the increased amount of pulmonary blood directed to the lungs determines a progressive increase in pulmonary vascular resistances with a higher risk to develop a pulmonary arterial hypertension over time (Fig. 2.2). In right-to-left shunts, a portion of deoxygenated blood

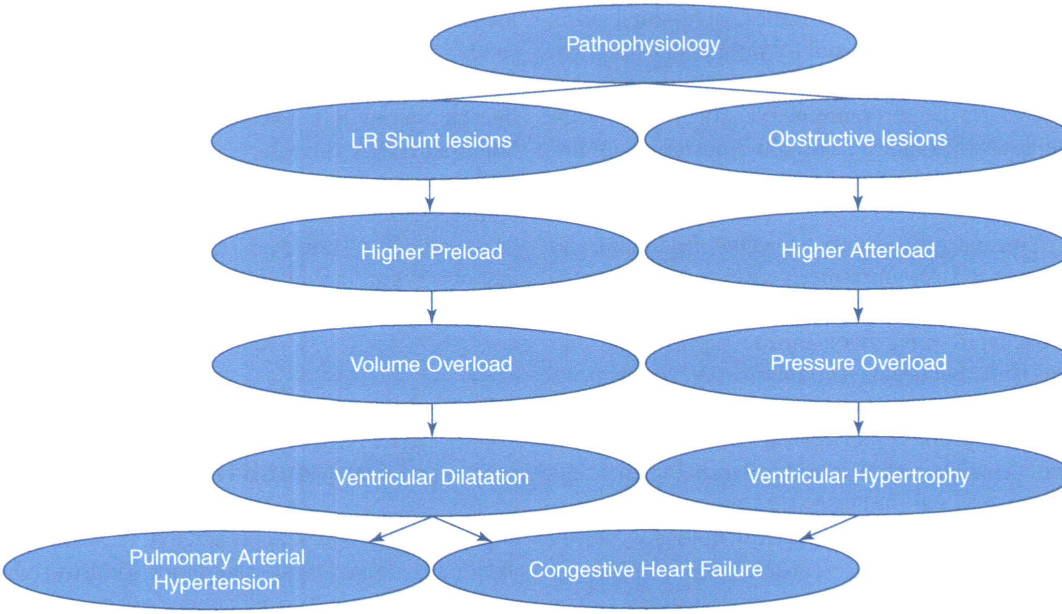

Fig. 2.2 Pathophysiology of shunt and obstructive lesions. Left-to-right (LR) shunts increase the preload with consequent volume overload and ventricular dilatation. Obstructive lesions increase the afterload with consequent pressure overload and ventricular hypertrophy. Without surgical or percutaneous correction, both conditions run up to congestive heart failure

flows from the pulmonary circulation to the systemic one and the most consequence is the onset of cyanosis. The development of a clinical evident cyanosis requires an amount of right-to-left shunt determining an increase in deoxygenated hemoglobin ≥5 g/dl; however, even small portions of right-to-left shunts may cause cerebral/systemic ischemic lesions due to episodes of paradoxical embolization. The main left-to-right lesions are as follows: partial anomalous pulmonary venous return, atrial septal defects, ventricular septal defects, atrioventricular septal defects, patent arterial duct, aortopulmonary collateral arteries, coronary fistula, aortopulmonary window, and truncus arteriosus. The amount of pulmonary hyperflow is related to the impedance through the defect and the pressure gradient between the connected chambers. Impedance is greatly associated with the dimension of defect even if the length of an anomalous vessel may have a non-negligible impact such as in the case of patent arterial duct or aortopulmonary collateral arteries. The main right-to-left lesions are as follows: total anomalous pulmonary venous return, pulmonary arteriovenous malformation, tetralogy of Fallot, pulmonary stenosis/atresia with ventricular septal defect, and Eisenmenger syndrome. In total anomalous pulmonary venous return, both systemic and pulmonary veins return into the right atrium and the mixed-blood shunts from this atrium to an "empty" left atrium realizing the desaturation of systemic blood. The Eisenmenger syndrome is the last step of left-to-right shunt natural history. The progressive increase in pulmonary vascular resistances due to pulmonary hyperflow determines a shunt inversion through the defect when these ones exceed the systemic vascular resistances. The development of Eisenmenger syndrome is a prognostically unfavorable event that limits the feasibility of CHD correction. In this stage, the defect is necessary to achieve a decompression of right ventricle and a pharmacologic therapy addressed to a drop of pulmonary vascular resistances is the only feasible strategy [6].

Obstructive lesions. The normal human circulation is characterized by a free flow from the right and the left ventricle to the pulmonary artery and the aorta, respectively. The pulmonary valve and the aortic valve allow the blood output during systole and avoid significant regurgitation from great arteries to ventricles with their closure during diastole. The presence of an obstructive lesion might compromise the normal function of ventricles and the pulmonary/systemic output. Congenital obstruction causes an increase in afterload with consequent systolic ventricular overload. A ventricular hypertrophy occurs with a consequent reduction in compliance and increased filling pressures. In this phase, the diastolic dysfunction determines a pulmonary congestion in left-sided obstruction and a systemic congestion (above all with a hepatic involvement) in right-sided obstruction. If the obstruction persists, the natural history of disease evolves to a systolic dysfunction with reduced stroke volume and pulmonary or systemic hypoperfusion (Fig. 2.2). More distal obstructions cause further hemodynamic changes due to a maldistribution of blood flow (like in a single pulmonary artery stenosis). The pressure gradient across the lesion reflects the degree of obstruction; however, a complete approach required the evaluation both of cardiac output and of a potential blood maldistribution for the distal lesions. In cases with low stroke volume and/or non-negligible blood flow maldistribution, the pressure gradient might underestimate the importance of lesion. This mistake is frequent with a significant systolic dysfunction and/or with a single pulmonary artery stenosis. In this last case, the unbalanced flow through the pulmonary arteries may determine an underestimation of pressure gradient across the obstructed vessel because of the less blood flow than the contralateral one. Conditions with increased right ventricle afterload are as follows: subvalvular, valvular, and supravalvular pulmonary stenosis, and pulmonary branch stenosis. The increased right ventricle afterload determines an increase in end-diastolic pressure. The relationship between right and left ventricle contributes to a consequence progressive increase in left ventricular end-diastolic pressure and pulmonary wedge pressure. These anomalies may reflect pulmonary circulation with the development of a paradoxical postcapillary pulmonary hypertension. Another consequence of right afterload is a right ventricle myocardial ischemia. The hypertrophy causes both higher oxygen requirements and lower myocardial perfusion with consequent ischemia and augmented risk of ventricular arrhythmia, syncope, and sudden cardiac death. Conditions with increased left ventricle afterload are as follows: subvalvular, valvular, and supravalvular aortic stenosis, and aortic coarctation/interruption. As right ventricle systolic overloads, the left ventricle replies to obstruction with a hypertrophy and increased filling pressures. These hemodynamic changes have a direct effect on pulmonary circulation with a high risk of pulmonary edema, particularly in higher output states. A left ventricle ischemia is possible even in left-sided obstruction with the same previous mechanisms. Right and left ventricles are different chambers. Right ventricle may be considered a chamber with a "reservoir" function, and it shows a good tolerance to volume overload (high preload) and a worse tolerance to pressure overload (high afterload); contrary, left ventricle has a "pump" function and it shows a good tolerance to pressure overload (high afterload) and a worse tolerance to volume overload (high preload). These different hemodynamic and pathophysiological evidences highlight the reasons of the better right ventricle adaptation to pre-tricuspid shunt lesions rather than to obstructive lesions. Cyanosis is not a typical sign of right obstructive lesions; however, if a site of shunt (atrial septal defect, patent foramen ovale, etc.) is associated with obstruction, the high right ventricle pressures may provoke a right-to-left shunt with consequent cyanosis [7].

Complex congenital heart diseases. Some complex CHDs may be a various association of shunt lesion, severe obstruction, chamber hypoplasia, and AV or VA discordance. Every patient should be considered differently because small variation of lesions might determine a significant change in hemodynamic and pathophysiology. The most complex CHDs are as follows: transposition of great artery (TGA), congenitally corrected transposition of great artery (ccTGA), Ebstein's anomaly, and univentricular heart (UVH) diseases.

In TGA, pulmonary artery and aorta arise from left ventricle and right ventricle, respectively. The consequence is the development of two divided and parallel circulation: the pulmonary circulation with oxygenated blood and the systemic one with deoxygenated blood. The patient survival is ensured by the presence of shunt sites that allow a mixing of circulations. In this CHD, the concept of anatomical shunt should be used to avoid misunderstanding. The term anatomical left-to-right shunt involves the amount of blood, which passes from the left to the right side of the heart. During fetal life, the most oxygenated and glucose-rich blood catches up the ascending aorta and then the brain through the foramen ovale right-to-left shunt. With TGA, this oxygenated blood delivers the splanchnic system through the arterial duct, whereas the less oxygenated blood is directed to brain. This pathophysiology might explain the lower head circumference, the altered brain metabolism, and the tendency to be rather large and heavy at birth [8–10]. In the postnatal period, as the two circulations run in parallel, a shunting between them allows an adequate mixing. The rule of interatrial shunt and arterial duct is crucial in TGA with intact ventricular septum. At birth, a bidirectional shunting is through interatrial septum, with a right-to-left shunt during atrial systole and a left-to-right shunt during early diastole. Also, breathing phases influence the direction of shunt with a predominant right-to-left shunt during inspiration and a predominant left-to-right shunt during expiration. Arterial duct shows an exclusive shunting mechanism in postnatal period. Immediately after birth, arterial duct has a bidirectional shunting (pulmonary to aorta in systole and aorta to pulmonary in diastole). The high saturation of pulmonary blood realizes a rapid fall of vascular pulmonary resistances with consequent reduction in pulmonary pressure. This mechanism determines a complete aorta-to-pulmonary shunt through the arterial duct. This total aorta-to-pulmonary shunt results in a large volume of pulmonary flow with consequent increase in left atrial pressure, which might force the flap of septum primum to septum secundum with progressive reduction until elimination of interatrial shunting through the foramen ovale. This mechanism is valid in neonates with a patent foramen ovale and a percutaneous atrioseptostomy is required to avoid the loss of an important source of shunt, while in neonates with a large atrial septum defect, the higher left atrial pressures may just increase the interatrial left-to-right shunt and the systemic blood saturation without the risk of an interatrial shunting reduction [11]. In TGA with ventricular septal defect (VSD), representing 40% of cases [12], interventricular shunting is an important source of mixing and it is often enough to achieve an acceptable systemic blood saturation. After the falling of vascular pulmonary resistances, VSD realizes a systolic right-to-left shunt and diastolic left-to-right. In some cases, VSD is undersized and an adjunctive source of shunting (atrial septum) is required to catch up adequate systemic blood saturation [13].

TGA is a complex CHD requiring a surgical correction within neonatal age. Differently, ccTGA is often diagnosed in adult age occasionally. This CHD is characterized by in-series circulations with a normal physiology; however, the aorta arises from a morphologically right ventricle left-sided and the pulmonary artery from a morphologically left ventricle right-sided. Since the initial physiology is preserved, patients do not have cardiac symptoms or signs until the onset of a late failure of the right ventricle and/or the development of significant tricuspid valve regurgitation. In this subset of patients, the right ventricle loses progressively the capability to sustain the systemic circulation characterized by high vascular resistances and progressive right chamber dilation occurs. The right ventricle dilation concurs with an enlargement of the tricuspid valve annulus and worsening of the tricuspid regurgitation [14]. The significant tricuspid regurgitation produces an increase in right ventricle preload, which worsens the right chamber dilation, which further stretches the tricuspid annulus. This "vicious circle" determines the progressive worsening of systemic right ventricle function and of hemodynamic of patient. Associated anomalies occur in well over 90% of cases and the most frequent are as follows: ventricular septal defect, left ventricular outflow tract obstruction (subpulmonary obstruction), and anomalies of the left-sided tricuspid valve [15]. Within 45 years old, 67% of patients with ccTGA and associated lesions develops a congestive heart failure, whereas just 25% of patients without associated lesions show this complication, highlighting the influence of the other anomalies on the prognosis of this population [16].

Ebstein's anomaly is characterized by a significant apical dislocation ($\geq$0.8 cm/m^2) of septal leaflet of the tricuspid valve. Tricuspid valve coaptation is rarely preserved with consequent at least moderate tricuspid regurgitation. The septal leaflet dislocation determines an atrialization of part of right ventricle with consequent huge right atrium and small right ventricle, which diminishes diastolic compliance with increased right filling pressure. Right atrium pressure is higher and worsened by tricuspid regurgitation. This hemodynamic setting produces a right congestive heart failure with hepatic congestion. A concomitant atrial septal defect may determine a right-to-left shunting with consequent cyanosis (above all on effort, when cardiac output is increased and the right ventricle has to accept more volume) [12].

The concept of "UVH physiology" (or "functionally UVH") concerns both the hearts with only one ventricle and the hearts characterized by an adequately developed ventricle and one hypoplastic ventricle unable to sustain either the systemic or the pulmonary circulation [17]. By this definition, CHDs such as hypoplastic left heart syndrome, pulmonary atresia with intact ventricular septum, and severely reductive right ventricle or Ebstein's anomaly with extreme atrialization of the right ventricle are considered functionally UVH [18, 19]. Every CHDs with a functionally UVH may be presented with various pathophysiologic mechanisms according to the different anatomic settings and associated lesions. UVH is characterized by just one functionally active ventricle either with right morphology or with left morphology. At birth, this ventricle received a mixed blood that drains both to pulmonary circulation and to systemic circulation. In this phase, the aim is to achieve balanced pulmonary and systemic circulation (QP:QS $\approx$ 1.00). Systemic saturations of 75–85% suggest an adequate balancing between systemic and pulmonary blood flow [19, 20]. When systemic saturation is over 90%, the systemic blood flow may be reduced (with tissue hypoperfusion, metabolic acidosis, and a low cardiac output state), whereas the pulmonary blood flow may be increased with signs of congestive heart failure. These patients usually show a well-balanced circulation in the first days of life; however, when the pulmonary vascular resistances fall, a pulmonary hyperflow status compares. In these cases, a palliative surgical procedure (banding of pulmonary artery) is often necessary to decrease the pulmonary blood flow. Reversely, associated pulmonary obstructive lesions may reduce the pulmonary flow unbalancing the blood flow to systemic circulation, with an oxygen saturation $\leq$70% and cyanosis. In this setting, additional sources of pulmonary blood flow are required to increase oxygen saturation and to achieve an adequate grow-up of pulmonary arteries over time. Further source of pulmonary flow may be achieved either with a surgical systemic-to-pulmonary shunt [21] or with a percutaneous stenting of arterial duct [22].

Over years, the aim of surgical management of UVH is to achieve a complete separation of circulations. Within the first 6–12 months of life, a direct anastomosis of superior vena cava to pulmonary artery (bidirectional Glenn shunt) is performed; then, an anastomosis of inferior vena cava (Fontan operation) complete the surgical palliation, usually 4–6 years later [23]. In this setting, pulmonary blood flow is not supported by a ventricular pump since the systemic veins are directly connected to the pulmonary arteries (Fig. 2.3). A well function of Fontan circuit needs a normal ventricular function and low resistances in Fontan circuit (systemic vein and pulmonary artery) [24, 25]. A normal ventricular function provides the driving force for the circuit allowing a normal cardiac output. During diastole, the suc-

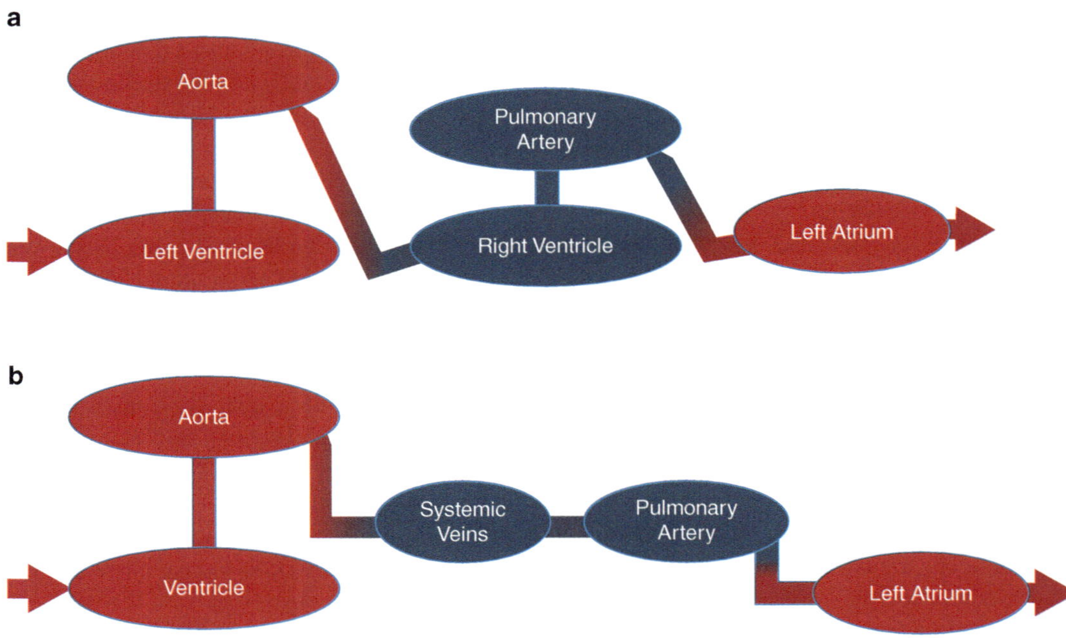

Fig. 2.3 Scheme of the normal cardiovascular circulation (**a**) and the Fontan circulation (**b**). (**a**) Normal biventricular circulation: Both pulmonary circulation and systemic one are supported by a ventricular pump. (**b**) Fontan circuit: The single ventricle supports both the systemic circulation (directly) and the nonpulsatile pulmonary circulation (indirectly) through the ventricular suction effect during diastolic phase. The line color reflects oxygen saturation

tion effect has an important rule providing a better blood flow to pulmonary vascular bed. In these ways, the ventricular contraction has both a direct rule and an indirect rule to allow a good functioning of Fontan circuit. Low resistances through the Fontan system are another key point. The vascular impedance exerted by every element of Fontan circuit is crucial to obtain efficient pulmonary blood flow. The most important elements to preserve are as follows: the surgical cavopulmonary anastomosis, the pulmonary arteries, the pulmonary capillary network, and the pulmonary veins with their atrial connections. Every narrowing of one or more elements might increase flow resistance contributing to a higher risk of Fontan failure. Extremely significant is the rule of pulmonary vascular resistances. Fontan circulation is a low-output system, and the pulmonary vascular resistances react to this low-flow status with a progressive increase up to the failing of circuit. The combination of high pulmonary vascular resistance index (>2 WU·m^2) and low cardiac index (<2.5 L/min/m^2) is an independent risk factor for Fontan failure [26].

2.3 Congenital Heart Diseases with Duct-Dependent Circulation

Arterial duct is a residue of sixth left aortic arch, which supports 80–90% of systemic output during the fetal life. With the onset of birth, the inflated lungs allowed a progressive complete separation of two circulations. The higher pulmonary blood flow increases the left atrial pressure with consequent juxtaposition of septum primum to septum secundum, with closure of foramen ovale. Within the first 15 h of life, arterial duct meets a functional closure as rapid vasoconstriction due to high blood oxygen tension. After several weeks, an intraluminal thrombosis provides an anatomical closure. A true persistence of arterial duct is defined by its presence beyond 3 months of age [27, 28].

Some CHDs need the persistence of arterial duct so that the patient's survival is guaranteed. These duct-dependent diseases are both isolated obstructive lesions (aortic coarctation/interruption, aortic stenosis, pulmonary atresia, etc.) and complex anatomical anomalies (hypoplasic left heart syndrome, Ebstein's anomaly, etc.). Three categories of duct-dependent CHDs may be identified: duct-dependent pulmonary circulation, duct-dependent systemic circulation, and duct-dependent TGA for blood mixing. In duct-dependent pulmonary circulation, lung perfusion is provided by arterial duct. A large arterial duct allows to achieve acceptable oxygen saturation. Severely hypoxic newborns might be useful to administer 100% oxygen since it is unlikely that these patients reach level of oxygen pressure such as to determine ductal vasoconstriction. Duct-dependent systemic circulations are a complex setting of CHDs wherein the systemic output is guaranteed by a persistent arterial duct. In newborns with critical aortic stenosis or hypoplastic left heart syndrome, a retrograde perfusion of aortic arch and ascending aorta is characteristic features of duct dependency. In such cases of TGA, arterial duct shows a double mechanism of blood mixing since it redirects desaturated systemic blood to lungs to improve oxygenation and it provides further source of mixing in the cases of high pulmonary pressure with consequent bidirectional shunt [29].

A strategy to achieve a ductal opening is the intravenous infusion of prostaglandin E1 (starting dose: 0.05–0.1 mcg/kg/min, maintenance dose: 0.002–0.05 mcg/kg/min); however, prostaglandins have several dose-related side effects (apnea, hyperthermia, and cutaneous vasodilatation are the most frequent) [30, 31]. An alternative change to achieve a long-lasting opening of arterial duct is to perform a percutaneous stenting. This procedure may be useful to maintain the ductal flow, even if a progressive neointimal hyperplasia realizes a closure of arterial duct over time. These patients need an antiplatelet therapy (ac. acetylsalicylic 3–5 mg/kg/die).

References

1. Fulcher AS, Turner MA. Abdominal manifestations of situs anomalies in adults. Radiographics. 2002;22:1439–56.
2. Van Praagh R, Van Praagh S. Atrial isomerism in the heterotaxy syndromes with asplenia, or polysplenia, or normally formed spleen: an erroneous concept. Am J Cardiol. 1990;66:1504.
3. Van Praagh R, Vlad P. Dextrocardia, mesocardia and levocardia: the segmental approach in congenital heart disease. In: Keith JD, Rowe RD, Vlad P (eds). Heart disease in infancy and childhood, 3rd edn. Macmillan, New York, 1978.
4. Jacobs JP, Anderson RH, Weinberg PM, Walters HL 3rd, Tchervenkov CI, Del Duca D, Franklin RC, Aiello VD, Béland MJ, Colan SD, Gaynor JW, Krogmann ON, Kurosawa H, Maruszewski B, Stellin G, Elliott MJ. The nomenclature, definition and classification of cardiac structures in the setting of heterotaxy. Cardiol Young. 2007;17(Suppl. 2):1–28.
5. Milo S, Ho SY, Macartney FJ, Wilkinson JL, Becker AE, Wenink AC, Gittenberger de Groot AC, Anderson RH. Straddling and overriding atrioventricular valves: morphology and classification. Am J Cardiol. 1979;44(6):1122–34.
6. Sommer RJ, Hijazi ZM, Rhodes JF Jr. Pathophysiology of congenital heart disease in the adult. Part I: Shunt lesions. Circulation. 2008;117(8):1090–9.
7. Rhodes JF, Hijazi ZM, Sommer RJ. Pathophysiology of congenital heart disease in the adult. Part II. Simple obstructive lesions. Circulation. 2008;117(9):1228–37.
8. Rosenthal GL. Patterns of prenatal growth among infants with cardiovascular malformations: possible fetal hemodynamic effects. Am J Epidemiol. 1996;143:505–13.
9. Limperopoulos C, Tworetzky W, McElhinney DB, et al. Brain volume and metabolism in fetuses with congenital heart disease: evaluation with quantitative magnetic resonance imaging and spectroscopy. Circulation. 2010;121:26–33.
10. Manzar S, Nair AK, Pai MG, Al-Khusaiby SM. Head size at birth in neonates with transposition of great arteries and hypoplastic left heart syndrome. Saudi Med J. 2005;26:453–6.
11. Rudolph AM. Congenital diseases of the heart: clinical- physiological considerations. Chichester, UK: Wiley-Blackwell; 2009.
12. Wernosvsky G. Transposition of the great arteries. In: Moss AJ, Allen HD, editors. Moss and Adams' Heart disease in infants, children, and adolescents including the fetus and young adult. 7th ed. Philadelphia, PA: Lippincott William & Wilkins; 2008. p. 1038–87.
13. Files MD, Arya B. Preoperative physiology, imaging, and management of transposition of the great arteries. Semin Cardiothorac Vasc Anesth. 2015;19(3):210–22.
14. Prieto LR, Hordof AJ, Secic M, Rosenbaum MS, Gersony WM. Progressive tricuspid valve disease in

patients with congenitally corrected transposition of the great arteries. Circulation. 1998;98:997–1005.
15. Presbitero P, Somerville J, Rabajoli F, et al. Corrected transposition of the great arteries without associated defects in adult patients: clinical profile and follow up. Br Heart J. 1995;74:57–9.
16. Graham TP Jr, Bernard YD, Mellen BG, Celermajer D, Baumgartner H, Cetta F, Connolly HM, Davidson WR, Dellborg M, Foster E, Gersony WM, Gessner IH, Hurwitz RA, Kaemmerer H, Kugler JD, Murphy DJ, Noonan JA, Morris C, Perloff JK, Sanders SP, Sutherland JL. Long-term outcome in congenitally corrected transposition of the great arteries: a multiinstitutional study. J Am Coll Cardiol. 2000;36:255–61.
17. Jacobs ML, Anderson RH. Nomenclature of the functionally univentricular heart. Cardiol Young. 2006;16(Suppl. 1):3–8.
18. Anderson RH, Cook AC. Morphology of the functionally univentricular heart. Cardiol Young. 2004;14(Suppl. 1):3–12.
19. Graham EM, Bradley SM, Atz AM. Preoperative management of hypoplastic left heart syndrome. Expert Opin Pharmacother. 2005;6(5):687–93.
20. Diller GP, Uebing A, Willson K, Davies LC, Dimopoulos K, Thorne SA, Gatzoulis MA, Francis DP. Analytical identification of ideal pulmonary systemic flow balance in patients with bidirectional cavopulmonary shunt and univentricular circulation: oxygen delivery or tissue oxygenation? Circulation. 2006;114:1243–50.
21. Salmon GW, Barkley HT. The Blalock-Taussig operation. J Pediatr. 1947;31(1):54–9.
22. Santoro G, Gaio G, Capozzi G, Giugno L, Palladino MT, Capogrosso C, D'Aiello AF, Caianiello G, Russo MG. Fate of hypoplastic pulmonary arteries after arterial duct stenting in congenital heart disease with duct-dependent pulmonary circulation. JACC Cardiovasc Interv. 2015;8(12):1626–32.
23. Castaneda AR. From Glenn to Fontan. A continuing evolution. Circulation. 1992;86(Suppl. 5):II80–4.
24. Gewillig M, Goldberg DJ. Failure of the Fontan circulation. Heart Fail Clin. 2014;10:105–16.
25. Gewillig M, Brown SC. The Fontan circulation after 45 years: update in physiology. Heart. 2016;102(14):1081–6.
26. Egbe AC, Connolly HM, Miranda WR, Ammash NM, Hagler DJ, Veldtman GR, Borlaug BA. Hemodynamics of Fontan failure: the role of pulmonary vascular disease. Circ Heart Fail. 2017;10:e004515.
27. Benson LN. The arterial duct: its persistence and its patency. In: Anderson RH, Baker EJ, Penny DJ, Redington AN, Rigby ML, Wernovsky G, editors. Paediatric cardiology. 3rd ed. London: Churchill Livingstone; 2010. p. 875–93.
28. Cassels DE, Bharati S, Lev M. The natural history of the ductus arteriosus in association with other congenital heart defects. Perspect Biol Med. 1975;18(4):541–72.
29. Costello JM, Franklin WF. Preoperative and postoperative care of the infant with critical congenital heart disease. In: MG MD, MMK S, Mullett MD, editors. Avery's neonatology: Pathophysiology & management of the newborn. 6th ed. Philadelphia (USA): Lippincott Williams & Wilkins; 2005. p. 710–67.
30. Lewis AB, Freed MD, Heymann MA, Roehl SL, Kensey RC. Side effects of therapy with prostaglandin E1 in infants with critical congenital heart disease. Circulation. 1981;64(5):893–8.
31. Holt L, McIntosh R. Congenital anomalies of the heart. In: Holt L, McIntosh R, editors. Holt's diseases of infancy and childhood: a textbook for the use of students and practitioners. New York: D. Appleton-Century Co.; 1933. p. 501–10.

Part II

Diseases: Modelling and Applications

3 Septal Defects: Clinical Concepts, Engineering Applications, and Impact of an Integrated Medico-Engineering Approach: Occluder Devices

Mario Giordano, Giorgia Bosi, and Gianfranco Butera

3.1 Introduction

The incidence of congenital heart disease (CHD) changes from about 4/1000 to 50/1000 live births, and it is estimated to be 6/1000 live births for moderate and severe forms of CHD. Septal defects represent the most common CHD. Isolated ventricular septal defect (VSD, incidence ≈ 4/1000) and isolated atrial septal defect (ASD, incidence ≈ 1/1000) are the most frequent of all CHD [1]. In some circumstances, small septal defects require neither surgical dealing nor transcatheter dealing; however, in other cases, they may produce symptoms or heart modeling with consequent necessity of closure. A comprehensive approach to the septal defects is necessary for a correct management and to choose the better therapeutic strategy available.

M. Giordano
Department Pediatric Cardiology, Monaldi Hospital, University of Campania L. Vanvitelli, Naples, Italy

G. Bosi
BioEngineering Department, University College of London, London, UK
e-mail: g.bosi@ucl.ac.uk

G. Butera (✉)
Department of Pediatric Cardiology, Cardiac Surgery and Heart Lung Transplantation - ERN GUARD HEART, Bambin Gesù Hospital and Research Institute, Rome, Italy

3.2 Ventricular Septal Defect

VSD is a direct communication between the right and left ventricles. Interventricular septum is a complex curvilinear structure divided into four main regions [2]:

- membranous septum, a thin fibrous septum between the septal leaflet of tricuspid valve and the right coronary/noncoronary cusps,
- inlet muscular septum, from the septal attachment of the tricuspid valve to the distal attachments of the tricuspid tensor apparatus,
- trabecular muscular septum, which corresponds to the septal triangle defined by the attachments of tricuspid leaflets, the apex, and the supraventricularis crista,
- outlet muscular septum (or infundibular septum), from the supraventricularis crista to the pulmonary valve.

Ventricular septum is crossed by the conduction system of heart. Atrioventricular node is placed at the apex of triangle of Todaro (formed by the tendon of Todaro, the ostium of coronary sinus, and the septal leaflet of tricuspid valve). From the atrioventricular node, the bundle of His crosses the membranous septum up to the crest of muscular septum, posteriorly to the anterior and the septal leaflets of tricuspid valve. At this level, the His divides into two bundles: left bundle branch and right bundle branch. The knowledge

G. Butera et al. (eds.), *Modelling Congenital Heart Disease*,
https://doi.org/10.1007/978-3-030-88892-3_3

of physiologic course of conduction system is crucial for a right approach to VSD. It is important to highlight that VSD may determine different anatomical patterns of conduction system according to the geography of the defect. Generally, the conduction system is placed along the posterior–inferior rim of all perimembranous defects, whereas it is far away from the inferior rim of other kinds of VSDs.

The complex anatomy of interventricular septum complicates the anatomical classification of the various VSD. Historically, different classifications were executed [3, 4]; however, recently, the International Society for Nomenclature of Pediatric and Congenital Heart Disease (ISNPCHD) develops a novel anatomical classification using both the anatomical location of the defect and the structures along the borders [5]. The goal of this categorization is to achieve a common VSD classification useful to describe every phenotype incorporating both anatomical geography and border features, giving an immediate crucial information to heart surgeons and interventional cardiologists. ISNPCHD identifies four main forms of VSD: perimembranous central, muscular trabecular, inlet VSD without a common atrioventricular valve/junction, and outlet VSD.

Perimembranous central VSD is located within the thin fibrous septum behind the septal leaflet of tricuspid valve and below the commissure between the right coronary and noncoronary cusps of aortic valve. Among the four main forms, this is the only type of VSD classified with the term "perimembranous" highlighting particularly the borders of defect rather than the geographic location. Perimembranous central VSD may have inlet or outlet extension. In the first case, the defect is classified as perimembranous central (perimembranous inlet type, to differentiate this kind of defect from the inlet canal-type VSD), whereas in the second case, the defect is classified as outlet VSD (outlet perimembranous type). This kind of defect is close to conduction system with high risk of injury to atrioventricular node and bundle of His with consequent atrioventricular block.

Inlet VSD interests the inlet interventricular septum, along the septal leaflet of tricuspid valve. This kind of defect is located between the medial papillary muscle, the posterior–inferior limb of the septal band, and the anteroseptal commissure of the tricuspid valve. Inlet VSDs are divided into perimembranous and muscular. Inlet perimembranous VSD involves the membranous septum, with a significant posterior–inferior dislocation of the atrioventricular node and the conduction system (this feature explains the basal anomalies of atrioventricular conduction at electrocardiogram). Since the conduction system runs along the posterior–inferior edge of the defect, the surgical suture line must be executed along the annulus of tricuspid septal leaflet to avoid an atrioventricular block. The inlet perimembranous defects are further subdivided into two different forms: with and without a malaligned atrioventricular septum. Inlet muscular defects have just muscular rims, without extension to membranous septum. The conduction system is not dislocated, and so, the risk of heart block is inferior. It is important to highlight that the inlet VSD with a common atrioventricular junction is considered an atrioventricular septal defect (and not an inlet VSD with a common atrioventricular valve).

Trabecular muscular VSD involves the trabecular muscular septum and it has always only muscular rims, without extension to membranous septum. This kind of defect undergoes often spontaneous closure. A complex anatomy may be present, with serpiginous course within the interventricular septum. Trabecular muscular VSDs are further subdivided into mid-septal, apical, posterior–inferior, and anterior–superior. Multiple trabecular muscular VSDs may coexist in different muscular sites. When numerous small trabecular muscular VSDs are present, the interventricular septum assumes a "Swiss cheese" aspect. Numerous authors agree that "Swiss cheese"-type VSDs are due to an embryologic lack of myocardial compaction, and for this reason, these defects do not undergo spontaneous closure [6].

Outlet VSD involves the interventricular septum between the limbs of septal band. This kind of defect is further classified as the malalignment (absent, anterior malalignment, posterior malalignment) and the borders of the defect:

muscular, perimembranous, and doubly committed juxtaarterial. Outlet muscular VSD has not a membranous extension with just muscular rims, whereas outlet perimembranous VSD is characterized by a membranous extension and a cranial fibrous rim. A particular form is the doubly committed juxtaarterial VSD, which is characterized by a subpulmonary extension.

Outside the aforementioned classification, a specific type is the "Gerbode VSD." This kind of perimembranous defect is a direct/indirect communication between the left ventricle and the right atrium. Gerbode VSD is often iatrogenic as a sequel of complete atrioventricular defect surgical correction; even if, it may also be congenital. This is a malignant form of VSD because the direct communication among two chambers with so different pressures produces a severe left-to-right shunt with early signs and symptoms of heart failure and/or pulmonary hypertension. In these cases, a surgical or percutaneous closure is usually necessary [7].

Isolated VSDs produce a post-tricuspid left-to-right shunt with consequent enlargement of left-sided chambers (with progressive left heart failure) and pulmonary hypertension. Differently from ASD, signs and symptoms of heart failure may arise in neonatal age when the defect is wide. At 4–5 days of life, the pulmonary resistance fall determines an increase in left-to-right shunt with the onset of heart failure (tachypnea, poor growth, fatigue in feeding, liver congestion, hypertension of fountain) in patient with large defects. An oral or intravenous therapy with diuretics is helpful to improve the hemodynamic of patient. In some cases, drug therapy is not enough and a palliative surgical intervention is required to limit the amount of shunt: pulmonary artery banding. The bandage of pulmonary trunk increases the right ventricle afterload limiting the interventricular left-to-right shunt with consequent improvement of symptoms. In some cases, the VSD is so much small that does not cause heart failure requiring just a cardiologic follow-up of the child. A spontaneous closure is a feasible chance, above all for small trabecular muscular VSDs. This event is rare for perimembranous defects, even if a redundant sub-tricuspid apparatus may create a "cushion," which limits partially the amount of shunting in some cases.

Surgical or percutaneous closure of VSD is recommended in patients with signs of left ventricular volume overload and hemodynamically significant shunts (pulmonary–systemic output ratio ≥ 1.5), if pulmonary–systemic systolic arterial pressure ratio is less than 0.50 and pulmonary–systemic vascular resistance ratio is less 0.33. Perimembranous central and outlet perimembranous defect may develop an aortic regurgitation. The subjacent VSD produces a prolapse of right coronary or noncoronary cusp due to Venturi effect (Laubry–Pezzi syndrome) with consequent eccentric regurgitation [8]. In these cases, it is reasonable to perform the closure of defect. The VSD closure may be reasonable in adults with a history of defect-related infective endocarditis (if not otherwise contraindicated) as well as in the presence of a hemodynamically significant left-to-right shunt when pulmonary–systemic systolic arterial pressure ratio is more than 0.50 and pulmonary–systemic vascular resistance ratio is more than 0.33 [9].

Although surgical strategy is valid and consolidated, percutaneous closure of VSD is a feasible choice in specific circumstances. Currently, muscular and perimembranous defects may be treated with a percutaneous approach, whereas surgery is again the only strategy to close inlet VSDs and perimembranous VSDs with aortic regurgitation. A transcatheter approach is feasible also to deal with Gerbode-type defects. A distance less than 4 mm between the defect and the atrioventricular or ventriculoarterial valves is a contraindication to percutaneous closure.

Several devices were designed for VSD closure. The most occluder devices are as follows: Amplatzer Muscular VSD Occluder (Abbott, Plymouth, MN, USA), Amplatzer Post-Infarct Muscular VSD (Abbott, Plymouth, MN, USA), Amplatzer Membranous VSD Occluder (Abbott, Plymouth, MN, USA), and Konar-MF VSD Occluder (Lifetech Scientific Corporation, Shenzhen, China).

Amplatzer Muscular VSD Occluder and Amplatzer Post-Infarct Muscular VSD Occluder are self-centering, self-expandable, double-disk devices consisting of a nitinol (nickel and titanium) wire framework. A central waist connects the two disks. A Dacron membrane covers the inner surface of the devices to aid the occlusion process. The size number of the device represents the waist diameter. Amplatzer Muscular VSD Occluder shows a central waist high 7 mm with both disks 8 mm larger than waist, whereas Amplatzer Post-Infarct Muscular VSD Occluder has a central waist high 10 mm with both disks 10 mm larger than waist. The size of Muscular and Post-Infarct Muscular devices varies from 4 mm to 18 mm and from 16 mm to 24 mm, respectively. Amplatzer Muscular VSD Occluder 4 mm is the only size characterized by both disks 5 mm larger than waist.

Amplatzer Membranous VSD Occluder is a self-centering and self-expandable device, characterized by two asymmetrical and decentered Nitinol disks. The central waist is very thin (high 1.5 mm), and its diameter corresponds to the size of device. The unique feature of this device is the presence of two asymmetrical and decentered disks. The left-sided disk has an asymmetrical structure with an aortic rim and an apical rim, which are 0.5 mm and 5.5 mm larger than waist, respectively, whereas the right-sided disk has a symmetrical structure in which both disks are 2 mm larger than waist. This conformation allows to reduce the risk of impingement on the aortic valve. The device is available in different sizes (from 4 mm to 18 mm).

Konar-MF VSD Occluder is a self-centering and self-expandable device, consisting of a nitinol wire framework. The device is characterized by a central waist with conal shape and two symmetrical retention disks. The conal waist is high 4 mm. It shows a larger left-sided diameter and a smaller right-sided diameter. These measures represent the sizes of the device, which may vary from 5–3 mm to 14–12 mm. The disks are 4 mm larger than the maximum diameter of the waist, except for the size 5–3 mm characterized by disks larger than 5 mm than waist. The major sizes of the device (from 9–7 mm to 14–12 mm) show an inner membrane to allow an immediate closure of VSD. The flexible waist is designed to conform the prosthesis to a great variety of defects, while the low profile minimizes the risks of injuries. A double-sided screw allows to deploy the device both for retrograde approach and for antegrade approach [10] (Fig. 3.1).

Transcatheter closure of VSD is guided by both the fluoroscopy and echocardiography. Left ventriculography in the hepatoclavicular view (35° left anterior oblique and 35° cranial) is optimal to delineate the interventricular septum; however, perimembranous and anterior muscular VSDs are better viewed with a 20° left anterior oblique and 20° cranial. The defect is usually crossed from the left side to reduce the risk of sheath trapping among the trabeculations of the right ventricle. A right side crossing is feasible even if complicated by the turbulence of the left-to-right shunt, which impairs the passage of wires and catheters. When the defect is crossed from the left side, it is necessary to capture the wire with a Goose Neck Snare (Microvena Corporation, White Bear Lake, MN, USA) and to pull back it outside from the venous vascular access. The arterial–venous loop created is necessary to allow a transcatheter closure from the venous side when the defect is crossed by the arterial side. The device chosen should be 1–2 mm larger than the defect measured at end diastole. In this phase, the transesophageal echocardiography is a fundamental tool both to achieve an optimal three-dimensional view of the defect and to check the route of wires and catheters avoiding interferences to valvular and subvalvular apparatus. At this time, a long delivery sheath is advanced onto the wire and the device chosen is deployed. During the placement of the device, it is necessary to open the left disk quite close to the defect to avoid injuries to aortic cusps or to subvalvular mitral apparatus. Konar-MF VSD Occluder has a double-sided screw, which allows to deploy the device also from the arterial side avoiding to create the arterial–venous loop. The only one problem of this retrograde approach is the need to use a large sheath from an arterial vascular access, and so, it should be discouraged for large defects in infants and small children. A hybrid approach was developed to improve the technique in infants and small children (less than 6 kg). This strategy allows both to reduce the vas-

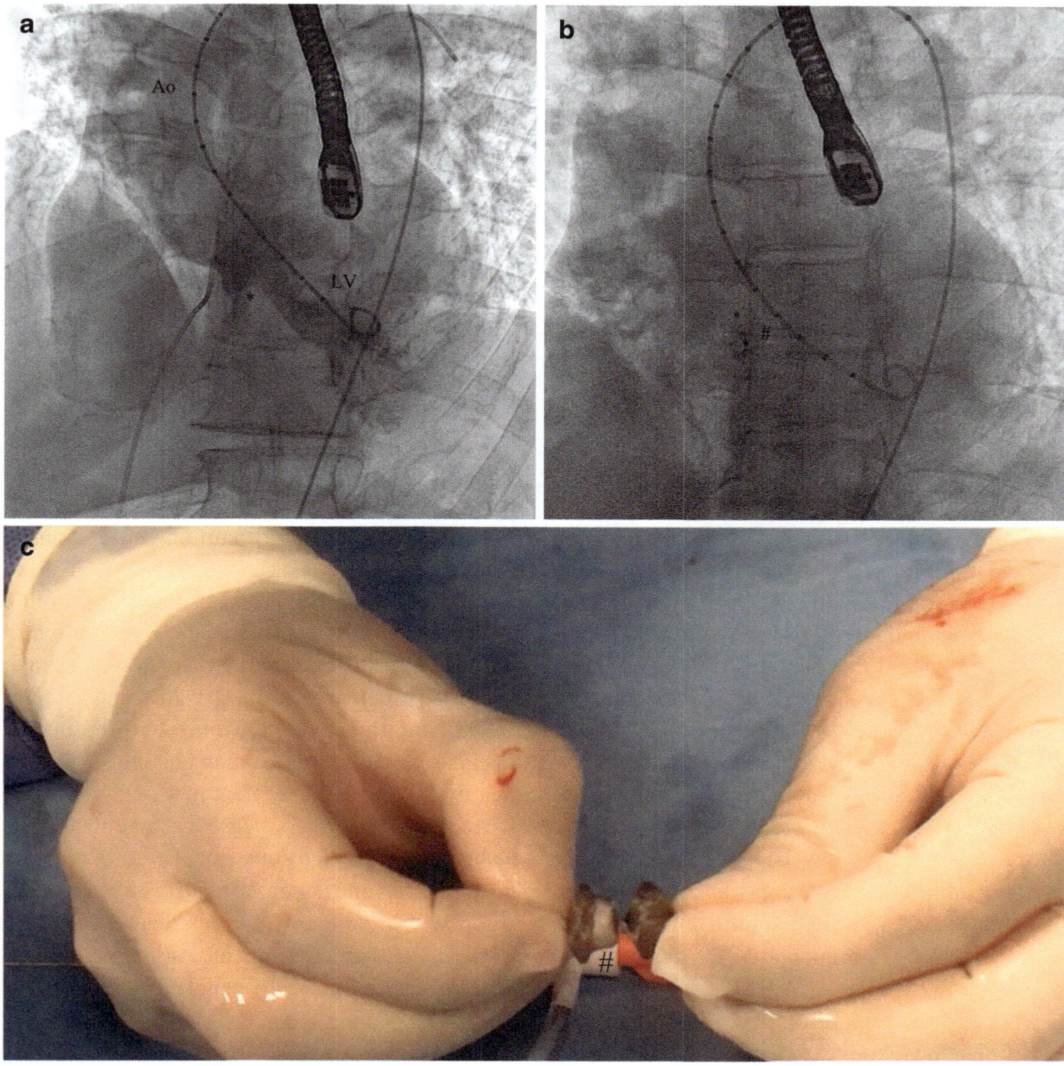

Fig. 3.1 Subaortic peri-membranous ventricular septal defect closure with a Konar-MF Ventricular Septal Occluder 10–8 mm. Abbreviations: aorta (Ao); left ventricle (LV); ventricular septal defect (*); Konar device (#)

cular complications of percutaneous approach and to avoid extracorporeal circulation necessary for surgery. The chest and the pericardium are exposed surgically, and a needle (usually 18-gauge) is used to puncture the free wall of the right ventricle. In this way, it achieved a direct access to the right ventricle. From this side, the VSD may be crossed and closed using the common devices [11]. Sometimes, an aneurysmatic septum may characterize the site of VSD. If the aneurysm is small, it may be covered completely by the device, whereas if the aneurysm is particularly large, the device may be deployed into the aneurysm. In the latter circumstance, ductal occluder devices may be considered to close the VSD, such as Amplatzer Duct Occluder (Abbott, Plymouth, MN, USA) type I, type II, and type II Additional Size or Amplatzer Vascular Plug type II (Abbott, Plymouth, MN, USA) [12, 13].

Procedure-related adverse events are similar to ASD closure: device embolization, cardiac erosion, arrhythmia, cobra shape of left disk, and pericardial effusion. However, some complications are typical of VSD closure. Hemolysis is rare and usually due to a residual shunt. The great turbulence through a residual shunt site may determine a direct damn to blood cells with consequent hemolysis. In this case, surgical transcatheter closure of the residual shunt is the only feasible strategy to solve the adverse event [14].

The onset of valvular regurgitation may be related to procedure. During the deployment of the left disk, the device may develop either an impingement of a valvular cusp and/or leaflet or an interference with the subvalvular mitral and/or tricuspid apparatus. These conditions may compromise the valvular closure with consequent regurgitation. In some circumstances, the passage of a large long delivery sheath through the trabeculations of right ventricle may determine tendon rope rupture with consequent flail of a tricuspid leaflet and significant regurgitation.

Heart block is the worst adverse event related to transcatheter closure of a perimembranous VSD (incidence 1–2%). The ventricular escape rate addresses the management of this complication. If the patient develops an acceptable ventricular heart rate, observation and oral therapy with corticosteroids may be performed waiting for the response; however, in the cases with an inadequate heart rate with significant symptoms (e.g., syncope), the implantation of a pacemaker is necessary [15].

All devices have a thrombogenic structure, so an antiplatelet therapy (oral acetylsalicylic acid 5 mg/kg die, maximum dosage 100 mg die) is recommended up to 6 months from procedure, at least.

3.3 Atrial Septal Defect

The interatrial septum is a nonlinear structure, which separates the left atrium from the right-sided one. During fetal life, the overlap of the thin anterior–inferior septum primum with the thick posterior–superior septum secundum creates a tunnel (foramen ovale) necessary for fetus' survivor. At birth, the first breathing acts to produce an important lung expansion with consequent higher pulmonary blood flow. This hemodynamic change causes an increase in pulmonary venous drainage into the left atrium. The higher pressure of this chamber pushes the septum primum to the septum secundum realizing an initial closure of patent foramen ovale. Usually, this tunnel is patent in the first year of life (50% of cases), but it may undergo spontaneous closure up to 5 years old. In a 25% of cases, this communication rests patent in the adult age and it is called patent foramen ovale [16].

ASD is a direct communication between right atrium and left atrium. Different interatrial defects are possible: ostium secundum (OS), ostium primum (OP), sinus venosus, and sinus coronary type.

ASD-OS is the most common cause of interatrial shunt (70% of cases). The defect involves the area of fossa ovale, and usually, the septum secundum is well-represented. The evaluation of periorificial rims is necessary to know the feasibility of percutaneous closure. In some circumstances, multiple ASD-OS may be present, but this anatomic finding is not an absolute contraindication to transcatheter treatment.

ASD-OP (18–25% of cases) is an endocardial cushion defect with a consequent communication close to atrioventricular valves. It is always associated with a defect of anterior leaflet of mitral valve (cleft) with consequent non-negligible valvular regurgitation, in the panel of a partial atrioventricular septal defect.

Sinus venosus ASD (4–11% of cases) is a specific type of interatrial communication through a pulmonary vein. It is usually due to an unroofing of a pulmonary vein with consequent left-to-right shunt. In the superior vena cava type (most common sinus venosus defect), the direct communication is between the right upper pulmonary vein and the superior vena cava, whereas in the inferior vena cava type, a direct communication is developed between the middle and/or the lower right pulmonary vein and the posterior wall of right atrium (a direct communication with the inferior vena cava is very rare). Sinus venosus defects are always associated with a partial anomalous drainage of pulmonary veins, even if with a normal left atrial connection.

Coronary sinus ASD (<1% of cases) shows shunting mechanism similar to sinus venosus type. An unroofing of coronary sinus causes an interatrial left-to-right shunt through the coronary sinus. The association with a persistent left superior vena cava is termed Raghib syndrome [17].

All defects determine a pre-tricuspid left-to-right shunt with consequent volume overload of the right ventricle, which may cause either a pulmonary hypertension or a right-sided heart failure. In 5–10% of adults with a hemodynamically significant left-to-right shunt, a pulmonary hypertension may arise [18]. The higher right ventricle output determines an increase in pulmonary vascular resistances with consequent pulmonary hypertension. Right heart failure may arise with symptoms of liver congestion or dyspnea on effort. Atrial arrhythmias are a consequence of right atrial dilation, while paradoxical systemic embolism is due to an inversion of shunt in specific hemodynamic changes.

Small ASD-OS (≤5–8 mm) may go to spontaneous closure within the first 5 years of life, whereas the other types of ASD never reduce their selves during follow-up [19]. Moreover, sinus venosus ASD is characterized by a rapid and marked right-sided heart diastolic overload due both to the indirect interatrial shunt and to anomalous pulmonary vein drainage.

In 2018, American College of Cardiology and American Heart Association developed Clinical Practice Guidelines about Adults with Congenital Heart Diseases. ASD closure is recommended in patient with a right atrial and/or ventricular dilation associated with "net left-to-right shunt sufficiently large to cause physiological sequelae" (pulmonary–systemic blood flow ratio ≥ 1.50), with a pulmonary–systemic systolic arterial pressure ratio < 0.50 and a pulmonary–systemic vascular resistance ratio <0.33 [9]. Transcatheter treatment is the first choice to close ASD-OS. Absent or poor periorificial rims and/or a huge defect (>40 mm) are contraindication for percutaneous closure. Transcatheter closure is not indicated to deal with ASD-OP, sinus venosus, and coronary sinus defects; however, rare cases of percutaneous closure of coronary sinus ASD [20] and sinus venosus ASD (type superior vena cava) [21] are described.

Transesophageal echocardiography (TOE) is a crucial imaging to guide the procedure, allowing an optimal identification of the anatomy, the assessment of accessory defects, and a measure of periorificial rims [22]. The important rims to consider are as follows: anterior–superior rim (or retro-aortic rim), posterior–superior rim (toward superior vena cava), posterior rim (toward the roof of right atrium), posterior–inferior rim (toward inferior vena cava), and anterior–inferior rim (toward the atrioventricular valves) (Fig. 3.2). A periorificial rim may be adequate (≥5 mm), poor (<5 mm), and absent (when it may not be identified in all echocardiographic view). The retro-aortic rim is the less important, since many patients undergo effective transcatheter closure of the defect even without a retro-aortic rim; in fact, the device deploys itself embracing the aorta, without a significant risk of prolapse. Poor or absent others rims are contraindication to percutaneous closure. The intracardiac echocardiography may be a useful alternative when TOE is unfeasible.

Periorificial rims are often "floppy," and so, a stretch diameter is necessary to identify the maximum diameter of the defect. The most used sizing balloon catheters are as follows: Amplatzer Sizing Balloon II (Abbott, Plymouth, MN, USA) and Equalizer Occlusion Balloon Catheter (Boston Scientific Corporation, Marlborough, MA, USA). The first one is available in various nominal diameters: 18 mm (maximum diameter 20 mm), 24 mm (maximum diameter 27 mm), and 34 mm (maximum diameter 40 mm). The balloon is inflated into the defect, and it assumes a characteristic "hourglass shape." The central waist of balloon is measured both in fluoroscopy and in echocardiography, and it represents the maximum "stretch diameter" of defect. The Equalizer Occlusion Balloon Catheter is available in the following maximum diameters: 20 mm, 27 mm, 33 mm, and 40 mm. The balloon is inflated into the right atrium and pushed toward the defect. The stretch diameter of ASD is represented by the minimum diameter of the balloon, which does not cross the defect achieving a

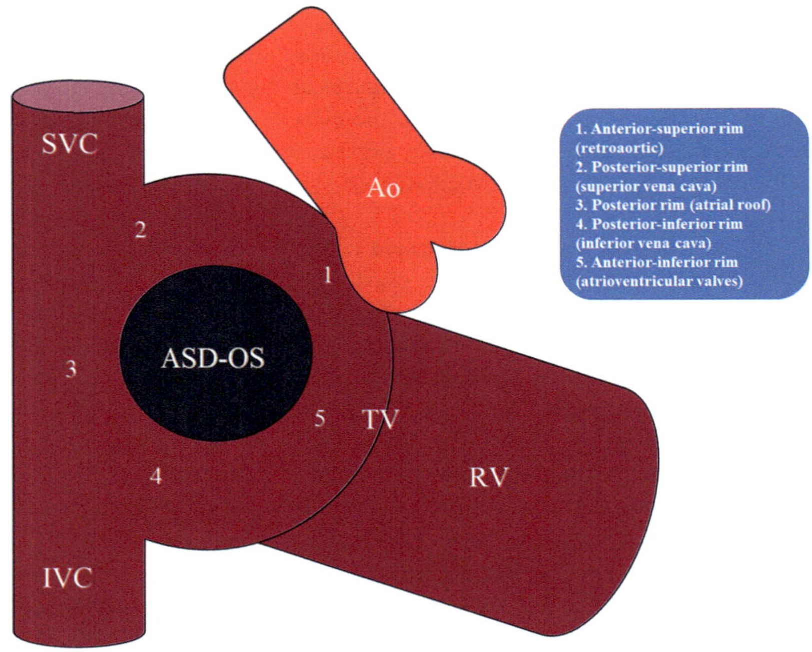

Fig. 3.2 Periorificial rims of atrial septal defect. Abbreviations: aorta (Ao); ostium secundum atrial septal defect (ASD-OS); inferior vena cava (IVC); right ventricle (RV); superior vena cava (SVC); tricuspid valve (TV)

complete "stop flow" through the hole. The stretch diameter represents the maximum diameter of the hole, and it guides the interventional cardiologist to choose the device. Usually, the central waist of the chosen device has a diameter similar to the stretch diameter. This technique ensures that the disks of the device are anchored to resistant periorificial rims, reducing the risks of prosthesis embolization.

The most occluder devices are as follows: Amplatzer Septal Occluder (Abbott, Plymouth, MN, USA), Figulla Flex II ASD Occluder (Occlutech GmbH, Jena, Germany), and GORE CARDIOFORM Septal Occluder device (W. L. Gore & Associates, Inc., Flagstaff, AZ, USA).

Amplatzer Septal Occluder is a self-centering, self-expandable, double-disk device consisting of a nitinol (nickel and titanium) wire framework. A central wide waist holds together the two disks. Inside the disks, a polyester fabric layer is inserted to increase the closing capability of the device. The number size of prosthesis identifies the diameter of central waist (high 3–4 mm), whereas the two disks have a variable diameter: 8–10 mm larger than waist for the right-sided disk and 12–16 mm larger than waist for the left-sided one. The prosthesis size varies from 4 mm to 40 mm, and for this reason, ASD >40 mm may be not suitable of closure (Fig. 3.3).

Figulla Flex II ASD Occluder is a self-centering, self-expandable, double-disk device composed of a nitinol wire framework ermined by a ceramic–titanium oxide surface, which gives a "golden" aspect to the device. Inside the disks, a polyethylene patch is placed to achieve an immediate closure. The delivery system is characterized by a bioptome, which hooks a sphere at the proximal end of the device. This system develops lower tension on the device allowing to see the final layout of the prosthesis immediately after the deployment and before to release it.

GORE CARDIOFORM Septal Occluder is a non-self-centering device, premounted on a specific delivery system. The device is characterized by an inner thin nitinol wire framework covered completely by an expanded polytetrafluoroethyl-

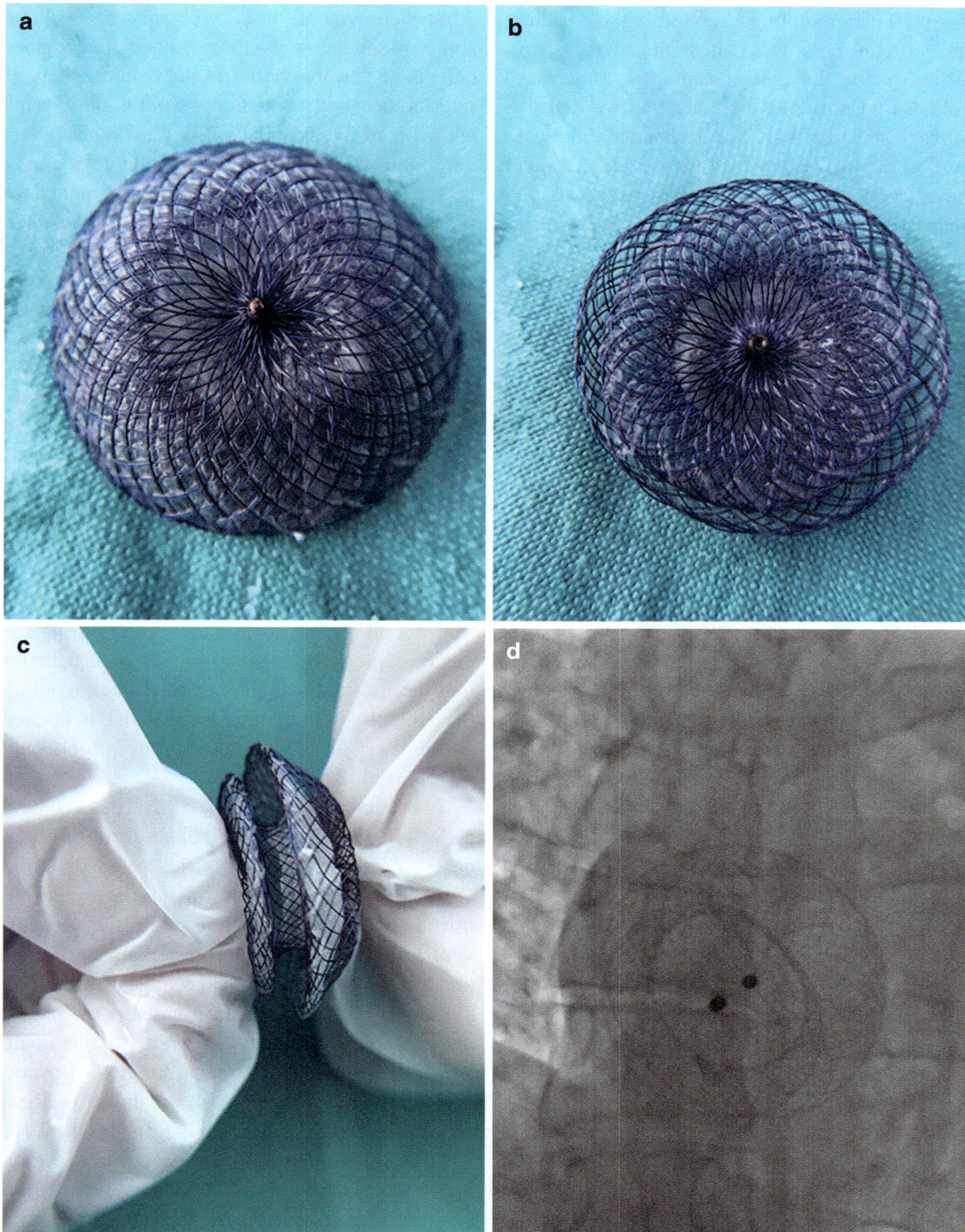

Fig. 3.3 Amplatzer Septal Occluder device: left disk view (**a**), right disk view (**b**), lateral view (**c**), fluoroscopic view (**d**)

ene (ePTFE) membrane (assuming a "folding fan" structure). For this reason, the device is very soft, reducing the risk of damages to the atrial wall. A thin pin holds together the two disks. A delivery sheath coupled to a "handle" allows to load, to deploy, and to lock the device. An inner retrieval cord is a safety system useful to pull back the device after the release. The device is available in four different sizes: 15 mm, 20 mm, 25 mm, and 30 mm. The number size of the prosthesis represents the diameter of both disks. The GORE instructions for use recommend a device–defect ratio at least of 1.75, because of the non-self-centering feature of the prosthesis. According to these instructions for the use, GORE CARDIOFORM Septal Occluder should not be used for ASD-OS larger than 17 mm at stretch diameter. This kind of device may be ideal in patients with nickel allergy, because of the absence of direct nitinol exposure to blood flow (Fig. 3.4).

CeraFlex (Lifetech Scientific Co., Shenzhen, China) and Cocoon Septal Occluder (Vascular Innovations Co., Nonthaburi, Thailand) are other ASD occluder devices. CeraFlex is a self-expanding, double-disk nitinol wire mesh device characterized by a wide central waist and a flexible connection between the device and delivery cable. The sizes of CeraFlex change from 6 to 42 mm (with 2 mm increments). The Cocoon Septal Occluder is a self-expanding, double-disk device characterized by a nano-platinum-coated nitinol framework filled with polypropylene fabric. The sizes of the occluder change from 8 to 40 mm (with 2-mm increments). Currently, the Cocoon Septal Occluder is considered the softest and lightest available device.

During percutaneous closure of ASD, the left-sided disk is opened into the left atrium and pulled back up to the interatrial septum. During this maneuver, a clockwise rotation of delivery sheath is useful to move the device far away from the aorta. When the interatrial septum is well-hooked, the right-sided disk may be opened. The periorificial rims have to be hooked adequately. After deployment, it is necessary to evaluate some crucial elements: the presence of accessory defects or periprosthetic residual shunt, signs of interference with systemic or pulmonary veins and with atrioventricular valves, and signs of aortic compression. The stability of the device may be checked with a push-and-pull maneuver ("Minnesota wiggle") and then realized. When the defect is very large and/or one or more rims are deficient, percutaneous closure may be very complex and some specific techniques are useful to solve these problems. The most diffused technique consists to use a pulmonary vein (frequently the right-upper one, but also the left-upper one) to improve the alignment of device toward the interatrial septum (Fig. 3.5). The left disk is opened partially into a pulmonary vein, pulled back up to the septum, and then the right disk is opened. In this way, the parallel orientation of left disk to interatrial septum reduces the risk of prolapse [23, 24]. Another feasible technique is to use a dilator or a balloon catheter. During the device deployment, a dilator or a balloon catheter may be useful to support the left disk avoiding the prolapse. When also the right disk is opened and the septum is well-hooked, these tools may be removed [25, 26].

In special situations, ASD closure is debated. Elderly patients are a particular subset. In these cases, numerous comorbidities (e.g., diastolic or systolic dysfunction, mitral regurgitation, mitral stenosis) may increase end-diastolic left ventricular pressure (and thereby also left atrial and pulmonary capillary wedge pressures). Temporary balloon occlusion test is a useful and reversible strategy to evaluate the acute hemodynamic changes during the ASD closure. A balloon is inflated through the defect to abolish the interatrial shunt, and left "in situ" for 15 min. Pulmonary pressures and end-diastolic pressure of left ventricle are recorded constantly. A persistent increase in left ventricular end-diastolic pressure (>20 mmHg and/or increase >50% compared to baseline) or a decrease ≥20% in systemic arterial pressure or

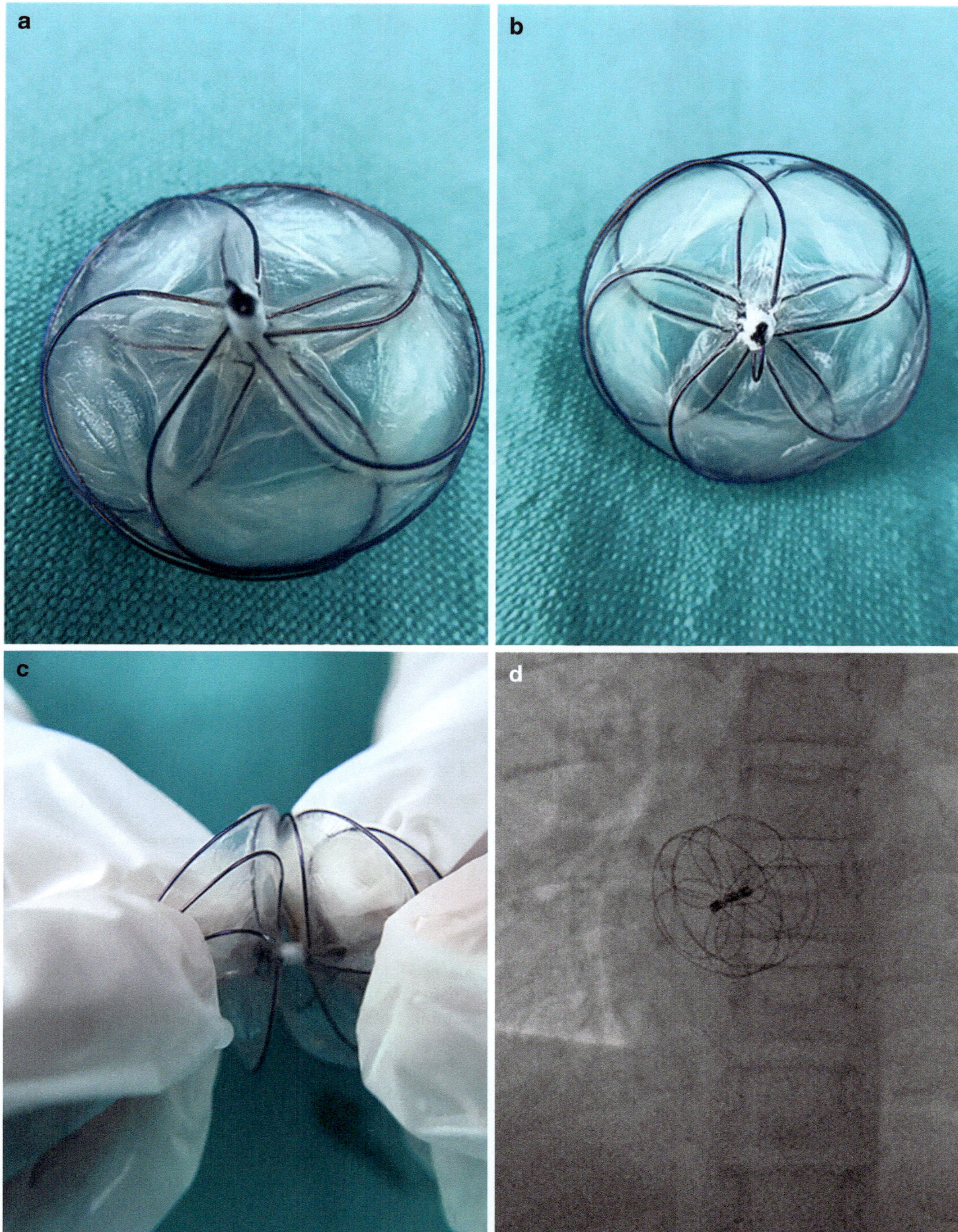

Fig. 3.4 Gore CARDIOFORM Septal Occluder device: left disk view (**a**), right disk view (**b**), lateral view (**c**), fluoroscopic view (**d**)

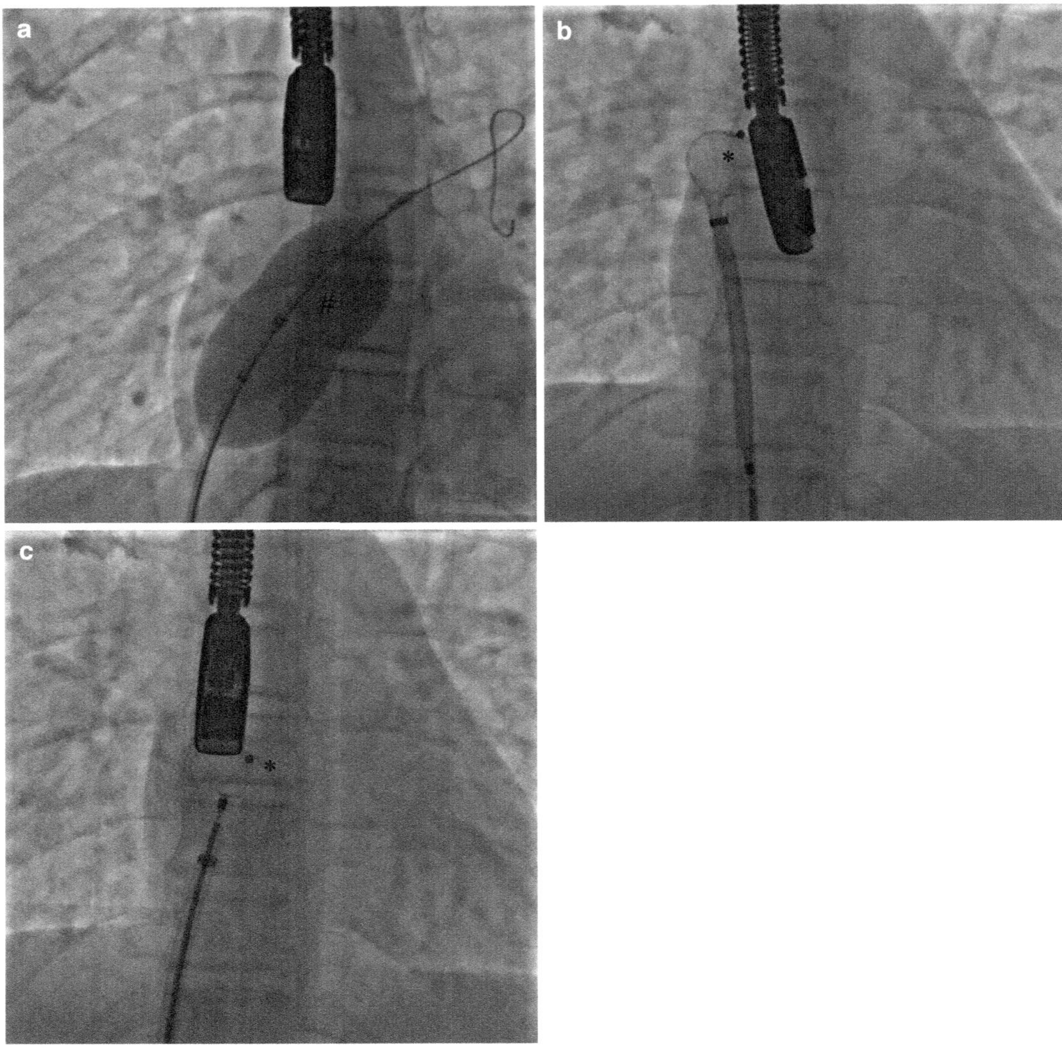

Fig. 3.5 Right pulmonary vein-assisted technique. The stretch diameter with the sizing balloon (#) highlighted a large atrial septal defect (**a**). The left disk is partially opened into the right upper pulmonary vein (**b**) to achieve an improved alignment of the prosthesis (*) with the atrial septum, with a final good deployment of the device (**c**)

the appearance of pulmonary edema is relative contraindication to ASD closure. In this subset, two different strategies are feasible: a pharmacologic preconditioning trial with oral diuretics and angiotensin-converting enzyme (ACE) inhibitors for 3 months followed by a novel right heart catheterization both in basal condition and during ASD balloon occlusion test or a partial ASD closure with a fenestrated device [27]. Fenestration may be performed with a 10–12 Fr dilator sheath. A peripheral angioplasty balloon may be used "in vitro" to increase the diameter of the hole up to 6–8 mm. The fate of fenestration is usually a rapid closure within few months; however, it is useful to tolerate the acute increase in left atrial pressure after ASD closure, avoiding the onset of pulmonary edema. In some cases, a post-dilation of fenestration may be required to improve the intra-prosthetic residual shunt. Tufaro and Butera described an "over-the-wire" implantation technique of the device to perform an "in vivo" post-dilation of fenestration [28].

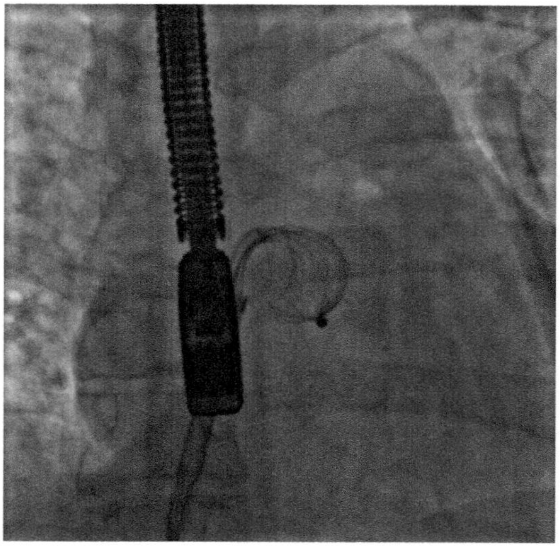

Fig. 3.6 Amplatzer Septal Occluder device with a "cobra head" shape

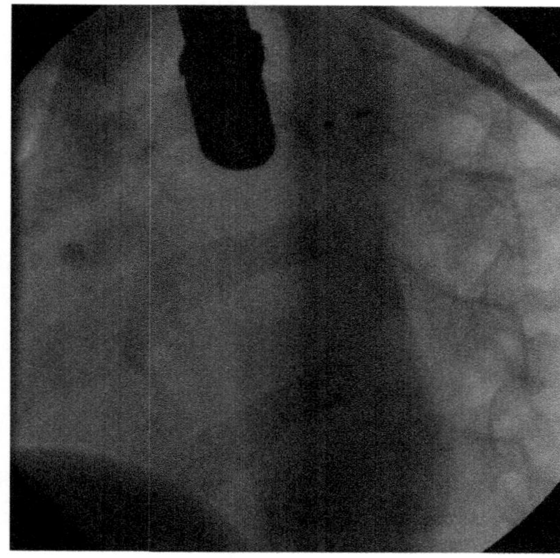

Fig. 3.7 Amplatzer Septal Occluder device embolization

During device deployment, various technical problems may arise. In some circumstances, the left disk assumes a "cobra head" shape (Fig. 3.6). This phenomenon occurs either for technical defects of device or because the device has been opened into a pulmonary vein (or in the left atrial appendage). The device has to be recaptured, and interventional cardiologists may try to place the device one more time. It is necessary to change the prosthesis, if the "cobra head" shape persists (never release a device that has a cobra head appearance of a disk) [29]. A redundant Eustachian valve is another factor, which may complicate the procedure. In this case, the device might capture the redundant valve; for this reason, a pigtail catheter is useful to move away the valve during device deployment [30].

Various and different procedure-related adverse events both major and minor are described. The most important and dangerous adverse events are as follows: device embolization and cardiac erosion. Device embolization (Fig. 3.7) is rare (about 1%) but very dangerous. The prosthesis migration may determine a hemodynamic collapse if a complete occlusion of pulmonary artery or aorta happens. This complication is typical of large ASD and/or deficient rims. The retrieval of device is performed usually percutaneously with a Goose Neck Snare (Microvena Corporation, White Bear Lake, MN, USA) and a larger sheath; however, when the transcatheter retrieval fails, it needed a surgical approach to remove the prosthesis.

Cardiac erosion is the most feared complication, even if very rare (about 0.1% of ASD closures). Erosion looks like be due to a direct lesion of device to the dome of the atria, close to the aortic root. Deficiency of any rim, device >5 mm larger than ASD basal diameter, and lower weight device ratio are associated with a high risk of cardiac erosion [31]. Most of the cardiac erosions occurred after the discharge. In literature, numerous cases about very late cardiac erosion are reported, and for this reason, it is recommended a long-lasting follow-up over time for these patients [32]. In literature, no cases of erosion with a GORE CARDIOFORM Septal Occluder are described; however, an early frame fracture with mitral valve perforation and partial prolapse of device is a possible complication [33].

Both tachyarrhythmias and bradyarrhythmias may arise after transcatheter ASD closure. Supraventricular tachyarrhythmias are the most frequent. All devices have a thrombogenic structure, so an antiplatelet therapy (oral acetylsalicylic acid 5 mg/kg die, maximum dosage 100 mg die) is recommended up to 6 months from procedure, at least. After this time, the endothelialization of device is partially completed and the risk of device thrombosis decreased. Nickel allergic reaction may arise in specific patients from 2 days up to 1 month after device implantation. Headaches, rash/urticaria, difficulty breathing, fever, and/or pericardial effusion are the most manifestation. Usually, the patients respond immediately to medical therapy; however, if the symptoms and signs persist, a surgical explant of the device is necessary [34].

3.4 Biodegradable New Devices

Due to the presented complications, permanent shape memory alloy occluders are acceptable for elderly patients to prolong life span or improve quality of life, but better solutions need to be developed for younger individuals, especially considering that a metallic device would prevent the access for future transseptal interventions, such as valve repair/replacement and left atrial appendage (LAA) obliteration [35–37].

Occluders made of materials able to be completely absorbed by the body, i.e., biodegradable, could be a desirable alternative so that the device provides a temporary "bridge" for cardiac self-repair of the defect [38]. Nevertheless, biodegradable materials possess much lower mechanical performances than widely used metals; therefore, partially biodegradable occluder was conceived by combining the advantages of nondegradable alloys and biodegradable materials [39].

Partially biodegradable occluders. The BioSTAR septal repair implant (NMT Medical, Boston, USA) is an umbrella-like structure with a nondegradable MP35N skeleton [40, 41] and a heparin-coated, acellular, porcine-derived collagen matrix that allows absorption and replacement with human tissue (95%) within 2 years [42], with the advantage of reducing thrombogenicity and improving biocompatibility [43–46]. An evaluation study was conducted in 58 patients and showed good results [47].

The Double BioDisk (Cook Medical, IN, USA) consists of two platinum-layered nitinol rings, connected with small cannulas, covered with porcine small intestinal submucosa and sutured with radio-opaque rings. It is a self-expanding and self-centering device, but so far it has only been implanted in adult sheep [47].

Carag Bioresorbable Septal Occluder (Carag AG, Baar, Switzerland) consists of two opposing foldable nonresorbable polyester fabrics and eight bioresorbable poly(lactic-co-glycolic acid) (PLGA) monofilaments, with Pt-Ir markers [48]. It has self-centering and complete retrieval/redeployment properties and was successfully implanted in seven patients in an initial first-in-human trial [49].

Fully biodegradable occluders. As partially biodegradable occluders have not been widely accepted due to unfavorable clinical results (12% implantation failure) and high early and late complication rates (9% and 12%, respectively) [50], fully degradable occluders started to be developed.

The most diffused structure for totally biodegradable devices is the "double umbrella."

The "Double Umbrella" occluder comprises two self-expandable disks connected with a stretchable stem. Each disk is made of four barium sulfate ($BaSO_4$)-doped poly(Ɛ-caprolactone) (PCL) spokes covered with polylactide-co-Ɛ-caprolactone (PLC) film. This device was successfully deployed in two pigs [51].

Other double-umbrella-like structures have been designed: (i) a patent ductus arteriosus (PDA) device build on PCL/PLC umbrellas, with the addition of radiopaque $BaSO_4$ [52]; (ii) a PCL skeleton components with PLGA/type I collagen blend nanofibrous membranes, with similar design of the BioSTAR occluder [53]; and (iii) a self-expandable poly(L-lactide) (PLLA) framework and two baffle PLLA membranes sewn interiorly (Lifetech Scientific Company) [54]. The last one is particularly innovative because of its locking components [55]. In the first animal experiments, the success rate of occlusion reached 100%, and in later animal experiments, 27 experimental piglets and 18 control piglets were all successfully implanted with the modified biodegradable and metal ASD devices, respectively, and both devices exhibited desirable sealing effects [54, 55].

The Chinese lantern occluder consists of a soft portion ("head," "waist," and "tail" films) made of the copolymer PLC and a hard structural skeleton (lock, and head tubes, and wires) made of pure PCL. It has a unique pull-fold mechanism for folding and sealing: On retraction of the loop wire, the head films and tail films will fold into the working structure with the waist film length being adjustable corresponding to the septum thickness [56]. The device can be repositioned and retrieved, and the preclinical study was conducted in swine models [56]. Importantly, the locking structure design of this device showed the possibility to replace the self-expansion feature of the shape memory alloys [39].

Another totally biodegradable lantern-like PLA-based occluder is composed of a PLLA skeleton, a PLLA locking tube, and two poly(D, L-lactide) (PDLLA) fabrics [57]. The two disks are connected by the pentagonal skeleton instead of the traditional "waist," making it suitable for narrow ASD closures. The occluder could be easily retrieved and implanted using only echocardiographic guidance (instead of fluoroscopy), thus not requiring metal markers. This device was successfully implanted in 18 sheep.

Finally, a self-expandable device was developed with similar design to the nondegradable Amplatzer septal occluder, but composed of a skeleton made of 0.298 mm poly(*p*-dioxanone) (PPDO) monofilaments, two PLA nonwoven fabric sewn interiorly, and two metal tantalum particles for X-ray marker [58]. The occlusion devices were successfully implanted in 16 healthy mongrels [59].

Future Directions for Biodegradable Devices. Surely, biodegradable polymer materials do not own the same elastic recovery/shape memory properties as Ni-Ti alloys, but structural design and processing technology of fully biodegradable occluders can play an important role in determining their performance. In the presented biodegradable occlusion devices, the elastic recovery was perfectly replaced by designing locking structures, obtaining satisfactory results [54–57]. Moreover, biodegradable shape memory polymers have been already applied to other medical devices, capable of recovering programmed permanent shape from a temporary one easy for delivery/implantation through a specific stimulus such as heat, alternating magnetic field, microwave radiation, and electricity [60–64]. Lastly, 3D printing technology could be employed to manufacture ASD occluders based on both complex locking components and biodegradable shape memory polymers, having the advantage of low cost, high reliability, simple operation, rapid prototyping, multipart structural designing, and personalized customization [65, 66].

3.5 3D Printing Techniques for Complex Cases

3D printing cannot only be used for device manufacturing, but also for interventional planning of complex cases, for which conventional imaging techniques are not sufficient for optimal preoperative preparation, especially when percutaneous closure would be contraindicated due to rim deficiency, large or multiple defects.

The 3D printing method was developed in the 1980s in the USA; it creates a hollow three-dimensional physical object from the virtual computer-aided design (CAD) model by successively adding material layer by layer under computer control. Compared to other manufacturing techniques, the 3D printing is able to produce extremely complex shapes in very short time. For this reason, it has found wide application in the medical field.

In cardiology, the first application dates to 2006, when Noecker and colleagues [67] 3D printed pediatric heart models using CT datasets and advocated how physical 3D models could assist in the positioning of medical devices [68]. In the last decade, additive manufacturing has experienced a fast development, and nowadays, the most common applications of 3D printing embrace procedural planning of complex cases, in vitro simulation for research purposes, and training and communication with patients and their families [69].

Current 2D and 3D clinical imaging techniques used for diagnosis and treatment of ASD often lack critical spatial information and are not intuitive or comprehensive [70]. In this light, 3D printed heart models allow clear visualization of the anatomy, the location and size of the defect, and the relationship between the lesion and surrounding structures (e.g., SVC, IVC, and aorta). Importantly, 3D printed models make possible to physically simulate transcatheter ASD closure on the model before the actual procedure, thus allowing more effective and individualized preoperative planning by trying different devices and different sizes of ASD occluders preoperatively.

For ASD occlusion applications, 3D printing was firstly employed by Kim et al. [71] in the pre-interventional assessment of the clinical case.

Patients with ASD are usually scanned using multi-slice CT, and the acquired clinical images are post-processed (i.e., segmented) to create a virtual 3D representation of the anatomical site of interest useful for the subsequent additive manufacturing.

Recently, 3D printed rigid phantoms were used to test preoperatively different sizes of the Beijing ASD occluder (Starway Medical Technology, Inc., Beijing, China) on a group of 6 patients with ASD-OS with rim deficiency less than 5 mm [68]. An occluder device was defined "appropriate" if it had no compression on adjacent tissues and did not obstruct the blood flow from the superior vena cava (SVC) or the inferior vena cava (IVC). In this study, the 3D printed aided planning showed to be a feasible approach for optimizing the choice of the occluder.

3D printing can be especially useful to assess feasibility of ASD occlusion that would be normally contraindicated for intervention [72–74]. An example of this application can be found in the work by Ref. [72] that investigated the possibility of transcatheter closure of multiple ASDs with an inferior sinus venosus ASD under the guidance of 3D printed models in 5 patients. In these cases, due to the absence of the residual edge of the inferior vena cava, an ASD occluder cannot be effectively fixed on the interatrial septum. Moreover, it can be extremely challenging to choose the correct position of one or more devices to be able to cover all the defects without leaving any residual shunt and avoiding interactions between occluders, which might lead to poor stability and difficulty in endothelialization. They tried different types and sizes of occluders on the 3D printed heart models to establish the best interventional approach; a device was defined "appropriate" if it was stable, not oppressed and exerted no compression on adjacent tissues. Ultimately, all of the cases were successfully treated with interventional therapy, as planned from the preoperative simulation tests [72].

The impact of 3D printing in preoperative planning can be further increased by exploring flexible materials that would allow to better replicate the human interatrial septum compliance [72–75], thus allowing to understand native tissue's displacement/deformation during the intervention [76].

Personalized heart models made of elastic rubber [75] were developed to find the appropriate candidate for percutaneous closure among 35 patients with ASD-OS with deficient posterior–inferior rim ($\leq$3 mm). The material was a mixture of three kinds of photosensitive resin materials: TangoPlus FLX930, VeroFlex Yellow RGD893, and VeroMagenta RGD851 (Stratasys, USA), and the 3D models were printed with an Objet350 Connex3 (Stratasys) 3D printer. The 3D printed models allowed to determine the size and the surrounding rim of the ASD, which is paramount for the implanted device stability. Moreover, the flexible material would make the in vitro ballooning sizing test significant [75].

Finally, 3D printing can aid procedural planning in case the ASD is combined with other structural heart defects, which are conventionally treated surgically, for example, built 3D printed flexible transparent models for 3 patients with sinus venosus ASD with partial anomalous pulmonary venous drainage to explore a possible interventional catheterization treatment in 3 patients [74]. After in vitro simulations in which computed tomography was performed on the instrumented model using 3D rotational X-ray acquisition on the cardiac catheterization table, a custom-made covered Cheatham–Platinum stent in the superior vena cava to right atrium junction to close the ASD while committing the anomalous pulmonary vein to the left atrium was successfully implanted.

3D printing may be more important for less experienced clinics and for more challenging cases; however, recently, its diagnostic value from 3D echocardiography against conventional three-dimensional transthoracic echocardiography (3DTTE) alone for the assessment of structural heart disease was investigated [58]; it was found that 3D printing provides essential information for preoperative evaluation and decision making for patients with structural heart diseases. A large-scale clinical trial with long-term follow-up may identify the positive/negative effects of 3D printing on patient outcomes; it is likely that it will reduce intraoperative time and postoperative complications, and increase success rates through a more efficient preoperative simulation [77].

3.6 Engineering Computational Models

Other advanced engineering techniques, such as computational simulations, can be used to shed light on the fluid-dynamic and biomechanic processes involved in congenital cardiovascular pathologies, including the ASD. However, although virtual simulations are widely used in the cardiovascular field, there are not many studies in the literature for the specific case of ASDs.

The first computational study on the topic was performed in 2006 [78], which focused on the pediatric proximal pulmonary vasculature obtained from X-ray angiogram images to simulate the congenital septal defect closure in patient-specific models of the pediatric pulmonary vasculature with fluid–structure interaction techniques.

Recently, a computational fluid-dynamic model has been developed to analyze the differences in the fluid-dynamic profile in the left atrium (LA) and left atrial appendage (LAA) of patients with mild/no right-to-left shunting (RLS) and permanent RLS [79]. After examining transesophageal echocardiography (TEE) and magnetic resonance (MR) data from 65 patients, they modeled both the right atrium (RA) and the LA, including the LAA, to prove that the vorticity magnitude is lower across the LA and LAA in patients with permanent RLS. This finding suggests possible higher blood stagnation in these anatomical locations, similarly as previously observed in atrial fibrillation (AF) patients.

A complex multiscale 2D model of the whole heart was developed to understand the hemodynamics of the circulation under normal conditions and ventricular septal defect (VSD) [80]. Simulation results showed that the VSD causes volume overload in the left ventricle and the pulmonary circulation because of the circuitous refluxing of blood. Importantly, this mathematical model could provide the measurements of the hole size, its flow, and stress around its rims during the entire cycle; this information will be paramount for VSD occluder selection. The same type of engineering models could be developed for analyzing the ASD case and thus aiding in pre-interventional planning.

References

1. Hoffman JIE, Kaplan S. The incidence of congenital heart disease. J Am Coll Cardiol. 2002;39:1890–900.
2. Soto B, Becker AE, Moulaert AJ, Lie JT, Anderson RH. Classification of ventricular septal defects. Br Heart J. 1980;43(3):332–43.
3. Donald DE, Edwards JE, Harshbarger HG, Kirklin JW. Surgical correction of ventricular septal defect: anatomic and technical considerations. J Thorac Surg. 1957;33(1):45–59.
4. Van Praagh R, Geva T, Kreutzer J. Ventricular septal defects: how shall we describe, name and classify them? J Am Coll Cardiol. 1989;14(5):1298–9.
5. Lopez L, Houyel L, Colan SD, Anderson RH, Béland MJ, Aiello VD, Bailliard F, Cohen MS, Jacobs JP, Kurosawa H, Sanders SP, Walters HL 3rd, Weinberg PM, Boris JR, Cook AC, Crucean A, Everett AD, Gaynor JW, Giroud J, Guleserian KJ, Hughes ML, Juraszek AL, Krogmann ON, Maruszewski BJ, St Louis JD, Seslar SP, Spicer DE, Srivastava S, Stellin G, Tchervenkov CI, Wang L, Franklin RCG. Classification of ventricular septal defects for the eleventh iteration of the international classification of diseases-striving for consensus: a report from the International Society for Nomenclature of Paediatric and congenital heart disease. Ann Thorac Surg. 2018;106(5):1578–89.
6. Agmon Y, Connolly HM, Olson LJ, Khandheria BK, Seward JB. Noncompaction of the ventricular myocardium. J Am Soc Echocardiogr. 1999;12(10):859–63.

7. Gerbode F, Hultgren H, Melrose D, Osborn J. Syndrome of left ventricular-right atrial shunt; successful surgical repair of defect in five cases, with observation of bradycardia on closure. Ann Surg. 1958;148(3):433–46.
8. Scott RC, McGuire J, Kaplan S, Fowler NO, Green RS, Gordon LZ, Shabetai R, Davolos DD. The syndrome of ventricular septal defect with aortic insufficiency. Am J Cardiol. 1958;2(5):530–53.
9. Stout KK, Daniels CJ, Aboulhosn JA, Bozkurt B, Broberg CS, Colman JM, Crumb SR, Dearani JA, Fuller S, Gurvitz M, Khairy P, Landzberg MJ, Saidi A, Valente AM, Van Hare GF. 2018 AHA/ACC guideline for the management of adults with congenital heart disease: a report of the American College of Cardiology/American Heart Association task force on clinical practice guidelines. Circulation. 2019;139:e698–800.
10. Schubert S, Kelm M, Koneti NR, Berger F. First European experience of percutaneous closure of ventricular septal defects using a new CE-marked VSD occlude. EuroIntervention. 2019;15:e242–3.
11. Haponiuk I, Chojnicki M, Jaworski R, Steffek M, Juscinski J, Sroka M, Fiszer R, Sendrowska A, Gierat-Haponiuk K, Maruszewski B. Hybrid approach for closure of muscular ventricular septal defects. Med Sci Monit. 2013;19:618–24.
12. El-Sisi A, Sobhy R, Jaccoub V, Hamza H. Perimembranous ventricular septal defect device closure: choosing between Amplatzer duct Occluder I and II. Pediatr Cardiol. 2017;38(3):596–602.
13. Hua N, Aquino P, Owada CY. Transcatheter closure of perimembranous ventricular septal defects with the Amplatzer vascular plug-II. Cardiol Young. 2016;26(6):1194–201.
14. Chang H, Hung CR, Huang FY, Wu GJ. Hemolysis following correction of ventricular septal defect. J Formos Med Assoc. 1990;89(11):1004–8.
15. Hong ZN, Chen Q, Huang LQ, Cao H. A meta-analysis of perventricular device closure of perimembranous ventricular septal defect. J Cardiothorac Surg. 2019;14(1):119.
16. Anderson RH, Brown NA, Webb S. Development and structure of the atrial septum. Heart. 2002;88:104–10.
17. Raghib G, Ruttenberg HD, Anderson RC, et al. Termination of left superior vena cava in left atrium, atrial septal defect, and absence of coronary sinus; a developmental complex. Circulation. 1965;31:906–18.
18. Saxena A, Divekar A, Soni NR. Natural history of secundum atrial septal defect revisited in the era of transcatheter closure. Indian Heart J. 2005;57:35–8.
19. Cockerham JT, Martin TC, Gutierrez FR, et al. Spontaneous closure of secundum atrial septal defect in infants and young children. Am J Cardiol. 1983;52:1267–71.
20. Santoro G, Gaio G, Russo MG. Transcatheter treatment of unroofed coronary sinus. Catheter Cardiovasc Interv. 2013;81(5):849–52.
21. Butera G, Sturla F, Pluchinotta FR, Caimi A, Carminati M. Holographic augmented reality and 3D printing for advanced planning of sinus venosus ASD/partial anomalous pulmonary venous return percutaneous management. JACC Cardiovasc Interv. 2019;12(14):1389–91.
22. Mazic U, Gavora P, Masura J. The role of transesophageal echocardiography in transcatheter closure of secundum atrial septal defects by the Amplatzer septal occluder. Am Heart J. 2001;142:482–8.
23. Berger F, Ewert P, Abdul-Khaliq H, et al. Percutaneous closure of large atrial septal defects with the Amplatzer septal occluder: technical overkill or recommendable alternative treatment? J Interv Cardiol. 2001;14:63–7.
24. Varma C, Benson LN, Silversides C, et al. Outcomes and alternative techniques for device closure of the large secundum atrial septal defect. Cathet Cardiovasc Interven. 2004;61:131–9.
25. Wahab HA, Bairam AR, Cao QL, Hijazi ZM. Novel technique to prevent prolapse of the Amplatzer septal occluder through large atrial septal defect. Cathet Cardiovasc Interven. 2003;60:543–5.
26. Dalvi BV, Pinto RJ, Gupta A. New technique for device closure of large atrial septal defects. Cathet Cardiovasc Interven. 2005;64:102–7.
27. Santoro G, Giordano M, Palladino MT, et al. ASD closure in special situations: elderly, PA-IVS. In: Butera G, Chessa M, Eicken A, Thomson J, editors. Atlas of cardiac catheterization for congenital heart disease. Cham: Springer; 2019.
28. Tufaro V, Butera G. Fenestrated ASD device "angioplasty": how to adjust a "pop-off" mechanism when needed. Catheter Cardiovasc Interv. 2018;92(7):1329–33.
29. Mazic U, Gavora P, Masura J. "Cobra-like" deformation of an Amplatzer septal occluder. Pediatr Cardiol. 2001;22(3):253–4.
30. Butera G, Montinaro A, Carminati M. The "pull-push" technique to deal with a redundant Eustachian valve interfering with placement of a PFO occluder. Catheter Cardiovasc Interv. 2006;68:961–4.
31. McElhinney DB, Quartermain MD, Kenny D, Alboliras E, Amin Z. Relative risk factors for cardiac erosion following transcatheter closure of atrial septal defects a case–control study. Circulation. 2016;133:1738–46.
32. Scognamiglio G, Barracano R, Colonna D, Mattera Iacono A, Santoro G, Spadafora A, Nappi G, Russo MG, Sarubbi B. A very late life-threatening complication after percutaneous closure of an atrial septal defect. Can J Cardiol. 2017;33(2):293.e1–2.
33. Qureshi AM, Mumtaz MA, Latson LA. Partial prolapse of a HELEX device associated with early frame fracture and mitral valve perforation. Catheter Cardiovasc Interv. 2009;74(5):777–82.
34. Singh HR, Turner DR, Forbes TJ. Nickel allergy and the amplatzer septal occluder. J Invasive Cardiol. 2004;16(11):681–2.

35. Bissessor N. Current perspectives in percutaneous atrial septal defect closure devices. Med Dev (Auckl). 2015;8:297–303.
36. Sigler M, Jux C. Biocompatibility of septal defect closure devices. Heart. 2007;93:444–9.
37. Tang B, Su F, Sun X, Wu Q, Xing Q, Li S. Recent development of transcatheter closure of atrial septal defect and patent foramen ovale with occluders. J Biomed Mater Res B Appl Biomater. 2018;106(1): 433–43.
38. Zhu YF, Huang XM, Cao J, Hu JQ, Bai Y, Jiang HB, Li ZF, Chen Y, Wang W, Qin YW, Zhao XX. Animal experimental study of the fully biodegradable atrial septal defect (ASD) occluder. J Biomed Biotechnol. 2012;2012:735989.
39. Shi D, Kang Y, Zhang G, Gao C, Lu W, Zou H, Jiang H. Biodegradable atrial septal defect occluders: a current review. Acta Biomater. 2019;96:68–80.
40. Carminati M, Chessa M, Butera G, Bini RM, Giusti S, Festa P, Spadoni I, Redaelli S, Hausdorf G. Transcatheter closure of atrial septal defects with the STARFlex device: early results and follow-up. J Interv Cardiol. 2001;14(3):319–24.
41. Morgan G, Lee KJ, Chaturvedi R, Benson L. A biodegradable device (BioSTAR) for atrial septal defect closure in children. Catheter Cardiovasc Interv. 2010;76(2):241–5.
42. Huang Y, Wong YS, Ng HCA, Boey FYC, Venkatraman S. Translation in cardiovascular stents and occluders: from biostable to fully degradable. Bioeng Transl Med. 2017;2(2):156–69.
43. Hoehn R, Hesse C, Ince H, Peuster M. First experience with the BioSTAR-device for various applications in pediatric patients with congenital heart disease. Catheter Cardiovasc Interv. 2010;75(1):72–7.
44. Jux C, Bertram H, Wohlsein P, Bruegmann M, Paul T. Interventional atrial septal defect closure using a totally bioresorbable occluder matrix: development and preclinical evaluation of the BioSTAR device. J Am Coll Cardiol. 2006;48(1):161–9.
45. Mullen MJ, Hildick-Smith D, De Giovanni JV, Duke C, Hillis WS, Morrison WL, Jux C. BioSTAR evaluation STudy (BEST): a prospective, multicenter, phase I clinical trial to evaluate the feasibility, efficacy, and safety of the BioSTAR bioabsorbable septal repair implant for the closure of atrial-level shunts. Circulation. 2006;114(18):1962–7.
46. Van den Branden BJ, Post MC, Plokker HW, ten Berg JM, Suttorp MJ. Patent foramen ovale closure using a bioabsorbable closure device: safety and efficacy at 6-month follow-up. JACC Cardiovasc Interv. 2010;3(9):968–73.
47. Pavcnik D, Tekulve K, Uchida BT, Luo ZH, Jeromel M, Van Alstine WG, Keller FS, Rösch J. Double BioDisk: a new bioprosthetic device for transcatheter closure of atrial septal defects—a feasibility study in adult sheep. Radiol Oncol. 2012;46(2):89–96.
48. Sigler M, Söderberg B, Schmitt B, Mellmann A, Bernhard J. Carag bioresorbable septal occluder (CBSO): histopathology of experimental implants. EuroIntervention. 2018;13(14):1655–61.
49. Sievert H, Söderberg B, Mellmann A, Bernhard J, Gafoor S, EuroIntervention first human use and intermediate follow-up of a septal occluder with a bioresorbable framework. EuroPCR, 2015; Paris.
50. Kenny DP, Hijazi ZM. Current status and future potential of transcatheter interventions in congenital heart disease. Circ Res. 2017;120(6):1015–26.
51. Duong-Hong D, Tang YD, Wu W, Venkatraman SS, Boey F, Lim J, Yip J. Fully biodegradable septal defect occluder-a double umbrella design. Catheter Cardiovasc Interv. 2010;76(5):711–8.
52. Huang Y, Wong YS, Wu J, Kong JF, Chan JN, Khanolkar L, Rao DP, Boey FY, Venkatraman SS. The mechanical behavior and biocompatibility of polymer blends for patent ductus arteriosus (PDA) occlusion device. J Mech Behav Biomed Mater. 2014;36:143–60.
53. Liu SJ, Peng KM, Hsiao CY, Liu KS, Chung HT, Chen JK. Novel biodegradable polycaprolactone occlusion device combining nanofibrous PLGA/collagen membrane for closure of atrial septal defect (ASD). Ann Biomed Eng. 2011;39(11):2759–66.
54. Xie ZF, Wang SS, Zhang ZW, Zhuang J, Liu XD, Chen XM, Zhang G, Zhang D. A novel-design poly-L-lactic acid biodegradable device for closure of atrial septal defect: long-term results in swine. Cardiology. 2016;135(3):179–87.
55. Li BN, Xie YM, Xie ZF, Chen XM, Zhang G, Zhang DY, Liu XD, Zhang ZW. Study of biodegradable occluder of atrial septal defect in a porcine model. Catheter Cardiovasc Interv. 2019;93(1):E38–45.
56. Wu W, Yip J, Tang YD, Khoo V, Kong JF, Duong-Hong D, Boey F, Venkatraman SS. A novel biodegradable septal defect occluder: the "Chinese lantern" design, proof of concept. Innovations (Phila). 2011;6(4):221–30.
57. Lu W, Ouyang W, Wang S, Liu Y, Zhang F, Wang W, Pan X. A novel totally biodegradable device for effective atrial septal defect closure: a 2-year study in sheep. J Interv Cardiol. 2018;31(6):841–8.
58. Zhu Y, Liu J, Wang L, Guan X, Luo Y, Geng J, Geng Q, Lin Y, Zhang L, Li X, Lu Y. Preliminary study of the application of transthoracic echocardiography-guided three-dimensional printing for the assessment of structural heart disease. Echocardiography. 2017;34(12):1903–8.
59. Huang XM, Zhu YF, Cao J, Hu JQ, Bai Y, Jiang HB, Li ZF, Chen Y, Wang W, Qin YW. Development and preclinical evaluation of a biodegradable ventricular septal defect occluder. Catheter Cardiovasc Interv. 2013;81(2):324–30.
60. Balk M, Behl M, Wischke C, Zotzmann J, Lendlein A. Recent advances in degradable lactide-based shape-memory polymers. Adv Drug Deliv Rev. 2016;107:136–52.
61. Cha KJ, Lih E, Choi J, Joung YK, Ahn DJ, Han DK. Shape-memory effect by specific biodegradable polymer blending for biomedical applications. Macromol Biosci. 2014;14(5):667–78.

62. Hardy JG, Palma M, Wind SJ, Biggs MJ. Responsive biomaterials: advances in materials based on shape-memory polymers. Adv Mater. 2016;28(27):5717–24.
63. Xu J, Song J. Polylactic acid (PLA)-based shape-memory materials for biomedical applications. In: Shape memory polymers for biomedical applications. Elsevier Ltd.; 2015. p. 197–217.
64. Zheng Y, Li Y, Hu X, Shen J, Guo S. Biocompatible shape memory blend for self-expandable stents with potential biomedical applications. ACS Appl Mater Interfaces. 2017;9(16):13988–98.
65. Jia H, Gu S-Y, Chang K. 3D printed self-expandable vascular stents from biodegradable shape memory polymer. Adv Polym Technol. 2018;37:3222–8.
66. Zarek M, Layani M, Cooperstein I, Sachyani E, Cohn D, Magdassi S. 3D printing of shape memory polymers for flexible electronic devices. Adv Mater. 2016;28(22):4449–54.
67. Noecker AM, Chen JF, Zhou Q, White RD, Kopcak MW, Arruda MJ, Duncan BW. Development of patient-specific three-dimensional pediatric cardiac models. ASAIO J. 2006;52(3):349–53.
68. Wang Z, Liu Y, Xu Y, Gao C, Chen Y, Luo H. Three-dimensional printing-guided percutaneous transcatheter closure of secundum atrial septal defect with rim deficiency: first-in-human series. Cardiol J. 2016;23(6):599–603.
69. Milano EG, Capelli C, Wray J, Biffi B, Layton S, Lee M, Caputo M, Taylor AM, Schievano S, Biglino G. Current and future applications of 3D printing in congenital cardiology and cardiac surgery. Br J Radiol. 2019;92(1094):20180389.
70. Olivieri LJ, Krieger A, Loke YH, Nath DS, Kim PC, Sable CA. Three-dimensional printing of intracardiac defects from three-dimensional echocardiographic images: feasibility and relative accuracy. J Am Soc Echocardiogr. 2015;28(4):392–7.
71. Kim MS, Hansgen AR, Wink O, Quaife RA, Carroll JD. Rapid prototyping: a new tool in understanding and treating structural heart disease. Circulation. 2008;117(18):2388–94.
72. He L, Cheng GS, Du YJ, Zhang YS. Feasibility of device closure for multiple atrial septal defects with an inferior sinus venosus defect: procedural planning using three-dimensional printed models. Heart Lung Circ. 2020;29(6):914–20.
73. Li P, Fang F, Qiu X, Xu N, Wang Y, Ouyang WB, Zhang FW, Hu HB, Pan XB. Personalized three-dimensional printing and echoguided procedure facilitate single device closure for multiple atrial septal defects. J Interv Cardiol. 2020;2020:1751025.
74. Velasco Forte MN, Byrne N, Valverde I, Gomez Ciriza G, Hermuzi A, Prachasilchai P, Mainzer G, Pushparajah K, Henningsson M, Hussain T, Qureshi S, Rosenthal E. Interventional correction of sinus venosus atrial septal defect and partial anomalous pulmonary venous drainage: procedural planning using 3D printed models. JACC Cardiovasc Imaging. 2018;11(2 Pt 1):275–8.
75. Yan C, Wang C, Pan X, Li S, Song H, Liu Q, Xu N, Wang J. Three-dimensional printing assisted transcatheter closure of atrial septal defect with deficient posterior-inferior rim. Catheter Cardiovasc Interv. 2018;92(7):1309–14.
76. Mashiko T, Otani K, Kawano R, Konno T, Kaneko N, Ito Y, Watanabe E. Development of three-dimensional hollow elastic model for cerebral aneurysm clipping simulation enabling rapid and low cost prototyping. World Neurosurg. 2015;83(3):351–61.
77. Schmauss D, Haeberle S, Hagl C, Sodian R. Three-dimensional printing in cardiac surgery and interventional cardiology: a single-Centre experience. Eur J Cardiothorac Surg. 2015;47(6):1044–52.
78. Hunter KS, Lanning CJ, Chen SY, Zhang Y, Garg R, Ivy DD, Shandas R. Simulations of congenital septal defect closure and reactivity testing in patient-specific models of the pediatric pulmonary vasculature: a 3D numerical study with fluid-structure interaction. J Biomech Eng. 2006;128(4):564–72.
79. Rigatelli G, Zuin M, Fong A. Computational flow dynamic analysis of right and left atria in patent foramen Ovale: potential links with atrial fibrillation. J Atr Fibrillation. 2018;10(5):1852.
80. Lee W, Jung E. A multiscale model of cardiovascular system including an immersed whole heart in the cases of normal and ventricular septal defect (VSD). Bull Math Biol. 2015;77(7):1349–76.

4 Aortic Coarctation: Clinical Concepts, Engineering Applications, and Impact of an Integrated Medico-Engineering Approach

Damien P. Kenny and John F. LaDisa Jr

4.1 Introduction (Disease, Pathophysiology, Natural and Unnatural History, Possible Therapies) Developed by the Clinician (Dr Kenny)

Coarctation of the aorta (CoA) is the fifth most common congenital heart defect, accounting for 6–8% of live births with congenital heart disease, with an estimated incidence of 1 in 2500 births [1]. It is likely that the incidence is higher in stillborn babies. There is preponderance in the male sex with a reported ratio of between 1.27:1 and 1.74:1. It is usually manifested by a discrete constriction of the aortic isthmus; however, it is more likely to represent a spectrum of aortic narrowing from a discrete entity to tubular hypoplasia with many variations seen in between these two extremes. Morphologists argue that tubular hypoplasia although it may coexist with discrete coarctation should be considered as a separate entity although this has not been proven. The presence of associated hypoplasia is relevant to longer-term risk for the development of hypertension.

The etiology of the discrete isthmic constriction of the aorta seen in patients with CoA remains very much in dispute. Although familial cases have been reported [2], and association with various gene deletions described [3], there is no experimental evidence to support a unifying theory, not helped by the absence of naturally occurring animal models. Developmental theories have focused on abnormalities of blood flow [4], abnormal migration patterns of the developing aortic arch, and excessive distribution of arterial duct-like tissue around the aortic isthmus [5]. Such singular mechanistic views do not reflect the widespread changes seen in both left heart structures (mitral valve abnormalities, bicuspid aortic valve) and upper body vascular structure (cerebral aneurysms), commonly associated with CoA. Peterson et al. [3] have demonstrated that changes induced by a gridlock mutation in the hey2 gene in the zebra fish lead to changes mimicking CoA in this species. Interestingly, inducing upregulation of vascular endothelial growth factor (VEGF) early in development is sufficient to suppress the gridlock phenotype and aortic abnormality in this model. VEGF plays a vital

D. P. Kenny
Department of Paediatric Cardiology, Our Lady's Children's Hospital, Crumlin, Dublin, Ireland
e-mail: damien.kenny@olchc.ie

J. F. LaDisa Jr (✉)
Department of Biomedical Engineering, Marquette University and the Medical College of Wisconsin, Milwaukee, WI, USA

Department of Pediatrics - Section of Pediatric Cardiology, Medical College of Wisconsin and the Herma Heart Institute, Children's Wisconsin, Milwaukee, WI, USA
e-mail: jladisa@mcw.edu

G. Butera et al. (eds.), *Modelling Congenital Heart Disease*,
https://doi.org/10.1007/978-3-030-88892-3_4

role in aortic development, acting as a chemoattractant, stimulating angioblast migration toward the midline before formation of the aorta [6]. Indeed, targeted disruption of VEGF in mice leads to significant disruption of the developing aorta [7]. VEGF is also involved in stimulating generalized arterial differentiation through its effect on angioblast migration. Whether an initial mutation leads to secondary effects on VEGF or on other signaling systems involved in recruiting mural cells in fetuses, leading to CoA, is unknown. However, a more widespread vascular abnormality might be expected if this were so, and numerous studies have demonstrated normal lower limb vascular structure and function both before and after early coarctation repair [8, 9]. The regional vasculopathy in patients with CoA is therefore more likely to be due to an effect of abnormal hemodynamics as a consequence of isthmic narrowing, but whether this is in response to reduced upper body flow dynamics or to increased intra-arterial pressure load is unclear. An increase in collagen and decrease in smooth muscle content of the pre-coarctation aorta in humans have been demonstrated in comparison with post-coarctation aorta or to proximal aorta of young transplant donors [10]. This is consistent with a rabbit model of CoA, where increased gene expression for collagen types I and III has been demonstrated in the aorta proximal to the coarctation site [11]. These investigators postulated that the mechanical stress associated with increased pressure load may initiate rapid gene expression for collagen production, leading to re-enforcement and reorganization of the vessel musculo-elastic fascicle, and thereby reducing the degree of pressure-induced aortic dilatation. However, a clear disadvantage of this is that the resultant stiffer vessel will lead to augmented central aortic systolic pressure and systolic hypertension.

The most notable early abnormality on fetal echocardiography of potential aortic coarctation is an imbalance in ventricular size with right ventricular dominance. This is likely to be due to increased left ventricular diastolic pressures secondary to the coarctation, reduced right-to-left flow across the foramen ovale, and an increased volume load to the right ventricle. It is possible that these flow abnormalities may have a long-term effect on ventricular development and function, which in turn may affect blood pressure control. For example, in the fetal sheep an increase in the load on the heart leads to an increase in the number of binucleated myocardial cells [12], which unlike their mononuclear counterparts are not able to divide. Although this has implications for the total number of myocardial cells in the heart, it also may influence how the heart will respond to pressure load in the future. Numerous studies have demonstrated increased ventricular mass in normotensive coarctectomy patients associated with changes in aortic arch morphology [13, 14] or linked to angiotensin-converting enzyme polymorphism. A recent report showed that increased ventricular systolic stiffness, coupled to arterial stiffness, may be implicated in hypertension in post-coarctectomy patients [15], and thus, abnormalities in ventricular development may be important in the hypertensive response.

4.1.1 Natural and Unnatural History

Untreated CoA has a significant early mortality with one report identifying CoA in 17% of neonates dying from congenital heart disease [16]. Most of those who survive infancy reach adult life; however, the mean age of death has been reported as low as 31 years for those surviving the first year without an operation. Even in the treated population, there is significant early morbidity and mortality with Cohen et al. reporting estimated mortality of almost 30% at 30 years following repair with average age of death of 38 years (Fig. 4.1) [17].

These outcomes are substantiated by other studies with Clarkson et al. reporting on 160 patients with 10–28 years of follow-up, and observing that only 20% of patients were alive without complications and with a normal blood pressure at 25 years of follow-up [18]. To some extent, these are historical reports and the identification and treatment outcomes of patients with congenital heart disease have improved significantly over the past two decades, and early to

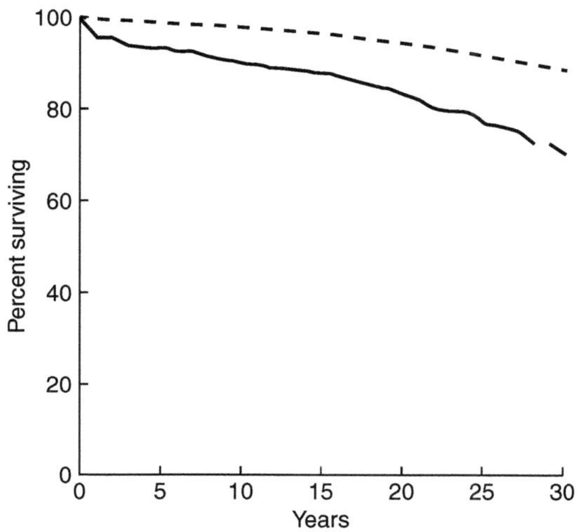

Fig. 4.1 from Cohen et al. [17] observed survival curves to 30 years of 588 surgically treated patients (solid line) and the expected survivorship of an age- and sex-matched population based on cohort life tables (dashed line). Reproduced from the American Heart Association under license 4,713,071,232,082 for Fig. 1 on page 842 of Marc Cohen, Valentin Fuster, Peter M Steele, David Driscoll, Dwight C McGoon. Coarctation of the aorta. Long-term follow-up and prediction of outcome after surgical correction in Circulation. 1989 Oct;80 (4):840–5 first accepted for publication May 23, 1989

midterm outcomes for patients with CoA are excellent with early mortality rates as low as 2% [19]. Significant longer-term morbidity remains, however, with more recent reports observing less than 50% of patients with normal blood pressure 1–27 years following CoA repair (Fig. 4.2) [20].

The average age at operation of this cohort was 9 years, and there are evolving data to suggest that earlier age at surgery may be protective against the development of hypertension [21, 22]. Although early surgery may prevent or delay the onset of hypertension, approximately 30% of children will be hypertensive by adolescence despite early surgery [23] and it is now arguable that hypertension is the single most important outcome variable in patients with repaired CoA. Most studies report resting blood pressure, and it is well accepted that a significant number of patients with normal resting blood pressure in the setting of CoA have an exaggerated blood pressure response to exercise, which may predict the onset of established hypertension [24].

4.1.2 Hypertension

Hypertension is likely to be involved in all of the end-organ dysfunction seen in patients with CoA, namely left heart failure, accelerated coronary artery disease, and cerebral aneurysm formation so much so that Cohen et al. [17] state that "the higher the postoperative systolic pressure, the higher the probability of death." The mechanisms driving this hypertensive response are poorly understood but are likely to involve vascular, central neural, and renal control mechanisms [25]. Investigators to date have tended to concentrate on single systems to evaluate their contribution to hypertension in CoA patients with reports indicating that reduced baroreceptor sensitivity, increased arterial stiffness, reduced endothelial function, and hyperactivation of the renin–angiotensin system secondary to reduced renal blood flow are implicated. These systems are widely regarded as the main regulators of blood pressure control and have been extensively researched in the pathogenesis of hypertension in adults.

1. ***Autonomic Function:*** Sealy et al. in 1957 suggested that initial rises in systolic blood pressure following repair of CoA may be due to resetting of the systemic arterial baroreceptors to chronically elevated pressure [26]. This author went on to demonstrate an increased release of noradrenaline in postoperative CoA patients and suggested that following CoA repair proximal arterial baroreceptors compensate for the reduction in proximal arterial pressure by increasing sympathetic nerve activity [27]. Later work by Inger et al. strengthened this postulate demonstrating resetting of arterial baroreceptors to operate at higher pressures in a canine model. Interestingly, in this study repair of the coarctation returned baroreceptor function toward normal [28]. Reports in older children with later repair and hypertension have also demonstrated baroreceptor dysfunction, and whether this is related to age of repair and potential chronic changes in aortic wall compliance, which houses arterial baroreceptor afferent fibers, is not clear.

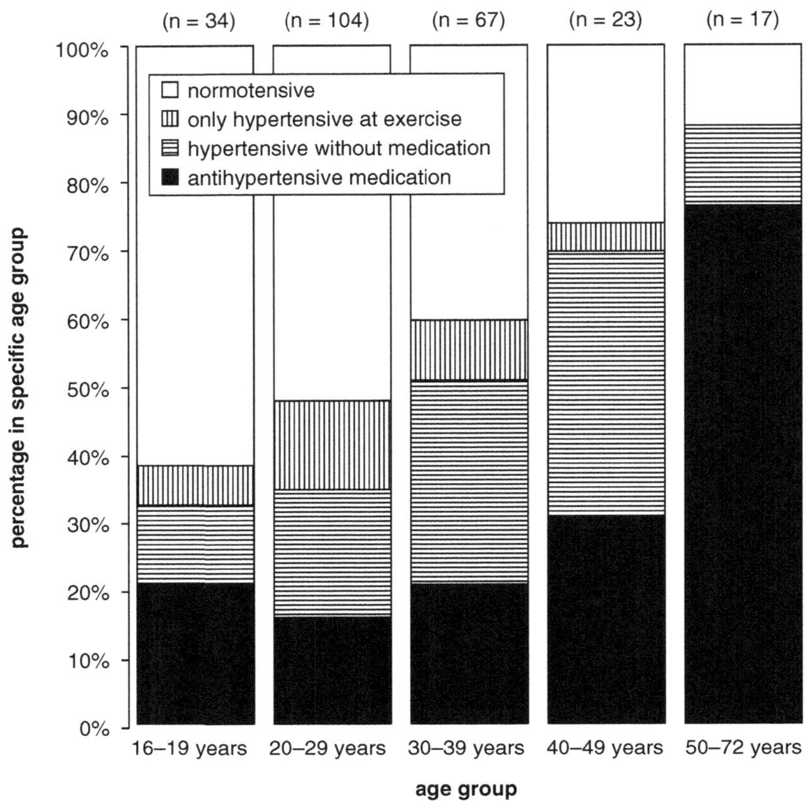

Fig. 4.2 From Hager et al. [20] provides a report of the hypertensive status and its management for CoA patients in the decades following repair. Reproduced from Elsevier under license 4,713,070,971,203 for Fig. 1 on page 742 of Alfred Hager, Simone Kanz, Harald Kaemmerer, Christian Schreiber, John Hess. Coarctation Long-term Assessment (COALA): significance of arterial hypertension in a cohort of 404 patients up to 27 years after surgical repair of isolated coarctation of the aorta, even in the absence of restenosis and prosthetic material in The Journal of Thoracic and Cardiovascular Surgery. 2007 Sep; 134 (3):738–45 first accepted for publication April 26, 2007

More recently, changes in baroreceptor reflex function have been demonstrated in preoperative neonates with CoA. Polson et al. evaluated autonomic control of blood pressure and heart rate in a group of neonates with CoA, prior to surgery ($n = 8$), and compared these findings to normal age- and sex-matched controls ($n = 13$) [29]. The study demonstrated a reduction in baroreceptor gain and heart rate variability of approximately 40% in the neonates with CoA, suggesting early maladaptive autonomic control of blood pressure. The authors postulated at the time that failure of these control mechanisms to normalize in some patients could cause long-term impediments to normal blood pressure control, possibly leading to hypertension. Reports on animals, however, have demonstrated normalization of these autonomic variables indicating that CoA repair offers good recovery of autonomic function. The development of autonomic control of blood pressure in patients with CoA throughout childhood has not been examined. How this control results relate to risk of hypertension in later years is unclear.

2. ***Vascular function:*** Systolic blood pressure is dependent on both large artery stiffness and cardiac output, whereas diastolic blood pressure is dependent on peripheral vascular resistance [29]. Increased arterial stiffness has been strongly linked with the development of

hypertension, and it is an independent predictor of longitudinal increases in blood pressure with aging. Numerous investigators have demonstrated a link between increased arterial stiffness and hypertension in post-coarctectomy patients [30]. These changes, however, appear to be restricted to the upper limb arteries and are more pronounced in those with later repair suggesting both a regional effect and temporal effect of the coarctation large artery stiffness. More recently, in order to examine the effects of CoA on fetal and early neonatal arterial stiffness, Vogt et al. measured local arterial stiffness indices and distensibility in the ascending and descending aortas of pre- and postoperative CoA neonates ($n = 17$) and compared these values to matched controls ($n = 17$) [8]. The authors demonstrated significantly reduced distensibility and increased stiffness indices in the ascending aortas of the pre- and postoperative group when compared to controls. There was no difference in the elastic properties of the descending aorta between the two groups. The same group ($n = 15$) were prospectively re-evaluated at 3 years of age, and aortic elastic properties were measured in a similar fashion. Persisting impairment of local elastic properties of the ascending aorta was noted in the CoA group when compared to controls. Longer-term evaluation of these parameters is required to determine if arterial properties normalize in those with normal blood pressure and remain abnormal in those with evolving hypertension. There was no indication of individual measure of arterial stiffness either initially or at follow-up, and therefore, a quantitative or predictive quality to the values cannot be commented on.

Reduced endothelial function has also been demonstrated in post-coarctectomy patients and has been suggested as a cause of hypertension. Reduced vascular reactivity appears to be restricted to the prestenotic arterial tree, and subsequent studies have demonstrated that these changes do not appear to be related to timing of surgery suggesting early changes in control of vascular reactivity. Whether this represents early programming is unclear as this mechanism has not been studied in young preoperative patients, and adult studies generally recognize endothelial dysfunction as a consequence rather than a cause of hypertension. Diffuse endothelial dysfunction is also likely to affect peripheral vascular resistance, which has most profound effects on mean and diastolic blood pressure values rather than systolic values and pulse pressure that are commonly raised in hypertensive CoA patients.

3. ***Renin–angiotensin system:*** Early studies examining the impact of the renin–angiotensin system (RAS) on hypertension in CoA concentrated on plasma renin levels with equivocal results, mostly demonstrating no significant increases in plasma renin levels in these patients [31, 32]. Subsequent studies have evaluated these levels following alterations to resting homeostasis such as fluid depletion or exercise. Although Parker et al. [33] demonstrated increased preoperative plasma renin levels following significant volume depletion, the subjects studied were older children (5–16 years) and the values normalized postoperatively including in those who remained hypertensive, thereby making a causal link for sustained long-term hypertension unlikely. Currently, most centers repair coarctation of the aorta in neonatal life, and thus, a prolonged period of renal hypoperfusion is not normal with transductal flow ensuring adequate renal perfusion before birth. In patients with late presentation of CoA, there is often significant collateral circulation ensuring that renal perfusion is not significantly affected; however, hypertension remains common suggesting that persistently elevated levels of renal renin or angiotensin although perhaps involved are not the primary mechanism involved in the development of long-term hypertension in these patients. It is less clear whether upper body blood pressure increases seen with developing coarctation induce changes in the overall number or sensitivity of angiotensin II type I receptors (AT1) in the brain. Sangaleti et al. [34] have

demonstrated that coarctation hypertension in the rat is associated with hyperactivity of the brain RAS as indicated by increased expression of AT1 mRNA in brainstem areas known to participate in cardiovascular control. It is possible that these receptors are involved in the progression of hypertension in post-coarctectomy patients involving the cardiac baroreceptor. This is more likely than a direct effect of angiotensin II on the arteriolar bed as this would not explain the differential changes seen in the upper and lower body and the expected effect of increased peripheral vascular resistance with angiotensin II is not typical of the systolic hypertension seen in CoA.

It is possible that more than one of the systems described above are involved in the delayed hypertensive response seen in some patients following CoA repair. Initial evidence for links between abnormal arterial structure and baroreceptor functioning was suggested by Sehested et al. [10]. The authors examined freshly resected coarctation tissue and demonstrated reduced isometric tension induced by potassium, noradrenaline, and prostaglandin in the prestenotic aortic tissue compared to the poststenotic area indicating reduced contractility of the prestenotic aorta. This was associated with increased collagen and reduced smooth muscle content of the prestenotic aortic wall. The authors postulated that aortic arch baroreceptors in this prestenotic area may be activated less at a given pressure than receptors placed in a vessel with normal distensibility, thus allowing a higher pressure to be tolerated by the cardiac baroreflex. The inter-relationship between reduced arterial compliance and a less sensitive baroreceptor reflex has been reported in other forms of secondary hypertension but has yet to be examined in patients with CoA. The progression in this inter-relationship throughout childhood when control mechanisms are potentially set for life is also of significant interest. For example, it is unclear, however, whether the baroreceptor reflex is dysfunctional from an early age as suggested by some investigators [30] and is therefore predictive of those likely to develop hypertension, or whether it becomes progressively less sensitive throughout childhood in conjunction with changes in vascular function.

4.1.3 Therapies

Currently, therapeutic options for CoA include (i) anatomical relief of the obstruction and (ii) management of hypertension. These therapies have focused on easily measurable markers such as Doppler gradient on echocardiography and peripheral blood pressure measurement. However, neither of these measurement variables provide data on the potential pathophysiological factors driving the hypertensive response as discussed above. Much debate exists in relation to optimal approach for relief of the obstruction with surgical repair preferred to ballooning or stenting in early infancy and stenting preferred to surgery in older children. However, the effect of these approaches has not been compared in relation to flow dynamics and vascular remodeling as discussed in the following section. Most studies evaluating treatment of hypertension have looked at peripheral blood pressure response to therapy rather than central blood pressure response. Indeed, outcomes are more closely related to central blood pressure measurements, responsible for perfusion pressures to the major organs.

4.2 Engineering Applications: Developed by the Engineer (Dr LaDisa)

4.2.1 Simulation-Based Modeling

Mechanical stimuli such as pressure and strain have been shown to influence the onset and progression of cardiovascular diseases. For example, wall tension can be estimated as the product of vessel radius and blood pressure. Chronic changes in wall tension initially driven by increases in blood pressure are believed to be the stimuli for vessel thickening, which then restores wall stress to a preferred operating range via changes in vessel thickness [35]. Strain similarly reflects changes in aortic deformation (e.g., pressure-induced dilation) as a response to vascular adaptation presenting in conditions like hypertension [36]. One of the most commonly reported indices from computational modeling studies is wall shear stress (WSS), which can be generally defined as the frictional force exerted on the walls of a vessel as a result of flowing blood. Areas of low time-averaged WSS (TAWSS) and elevated oscillatory shear index (OSI) are known to correlate with sites of atherogenesis and inflammation [37, 38] have also been found in a rotating pattern down the descending aorta [35], correlate with areas of plaque [39], and are accentuated after correction of CoA [40]. The associations above suggest that specific alterations in mechanical stimuli presenting from a given congenital heart disease (CHD) may serve as stimuli ultimately contributing to morbidity. Hence, there is value in further characterizing mechanical stimuli from CoA, as well as knowing how such stimuli lead to structural, functional, and mechanistic vascular alterations.

Investigation into the mechanical contributors to morbidity in CoA is exciting considering recent advancements in computational modeling. Patient-specific anatomy can be extracted, and representative models of the vasculature can be created using data obtained during a routine clinical imaging session. This anatomic data, together with physiological measurements, can be used to create 3D patient-specific representations of hemodynamics that consider vascular properties associated with a patient's current state. Alternatively, data from multiple patients may be used to create idealized or aggregate representations of a patient population. These approaches have been applied with single ventricle congenital defects where computational fluid dynamic (CFD) simulations of the Fontan procedure have led to several technical modifications demonstrated to be hemodynamically superior to previous surgical techniques [41]. Researchers employing computational modeling for CoA are eager to achieve similar benefits, but there are several considerations that require careful attention in order to make such studies transferable to the clinical condition.

The general requirements for computational vascular modeling related to CHD include first creating a model of the vessel geometry from three-dimensional (3D) medical imaging data. Volumetric data such as cardiac magnetic resonance (CMR) or computed tomographic angiography (CTA) data that provide clear definition of anatomy are readily available as they are often clinically indicated in many conditions. CFD also requires prescription of information for the entrance and exit of vessels. In many cases, inflow waveforms are obtained from phase-contrast (PC) velocity encoded CMR data acquired at these sites. It is also necessary to prescribe the hemodynamic state beyond the borders of the 3D imaging dataset (i.e., computational model) in order to obtain physiologic results (e.g., setting downstream resistance to obtain a realistic range of pressure). These inlet and outlet prescriptions are often referred to as boundary conditions. Rheological properties, such as blood density and viscosity, are then assigned. The last step in the process entails the use of a powerful computer or cluster of computers to solve the governing equations for fluid flow throughout a version of the vessel geometry that is represented as a computational mesh. When considering these requirements relative to CoA, the complexity of aortic flow patterns, selection of physiologic boundary conditions, replicating vascular compliance, including collateral arteries and vascular tethering, and the motion of the aorta all contribute to the difficulty of the problem.

In 1971, O'Rourke and Cartmill suggested that much of the morbidity observed in patients with CoA can be explained on the basis of abnormal hemodynamics through the ascending aorta and its branches [42]. The authors specifically described what they referred to as the conduit (i.e., blood flow) and cushioning (i.e., capacitance) function in the aortic arch being altered by CoA. They described how the presence of CoA introduces a pressure wave reflection site near the heart causing drastic reductions in aortic capacitance and elevates pulse pressure. These findings are consistent with hypertension often observed during rest and elevated blood flow conditions. Moreover, associated reductions in diastolic blood pressure observed from CoA [42] can theoretically alter coronary artery perfusion, thereby accentuating the likelihood for adverse hemodynamics associated with premature CAD, such as low WSS and OSI. Concomitant increases in afterload from the coarctation also explain the propensity for heart failure in untreated CoA patients. We subscribe to the notion suggested by O'Rourke and Cartmill, which suggests computational modeling studies of CoA that employ physiologic outlet boundary conditions and deformable walls have the greatest potential to replicate the clinical condition and foster translation. We therefore only focus our attention in the current work on studies with realistic inlet and outlet boundary conditions. We also primarily highlight studies that used deformable walls via fluid–structure interaction (FSI), although there are data suggesting rigid wall CFD simulations may be reasonable given the stiffness that ensues following pronounced exposure to CoA [43, 44].

As alluded to in the discussion above, mechanical indices involving vascular loading upstream of the coarctation seem to be principally involved in the morbidity of CoA [43]. For example, although WSS may be altered downstream of the coarctation and promote intimal thickening, WSS seems to be of secondary importance since severe stenosis from intimal thickening is unlikely in an adult aorta (~2.5 cm in diameter). In contrast, other mechanical stimuli influencing stiffening can theoretically explain aneurysm formation, early-onset coronary artery disease, and hypertension in regions above the coarctation.

One of the first computational modeling studies of CoA potentially relatable to the clinical condition was published by LaDisa and Figueroa et al. in 2011 [45]. The goals of this work were to use clinical imaging data and FSI to further understand hemodynamic alterations under resting and elevated (i.e., simulated exercise) blood flow conditions, uniquely investigate potential sources of morbidity through the use of patient-specific computational modeling, and evaluate treatment outcomes and coarctation severity in a manner not possible using current imaging modalities alone. More specifically, results from four CoA patients were contrasted against those from a normal subject. The four CoA patients studied were selected to provide initial results for a range of patients seen clinically. Patients selected for study included those with native CoA having modest (12 mmHg) and severe (25 mmHg) blood pressure gradients (BPG) as well as postoperative CoA patients treated by resection by end-to-end anastomosis and end-to-side anastomosis having no residual BPG. Simulations incorporated vessel deformation, downstream vascular resistance, and compliance through the use of 3-element Windkessel approximations. The use of a viscoelastic boundary condition also accounted for in vivo tissue tethering and helped dampen high-frequency fluttering modes of vessel wall motion that occurred from severe CoA and from the implementation of simulated exercise boundary conditions. The importance of simulating exercise, despite its challenges computationally, was highlighted in the historical review of CoA above. Outlet boundary conditions were tuned to match available clinical data using ranges for all parameters from the literature. Indices including cyclic strain, TAWSS, and OSI were then quantified and reported in intuitive ways that extracted as much data from the results as possible at the time. The main findings of the study were that the methods employed replicated values of indices like strain consistent with the pathophysiology seen in CoA patients and that simulated exercise could be used to further assess important mechanical indices for borderline patients having BPG near the current clinical treatment guidelines. Unwrapping of WSS results together with local quantification also helped elucidate the range of normal to potentially worst-case values for the first time.

More recently, Zhu et al. [46] created patient-specific CFD models of 25 CoA patients after multidetector CTA. The authors used velocity from transthoracic echocardiography (TTE) as input boundary conditions and physiologic outlet boundary conditions (i.e., 3-element Windkessel representations). There was good agreement between CTA and computational model dimensions when compared at the ascending aorta, level of the coarctation, and descending aorta. The peak systolic pressure values reported from simulation results were also in good agreement with catheter-based measurements (i.e., $R^2 = 0.918$), but this is not necessarily surprising given that boundary condition parameters are often tuned to match aimed clinical values such as those from catheterization. Velocity patterns and localized WSS distributions were similar to results represented previously [40]. Interestingly, the authors only reported WSS values during peak systole despite conducting pulsatile simulations. Notwithstanding some potential limitations, to our knowledge this is the largest computational study of CoA patients to date.

There have also been studies with physiologic boundary conditions aimed as characterizing turbulence intensity in CoA [47–49]. Pointing to the importance of turbulence in biological processes such as platelet activation, atherosclerosis, and intimal hyperplasia, Arzani et al. created a subject-specific model of 60-year-old patient previously repaired by resection with end-to-end anastomosis [47]. Turbulent kinetic energy (TKE) was calculated from the simulation and validated to that from the CMR session used for computational modeling. A follow-up study by a subset of the original authors used similar methods to further quantify TKE using vessel morphology after balloon angioplasty [49]. With a technical approach in place for the calculation of TKE based on CFD models of CoA, larger-scale studies linking this index to coarctation-induced pathology are likely forthcoming.

CoA rarely presents in isolation. For example, hypoplastic left heart syndrome (HLHS) often includes CoA (67–80%) [50]. Biglino et al. constructed a multiscale CFD model of the stage 1 Norwood circulation that included a modified Blalock–Taussig shunt in a patient with CoA and validated results to benchtop data [51]. Several computational studies, both idealized and patient-specific, have also characterized the combined impact of CoA in the setting of a bicuspid aortic valve (BAV), which is a concomitant occurrence in 50–80% of CoA patients [52]. Specifically, Wendell et al. implemented physiologic outlet boundary conditions along with computational masking at the inlet of a patient-specific CFD model of a 34-year-old male CoA patient with a BAV [48]. The masking served to constrain inlet velocity distributions to the orifice area of the BAV segmented from CMR. TKE, TAWSS, and OSI results were compared to equivalent simulations using a plug inlet profile. The plug inlet greatly underestimated TKE and TAWSS differences extended throughout the thoracic aorta when compared to those from the BAV simulation. OSI differences existed mainly in the ascending aorta. Importantly, differences in TAWSS from the use of the BAV vs plug inlet velocity profiles were greater than the interobserver variability in defining local geometry from a prior study [53] for nearly the entire thoracic aorta and its branches. These findings underscore the importance of inlet boundary conditions and including valve morphology in patient-specific models for CoA.

4.2.2 Use of Velocity in Estimating Coarctation-Induced Pressure Gradients

As alluded to above, a patient's BPG currently plays an important role in assessing the severity and potential treatment options for CoA. In many centers, ultrasound-based assessment of the BPG has become the first-line imaging modality for the assessment protocol applied with CoA patients [54]. This method relies on BPG estimated from peak systolic velocity using the simplified Bernoulli equation [55]. This simplification ignores contributions from the pressure recovery and viscous losses terms within the Bernoulli equation [56]. Perhaps more importantly, accurate assessment of peak velocity by ultrasound assumes that values obtained by the sonographer are measured at the specific location of maximum velocity within the extended velocity profile resulting from convective acceleration through the coarctation. In practice, difficulties can be present. For example, transmission of the ultrasound signal can be hindered by anatomical structures and depends on the angle of correction implemented. The potential for error with this approach can therefore be substantial depending on the patient and if measurements are not made carefully.

Perhaps not surprisingly, many of the image-based computational modeling studies conducted to date for CoA are therefore centered around improved estimation of BPG. For example, Ralovich et al. [57] and Shi et al. [58] have described their approaches to rapidly creating reduced order models that include physiologic boundary conditions and the impact of local aortic compliance. BPG from simulations was in good agreement with available clinical data, and run times were a fraction of those seen with full-scale CFD or FSI studies. One potential drawback of such reduced order models is that spatiotemporal hemodynamic results such as indices of WSS are not available with this approach.

The work of Zhu et al. mentioned above reports peak systolic velocity measured by TTE relative to that extracted from CFD simulations using physiologic boundary conditions. CFD results represent a surrogate version of the flow domain for each patient assuming careful segmentation of imaging data and proper study of mesh independence. In contrast to TTE results, CFD results can be searched throughout the computational flow domain for values of interest, such as peak systolic velocity. There was good agreement between velocity from the two approaches (i.e., $R^2 = 0.968$) but inconsistency in which measurement type is greater in value. For example, in Case 01 peak systolic velocity is reported as 160 cm/s by TTE and lower (i.e., 151 cm/s) by CFD, but for Case 2 peak systolic velocity was 85 cm/s by TTE and higher (i.e., 100 cm/s) by CFD. Although the authors' use of rigid walls introduces some error in their CFD results, this finding supports the point that ultrasound results are frequently taken within the coarctation-induced velocity profile, but not necessarily directly at the location of peak velocity.

In an effort to mitigate pressure drop overestimation by simplified Bernoulli equation [55] and the potential for operator-dependent error during Doppler ultrasound measurement, Saitta et al. recently published initial findings from noninvasive estimation of blood pressure using an algorithm that applies the Poisson pressure equation to 4D flow from CMR (4DF-FEPPE) [59]. Of note, the authors compared instantaneous and time-averaged 4DF-FEPPE results at 19 different cross-sections throughout the thoracic aorta of a 57-year-old CoA patient to spatially equivalent locations from an FSI model created from the same imaging data. The model applied many of the best practices desirable in current patient-specific simulations including use of 4D flow data to proscribe pulsatile flow at the inlet, use of deformable walls based on available material properties in the literature, careful consideration of mesh density, and use of outlet boundary conditions resulting in physiologic values (i.e., 3-element Windkessel representations) despite a lack of pressure data for the patient studied. Their results showed excellent agreement between pressure assessment calculated by 4DF-FEPPE vs that extracted from the patient-specific FSI simulation both spatially and temporally. Specifically, the Bland–Altman analysis of peak systolic, end-diastolic, and time-averaged values yielded biases (means of differences) of +0.4, +1.1, and + 0.6 mmHg, respectively. The corresponding limits of agreement (2 standard deviation of differences) were ± 0.978 mmHg, ±1.97 mmHg, and ± 1.06 mmHg, respectively. Agreement between peak-to-peak (ΔPpp) and maximum pressure drops from two cross-sectional planes located immediately proximally and distally to the coarctation was excellent between the two approaches (ΔPpp: 4DF-FEPPE = 17.6 vs FSI = 16.9 mmHg). These exciting methodological advancements have potential for clinical translation following application to a larger group of CoA patients of various ages and disease severities, hopefully with corresponding catheterization data.

4.2.3 Experimental Models

The etiology of long-term morbidity in CoA patients seems to stem from persistent arterial pathology [43, 44]. In contrast to the secondary consequences of exposing the aortic arch and arteries above the coarctation to mechanical stimuli including elevated blood pressure and a gradient across the narrowing upon closure of the ductus arteriosus, there putatively is a causal genetic basis for this pathology that may relate to the presentation of the initial narrowing. As mentioned in the introduction, the prevailing causal hypothesis is based on histology showing that tissue with features similar to the ductus arteriosus also exists near the coarctation, suggesting CoA may be created during closure of the DA in the first week of life [60]. Although the subsequent BPG has become a primary indication for treatment, there is a lack of data for the current ≥20 mmHg BPG treatment value. A recent review [61] called the evidence (Level C) for this BPG in the treatment guidelines [52] suboptimal. This suggests the putative BPG guideline reflects the opinion of many clinicians, but there are no directly applicable studies or randomized clinical trials. Although the evidence level is likely to improve over time, the putative BPG dates back to surgical outcomes from several decades ago [62] and seems to persist despite notable surgical advances.

The ability of CoA researchers to elucidate an ideal BPG for treatment and associated mechanisms of morbidity in humans with CoA is complicated by a relatively small number of patients at each center annually, their heterogeneity from confounding variables (age prior to surgery, time to follow-up, severity of CoA before correction), and concomitant anomalies such as BAV and septal defects. Mechanistic understanding of long-term morbidity in CoA, such as hypertension, is also precluded by the difficulty in separating causal genetic contributions from changes in gene expression associated with mechanical stimuli after closure of the ductus arteriosus. Recently, animal models of CoA have been described [63] to control for the variability and limitations seen in humans (CoA severity, duration, and subject age) without concomitant anomalies, and to study mechanisms of persistent cardiovascular dysfunction. The use of CoA to study hypertension in animals is not new, but at least some of these more recent models are different in that they use lower BPG that are representative of the condition seen in pediatrics and have allowed for the recovery of vasculature (i.e., plasticity) by simulating treatment with the use of absorbable suture. Importantly, blood pressure is similar between animals and humans [64], suggesting that pressure-based results from a rabbit model of CoA (for example) have potential for translation back to humans.

Recent experimental results from a rabbit model of CoA implementing a 20 mmHg BPG treatment value revealed arterial changes within the region exposed to high blood pressure. These changes included medial thickening and stiffening, and endothelial dysfunction, which were reviewed in the prior section and all persisted after treatment [43]. This persistence is a paradigm shift from the dogma of complete vascular plasticity and indicates a 20 mmHg BPG is not the ideal treatment threshold since it may contribute to long-term HTN via central aortic stiffening. Computational modeling using data from this rabbit model with a 20 mmHg BPG confirmed that the mechanical stimuli for remodeling above the level of the coarctation are blood pressure, wall tension, and strain. In contrast, WSS primarily contributed to persistent arterial dysfunction distal to untreated CoA [43] and has been studied elsewhere [40, 45]. CoA-induced increases in wall tension were also associated with increased nonmuscle myosin confirming a phenotypic shift in smooth muscle cells to the de-differentiated proliferating state, likely to increase medial thickness and restore stress within the aortic wall to a preferred level. Work is now underway using this model to identify and translate thresholds for arterial stiffening and dysfunction identified in rabbits for severity, duration, and age to the human condition of CoA.

4.2.4 Relationship with Genetic Markers

Genetic factors are believed to be associated with CoA, and may contribute to persistent morbidity after treatment [2, 65]. As mentioned above, the specific genes contributing to the pathology of CoA, and their functional relevance, remain unknown. Microarray analysis of aortic tissue from the animal model of CoA discussed above identified several differentially expressed genes in the region exposed to pronounced mechanical stimuli developing from CoA [66]. More recently, RNA sequencing (RNAseq) of aortic tissue from humans with CoA who had upper extremity systolic blood pressure > 99th percentile for their gender, height, and age to focus on mechanical consequences of the coarctation was used in concert with the prior microarray data from the animal model above to identify one or more candidate genes for use in future mechanistic studies of coarctation-induced arterial dysfunction. RNAseq from humans treated for CoA revealed downregulation of *natriuretic peptide receptor C (NPR-C; also known as NPR3)* in proximal sections of the thoracic aorta subjected to high blood pressure when compared to sections from regions in the distal thoracic aorta exposed to normal blood pressure. Importantly, microarray data from the experimental rabbit model of CoA also showed downregulation of *NPR-C* in proximal aortas from both CoA and treated rabbits experiencing high blood pressure, as compared to controls experiencing normal blood pressure. NPR-C is one of three receptors for natriuretic peptides that include atrial (ANP), brain (BNP), and C type (CNP). ANP and BNP are mostly found in the atria and ventricles, whereas CNP is abundantly expressed in vascular endothelial cells [67]. NPR-C is found in many tissues including smooth muscle cells where it has been shown to play a role in inhibiting proliferation and endothelial cells where recent literature suggests a role in re-endothelialization and viability under healthy conditions [68]. The normal response of a health aorta and human aortic endothelial cells decreased intracellular calcium ($[Ca^{2+}]_i$) activity in response to CNP administration. However, this normal activity and vascular relaxation induced by CNP, as well as ANP, were impaired for aortic segments exposed to elevated blood pressure from CoA. Inhibition of NPR-C (M372049) also impaired aortic relaxation and $[Ca^{2+}]_i$ activity. Prior studies point to a role for NPR-C in regulating blood pressure as systemic administration of an NPR-C agonist attenuated high blood pressure in spontaneously hypertensive rats by inhibiting enhanced levels of inhibitory guanine nucleotide regulatory protein (Gi) [69]. Activation of NPR-C increases eNOS activity resulting in the formation of nitric oxide and vasorelaxation via cGMP. While this prior literature implicated NPR-C in arterial dysfunction, the more recent study [70] seems to be the first to show altered *NPR-C* transcript levels in aortic tissue from human CoA patients, and promising evidence of functional ramifications to altered NPR-C activity from pronounced mechanical stimuli.

4.3 Potential Impact on Clinical Arena Now and in the Future (Clinician's Perspective) (Dr Kenny)

4.3.1 Relief of Obstruction

As we move toward demand for less invasive solutions to congenital cardiovascular lesions such as CoA, greater understanding of the longer-term physiological effect of a metal stent in the descending aorta is required. This has not been widely studied; however, evaluation in a small cohort of 12 adults pre- and poststenting demonstrated improvement in central aortic function poststenting but no change in peripheral vascular dysfunction [71]. Modeling techniques described above may facilitate preprocedural simulation to optimize hemodynamics. Indeed, several studies of stenting for CoA using realistic boundary conditions and, in some cases, commercially available stent designs have already been conducted [72, 73]. The ultimate stenting solution may be a bioresorbable scaffold implanted in neonatal life to provide early aortic remodeling and minimize scar or metal within the descending aorta. Aortic remodeling may be augmented by anti-inflammatory drug-eluting properties within the bioresorbable material, possibly involving one of the targets mentioned above. The optimal bioresorbable material and scaffold design have yet to be elucidated but will require sufficient radial strength despite requirements for low profile delivery.

4.3.2 Pharmacological Management of Hypertension in CoA

Several therapeutic agents are prescribed to children with hypertension after correction for CoA (e.g., ramipril, losartan, amlodipine, and propranolol). Beta blockers, ACE inhibitors, and angiotensin receptor blockers are recommended first-line agents [52]. In reality, the choice of agent depends on patient age, associated pathology, and personal bias since there is a paucity of data comparing the various agents in children with or without CoA. Despite different mechanisms of action for each agent, their common end result is to reduce blood pressure, thereby making it reasonable to surmise some may have a secondary effect of mitigating the specific persistent vascular changes observed after correction as discussed above. For example, endothelial dysfunction in CoA may result from increased vascular inflammation and Brill et al. [74] have reported that the ACE inhibitor ramipril decreases expression of pro-inflammatory cytokines in CoA patients in addition to reducing blood pressure.

Agents that agonize or antagonize specific proteins and pathways that are already altered by CoA may either further exacerbate the apparent irreversibility of vascular changes, or facilitate restoring vascular function closer to normal. Interestingly, a genomic data mining tool [75] revealed an association of differentially expressed genes from the CoA model above with nifedipine, an L-type calcium channel blocker. A brief review of the literature indicates no systematic reviews on the use of nifedipine for hypertension following correction of CoA, but it appears it may be prescribed for this purpose in children. This example illustrates continued movement toward personalized medicine based on genetic testing and the ability of such data in concert with those related to mechanical stimuli to revise current treatment guidelines such as those related to medical therapy. The potential for impact from other advancements reviewed above is further discussed below.

4.4 Potential Impact on Clinical Arena Now and in the Future (Engineer's Perspective) (Dr LaDisa)

The collection of advancements summarized above is likely to manifest in several ways clinically. We expect to see more computational models like that presented by Coogan, Humphrey, and Figueroa [76]. The authors implemented a growth and remodeling framework to update an extensive FSI simulation that included the cerebral vasculature based on changes resulting from a narrowing in the distal descending thoracic aorta. The use of computational modeling for use in clinical assessment is also likely to increase. In the past decade, several exceptional open-source software packages have emerged [77, 78]. These packages have features specifically designed to create idealized and patient-specific models based on medical imaging data, create complex computational meshes, perform simulations with rigid or deformable walls, and visualize results. These tools are being downloaded and used by emerging and established researchers at an exciting rate. When coupled with existing commercially available software packages and the proliferation of researchers interested in computational modeling, it is our opinion that the number of computational modelers will grow exponentially in the next decade. The ability to run publication quality simulations with mesh densities sufficient to provide confidence in mechanical indices of interest continues to grow with recent movement of high-performance computing to the cloud. This spreads the opportunity for clinical and remote collaboration beyond those institutions or organizations with dedicated high-performance computing equipment and personnel.

Perhaps most importantly, as mentioned in a recent review by Biglino and colleagues [79], there is a need for studies linking computational simulation results from CoA to improvements in patient outcomes. This means not only accurate and precise noninvasive estimation of BPG from imaging, for example, but also linking changes in stimuli elucidated from modeling to long-term morbidity in CoA. Such progress will involve merging of the advances above, such as those showing the relationship between coarctation-induced stimuli and arterial thickening. These relationships between stimuli and response will have to go beyond application in animals to also being applied with humans using available techniques that are perhaps not yet commonly applied to CoA. For example, optical coherence tomography (OCT; Fig. 4.3) and intravascular ultrasound have the potential to assess temporal vascular changes that are a manifestation of stimuli delineated by modeling. While aortic size may limit the use of OCT in older CoA patients due to aortic diameter, branch arteries from the aorta are likely more appropriate in their diameter, making assessment by OCT tractable for any CoA patients undergoing catheterization. The use of other clinical tools is particularly important as Astengo recently reported that Class I European Society of Cardiology recommendations, alone or in concert with Class IIa recommendations, had a relatively poor ability to predict whether CoA patients were to be hypertensive [80]. More accurate models, perhaps using results from OCT or other imaging modalities, will allow for the inclusion of additional predictors and vessel thickness is likely to increase the specificity of such approaches. Toward this end, the American College of Cardiology/American Heart Association 2008 Guidelines for the Management of Adults with Congenital Heart Disease recommends that every patient with CoA, whether repaired or not, has at least one complete evaluation of the thoracic aorta and intracranial vessels at least by CMR or CTA [52]. Evaluation of the coarctation repair site by CMR or CTA is also recommended at most every 5 years. The functional assessment that is possible with MRI, such as BPG estimates via calculation of 4DF-FEPPE from 4D flow, provides a strong argument for the use of CMR, rather than echocardiography [59], as the first-line diagnostic imaging tool for CoA. Computational models to emerge in the future may also ultimately predict vascular changes (including gene and associated protein expression) resulting from image-based patient-specific indices using additional novel methods yet to be derived or applied to CoA. Continued concerted efforts between clinicians and engineers will be required to increase the number of CoA patients studied and reported using such methods, likely by leveraging existing registries that include CoA patients.

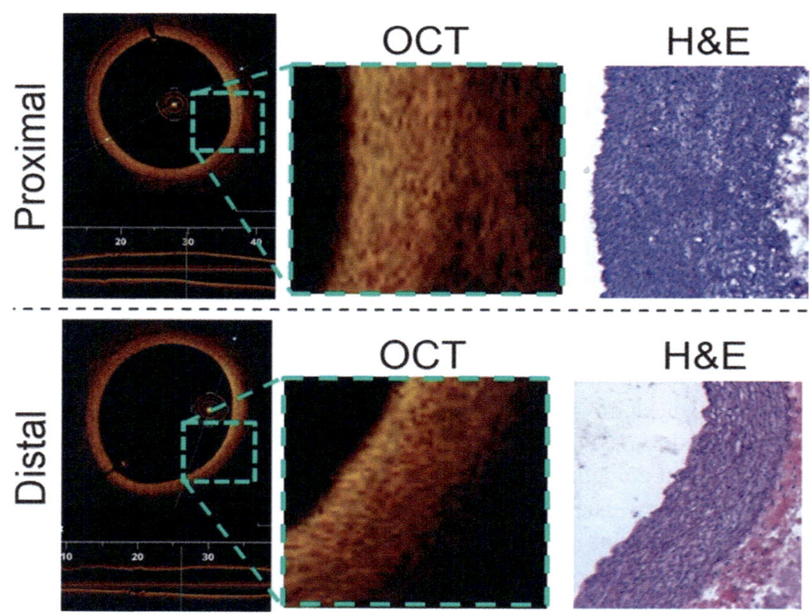

Fig. 4.3 OCT, which is clinically available but not often used in the study of CoA, may permit the study of temporally changes in vasculature in response to mechanical stimuli. The example here shows thickness and remodeling quantified at spatially equivalent locations for OCT and histology in response to a 20 mmHg BPG imposed in a rabbit

Acknowledgements Support was provided by the National Institutes of Health (NIH) grants R01HL142955 and R15HL096096 and American Heart Association Grant-In-Aid Award 15GRNT25700042. Portions of the data discussed were also supported by the National Center for Advancing Translational Sciences, National Institutes of Health, through Grant Number UL1TR001436. Contents are solely the responsibility of the authors and do not necessarily represent the official views of the NIH.

References

1. Anderson RH. Paediatric cardiology. Churchill Livingstone; 2002.
2. Sehested J. Coarctation of the aorta in monozygotic twins. Br Heart J. 1982;47:619–20.
3. Peterson RT, Shaw SY, Peterson TA, Milan DJ, Zhong TP, Schreiber SL, MacRae CA, Fishman MC. Chemical suppression of a genetic mutation in a zebrafish model of aortic coarctation. Nat Biotechnol. 2004;22:595–9.
4. Rudolph AM, Spitznas U, Heymann MA. Hemodynamic considerations in development of narrowing of aorta. Am J Cardiol. 1972;30:514.
5. Russell GA, Berry PJ, Watterson K, al. e. Patterns of ductal tissue in coarctation of the aortic arch. J Cardiovasc Surg. 1991:102, 596.
6. Cleaver O, Krieg PA. Vegf mediates angioblast migration during development of the dorsal aorta in xenopus. Development. 1998;125:3905–14.
7. Carmeliet P, Ferreira V, Breier G, Pollefeyt S, Kieckens L, Gertsenstein M, Fahrig M, Vandenhoeck A, Harpal K, Eberhardt C, Declercq C, Pawling J, Moons L, Collen D, Risau W, Nagy A. Abnormal blood vessel development and lethality in embryos lacking a single vegf allele. Nature. 1996;380:435–9.
8. Vogt M, Kuhn A, Baumgartner D, Baumgartner C, Busch R, Kostolny M, Hess J. Impaired elastic properties of the ascending aorta in newborns before and early after successful coarctation repair: proof of a systemic vascular disease of the prestenotic arteries? Circulation. 2005;111:3269–73.
9. de Divitiis M, Pilla C, Kattenhorn M, Zadinello M, Donald A, Leeson P, Wallace S, Redington A, Deanfield JE. Vascular dysfunction after repair of coarctation of the aorta: impact of early surgery. Circulation. 2001;104:I165–70.
10. Sehested J, Baandrup U, Mikkelsen E. Different reactivity and structure of the pre-stenotic and post-stenotic aorta in human coarctation: implications for baroreceptor function. Circulation. 1982;65:1060–5.
11. Xu CP, Zarins CK, Bassiouny HS, Briggs WH, Reardon C, Glagov S. Differential transmural distribution of gene expression for collagen types i and iii proximal to aortic coarctation in the rabbit. J Vasc Res. 2000;37:170–82.
12. Barbera A, Giraud GD, Reller MD, Maylie J, Morton MJ, Thornburg KL. Right ventricular systolic pressure load alters myocyte maturation in fetal sheep. Am J Physiol-Regul Integr Comp Physiol. 2000;279:R1157–64.
13. Ou P, Celermajer DS, Raisky O, Jolivet O, Buyens F, Herment A, Sidi D, Bonnet D, Mousseaux E. Angular (gothic) aortic arch leads to enhanced systolic wave reflection, central aortic stiffness, and increased left ventricular mass late after aortic coarctation repair: evaluation with magnetic resonance flow mapping. J Thorac Cardiovasc Surg. 2008;135:62–8.
14. Ou P, Mousseaux E, Celermajer DS, Pedroni E, Vouhe P, Sidi D, Bonnet D. Aortic arch shape deformation after coarctation surgery: effect on blood pressure response. J Thorac Cardiovasc Surg. 2006;132:1105–11.
15. Zadinello M, Greve G, Liu XQ, Barbosa JR, Schulze-Neick I, Wilkinson JL, Redington AN. Angiotensin I converting enzyme genotype affects ventricular remodelling in children with aortic coarctation. Heart. 2005;91:367–8.
16. Campbell M. Natural history of coarctation of the aorta. Br Heart J. 1970;32:633–40.
17. Cohen M, Fuster V, Steele PM, Driscoll D, McGoon DC. Coarctation of the aorta - long-term follow-up and prediction of outcome after surgical-correction. Circulation. 1989;80:840–5.
18. Clarkson PM, Nicholson MR, Barrattboyes BG, Neutze JM, Whitlock RM. Results after repair of coarctation of the aorta beyond infancy – a 10 year to 28 year follow-up with particular reference to late systemic hypertension. Am J Cardiol. 1983;51:1481–8.
19. Kaushal SBC, Patel JN, Patel SK, Walker BL, Weigel TJ, Randolph G, Wax D, Mavroudis C. Coarctation of the aorta: midterm outcomes of resection with extended end-to-end anastomosis. Ann Thorac Surg. 2009;88:1932–8.
20. Hager A, Kanz S, Kaemmerer H, Schreiber C, Hess J. Coarctation long-term assessment (coala): significance of arterial hypertension in a cohort of 404 patients up to 27 years after surgical repair of isolated coarctation of the aorta, even in the absence of restenosis and prosthetic material. J Thorac Cardiovasc Surg. 2007;134:738–U712.
21. Seirafi PA, Warner KG, Geggel RL, Payne DD, Cleveland RJ. Repair of coarctation of the aorta during infancy minimizes the risk of late hypertension. Ann Thorac Surg. 1998;66:1378–82.
22. Daniels SR. Repair of coarctation of the aorta and hypertension: does age matter? Lancet. 2001;358:89.
23. O'Sullivan JJ, Derrick G, Darnell R. Prevalence of hypertension in children after early repair of coarctation of the aorta: a cohort study using casual and 24 hour blood pressure measurement. Heart. 2002;88:163–6.
24. Ou P, Celermajer DS, Mousseaux E, Pedroni E, Brunelle F, Vouhe P, Sidi D, Bonnet D. Exercise-related hypertension predicts resting hypertension, left ventricular and vascular hypertrophy late after coarctation repair. Circulation. 2006;114:504.

25. de Divitiis M, Rubba P, Calabro R. Arterial hypertension and cardiovascular prognosis after successful repair of aortic coarctation: a clinical model for the study of vascular function. Nutr Metab Cardiovasc Dis. 2005;15:382–94.
26. Sealy WC, Harris JS, Young WG, Callaway HA. Paradoxical hypertension following resection of coarctation of aorta. Surgery. 1957;42:135–47.
27. Sealy WC, Goodall MC. Paradoxical hypertension following resection of coarctation of aorta - possible relationship of increased epinephrine + norepinephrine secretion to its development. Circulation. 1964;30:155.
28. Polson JW, McCallion N, Waki H, Thorne G, Tooley MA, Paton JF, Wolf AR. Evidence for cardiovascular autonomic dysfunction in neonates with coarctation of the aorta. Circulation. 2006;113:2844–50.
29. McEniery CM, Yasmin WS, Maki-Petaja K, McDonnell B, Sharman JE, Retallick C, Franklin SS, Brown MJ, Lloyd RC, Cockcroft JR, Wilkinson IB. Increased stroke volume and aortic stiffness contribute to isolated systolic hypertension in young adults. Hypertension. 2005;46:221–6.
30. Brili S, Tousoulis D, Antoniades C, Aggeli C, Roubelakis A, Papathanasiu S, Stefanadis C. Evidence of vascular dysfunction in young patients with successfully repaired coarctation of aorta. Atherosclerosis. 2005;182:97–103.
31. Werning C, Schonbeck M, Weidmann P, Baumann K, Gysling E, Wirz P, Siegenthaler W. Plasma renin activity in patients with coarctation of the aorta. A comment of the pathogenesis prestenotic hypertension. Circulation. 1969;40:731–7.
32. Markiewicz A, Wojczuk D, Kokot F, Cicha A. Plasma renin acitivity in coarctation of aorta before and after surgery. Br Heart J. 1975;37:721–5.
33. Parker FB, Farrell B, Streeten DHP, Blackman MS, Sondheimer HM, Anderson GH. Hypertensive mechanisms in coarctation of the aorta - further-studies of the renin-angiotensin system. J Thorac Cardiovasc Surg. 1980;80:568–73.
34. Sangaleti CT, Crescenzi A, Michelini LC. Endogenous angiotensin and pressure modulate brain angiotensinogen and AT (1A) mRNA expression. Hypertension. 2004;43:317–23.
35. Frydrychowicz A, Stalder AF, Russe MF, Bock J, Bauer S, Harloff A, Berger A, Langer M, Hennig J, Markl M. Three-dimensional analysis of segmental wall shear stress in the aorta by flow-sensitive four-dimensional-MRI. J Magn Reson Imaging. 2009;30:77–84.
36. Xu CP, Glagov S, Zatina MA, Zarins CK. Hypertension sustains plaque progression despite reduction of hypercholesterolemia. Hypertension. 1991;18:123–9.
37. Malek AM, Alper SL, Izumo S. Hemodynamic shear stress and its role in atherosclerosis. JAMA. 1999;282:2035–42.
38. Ku DN, Giddens DP, Zarins CK, Glagov S. Pulsatile flow and atherosclerosis in the human carotid bifurcation. Positive correlation between plaque location and low oscillating shear stress. Arteriosclerosis. 1985;5:293–302.
39. Wentzel JJ, Corti R, Fayad ZA, Wisdom P, Macaluso F, Winkelman MO, Fuster V, Badimon JJ. Does shear stress modulate both plaque progression and regression in the thoracic aorta? Human study using serial magnetic resonance imaging. J Am Coll Cardiol. 2005;45:846–54.
40. LaDisa JF Jr, Dholakia RJ, Figueroa CA, Vignon-Clementel IE, Chan FP, Samyn MM, Cava JR, Taylor CA, Feinstein JA. Computational simulations demonstrate altered wall shear stress in aortic coarctation patients treated by resection with end-to-end anastomosis. Congenit Heart Dis. 2011;6:432–43.
41. Pizarro C, De Leval MR. Surgical variations and flow dynamics in cavopulmonary connections: a historical review. Semin Thorac Cardiovasc Surg Pediatr Card Surg Annu. 1998;1:53–60.
42. O'Rourke MF, Cartmill TB. Influence of aortic coarctation on pulsatile hemodynamics in the proximal aorta. Circulation. 1971;44:281–92.
43. Menon A, Eddinger TJ, Wang H, Wendell DC, Toth JM, LaDisa JF Jr. Altered hemodynamics, endothelial function, and protein expression occur with aortic coarctation and persist after repair. Am J Physiol Heart Circ Physiol. 2012;303:H1304–18.
44. Niwa K, Perloff JK, Bhuta SM, Laks H, Drinkwater DC, Child JS, Miner PD. Structural abnormalities of great arterial walls in congenital heart disease: light and electron microscopic analyses. Circulation. 2001;103:393–400.
45. LaDisa JF Jr, Figueroa CA, Vignon-Clementel IE, Kim HJ, Xiao N, Ellwein LM, Chan FP, Feinstein JA, Taylor CA. Computational simulations for aortic coarctation: representative results from a sampling of patients. J Biomech Eng. 2011;133:091008.
46. Zhu Y, Chen R, Juan YH, Li H, Wang J, Yu Z, Liu H. Clinical validation and assessment of aortic hemodynamics using computational fluid dynamics simulations from computed tomography angiography. Biomed Eng Online. 2018;17:53.
47. Arzani A, Dyverfeldt P, Ebbers T, Shadden SC. In vivo validation of numerical prediction for turbulence intensity in an aortic coarctation. Ann Biomed Eng. 2012;40:860–70.
48. Wendell DC, Samyn MM, Cava JR, Ellwein LM, Krolikowski MM, Gandy KL, Pelech AN, Shadden SC, LaDisa JF Jr. Including aortic valve morphology in computational fluid dynamics simulations: initial findings and application to aortic coarctation. Med Eng Phys. 2013;35:723–35.
49. Lantz J, Ebbers T, Engvall J, Karlsson M. Numerical and experimental assessment of turbulent kinetic energy in an aortic coarctation. J Biomech. 2013;46:1851–8.
50. Fruitman DS. Hypoplastic left heart syndrome: prognosis and management options. Paediatr Child Health. 2000;5:219–25.

51. Biglino G, Corsini C, Schievano S, Dubini G, Giardini A, Hsia TY, Pennati G, Taylor AM. Computational models of aortic coarctation in hypoplastic left heart syndrome: considerations on validation of a detailed 3d model. Int J Artif Organs. 2014;37:371–81.
52. Warnes CA, Williams RG, Bashore TM, Child JS, Connolly HM, Dearani JA, Del Nido P, Fasules JW, Graham TP Jr, Hijazi ZM, Hunt SA, King ME, Landzberg MJ, Miner PD, Radford MJ, Walsh EP, Webb GD. Acc/aha 2008 guidelines for the management of adults with congenital heart disease: executive summary: a report of the American college of cardiology/American heart association task force on practice guidelines (writing committee to develop guidelines for the management of adults with congenital heart disease). Circulation. 2008;118:2395–451.
53. Bieging ET, Frydrychowicz A, Wentland A, Landgraf BR, Johnson KM, Wieben O, Francois CJ. In vivo three-dimensional mr wall shear stress estimation in ascending aortic dilatation. J Magn Resonance Imaging. 2011;33:589–97.
54. Baumgartner H, Bonhoeffer P, De Groot NM, de Haan F, Deanfield JE, Galie N, Gatzoulis MA, Gohlke-Baerwolf C, Kaemmerer H, Kilner P, Meijboom F, Mulder BJ, Oechslin E, Oliver JM, Serraf A, Szatmari A, Thaulow E, Vouhe PR, Walma E. Esc guidelines for the management of grown-up congenital heart disease (new version 2010). Eur Heart J. 2010;31:2915–57.
55. Itu L, Sharma P, Ralovich K, Mihalef V, Ionasec R, Everett A, Ringel R, Kamen A, Comaniciu D. Non-invasive hemodynamic assessment of aortic coarctation: validation with in vivo measurements. Ann Biomed Eng. 2013;41:669–81.
56. Donati F, Myerson S, Bissell MM, Smith NP, Neubauer S, Monaghan MJ, Nordsletten DA, Lamata P. Beyond bernoulli: improving the accuracy and precision of noninvasive estimation of peak pressure drops. Circ Cardiovasc Imaging. 2017;10
57. Ralovich K, Itu L, Vitanovski D, Sharma P, Ionasec R, Mihalef V, Krawtschuk W, Zheng Y, Everett A, Pongiglione G, Leonardi B, Ringel R, Navab N, Heimann T, Comaniciu D. Noninvasive hemodynamic assessment, treatment outcome prediction and follow-up of aortic coarctation from mr imaging. Med Phys. 2015;42:2143–56.
58. Shi Y, Valverde I, Lawford PV, Beerbaum P, Hose DR. Patient-specific non-invasive estimation of pressure gradient across aortic coarctation using magnetic resonance imaging. J Cardiol. 2019;73:544–52.
59. Saitta S, Pirola S, Piatti F, Votta E, Lucherini F, Pluchinotta F, Carminati M, Lombardi M, Geppert C, Cuomo F, Figueroa CA, Xu XY, Redaelli A. Evaluation of 4d flow mri-based non-invasive pressure assessment in aortic coarctations. J Biomech. 2019;94:13–21.
60. Kim JE, Kim EK, Kim WH, Shim GH, Kim HS, Park JD, Bae EJ, Kim BI, Noh CI, Choi JH. Abnormally extended ductal tissue into the aorta is indicated by similar histopathology and shared apoptosis in patients with coarctation. Int J Cardiol. 2010;145:177–82.
61. Torok RD, Campbell MJ, Fleming GA, Hill KD. Coarctation of the aorta: management from infancy to adulthood. World J Cardiol. 2015;7:765–75.
62. Rosenthal E. Stent implantation for aortic coarctation: the treatment of choice in adults? J Am Coll of Cardiol. 2001;38:1524–7.
63. Menon A, Wendell DC, Wang H, Eddinger TJ, Toth JM, Dholakia RJ, Larsen PM, Jensen ES, LaDisa JF Jr. A coupled experimental and computational approach to quantify deleterious hemodynamics, vascular alterations, and mechanisms of long-term morbidity in response to aortic coarctation. J Pharmacol Toxicol Methods. 2012;65:18–28.
64. Wolinsky H, Glagov S. A lamellar unit of aortic medial structure and function in mammals. Circ Res. 1967;20:99–111.
65. Hager A, Bildau J, Kreuder J, Kaemmerer H, Hess J. Impact of genomic polymorphism on arterial hypertension after aortic coarctation repair. Int J Cardiol. 2011;151:63–8.
66. LaDisa JF Jr, Bozdag S, Olson J, Ramchandran R, Kersten JR, Eddinger TJ. Gene expression in experimental aortic coarctation and repair: candidate genes for therapeutic intervention? PLoS One. 2015;10:e0133356.
67. Chen HH, Burnett JC Jr. C-type natriuretic peptide: the endothelial component of the natriuretic peptide system. J Cardiovasc Pharmacol. 1998;32:S22–8.
68. Khambata RS, Panayiotou CM, Hobbs AJ. Natriuretic peptide receptor-3 underpins the disparate regulation of endothelial and vascular smooth muscle cell proliferation by c-type natriuretic peptide. Br J Pharmacol. 2011;164:584–97.
69. Li Y, Sarkar O, Brochu M, Anand-Srivastava MB. Natriuretic peptide receptor-c attenuates hypertension in spontaneously hypertensive rats: role of nitroxidative stress and gi proteins. Hypertension. 2014;63:846–55.
70. LaDisa JF Jr, Tomita-Mitchell A, Stamm K, Bazan K, Mahnke DK, Goetsch MA, Wegter BJ, Gerringer JW, Repp K, Palygin O, Zietara AP, Krolikowski MM, Eddinger TJ, Alli AA, Mitchell ME. Human genotyping and an experimental model reveal npr-c as a possible contributor to morbidity in coarctation of the aorta. Physiol Genomics. 2019;51:177–85.
71. Chen SS, Donald AE, Storry C, Halcox JP, Bonhoeffer P, Deanfield JE. Impact of aortic stenting on peripheral vascular function and daytime systolic blood pressure in adult coarctation. Heart. 2008;94:919–24.
72. Coogan JS, Chan FP, Taylor CA, Feinstein JA. Computational fluid dynamic simulations of aortic coarctation comparing the effects of surgical- and stent-based treatments on aortic compliance and ventricular workload. Catheter Cardiovasc Interv. 2011;77:680–91.
73. Kwon S, Feinstein JA, Dholakia RJ, LaDisa JF Jr. Quantification of local hemodynamic alterations caused by virtual implantation of three commercially available stents for the treatment of aortic coarctation. Pediatr Cardiol. 2014;35:732–40.

74. Brili S, Tousoulis D, Antoniades C, Vasiliadou C, Karali M, Papageorgiou N, Ioakeimidis N, Marinou K, Stefanadi E, Stefanadis C. Effects of ramipril on endothelial function and the expression of proinflammatory cytokines and adhesion molecules in young normotensive subjects with successfully repaired coarctation of aorta: a randomized cross-over study. J Am Coll Cardiol. 2008;51:742–9.
75. Zhang F, Drabier R. Ipad: the integrated pathway analysis database for systematic enrichment analysis. BMC Bioinformatics. 2012;13:S7.
76. Coogan JS, Humphrey JD, Figueroa CA. Computational simulations of hemodynamic changes within thoracic, coronary, and cerebral arteries following early wall remodeling in response to distal aortic coarctation. Biomech Model Mechanobiol. 2013;12:79–93.
77. Figueroa CA. Crimson: cardiovascular integrated modeling and simulation; 2019.
78. Updegrove A, Wilson NM, Merkow J, Lan H, Marsden AL, Shadden SC. Simvascular: an open source pipeline for cardiovascular simulation. Ann Biomed Eng. 2017;45:525–41.
79. Biglino G, Capelli C, Bruse J, Bosi GM, Taylor AM, Schievano S. Computational modelling for congenital heart disease: how far are we from clinical translation? Heart. 2017;103:98–103.
80. Astengo M, Berntsson C, Johnsson AA, Eriksson P, Dellborg M. Ability of noninvasive criteria to predict hemodynamically significant aortic obstruction in adults with coarctation of the aorta. Congenit Heart Dis. 2017;12:174–80.

5 Tetralogy of Fallot, the Right Ventricular Outflow Tract: Clinical Concepts, Engineering Applications and Impact of an Integrated Medico-Engineering Approach

Laxmi Kaliyappan, Nithin Balasubramanian, Silvia Schievano, and Louise Coats

5.1 TOF-RVOT from a Clinical Aspect

5.1.1 Introduction

Tetralogy of Fallot (TOF) is the commonest form of cyanotic congenital heart disease (CHD). Cardinal morphological characteristics (right ventricular outflow obstruction (RVOTO), overriding aorta, right ventricular hypertrophy (RVH) and interventricular communication) occur due to the antero-cephalad deviation of the outlet septum. Whilst all these deformities can have long-lasting implications, the extent of RVOTO underpins the clinical course of a TOF patient [1]. For instance, infants with severe RVOTO and inadequate pulmonary flow can typically present at birth with profound central and peripheral cyanosis and require early intervention. TOF with pulmonary atresia and major aortopulmonary collateral arteries is the most extreme variant of TOF, in which complete atresia of the pulmonary valve replaces pulmonary stenosis [2]. While 80% of TOF patients have varying degrees of pulmonary stenosis, the remaining 20% have pulmonary atresia [3].

RVOTO can occur at multiple levels [4]:

- Hypoplastic pulmonary valve annulus.
- Bicuspid or dysplastic stenotic pulmonary valve.
- Subvalvar obstruction.
- Hypertrophy of RVOT muscle band.

5.1.2 Surgical Correction

5.1.2.1 Primary Repair

TOF patients with symptomatic RVOTO require surgical correction to enlarge the outflow tract and relieve the obstructed pulmonary flow. The nature of the surgery depends on the level and degree of

Laxmi Kaliyappan and Nithin Balasubramanian contributed equally to this work.

L. Kaliyappan · N. Balasubramanian
University College London, Institute of Cardiovascular Science, London, UK
e-mail: laxmi.kaliyappan.19@ucl.ac.uk; nithin.balasubramanian.19@ucl.ac.uk

S. Schievano
UCL Institute of Cardiovascular Science and Great Ormond St Hospital for Children, London, UK
e-mail: s.schievano@ucl.ac.uk

L. Coats (✉)
Freeman Hospital, Newcastle upon Tyne Hospitals NHS Foundation Trust and Population Health Sciences Institute, Newcastle University, Newcastle upon Tyne, UK
e-mail: louise.coats@newcastle.ac.uk

G. Butera et al. (eds.), *Modelling Congenital Heart Disease*,
https://doi.org/10.1007/978-3-030-88892-3_5

RVOTO. Patients with a functional pulmonary valve annulus are suitable for RVOT muscle bundle resection and annular sparing. Whereas those with severe pulmonary annular hypoplasia and infundibular stenosis may require a transannular patch (TAP) and muscle bundle resection for adequate RVOTO relief [5]. Patients with pulmonary atresia may require reconstruction of the continuity between the right ventricle and the main pulmonary artery using a conduit [6]. Over time primary repair surgeries have shifted away from a trans-ventricular to a transatrial or transatrial–transpulmonary approach. Ventriculotomy can cause transmural myocardial scarring and coronary artery damage, which can contribute to long-term implications for right ventricular function and an increased propensity for ventricular arrhythmias [7]. In addition, inadequate RVOT resection with a transatrial-transpulmonary approach can increase long-term risk of residual RVOTO [8]. With increasing recognition of the deleterious effects of pulmonary regurgitation in later life, there has also been an inclination towards a 'valve-sparing approach' to maintain pulmonary valve competence using techniques such as implantation of a monocusp valve, a valved right ventricle-to-pulmonary artery (RV-PA) conduit or a homograft valve. However, the efficacy of valve-sparing techniques is yet to be demonstrated [5].

5.1.2.2 Staged Repair

Staged repair is preferred in neonates for whom primary surgical repair may not be immediately feasible due to infant size (prematurity) or due to challenging anatomy. Staged repair involves initial palliation to establish adequate pulmonary blood flow in the neonatal period followed at a later stage by elective complete repair. Typically, a systemic to pulmonary shunt, like the modified Blalock-Taussig shunt which connects the subclavian artery and pulmonary artery, is used. It allows the growth of the pulmonary arteries and reduces hypoxaemia until an appropriate age and body weight suitable for complete repair is reached [9]. However, there have been some reports of post-procedural instability and early mortality associated with the use of the modified Blalock- Taussig shunt which has led to a search for safer alternatives.

In recent times, stenting of the ductus arteriosus to re-establish the connection between the aorta and pulmonary artery has been increasingly preferred over the modified Blalock-Taussig shunt as it provides greater post-procedural stability and improves patient survival to corrective surgery [10]. However, in cyanotic CHDs like TOF, the anatomy of the ductus arteriosus may be complex and unsuitable for stenting, which may only be achievable in selected cases. An alternative is balloon pulmonary valvuloplasty which can be used in infundibular obstruction. When performed early, this can rapidly increase systemic oxygen saturations and postpone reparative surgery. Balloon pulmonary valvuloplasty for TOF has shorter hospital stays when compared to ductus arteriosus stenting; however, long-term outcomes remain controversial [11]. Similarly, RVOT stenting is a novel technique that is sometimes considered as a bridge to repair in neonatal life. Limited studies have reported safety, lower complication rates and shorter hospital stays when compared to the modified Blalock-Taussig shunt [12].

5.1.2.3 Timing of Surgical Intervention

Most patients with TOF typically undergo complete repair between 6 months and 1 year of age. Correction at this stage limits prolonged right ventricular exposure to adverse pressure load and the effects of hypoxia, thus optimising the conditions for adequate myocardial development and function [13]. TOF patients with mild RVOTO can be managed medically and surgery can be deferred until after the neonatal period, providing time for pulmonary vascular resistance to fall and the infant to grow larger. However, TOF patients with severe RVOTO may require early surgical interventions in the form of primary or staged repair.

5.1.2.4 Primary Repair Versus Staged Repair

The benefits of an early corrective repair of TOF neonates versus a staged repair with palliative intervention remain unclear. Early primary correction minimises chronic hypoxia and RVH and promotes pulmonary artery growth, whilst the staged approach exposes the right ventricle to prolonged pressure overload and persistent cyanosis. There is also a risk of thrombosis of the shunt and pulmonary artery distortion [14]. However, complete repair procedures in the neonatal period are associated with a threefold increase in mortality and 50% increase in longer hospital stay. Additionally, neonatal repair is also associated with an increased use of TAP which will result in late pulmonary regurgitation and attendant long-term complications [15].

5.1.3 Survival and Long-Term Complications

Immediate and long-term survival continues to improve in repaired TOF (rTOF) patients with major congenital cardiothoracic surgical centres reporting operative mortality <2% in recent years [16]. Consequently, a 20- to 30-year survival rate of over 90% has been observed. Nevertheless, despite these successes, late survival (after 30 years) remains markedly lower than the general population, likely due to development of long-term complications like pulmonary regurgitation, residual RVOTO, aortic root dilation and arrhythmias, which can occur in isolation or concomitantly [17].

5.1.3.1 Pulmonary Regurgitation (PR)

Is a key haemodynamic lesion arising from RVOT dysfunction occurring at the expense of RVOTO relief. It is one of the most frequent complications seen in rTOF patients and use of the TAP technique and RV-PA conduits are associated with higher rates of chronic PR. Whilst PR is generally well-endured short-term [18], long-standing PR can lead to RV volume overload and RV dilatation due to the increased workload. This can attenuate tricuspid annulus dilation and tricuspid regurgitation (TR) which further increase RV overload [19].

This presents functionally with decreased exercise performance in patients with outcomes predicted by chronotropic incompetence [20]. Depending on the severity of volume overload, poor RV adaptation can worsen RVH, ultimately resulting in decompensation and RV failure (RVF) without intervention [18]. So, extent of PR is a vital consideration for timely re-intervention with pulmonary valve replacement (PVR). PVR has been shown to improve cardiac remodelling and patient quality of life [21] but survival benefit remains to be demonstrated [22]. Currently, PVR is indicated for severe PR in symptomatic rTOF patients with a regurgitant fraction >30-40% measured by cardiac MRI [23].

5.1.3.2 Residual or Recurrent RVOTO

Often associated with valve-sparing surgeries, can result in remnant hypertrophied subvalvular muscle, pulmonary valve stenosis or annular hypoplasia post-operatively. While mild obstruction is well-tolerated and does not require reintervention, severe obstruction can lead to long-lasting problems for the patient and is an important structural cause for reintervention. Severe obstruction can lead to extensive pulmonary stenosis and RVH which is a major risk factor for life-threatening arrhythmias. Re-intervention using balloon valvuloplasty or PVR to relieve the pressure-loaded right ventricle has been demonstrated to relieve symptoms. Consequently, significant RVOTO (peak velocity>3 m/s) is also an indication for PVR in symptomatic rTOF patients [23].

In asymptomatic patients, in addition to severe PR and/or significant RVOTO, one of the following criteria must be present: objective exercise tolerance reduction; progressive RV dilation monitored by repeated measurements to RV end systolic index (RVESVi)≥80 mL/m^2 and/or RV end diastolic index (RVEDVi) ≥160 mL/m^2 and/or progression of TR to at least moderate; right ventricular systolic pressure (RVSP)≥80 mmHg due to RVOTO or progressive RV systolic dysfunction (Fig. 5.1) [19, 23]. Surgical reintervention is more frequent in patients with a RV-PA conduit than TAP [24].

5.1.3.3 Arrhythmias

Are a relatively frequent late post-operative complication with variable prevalence rates (~20% for supraventricular tachycardias (SVT) and ~15% of rTOF patients developing ventricular tachycardias (VT)). The risk of developing arrhythmias also increases significantly after the age of 45 in rTOF patients [25]. Additionally, the INdiCaTOR study findings support RVH, due to increased mass-to-

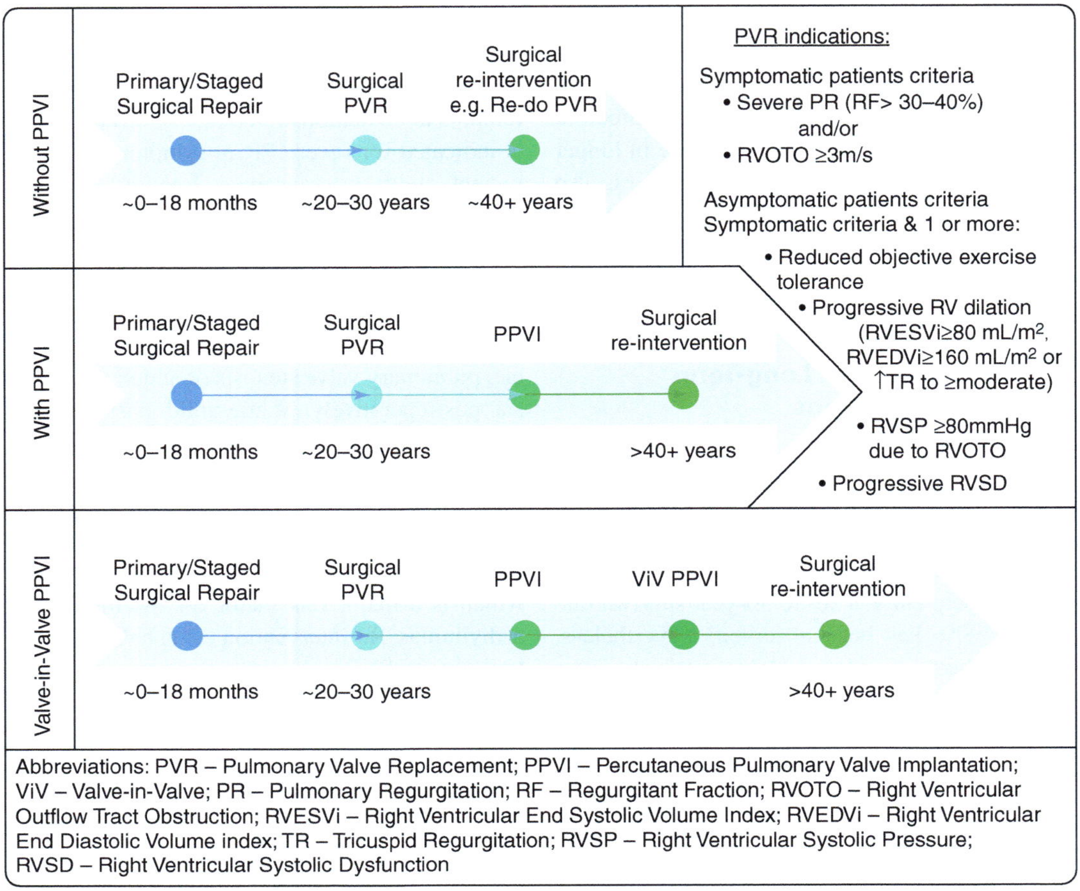

Fig. 5.1 Timeline schematic of cardiac operations in repaired TOF patients and the potential impact of PPVI in delaying surgical re-interventions. Stated age ranges may vary widely depending on multiple factors, e.g. disease severity and initial surgical repair technique (Tweddell et al. 2012 [30], Stout et al. 2018 [19] and Baumgartner et al. 2020 [23])

volume ratio, as an important long-term risk factor for VT and death [26]. Older age at repair, higher number of previous cardiac surgeries, LV dysfunction, use of TAP and prolongation of QRS duration are also important risk factors for the development of VT. Patients with sustained VT may require an implantable cardioverter defibrillator or radiofrequency ablation to prevent sudden cardiac death. Risk factors for SVT include the presence of tricuspid regurgitation and lower right and left ventricular ejection fractions; however, ambiguity remains around determining optimal thresholds for risk stratification [25].

5.2 Challenges

One of the main clinical challenges with rTOF patients is balancing the risk of pulmonary regurgitation or stenosis against the risk and benefits of reintervention. Definitive reintervention involves PVR and uses a prosthetic valve or surgical placement of a new conduit. However, both these options are limited by a finite lifespan [27]. Thus, during a lifetime, patients will require repeated conduit or valve replacement surgeries, each with the attendant risk of morbidity or mortality. Therefore, valve/conduit replacement should be optimally timed to prevent irreversible adverse remodelling and to limit the number of re-interventions [28] (Fig. 5.1). Percutaneous pulmonary valve implantation (PPVI), with devices such as the Medtronic Melody valve or the Edwards SAPIEN XT valve, can complement this strategy offering a less invasive alternative to open-heart surgery that can reduce the number of redo surgeries to the pulmonary valve needed over a lifetime [29]. Whilst clinical indications for PPVI are the same as for surgical PVR [19], PPVI is preferred in patients without native RVOT tissue [23]. PPVI has been associated with higher rates of infective endocarditis compared to surgical PVR (2.3–0.3%) [5].

5.3 Engineering Applications

The application of engineering methods to solve problems related to the RVOT in TOF has gained increasing popularity. From assessment of RVOT anatomy and haemodynamics to simulating mechanical effects of PPVI devices, we aim to present a summary of currently utilised techniques.

5.3.1 Rapid Prototyping and 3D Printing

3D printing is a rapid prototyping technique that involves building three dimensional objects from computer-aided models (CAD) by layering plastic materials. Research in the use of 3D-printed heart models in the management of TOF and CHD is popular as this technique provides a more accurate visualisation of the heart, exceeding the capacity of existing diagnostic tools and aiding surgical planning and education [31].

TOF models have been printed mainly from high resolution CT and MRI pictures with attempts also made from 3D echocardiography images and foetal ultrasound of the heart. Images are easily segmented using commercial 3D modelling software such as Mimics (Materialise) and ScanIP (Simpleware) or open source platforms like 3D slicer [32]. Different types of 3D printing technologies such as stereolithography (SLA), selective laser sintering (SLS), fused deposition modelling (FDM) and polyjet printing (PJ) have been used to produce TOF models, considering advantages and disadvantages of each technique [33].

In TOF patients, 3D-printed rigid models (Fig. 5.2) are used to accurately visualise the

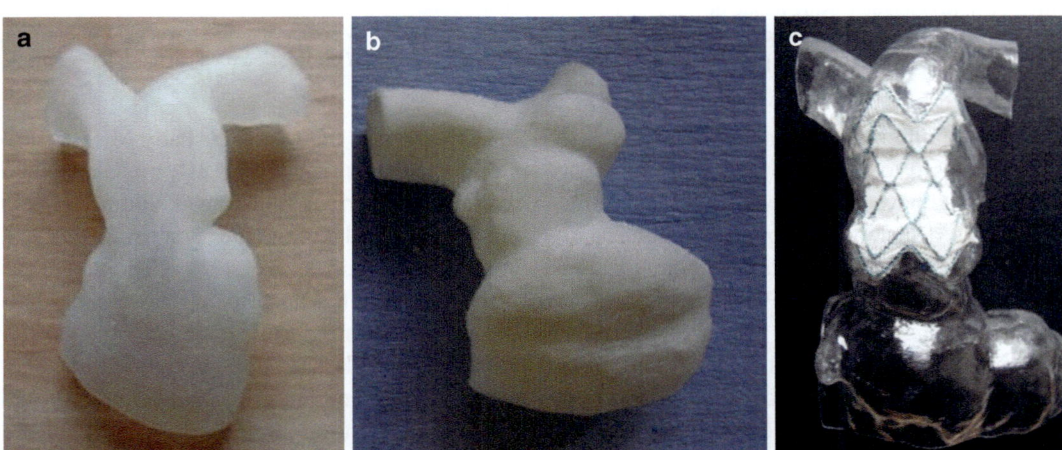

Fig. 5.2 3D-printed models of the right ventricular outflow tract and pulmonary arteries in (**a**) rubber-like material, (**b**) nylon and (**c**) hollow transparent resin implanted with a novel percutaneous pulmonary valve

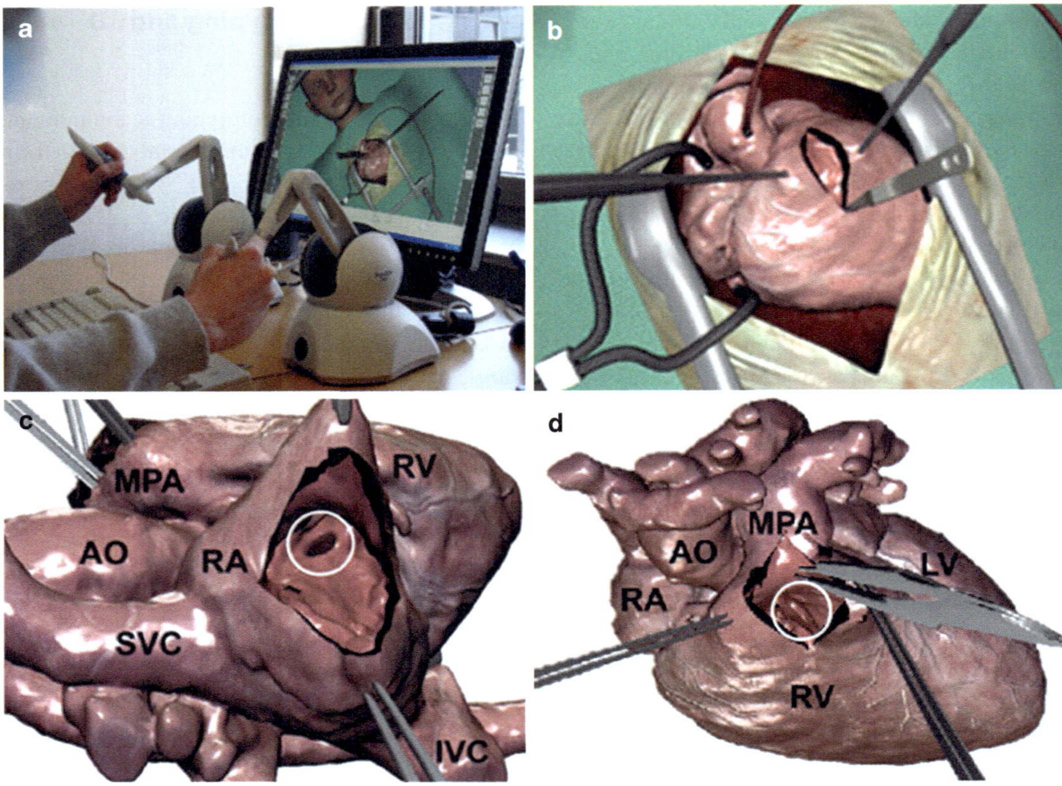

Fig. 5.3 Virtual surgical simulation enables interactive cardiac tissue manipulation, e.g. dissection, rotation and device employment, in 3-dimensional (3D) models reconstructed from patient cardiac magnetic resonance imaging (MRI) data. (**a**) Operator using haptic devices on a virtual patient heart to study cardiac anatomy. (**b**) Magnified view of surgical field in a virtual 3D patient heart. (**c**) Virtual simulation can also aid decisions for surgical approach; 3D cardiac model of a patient with a ventricular septal defect (*white circle*) approached via a trans-atrial incision or (**d**) a trans-ventricular approach (reproduced from Sørensen et al. 2006 [38]). Abbreviations: *MPA* main pulmonary artery, *AO* aorta, *SVC* superior vena cava, *RA* right atrium, *RV* right ventricle, *IVC* inferior vena cava, *LV* left ventricle

RVOT and pulmonary trunk anatomy. These models can be used to refine the selection of patients for PPVI and have been shown to be more effective in this decision-making process than using MRI scans [34]. The models also help surgeons train for surgical repair or PPVI in complex cases by familiarising themselves with a very accurate anatomy of the patient's heart defect. Additionally, they can aid patient-specific care by providing clinicians with a visual resource to support explanation of surgeries and interventions.

5.3.2 Virtual Simulation

Virtual simulation, much akin to 3D printing, but displayed on a computer screen, involves utilising 3D-imaging modalities, usually MRI, to create patient-specific virtual 3D models. These models have various functions including roles in pre-operative planning and practising surgical procedures (Fig. 5.3) [35].

Considering the anatomical and clinical heterogeneity amongst TOF patients, off-the-shelf devices are unlikely to provide optimal management. Virtual simulation in cardiac surgery and catheter interventions are relatively novel and aid optimisation and advancement of patient-tailored care.

Case reports of pre-operative virtual surgery detail using virtual 3D models which allow surgeons to simulate multiple RV-PA conduit implantations and to choose optimally sized conduits. This methodology enabled implantation of a larger conduit than initially intended, thereby minimising likelihood of future surgical or transcatheter re-intervention. Virtual cardiac surgery also has the potential to inform timing of RVOT intervention in elective cases [36].

Virtual 3D models can be rotated and cross-sectioned at any depth and plane offering enhanced spatial visualisation into intimate structures (Fig. 5.3) compared to standard imaging modalities [37]. This insight may be crucial for pre-operative planning, particularly for rare yet important anatomical anomalies to assess the spatial relationship of relevant cardiovascular structures. Demonstrated in a case report, precise knowledge of the location of the RVOT and an aberrant right coronary artery pre-operatively was vital for avoiding injury during RV-PA conduit placement in a patient with TOF and absent pulmonary valve [37].

Whilst PPVI is a minimally invasive solution for PVR, it poses its own inherent challenges with the potential for insecure implantation in patients with complex RVOTs. A study comparing implantation in ovine RVOT virtual models and actual PPVI device contact demonstrated that virtual pre-implant modelling mimicked actual device implantation. However, it did not accurately predict distal RVOT expansion perhaps due to the lack of deformation of the RVOT virtual anatomy which was modelled as a rigid vessel. Nonetheless, it showed potential to be a useful tool in combination with other modelling techniques to facilitate PPVI patient selection [39]. To overcome the limitations of virtual image overlay in PPVI assessment; finite element modelling, another engineering methodology which has been used over the past decade, allows in-depth study of the physical interactions between structures.

5.3.3 Finite Element Analysis

PPVI is associated with device-related complications such as stent fracture, anchoring issues and utility in a limited number of patients with specific anatomy. Computational simulations such as finite element (FE) analysis (Fig. 5.4) can assist in understanding the limitations of the prosthetic valve in order to optimise the design for better patient outcomes. Though in vivo and in vitro experiments can provide valuable data, FE analysis can predict patient outcomes more accurately. Modelling tools can contribute to improved procedures (e.g. design modifications

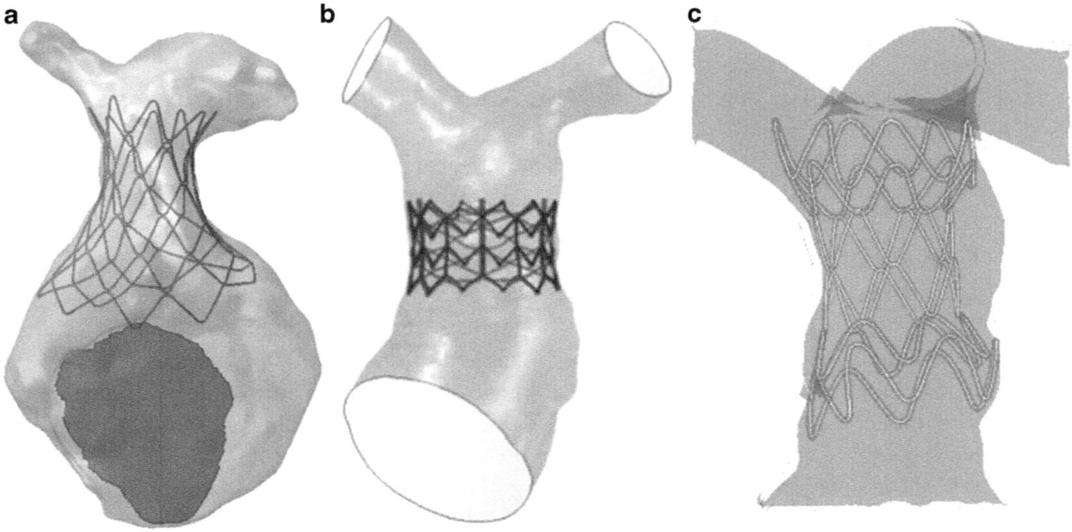

Fig. 5.4 Patient-specific right ventricular outflow tract (RVOT) finite element models of different cases implanted with (**a**) Melody valve, (**b**) 27 mm SAPIEN valve pre-stented with a bare metal stent and (**c**) a novel shape memory allow device

or optimal route identification for device insertion) and increase patient safety in the early introduction of these devices into clinical practice. FE models can also help test realistic loading scenarios for accurate simulation of 'worst case' mechanical behaviour [40].

In refining the design of PPVI stents, MRI data from patients who underwent PPVI and had subsequent stent fractures were used to create a FE model of the implantation site. These models were used to simulate the expansion of the PPVI stent into the RVOT geometry and compared with free expansions of the stent up to uniformly deployed configuration. The data acquired from these models provided useful information about the influence of the RVOT on the final geometry and mechanical performance of the stent [41]. FE models have also been used to retrospectively assess stent deformations and identify RVOT anatomical risk factors that are associated with higher risk of stent fracture [42].

Whilst RVOT stenting and PPVI can relieve RVOTO, accurate pre-procedural evaluation is required to minimise procedural complications such as coronary artery compression, conduit injury or arterial distortion. Caimi et al. used patient-specific FE modelling as a tool to assess stenting feasibility and investigate clinically relevant risks associated to percutaneous intervention, which showed results consistent with in-vivo evidence. The FE models yielded useful data on stent distortion and stresses, thus highlighting that the numerical approach could be used to understand the role of patient-specific anatomical features [43].

5.3.4 Computational Fluid Dynamics

Computational fluid dynamics (CFD) is a tool that allows for fluid flow evaluation, often applied clinically to model and investigate haemodynamic interactions [44]. CFD modelling to evaluate pressure, shear stress and flow profiles in the RVOT and pulmonary trunk (Fig. 5.5) has demonstrated low flow with elevated pressure between the pulmonary trunk and bifurcation into PAs. Regions with low flow consequently have low shear stress which leads to activation of proatherogenic genes implicated in intimal hyperplasia. Thus, CFD modelling provides important fluid dynamics information that could characterise graft stenosis in rTOF [45].

It can also be used to develop and inform surgical techniques. Miyaji et al. investigated the efficacy of virtual surgery based on haemodynamic analyses from CFD models, in complicated RVOT reconstruction and PA-plasty. This novel surgical strategy provided surgeons with drawings that indicated the shape required for the graft to achieve efficient flow instead of solely depending on surgical experience. These grafts were made intra-operatively and aided creating ideal PA form with laminar flow. This system provides robust constructs that guide surgeons in conjunction with their own expertise [44]. Rao et al. also found virtual surgery with subsequent CFD evaluation enabled a more objective evaluation of surgical strategy accounting for patient-specific morphology [46].

Therefore, studies suggest CFD modelling informs pre-operative plans and promotes morphology-informed decisions. There are limitations to consider as these models were rigid and used steady-state flow with an always-open valve. Conversely, pulsatile flow with an active valve is present in the RVOT in vivo [44, 45].

5.3.5 Advanced Image Analysis

Statistical shape analysis (SSA) is a tool that allows for consistent quantitative description of complex shapes (Fig. 5.6). Cardiac shapes can be correlated with known functional parameters, potentially enabling the discovery of novel biomarkers for use as predictive tools [47]. SSA applied to RVOTs in TOF patients with late PR post-surgical repair aids identifying anatomical features vital to guiding optimisation of novel PPVI device designs [48].

Pre-operative geometry descriptors of the PAs and the RVOT were also shown to predict late pulmonary valve dysfunction. Dilated RVOT circumference and acute bifurcation angle of the

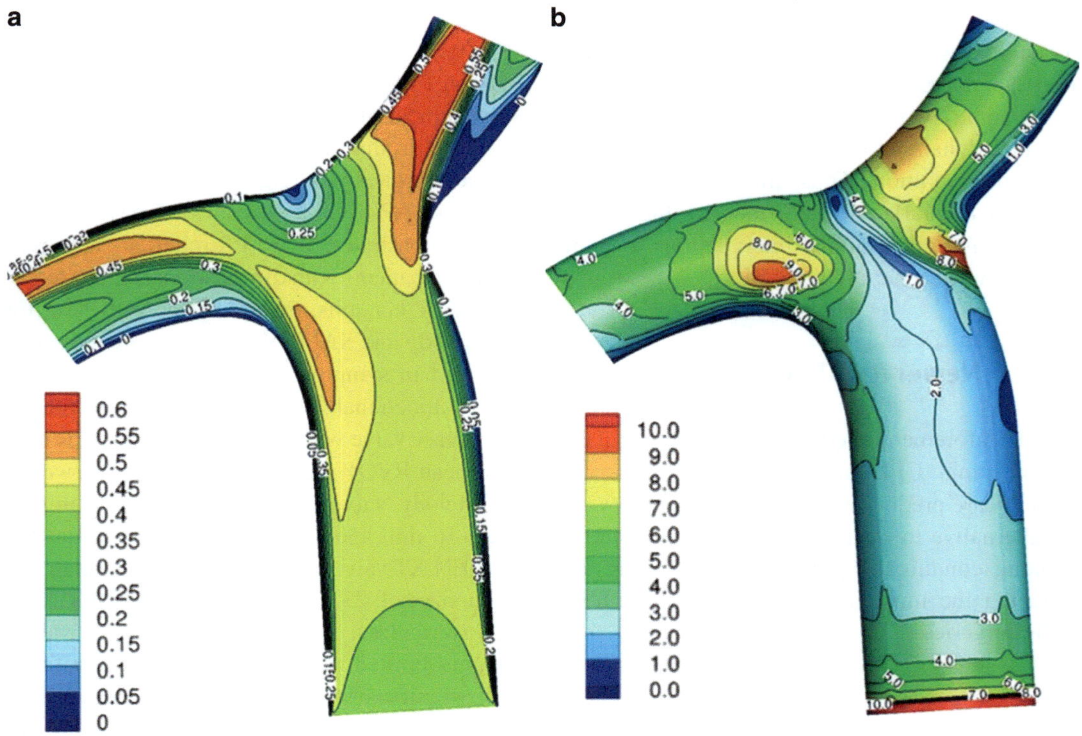

Fig. 5.5 Right Ventricular Outflow Tract Computational Fluid Dynamic Models of Wall (**a**) Velocities and (**b**) Shear-Stress in a Non-pulsatile Flow Rate of 2.5 (L/min). Higher Velocities Seen Centrally and the Ostia of the Pulmonary Arteries and Low Shear Stress Visible at the Anterior Walls (Images Reproduced from Mosbahi et al. 2014 [45])

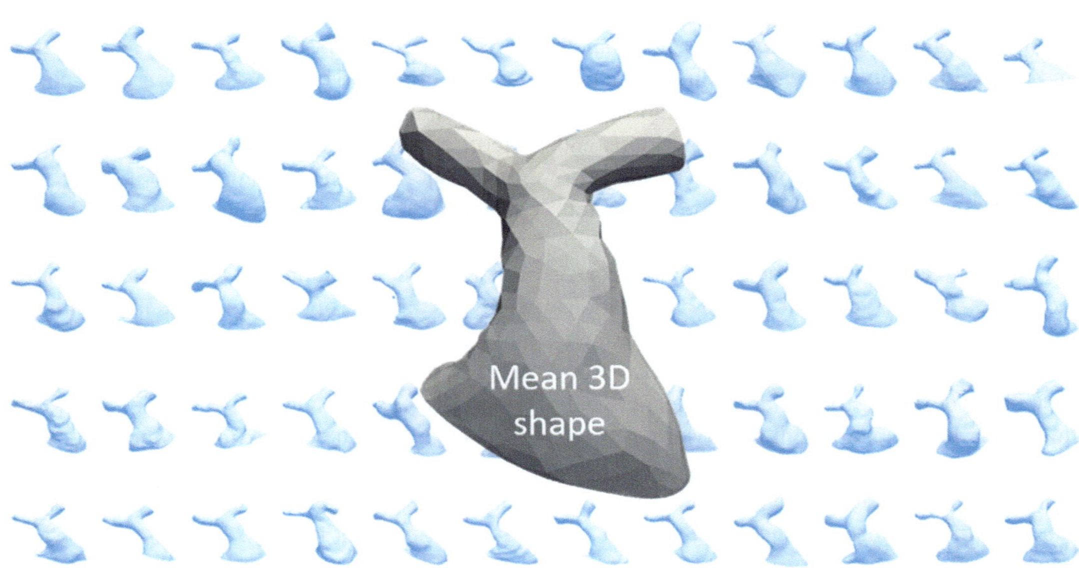

Fig. 5.6 Statistical shape analysis of three-dimensional RVOT models (*blue*) computed to produce a mean 3D shape (*model overlaid in grey*) of the population

pulmonary trunk and LPA were significantly associated with valve degeneration. RVOT dilation can increase valvular shear stress and an acute angle disrupts laminar flow, compounding valvular dynamic load all of which promote valvular degeneration. Thus, this information can act as a biomarker to help identify higher-risk patients and guide aimed interventions to minimise valve degeneration risk [49].

5.4 Newer RVOT Treatments

All the aforementioned techniques, particularly 3D printing, CFD and FEA, have influenced patient care predominantly via enabling PPVI as an alternative to treat RVOT dysfunction and to prolong conduit lifespan; thereby delaying and reducing the number of subsequent open-heart PVR surgeries.

5.4.1 Conduits

Conduit selection in PVR remains a significant challenge due to recognised complications with shortened lifespan from calcification and poor haemodynamic utility. With new conduit designs emerging, the importance of modelling techniques is highlighted in assessing performance and clinical viability. Dur et al. evaluated the performance of a new bicuspid polytetrafluoroethylene valve design, using in vitro flow loops and probes to assess valve motion. CFD was also used to assess the effect of geometric parameters like conduit orientation and pulmonic curvature on haemodynamics inside the lesser and greater curvature of the conduit. CFD simulations showed reduced flow in the lesser curvature, which correlated with clinical experience from previously designed tricuspid valve leaflets getting stuck due to insufficient flow over the lesser curvature. Regardless, valve competence was independent of conduit curvature. These findings can be utilised to further aid conduit designs and improve valve competency in an effort to minimise reoperation [50].

5.4.2 Valves and Stents

Balloon PPVI has increasingly been favoured as the treatment of choice in rTOF patients with PV dysfunction due its ease of implantation and reduced surgical risk. The Medtronic Melody and Edwards SAPIEN XT valves, which have successfully demonstrated robust PR relief, are approved for use in patients with dysfunctional circumferential RVOT conduits. Patient selection criteria for the Melody and SAPIEN XT valves are the same, indicated in symptomatic patients with evidence of a dysfunctional RVOT conduit or bioprosthetic pulmonary valve with moderate or severe PR and/or a mean RVOT gradient >35 mmHg. However, the Melody valve is currently available in two sizes: 16 mm and 18 mm diameter; whereas the SAPIEN XT valve, which is shorter, is available in larger sizes of: 23 mm, 26 mm and 29 mm [1].

Nevertheless, application of these valves remains limited in the large subset of patients with significant PR who have native RVOT lesions and complex RVOT geometry as a larger diameter valve rather than a balloon-expandable valve is required to maintain a stable device position. Although off-label use in selected patients has shown limited feasibility, many patients with larger RVOTs remain unsuitable candidates for PPVI with existing technology [1].

Next-generation self-expandable delivery systems to implant the PV have shown great potential and are undergoing human clinical trials. The Medtronic Harmony valve, a self-expanding porcine tissue valve mounted in a nitinol frame, was designed to fulfil the unmet need for effective percutaneous intervention in patients with dysfunctional non-conduit RVOTs. Initial feasibility study, recording 6-month outcomes in 20 patients, showed the valve continued to function well with trace or no PR in 94% of patients and mild PR in 6%. Two patients had the valve surgically explanted, one due to proximal migration of the valve and the other due to stent fracture. Notably, out of the 66 patients initially enrolled for the study, 46 patients (~70%) did not get the valve implanted due to inappropriate anatomy [51]. Recently published 3-year outcomes are encouraging with no deaths, stable device position and good function with only one patient developing mild PR and another that had a mild paravalvular leak. However, two patients developed significant neointimal proliferation and further long-term follow-up will be required to assess duration of RVOTO relief [52].

The Pulsta valve (TaeWoong Medical Co) is a self-expandable valve (three leaflets of treated porcine pericardial tissue) with flared-ends to adapt to the larger native RVOT and is positioned by a delivery catheter made of a knitted nitinol wire backbone. A recent human feasibility study in the native RVOT in rTOF using the Pulsta valve demonstrated good short-term effectiveness without serious adverse events [53]. Similarly, devices such as the Alterra Adaptive Prestent are designed to internally remodel a wide variety of RVOT morphologies, thereby creating a suitable site for implantation of standard balloon expandable valves (SAPIEN S3) in patients who had previous surgery and are left with large, compliant, irregular RVOTs. Various early case studies have demonstrated the feasibility of implantation of this device; however, larger trials are required to test for its efficacy and safety [54].

5.5 Overview

Application of medico-engineering methods to the RVOT in TOF is a growing venture requiring continual interaction between clinicians, engineers and industry. Whilst there has been valuable work in RVOT characterisation, pre-operative planning and PPVI device validation; there is a growing need to develop devices suitable for complex RVOT anatomy commonly seen in these patients and to assess how long-term survival can be improved.

References

1. Balzer D. Pulmonary valve replacement for tetralogy of Fallot. Methodist Debakey Cardiovasc J. 2019;15(2):122–32. https://doi.org/10.14797/mdcj-15-2-122.
2. Liao P-K, Edwards WD, Julsrud PR, Puga FJ, Danielson GK, Feldt RH. Pulmonary blood supply in patients with pulmonary atresia and ventricular septal defect. J Am Coll Cardiol. 1985;6(6):1343–50.
3. Ferencz C. Epidemiology of congenital heart disease: the Baltimore-Washington infant study 1981–1989. Futura Publishing Company; 1993.
4. Doyle T, Kavanaugh-McHugh A. Pathophysiology, clinical features, and diagnosis of tetralogy of Fallot; 2020.
5. van der Ven JPG, van den Bosch E, Bogers AJCC, Helbing WA. Current outcomes and treatment of tetralogy of Fallot. F1000Res. 2019;8 https://doi.org/10.12688/f1000research.17174.1.
6. Pluchinotta FR, Bussadori C, Butera G, Piazza L, Chessa M, Saracino A, et al. Treatment of right ventricular outflow tract dysfunction: a multimodality approach. Eur Heart J Suppl. 2016;18(Suppl. E):E22–6.
7. Luijten LW, van den Bosch E, Duppen N, Tanke R, Roos-Hesselink J, Nijveld A, et al. Long-term outcomes of transatrial-transpulmonary repair of tetralogy of Fallot. Eur J Cardiothorac Surg. 2015;47(3):527–34. https://doi.org/10.1093/ejcts/ezu182.
8. Giannopoulos N, Chatzis A, Karros P, Zavaropoulos P, Papagiannis J, Rammos S, et al. Early results after transatrial/transpulmonary repair of tetralogy of Fallot. Eur J Cardiothorac Surg. 2002;22(4):582–6.
9. Kiran U, Aggarwal S, Choudhary A, Uma B, Kapoor PM. The Blalock and taussig shunt revisited. Ann Card Anaesth. 2017;20(3):323.
10. Bentham JR, Zava NK, Harrison WJ, Shauq A, Kalantre A, Derrick G, et al. Duct stenting versus modified Blalock-Taussig shunt in neonates with duct-dependent pulmonary blood flow: associations with clinical outcomes in a multicenter national study. Circulation. 2018;137(6):581–8.
11. Muneuchi J, Watanabe M, Sugitani Y, Kawaguchi N, Matsuoka R, Ando Y, et al. Early palliative balloon pulmonary valvuloplasty in neonates and young infants with tetralogy of Fallot. Heart Vessel. 2020;35(2):252–8.
12. Quandt D, Ramchandani B, Penford G, Stickley J, Bhole V, Mehta C, et al. Right ventricular outflow tract stent versus BT shunt palliation in tetralogy of Fallot. Heart. 2017;103(24):1985–91.
13. Daliento L, Mapelli D, Russo G, Scarso P, Limongi F, Iannizzi P, et al. Health-related quality of life in adults with repaired tetralogy of Fallot: psychosocial and cognitive outcomes. Heart. 2005;91(2):213–8.
14. Kantorova A, Zbieranek K, Sauer H, Lilje C, Haun C, Hraska V. Primary early correction of tetralogy of Fallot irrespective of age. Cardiol Young. 2008;18(2)
15. Loomba RS, Buelow MW, Woods RK. Complete repair of tetralogy of Fallot in the neonatal versus non-neonatal period: a meta-analysis. Pediatr Cardiol. 2017;38(5):893–901.
16. Townsley MM, Windsor J, Briston D, Alegria J, Ramakrishna H. Tetralogy of Fallot: perioperative

management and analysis of outcomes. J Cardiothorac Vasc Anesth. 2019;33(2):556–65. https://doi.org/10.1053/j.jvca.2018.03.035.
17. Dłużniewska N, Podolec P, Skubera M, Smaś-Suska M, Pająk J, Urbańczyk-Zawadzka M, et al. Long-term follow-up in adults after tetralogy of Fallot repair. Cardiovasc Ultrasound. 2018;16(1):28. https://doi.org/10.1186/s12947-018-0146-7.
18. Bouzas B, Kilner PJ, Gatzoulis MA. Pulmonary regurgitation: not a benign lesion. Eur Heart J. 2005;26(5):433–9. https://doi.org/10.1093/eurheartj/ehi091.
19. Stout KK, Daniels CJ, Aboulhosn JA, Bozkurt B, Broberg CS, Colman JM, et al. 2018 AHA/ACC guideline for the Management of Adults with Congenital Heart Disease: a report of the American College of Cardiology/American Heart Association task force on clinical practice guidelines. Circulation. 2019;139(14):e698–800. https://doi.org/10.1161/CIR.0000000000000603.
20. Carvalho JS, Shinebourne EA, Busst C, Rigby ML, Redington AN. Exercise capacity after complete repair of tetralogy of Fallot: deleterious effects of residual pulmonary regurgitation. Br Heart J. 1992;67(6):470–3. https://doi.org/10.1136/hrt.67.6.470.
21. McRae ME, Coleman B, Atz TW, Kelechi TJ. Patient outcomes after transcatheter and surgical pulmonary valve replacement for pulmonary regurgitation in patients with repaired tetralogy of Fallot: a quasi-meta-analysis. Eur J Cardiovasc Nurs. 2017;16(6):539–53. https://doi.org/10.1177/1474515117696384.
22. Mongeon F-P, Ben Ali W, Khairy P, Bouhout I, Therrien J, Wald RM, et al. Pulmonary valve replacement for pulmonary regurgitation in adults with tetralogy of Fallot: a meta-analysis—a report for the writing committee of the 2019 update of the Canadian cardiovascular society guidelines for the Management of Adults with Congenital Heart Disease. Can J Cardiol. 2019;35(12):1772–83. https://doi.org/10.1016/j.cjca.2019.08.031.
23. Baumgartner H, De Backer J, Babu-Narayan SV, Budts W, Chessa M, Diller G-P, et al. ESC guidelines for the management of adult congenital heart disease the task force for the management of adult congenital heart disease of the European Society of Cardiology (ESC). Eur Heart J. 2020;2020:50–3. https://doi.org/10.1093/eurheartj/ehaa554.
24. Balasubramanya S, Zurakowski D, Borisuk M, Kaza AK, Emani SM, del Nido PJ, et al. Right ventricular outflow tract reintervention after primary tetralogy of Fallot repair in neonates and young infants. J Thorac Cardiovasc Surg. 2018;155(2):726–34. https://doi.org/10.1016/j.jtcvs.2017.09.019.
25. Khairy P, Van Hare GF, Balaji S, Berul CI, Cecchin F, Cohen MI, et al. PACES/HRS expert consensus statement on the recognition and management of arrhythmias in adult congenital heart disease: developed in partnership between the Pediatric and Congenital Electrophysiology Society (PACES) and the Heart Rhythm Society (HRS). Endorsed by the governing bodies of PACES, HRS, the American College of Cardiology (ACC), the American Heart Association (AHA), the European Heart Rhythm Association (EHRA), the Canadian Heart Rhythm Society (CHRS), and the International Society for Adult Congenital Heart Disease (ISACHD). Heart Rhythm. 2014;11(10):e102–65.
26. Valente AM, Gauvreau K, Assenza GE, Babu-Narayan SV, Schreier J, Gatzoulis MA, et al. Contemporary predictors of death and sustained ventricular tachycardia in patients with repaired tetralogy of Fallot enrolled in the INDICATOR cohort. Heart. 2014;100(3):247–53. https://doi.org/10.1136/heartjnl-2013-304958.
27. Tweddell JS, Pelech AN, Frommelt PC, Mussatto KA, Wyman JD, Fedderly RT, et al. Factors affecting longevity of homograft valves used in right ventricular outflow tract reconstruction for congenital heart disease. Circulation. 2000;102(Suppl. 3):Iii-130–Iii-5.
28. Daily JA, Tang X, Angtuaco M, Bolin E, Lang SM, Collins RT. Transcatheter versus surgical pulmonary valve replacement in repaired tetralogy of Fallot. Am J Cardiol. 2018;122(3):498–504. https://doi.org/10.1016/j.amjcard.2018.04.028.
29. Bonhoeffer P, Boudjemline Y, Saliba Z, Merckx J, Aggoun Y, Bonnet D, et al. Percutaneous replacement of pulmonary valve in a right-ventricle to pulmonary-artery prosthetic conduit with valve dysfunction. Lancet. 2000;356(9239):1403–5.
30. Tweddell JS, Simpson P, Li S-H, Dunham-Ingle J, Bartz PJ, Earing MG et al., editors. Timing and technique of pulmonary valve replacement in the patient with tetralogy of Fallot. Seminars in Thoracic and Cardiovascular Surgery: Pediatric Cardiac Surgery Annual; 2012: Elsevier.
31. Vukicevic M, Mosadegh B, Min JK, Little SH. Cardiac 3D printing and its future directions. J Am Coll Cardiol Img. 2017;10(2):171–84. https://doi.org/10.1016/j.jcmg.2016.12.001.
32. Sun Z, Lau I, Wong YH, Yeong CH. Personalized three-dimensional printed models in congenital heart disease. J Clin Med. 2019;8(4):522. https://doi.org/10.3390/jcm8040522.
33. Anwar S, Singh GK, Miller J, Sharma M, Manning P, Billadello JJ, et al. 3D printing is a transformative technology in congenital heart disease. JACC Basic Transl Sci. 2018;3(2):294–312. https://doi.org/10.1016/j.jacbts.2017.10.003.
34. Schievano S, Migliavacca F, Coats L, Khambadkone S, Carminati M, Wilson N, et al. Percutaneous pulmonary valve implantation based on rapid prototyping of right ventricular outflow tract and pulmonary trunk from MR data. Radiology. 2007;242(2):490–7. https://doi.org/10.1148/radiol.2422051994.
35. Sørensen TS, Mosegaard J, Kislinskiy S, Greil GF. Virtual surgery in congenital heart disease. In: Cardiac CT and MR for adult congenital heart disease. Springer; 2014. p. 515–23.
36. Ong CS, Loke Y-H, Opfermann J, Olivieri L, Vricella L, Krieger A, et al. Virtual surgery for conduit recon-

struction of the right ventricular outflow tract. World J Pediatr Congenit Heart Surg. 2017;8(3):391–3. https://doi.org/10.1177/2150135117692777.
37. Kiraly L. Three-dimensional modelling and three-dimensional printing in pediatric and congenital cardiac surgery. Transl Pediatr. 2018;7(2):129–38.
38. Sørensen TS, Greil GF, Hansen OK, Mosegaard J. Surgical simulation–a new tool to evaluate surgical incisions in congenital heart disease? Interact Cardiovasc Thorac Surg. 2006;5(5):536–9.
39. Jolley MA, Lasso A, Nam HH, Dinh PV, Scanlan AB, Nguyen AV, et al. Toward predictive modeling of catheter-based pulmonary valve replacement into native right ventricular outflow tracts. Catheter Cardiovasc Interv. 2019;93(3):E143–E52. https://doi.org/10.1002/ccd.27962.
40. Capelli C, Biglino G, Petrini L, Migliavacca F, Cosentino D, Bonhoeffer P, et al. Finite element strategies to satisfy clinical and engineering requirements in the field of percutaneous valves. Ann Biomed Eng. 2012;40(12):2663–73.
41. Schievano S, Taylor AM, Capelli C, Lurz P, Nordmeyer J, Migliavacca F, et al. Patient-specific finite element analysis results in more accurate prediction of stent fractures: application to percutaneous pulmonary valve implantation. J Biomech. 2010;43(4):687–93.
42. Cosentino D, Zwierzak I, Schievano S, Díaz-Zuccarini V, Fenner JW, Narracott AJ. Uncertainty assessment of imaging techniques for the 3D reconstruction of stent geometry. Med Eng Phys. 2014;36(8):1062–8.
43. Caimi A, Sturla F, Pluchinotta FR, Giugno L, Secchi F, Votta E, et al. Prediction of stenting-related adverse events through patient-specific finite element modelling. J Biomech. 2018;79:135–46.
44. Miyaji K, Miyazaki S, Itatani K, Oka N, Kitamura T, Horai T. Novel surgical strategy for complicated pulmonary stenosis using haemodynamic analysis based on a virtual operation with numerical flow analysis. Interact Cardiovasc Thorac Surg. 2018;28(5):775–82. https://doi.org/10.1093/icvts/ivy326.
45. Mosbahi S, Mickaily-Huber E, Charbonnier D, Hullin R, Burki M, Ferrari E, et al. Computational fluid dynamics of the right ventricular outflow tract and of the pulmonary artery: a bench model of flow dynamics. Interact Cardiovasc Thorac Surg. 2014;19(4):611–6. https://doi.org/10.1093/icvts/ivu202.
46. Rao AS, Menon PG. Presurgical planning using image-based in silico anatomical and functional characterization of tetralogy of Fallot with associated anomalies. Interact Cardiovasc Thorac Surg. 2014;20(2):149–56. https://doi.org/10.1093/icvts/ivu368.
47. Biffi B, Bruse JL, Zuluaga MA, Ntsinjana HN, Taylor AM, Schievano S. Investigating cardiac motion patterns using synthetic high-resolution 3D cardiovascular magnetic resonance images and statistical shape analysis. Front Pediatr. 2017;5:34. https://doi.org/10.3389/fped.2017.00034.
48. Couvreur S, Nowacka A, Biffi B, Bruse J, Burriesci G, Taylor A et al., editors. Statistical shape analysis of the right ventricular outflow tract in patients with tetralogy of Fallot congenital heart defects and pediatric cardiology applications. 8th World Congress of Biomechanics; 2019; Dublin: J Vasc Endovasc Therapy.
49. Toro Arana V, Chan F, Shiavone N, McElhinney D, Reddy S, Marsden A. Pre-operative right ventricular outflow tract and pulmonary artery geometry predicts pulmonary valve replacement outcomes in patients with tetralogy of Fallot. Circulation. 2018;138(Suppl. 1):A17390.
50. Dur O, Yoshida M, Manor P, Mayfield A, Wearden PD, Morell VO, et al. In vitro evaluation of right ventricular outflow tract reconstruction with bicuspid valved polytetrafluoroethylene conduit. Artif Organs. 2010;34(11):1010–6.
51. Bergersen L, Benson LN, Gillespie MJ, Cheatham SL, Crean AM, Hor KN, et al. Harmony feasibility trial: acute and short-term outcomes with a self-expanding transcatheter pulmonary valve. J Am Coll Cardiol Intv. 2017;10(17):1763–73.
52. Benson LN, Gillespie MJ, Bergersen L, Cheatham SL, Hor KN, Horlick EM, et al. Three-year outcomes from the harmony native outflow tract early feasibility study. Circ Cardiovasc Interv. 2020;13(1):e008320.
53. Kim GB, Song MK, Bae EJ, Park E-A, Lee W, Lim H-G, et al. Successful feasibility human trial of a new self-expandable percutaneous pulmonary valve (Pulsta valve) implantation using knitted nitinol wire backbone and Trileaflet α-gal–free porcine pericardial valve in the native right ventricular outflow tract. Circ Cardiovasc Interv. 2018;11(6):e006494.
54. Zahn EM, Chang JC, Armer D, Garg R. First human implant of the Alterra adaptive PrestentTM: a new self-expanding device designed to remodel the right ventricular outflow tract. Catheter Cardiovasc Interv. 2018;91(6):1125–9.

Tetralogy of Fallot Ventricle: Clinical Concepts, Engineering Applications, and Impact of an Integrated Medico-Engineering Approach

6

Henrik Brun and Kristin McLeod

6.1 The Tetralogy of Fallot Ventricles

Tetralogy of Fallot (TOF) is the most common cyanotic congenital heart defect. Repair is normally performed during the first year of life with a mortality below 2%. This includes VSD closure, cutting of right ventricular outflow tract (RVOT) muscle, often augmenting RVOT with noncontractile patch material. This may extend through the pulmonary valvular annulus, creating pulmonary regurgitation (see Fig. 6.1). Despite good long-term survival with 90% of patients surviving to 30 years of age, TOF is associated with a significant increase in morbidity, reduction of physical capacity, and risk for cardiac adverse events such as life-threatening arrhythmias and death. Preserved biventricular function is a main key to a long and good life with the repaired TOF heart. This chapter will address novel computational modeling efforts to expand the toolbox for individualized assessment, risk stratification, and thus potentially preservation of ventricular function. TOF represents a wide spectrum of disease ranging from conditions that are best categorized as pulmonary atresia to those resembling VSD with moderate pulmonary stenosis. The morphological starting point with all details taken into account, sets limitations for the clinical trajectory and prognosis for every individual patient. This again is modified substantially by time and type of initial palliation and/or repair and redo surgery or interventions down the road. Emerging knowledge tells us that genetic predisposition plays a role for ventricular remodeling as well (Mital S). In the ideal world, the engineering efforts described below in combination with rigorously collected clinical follow-up data should lead into an algorithm for the choice of ideal therapy for each TOF patient with her characteristics, at any time point of encounter—including surgical, interventional/electrophysiological, and medical therapies.

Any decision-making tool in congenital heart disease must have a lifetime perspective—from the newborn patient to the aging grown-up congenital heart (GUCH) patient. A major challenge is making the best decisions early in life—sometimes with limited time and preoperative data available—with focus on avoiding early postoperative complications but taking into perspective that the solution we choose must be optimal for biventricular longevity.

H. Brun
Department for Pediatric Cardiology, Oslo University Hospital – Rikshospitalet, Oslo, Norway

The Intervention Centre, Oslo University Hospital—Rikshospitalet, Oslo, Norway
e-mail: hbrun@ous-hf.no

K. McLeod (✉)
GE Vingmed Ultrasound AS, GE Healthcare, Horten, Norway
e-mail: Kristin.Mcleod@ge.com

G. Butera et al. (eds.), *Modelling Congenital Heart Disease*,
https://doi.org/10.1007/978-3-030-88892-3_6

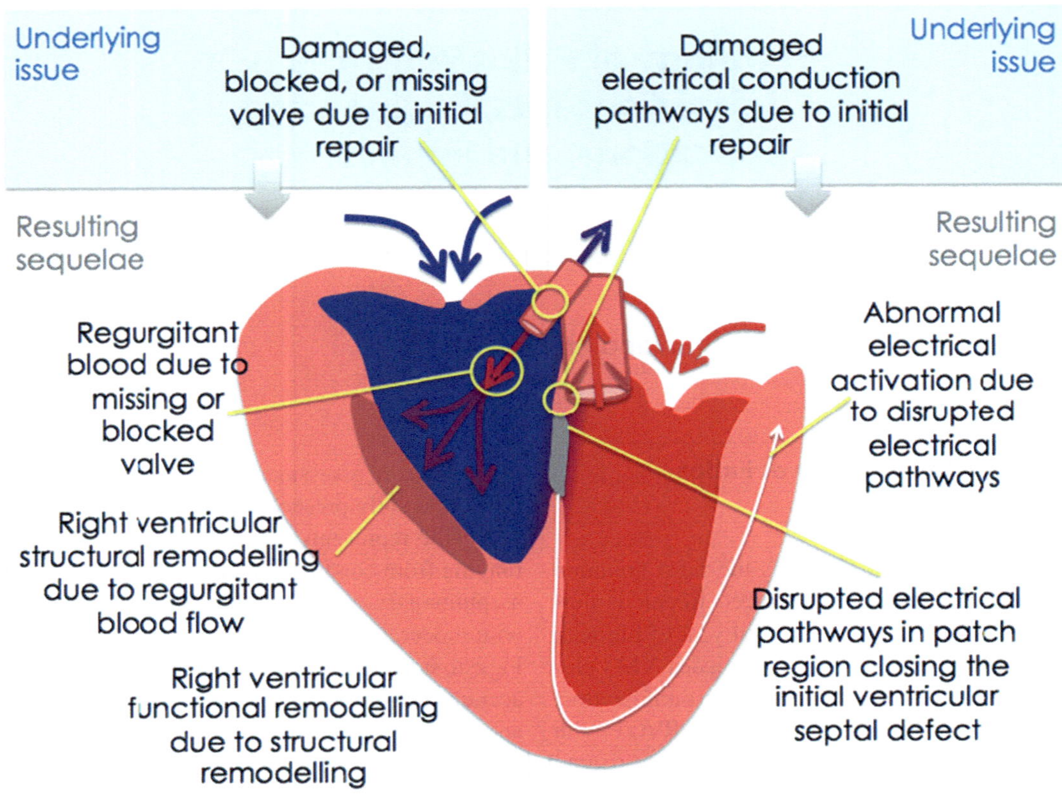

Fig. 6.1 Overview of the key underlying issues related to the Tetralogy of Fallot heart and their subsequent sequelae (Courtesy of Dr Krissy McLeod)

The first choice in early symptomatic TOF treatment is—albeit less frequent as surgical techniques are improved—whether to palliate with a shunt or RVOT stent or to do early primary repair. Ventricular structure and function has less of a place in these decisions but little is known about whether individual ventricular characteristics should be part of this treatment plan. Other structures such as pulmonary valve size and morphology, pulmonary artery and branch artery size, in relation to baby size make up the currently applied background information. Single factors such as a significant coronary artery crossing the RVOT may change treatment plan alone.

In cases that need surgery from approximately 3 months of age, a primary repair will most often be the choice. In this situation comes the choice of pulmonary valve sparing techniques/valvuloplasty versus valve destructive surgery. The optimal cut between residual RVOT and/or valve stenosis and regurgitation plus artery and branch size is sought, based on clinical consensus guidelines and the best subjective evaluation by the surgeon. With respect to the variation of RV muscle bundle morphology creating RVOT obstruction and pulmonary valve and artery size and morphology plus the increasing use of combinations of interventional and surgical strategies, this decision-making process is getting increasingly complex—calling for more objective individualized modeling and simulation support tools. Our future goal is to move decisions like these into the preoperative planning phase. How much muscle to be incised and/or removed in the RVOT during primary repair could be, determined based on patient-specific (simulation) models for how much pressure load the ventricle will tolerate without developing increased fibrosis and further including a determination of tolerable scars to contractile and conductive tissues from both a contractile and arrhythmogenic perspective.

The focus of clinical decision-making in repaired TOF has moved from right ventricular dilatation assessment and ECG changes, through an increasing focus on left ventricular function, at present appreciating the complex interplay between both ventricles, both structural, electrical, and mechanical. In everyday practice, it is hard to bring complicated biventricular functional evaluations into practice, and we still tend to lean on simple numbers such as QRS duration and MRI-derived ventricular volumes in combination with echocardiographic LV EF and RV FAC in our choice of therapies, omitting the impact of regional differences in ventricular shape and myocardial performance [1]. It is our belief that complex mathematical models based on large and detailed datasets, in combination with deep learning algorithms, may help bringing a more individualized approach to therapeutic choice into practical use.

The increasing application of percutaneous procedures for pulmonary valve replacement introduces a new challenge to clinical decision-making in TOF. The relative ease of PPVI as compared to open-heart PVR makes it tempting and possible to lower the intervention threshold without putting the patient at risks related to surgery and taking only a couple of days of his active life for hospital stay. An increased frequency of right-sided endocarditis after PPVI has moderated the eagerness to treat to some extent and brings another factor into the decision tree. This highlights the need for simulations and modeling efforts that take into account the differences between a surgical and interventional approach when it comes to effects on the ventricles and ventriculoarterial coupling mediated by stiffness of stents as compared to patches, expected residual obstructions, arrhythmogenicity of the changes made, and expected function over time of the valve replacement with its implications for future volume and pressure load on the ventricles. The effect of a planned intervention on a specific heart's creation of turbulent flows through a new pulmonary valve may be of particular interest to reduce risk for endocarditis.

All simulations and models for ventricular performance after TOF palliation, repair, and redo interventions or surgery must take into account the complex interplay between right and left ventricular tissue properties, morphology, size, regional and global activation, deformation, and timing of contractile and filling events.

In TOF, the right ventricle is challenged first with hypoxia, then by surgical scarring followed by pressure and or volume overload depending on the surgical result affecting both diastolic and systolic properties. The left ventricle isbeing a relatively frequent entity, TOF has the potential to become the first CHD in which true lifetime individualized and detailed treatment choices can be made—helped by comprehensive computer-modeled decision-making algorithms.

6.2 Engineering Applications

Computational techniques applied to the TOF heart have progressed significantly in the past decade. Given that this condition effects multiple physical phenomena of the heart, computational techniques to describe and simulate these different physical phenomena are being constantly refined as better techniques and computational resources improve. This includes techniques to study blood flow in the ventricles, structural remodeling of the ventricles, functional remodeling of the ventricles, fiber reorientation and tissue analysis in the ventricles, and electrophysiological modeling of the ventricles. In this chapter in order to simplify the discussion of engineering applications, we separately describe the analysis techniques (i.e., **data-driven**) and simulation techniques (**model-driven**), where we position analysis techniques as a means to **understand** the current state of a given patient and characterize it based on the data of the patient, whereas simulation techniques are more suited to **predict** the state given different clinical or biophysical variables and can be used to derive additional clinical metrics

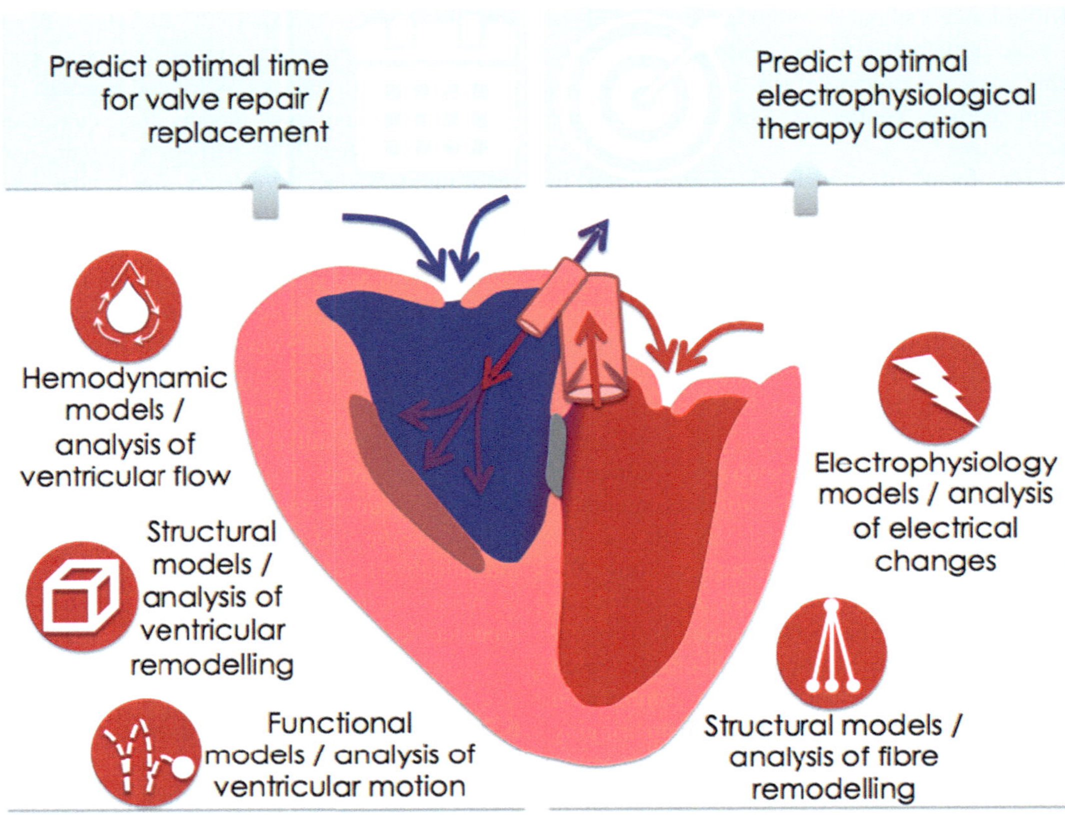

Fig. 6.2 Engineering techniques to assess the dominant issues of the TOF heart covering different biophysical modelling and analysis approaches (Courtesy of Dr Krissy McLeod

to support clinical decision-making. The two techniques, as such, are complementary and are not necessarily distinct, i.e., hybrid techniques also exist, and data-driven techniques are not only used to understand but also used for predict (similarly for model-driven approaches). As summarized in Fig. 6.2, engineering techniques are able to provide crucial insight to drive two key clinical decision-making challenges, namely; when to repair or replace the pulmonary valve, and to guide lead placement for cardiac resynchronization therapy. Examples of state-of-the-art techniques for studying cardiac phenomena are provided in the following subsections. Note that this is not a comprehensive review of current methods, the examples described here are used for the purpose of forming the discussion of engineering applications for TOF clinical decision-making guidance. A summary of the techniques and examples of clinical applications, as well as a summary of the current status of the different techniques in terms of clinical feasibility are provided in Table 6.1.

6.2.1 Blood Flow Distrubances in the Tetralogy of Fallot Heart

Analysis techniques for studying blood flow in the ventricles of TOF patients have been applied to assess the impact of the regurgitant blood flow from the pulmonary artery back to the right ventricle. 4D flow MRI is an imaging technique to characterize blood flow patterns and has been applied to the analysis of flow in the ventricles in TOF patients, for example by François et al. [2]. Kutty et al. [3], demonstrated the use of high-resolution ultrasound contrast particle imaging

Table 6.1 A summary of the physical phenomena affected in Tetralogy of Fallot hearts and how these phenomena can be represented by modern modeling/analysis techniques

Physical phenomena	Technique	Examples of methods	Examples of clinical application(s)	Current status
Blood flow/ hemodynamics	Analysis (data-driven)	4D flow MRI, contrast ultrasound, blood speckle imaging	Isolate vortex locations and relate these to increased wall shear stress	Methods quite evolved but limited studies of ventricular blood flow imaging
	Simulation (model-driven)	Computational fluid dynamics	Simulate the impact of varying degrees of regurgitation, correlate with the impact on wall shear stress	Methodology well developed for CFD simulations but currently limited applications to ventricular blood flow
Structure/ morphology	Analysis (data-driven)	Segmentation of MRI, ultrasound, CT images	Describe ventricular remodeling due to regurgitant blood flow	Well-established literature on the structural remodeling patterns in TOF ventricles. Mostly focused on the right ventricle to date
	Simulation (model-driven)	Statistical shape analysis	Determine structural features and link to clinical outcomes. Characterize ventricular structure by statistical shape features, simulate remodeling over time, estimate the degree of remodeling after which poor clinical outcomes increase, use these to predict the optimal time to intervene	Well-established methodology for studying ventricular morphology at the population level and preliminary studies on predicting the evolution of ventricular structure with aging. Large-scale clinical feasibility studies lacking to take the techniques from proof-of-concept to clinical usability
Function (motion)	Analysis (data-driven)	Speckle tracking ultrasound, tagged-MRI	Use increased strain and stress in different regions as indicators to support clinical decision-making	Speckle tracking and tagged MRI are well-established techniques for measuring cardiac function based on cardiac strain and work
	Simulation (model-driven)	Computational biomechanics modeling	Use computational models to simulate stress/strain dynamics to evaluate the use of biomechanical parameters for clinical decision-making	Prelimary applications analyzing stress and strain in TOF patients and comparing surgical techniques have been described, but are limited to proof-of-concept studies. Larger-scale clinical feasibility studies are lacking
Fiber structure	Analysis (data-driven)	Diffusion tensor MRI, ultrasound	Link patient-specific fiber structure (disarray) with clinical outcomes or use as input for further analysis	Diffusion tensor MRI is limited in patient-specific clinical care due to the time required to image the whole heart. Ultrasound imaging has been used to evaluate fiber orientation in animal hearts but clinical feasibility remains limited
	Simulation (model-driven)	Computational fiber generation	Estimate fiber orientation as an input to other simulation techniques (electrical, mechanical)	Rule-based and atlas-based methods for fiber generation are fairly well established. These methods are limited due to the fact that they are only approximations of patient-specific fiber orientation.

(continued)

Table 6.1 (continued)

Physical phenomena	Technique	Examples of methods	Examples of clinical application(s)	Current status
Electrical propogation	Analysis (data-driven)	Electroanatomical mapping, electrocardiogram	Guide ablation therapy based on patient-specific measurements of electrical activation. Real-time visualization of catheter location in the heart	Electroanatomical mapping is used clinically for therapy guidance
	Simulation (model-driven)	Computational electrophysiology	Simulate electrical propogation as a surrogate for invasive electrophysiology studies. Simulate ablation or CRT therapy to predict optimal regions of ablation/activation	Proof-of-concept studies suggest that electrophysiological models may be useful in stratifying risk of ventricular tachycardias, but clinical feasibility is still lacking

velocimetry to describe blood flow patterns in TOF patients, in particular showing the blood flow vortex concentrated at the right ventricular free wall as opposed to the outflow inflow tract. Blood speckle imaging is an alternative to constrast imaging for imaging blood flow [4], but has not to date been applied to studying the blood flow in TOF. Analysis of blood flow in TOF patients has the potential to shed insights into the clinical risk of patients.

Modeling techniques, such as computational fluid dynamics, have been primarily focused on the pulmonary artery and aorta in TOF patients. However, a recent study analyzing right ventricular blood flow from Wiputra et al. [5], in three fetal congenital heart disease patients demonstrated increased diastolic wall shear stress in areas with flow vortices. Increased wall stress can result in structural remodeling, as discussed in the next subsection.

6.2.2 Structural Remodeling of the Tetralogy of Fallot Ventricles

Analysis of structural remodeling is typically performed using image-processing techniques such as image segmentation to extract the object of interest from the image (the ventricles), providing a 2D or 3D representation of the structure from which clinical parameters such as volumes or ejection fraction can be computed. A recent study of CT images compared clinical parameters between patients pre and postrepair, finding statistically significant differences in right ventricular end-systolic volume and ejection fraction. The potential for such analyses as growing as fully automatic techniques are developing; however, a large-scale clinical evaluation of TOF patients has not to-date been performed since such tools are not as yet in the hands of the clinician.

Modeling of structural remodeling is emerging as a potential avenue to go beyond standard clinical metrics such as volumes and ejection fraction, to provide more local structural descriptors such as curvature [6]. Regional bulging (e.g., due to pulmonary regurgitation), among other structural remodeling patterns, was modeled over time in the right ventricle by Mansi et al. [7] to describe the progression of these structural remodeling patterns over time. A method to study the biventricular interaction in structural remodeling was described by Gilbert et al. [8] for the broader topic of congenital heart disease, and more specifically to TOF patients in McLeod et al. [9], as an extension of the methods described in Mansi et al. [7].

6.2.3 Functional Remodeling of the Tetralogy of Fallot Ventricles

Analysis of ventricular function via data-driven approaches uses imaging information to derive functional information. In the case of ultrasound or tagged-MRI, where physical markers in the image can be tracked over time, these markers are tracked across the cardiac cycle. Based on this tracking, parameters such as stress, strain, and cardiac work can be derived. Left ventricular strain is used in clinical practice; however, right ventricular strain is less common due to challenges in imaging the right ventricle either because of image acquisition window (in the case of ultrasound) or simply due to the thinner tissue compared to the left ventricle. Stress and strain and the comparison of these parameters between different surgical scenarios have been proposed by Zhou et al. [10].

Modeling of ventricular function using computational biophysical models enables the possibility to link ventricular functional parameters such as stress to clinical outcomes. A proof-of-concept study was recently described by Tang et al. [11], to relate mechanical stress to right ventricular function after pulmonary valve replacement.

6.2.4 Structural Remodeling of the Ventricular Fiber Structure

Analysis of fiber structure via high-resolution imaging is becoming increasingly important for understanding how macroscopic reorganization of cardiac tissue fibers in the electrical and mechanical activation of the heart. A recent review of imaging techniques is provided by Watson et al. [12], with the main focus on histology, or multimodal techniques using ultrasound and MR diffusion tensor imaging for imaging and visualization of cardiac fiber structure. Current imaging techniques are not able to provide sufficient resolution of fiber orientation in patient-specific cases.

Modeling of fiber sturcture has been proposed to address the short-comings of current imaging techniques, using rule-based techniques that describe the fiber structure in a given geometry based on some mathematical rules, or atlas-based techniques that map fiber structure from one subject to another, as summarized in Crozier et al. [13]. Both techniques have shown the ability to support subsequent simulation techniques but remain limited in the fact that they remain estimates and thus may not capture complex fiber disarray in a given patient.

6.2.5 Electrical Disturbances in the Tetralogy of Fallot Ventricles

Analysis of electrical disturbances is typically performed using electrocardiograms and electroanatomical mapping. Kimura et al. [14] reported disturbances in electrical propagation between the right atria and ventricle in relation to enlarged right ventricles. Electroanatomical mapping can be used during therapy for catheter guidance and to visualize scar locations, and may have implications in stratification for arrhythmia risk, as suggested by Pazzano et al. [15]. Cardiac resynchronization therapy in the right ventricle, while still an emerging procedure, can benefit from electroanatomical mapping to guide lead placement.

Modeling of electrical disturbances has been shown, in preliminary studies, to support therapy by stratifying patients at risk of arrhythmias and by providing suggestions for where to apply therapy. A feasibility study of electrophysiological simulations to stratify the risk of ventricular tachycardia was described by Shade et al. and found that these techniques may have the potential for predicting the risk ventricular tachycardia, compared to standard clinical predictors [16].

6.2.6 Summary

While engineering approaches to analyze and simulate cardiac phenomena have been increasing in recent years, these have been primarily limited to proof-of-concept studies and thus remain in their infancy with respect to clinical translation. There is a need to validate the different techniques in clinical feasibility studies. Methods to automate the processing steps will enable this potential for clinical translation, and are currently under development. A key bottleneck in the clinical translation of these techniques is the availability of such tools for clinical use, due to challenges in gaining regulatory approval for such tools. This creates a chicken and egg problem in which there is a lack of clinical studies to support the translation of engineering techniques and adoption in clinical tools, and at the same time a challenge of providing such tools in clinical tools due to the lack of clinical studies. Moreover, the analysis or simulation of cardiac phenomena in isolation of other phenomena neglects the important interplay between different phenomena (and additionally between different structures in the heart), therefore these models provide only a starting point for understanding the mechanisms driving remodeling in these patients.

6.3 Future Potential of Engineering Applications in Tetralogy of Fallot Care

6.3.1 From Models or Analysis Tools to Clinical Decision-Making Capabilities

A key component missing in existing engineering applications is clinical feasibility studies. The majority of articles referenced in this chapter describe methods applied to only 10's of patients and as such the true clinical impact is not clear.

6.3.2 Coupling Biophysical Models and Analysis to Build a Digital Twin of the Tetralogy of Fallot Heart

Due to the infancy of the described engineering applications, where most are still limited to proof-of-concept applications, the coupling of these phenomena has not been the first priority. However, hemodynamic load intrinsically effects structural remodeling which intrinsically effects functional remodeling, and so on. Coupling of the biophysical models (computational fluid dynamics, biomechanics, and electrophysiology) is a growing area of research but remains limited, due in part to the infancy of the respective models, but also due to challenges in computational requirements (these models are typically computationally intensive) and a lack of data for validation. Both computing power and data availability are improving with time, therefore the potential for coupled models will no doubt also improve with time. Pooling these techniques into a single representation, such as a digital twin (Fig. 6.3), can have significant potential to support clinical

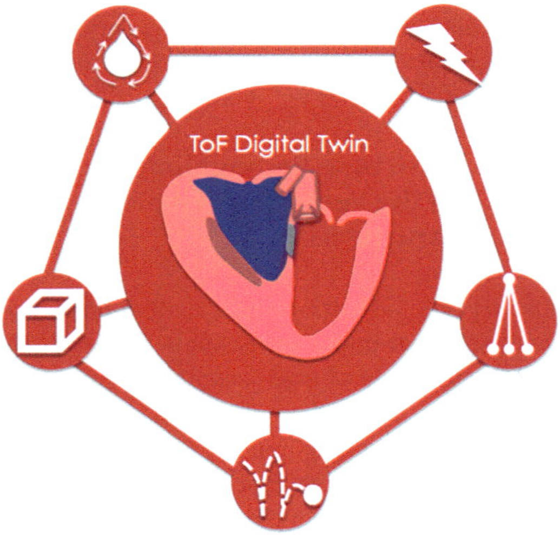

Fig. 6.3 The next generation of engineering applications will need to focus on the development of digital twins that incorporate multiple physical phenomena (Courtesy of Dr Krissy McLeod)

decision-making by reducing the need for invasive examinations, and by isolating biomarkers that help to identify the most at-risk patients.

6.3.3 Modeling Biophysical Changes in the Ventricles Over Aging

Current techniques have focused on analyzing or simulating the present state of the ventricles;however, the most interesting clinical question to solve is the optimal time of intervention for pulmonary valve repair. While there has been preliminary work on techniques to study the progression of the structural remodeling over time [7] and the functional remodeling over time [9], there are limited studies to describe the evolution of other cardiac phenomena over time. A key reason preventing the development of time-evolving models is a lack of longitudinal data following a patient over time (both Mansi et al. [7] and McLeod et al. [9] use cross-sectional data rather than longitudinal). A digital twin would ideally be a growing digital twin (Fig. 6.4).

6.3.4 The Role of Artificial Intelligence in Tetralogy of Fallot Decision-Making in the Future

Artificial intelligence (AI) is growing in the field of clinical decision-making, with the first few clinical tools gaining regulatory approval in recent years. Artificial intelligence has the potential to support both analysis and simulation techniques by improving the automation of data processing, enabling increased throughput of data which could support, for example, clinical feasibility studies. Furthermore, the potential to isolate unknown clinical features through large-scale data analysis could unlock the possibility to derive novel biomarkers that could be considered in combination with other biomarkers (existing and novel).

6.4 Potential Impact on Clinical Arena Now and in the Future, from the Clinician's Perspective

Integration and pattern analysis of big data supported by artificial intelligence bring expectations that clinicians will have new tools in the near future to make complex decisions for TOF patients in a personalized way. These solutions should ideally identify more exactly which patients will profit from what type of surgery or intervention at what time in life to ensure cardiac longevity. Contributions from computational ventricular modeling as described in this chapter may become an important part of new and comprehensive phenotyping systems that inform more complex decision-making algorithms than clinicians are used to. In early life-models that describe the native ventricular structure, function and development could potentially become part of the information that surgeons depend on, in order to decide when and how to do primary repair in stable TOF patients, aiming to optimize long-term preservation of biventricular function. New ways of analyzing and understanding ventricular shape, size, activation, and contraction pattern most obviously will have impact on decisions of how and when to intervene after primary surgery, when scars, artificial and potentially future contractile implants, changes in myocardial activation pattern, and ventricular function become important parts of the models.

6.4.1 The Potential Impact of Engineering Tools Over the TOF Lifecycle

The engineering tools described in this chapter could have potential impact over several stages in the care of TOF hearts:

1. **Antenatal decisions.** In the ideal world—before a TOF patient is born—genetic characteristics and fetal echocardiographic and fetal MRI images are harvested and fed into a

Fig. 6.4 As with all congenital heart diseases, analysis of TOF hearts through digital twins will need to incorporate the evolution of these models, to adequately represent the development of the heart over time as through ageing (Courtesy of Dr Krissy McLeod)

model that informs parents about their child's prognosis with optimal treatment. As standard TOF pregnancies are normally not advised for termination, this could be of special importance to identify pathologies that carry a higher risk of early death or increased morbidity such as the extreme morphologies resembling PA-VSD-MAPCAS or absent pulmonary valve TOF. Theoretically such models could be part of fetal interventions decision-making, such as for intrauterine pulmonary balloon valvuloplasty.

2. **Primary repair decisions.** After birth, echocardiographic, CT, and MRI data will increasingly be used to feed mathematical models that predict outcome for a specific child, following the available treatment options included the timing of these. Decisions such as whether to stent the RVOT, stent the duct or do surgical shunt palliation may become less a matter of heart conference likes and dislikes and more rely on calculations and predictions from large datasets.
3. **Redo interventional and surgical decisions.** This is probably the area where modeling of right and left ventricular size, shape, activation, and contraction pattern and ventricular interaction will become the basis for choice like what size and type of pulmonary valve the patient will benefit the most from, and when it is time to provide the new valve – be it interventional or surgical.
4. **Electrophysiology management.** One of the significant long-term risks in repaired TOF is ventricular arrhythmias that may potentially be lethal. Valid risk stratification to treat the right patient at the right time is an unresolved problem. Ventricular modeling may inform EP decision algorithms, including risk stratification for choice of ICD treatment and future prospected increased use of CRT devices for resynchronization in repaired TOF with ventricular failure, typically with right bundle branch block.

Arrhythmias occur as false and circular ways of activation signaling that obviously are 3D structures. Segmentation of scars in LV have been used for VT prediction TOF-like animal models with RBBB taking into account biventricular activation optimization by correct pacing strategy will hopefully teach us about which TOF patients to treat with CRT and how the activation can be normalized to reverse negative ventricular remodeling and preserve function.

5. **Medical treatment.** Lastly, ventricular modeling could also become informative to identify postrepair TOF ventricles that profit from specific cardiovascular drugs such as beta-blockers or ACE inhibitors. Global ventricular wall stress and work may be totally different for two ventricles with the same volume but different shapes. Rather than basing the drug treatment on ventricular size, volume, and ejection fraction, different ventricular shapes may need different pharmacological support, based on complex calculations of how the sum of myocardial fiber contraction and interaction can be optimally supported for each patient.

6.4.2 Engineering Tools that Could Address Key Challenges in the Care of TOF Patients

1. **Prediction of RV failure**: When it comes to early prediction of RV failure, possibly preventing by interventions to relieve loads, clearly we have not succeeded in finding the valid follow-up parameters to apply when assessing TOF patients for PVR need typically 10 years after last intervention. New ventricular descriptors such as shape analysis and specifically change of shapes with time in longitudinal studies will teach us what are the clinically important markers of deterioration that we should look for and act upon.
2. **Testing of new PPVI devices and also TVR devices** could be possible via simulation models. Development of new PPVI devices and techniques should be performed in simulations in silico and in vitro before application in vivo in animal models to save animals. Improved realism of simulations is needed to offer device developers a wide range of valid testing scenarios supported by new visualization technologies such as virtual or augmented reality.
3. **Prediction of RV CRT/biventricular pacing success** and the optimalization of lead(s) positioning could be supported by modeling through individualized medicine approaches. 3D models of both ventricles with representation of postoperative scars from MRI LGE imaging may become an important tool in TOF patients who need CRT pacing/resync therapy for HF. Patients with large RV's, RBBB, and wide QRS may profit from resyncronization and modeling that includes individual activation patterns expected from pacing at different sites could become effevtive tools in selecting a personalized pacing lead placement to optimize pacing therapy [17].
4. **Resting and stress 4D flow simulation** can be applied to determine optimal RVOT—PVR—PA -creation (intervention/surgery). There is little data on the blood flow and conservation of right ventricular pumping force/energy into blood transportation through the lungs after PVR in TOF. Computational flow modeling simulations of blood flow with vortex formation and turbulence/energy loss could be valuable in choosing personalized PVR solutions based on RVOT size, shape, and function/regional lack of function in repaired TOF to achieve optimized ventricular pulmonary artery coupling in the highly variable range of anatomies including special morphologies such as DORV – TOF or AVSD -TOF.
5. **Prediction of mildly hypoplastic RV growth** after RF PA opening or critical vPS balloon can be modeled by modern AI approaches to improve directed surgical RV overhaul.
6. **Different RV biomarkers** could be combined and analyzed using modern AI-based approaches—continuous parameters combined as big data possibly providing new treatment and management tools. RV shape, RV regional strain, CT, MRI, and echo data could be modeled together in one model.

7. **Aggressive resection or leave stenosis** to preserve valvar function is a core question in primary repair. Advanced right ventricle models including genetics, MRI/CT, and echo data including strain, flow, and pressure signals could help deciding the individual tolerance for residual pressure and/or volume load for each patient to preserve long-term ventricular function.

There is a clear need for advanced engineering techniques to support the care of TOF patients. While the engineering techniques have so far been primarily focused on generating representations of the biophysical processing underlying this condition, along with the remodeling that occurs as the heart develops under as a result of the sequalae of this condition, the tools and methods available today pave the way toward advanced applications that will enable and empower cardiologists in the care of these patients.

References

1. Stout KK, Daniels CJ, Aboulhosn JA, Bozkurt B, Broberg CS, Colman JM, Crumb SR, Dearani JA, Fuller S, Gurvitz M, Khairy P. 2018 AHA/ACC guideline for the management of adults with congenital heart disease: executive summary: a report of the American College of Cardiology/American Heart Association Task Force on Clinical Practice Guidelines. J Am Coll Cardiol. 2019;73(12):1494–563.
2. François CJ, Srinivasan S, Schiebler ML, Reeder SB, Niespodzany E, Landgraf BR, Wieben O, Frydrychowicz A. 4D cardiovascular magnetic resonance velocity mapping of alterations of right heart flow patterns and main pulmonary artery hemodynamics in tetralogy of Fallot. J Cardiovasc Magn Reson. 2012;14(1):16. https://www.researchgate.net/publication/221813931_4D_Cardiovascular_Magnetic_Resonance_Velocity_Mapping_of_Alterations_of_Right_Heart_Flow_Patterns_and_Main_Pulmonary_Artery_Hemodynamics_in_Tetralogy_of_Fallot
3. Kutty S, Li L, Danford DA, Houle H, Datta S, Mancina J, Xiao Y, Pedrizzetti G, Porter TR. Effects of right ventricular hemodynamic burden on intraventricular flow in tetralogy of Fallot: an echocardiographic contrast particle imaging velocimetry study. J Am Soc Echocardiogr. 2014;27(12):1311–8. http://www.dic.univ.trieste.it/perspage/Gianni/papers/2014jase2.pdf
4. Wigen MS, Fadnes S, Rodriguez-Molares A, Bjåstad T, Eriksen M, Stensæth KH, Støylen A, Lovstakken L. 4-D Intracardiac ultrasound vector flow imaging–feasibility and comparison to phase-contrast MRI. IEEE Trans Med Imaging. 2018;37(12):2619–29. https://www.researchgate.net/profile/Morten_Wigen/publication/325608907_4D_Intracardiac_Ultrasound_Vector_Flow_Imaging_-Feasibility_and_Comparison_to_Phase-Contrast_MRI/links/5d08c30c299bf1f539cbc4a7/4D-Intracardiac-Ultrasound-Vector-Flow-Imaging-Feasibility-and-Comparison-to-Phase-Contrast-MRI.pdf
5. Wiputra H, Lai CQ, Lim GL, Heng JJW, Guo L, Soomar SM, Leo HL, Biwas A, Mattar CNZ, Yap CH. Fluid mechanics of human fetal right ventricles from image-based computational fluid dynamics using 4D clinical ultrasound scans. Am J Physiol-Heart Circ Physiol. 2016;311(6):H1498–508. https://doi.org/10.1152/ajpheart.00400.2016.
6. Moceri P, Duchateau N, Baudouy D, Squara F, Ferrari E, Sermesant M. 3D right ventricular strain and shape in volume overload: comparative analysis of Tetralogy of Fallot and atrial septal defect patients. Arch Cardiovasc Dis Suppl. 2019;11(1):128–9. https://www.sciencedirect.com/science/article/pii/S1878648018305378
7. Mansi T, Voigt I, Leonardi B, Pennec X, Durrleman S, Sermesant M, Delingette H, Taylor AM, Boudjemline Y, Pongiglione G, Ayache N. A statistical model for quantification and prediction of cardiac remodelling: application to tetralogy of fallot. IEEE Trans Med Imaging. 2011;30(9):1605–16. http://citeseerx.ist.psu.edu/viewdoc/download?doi=10.1.1.221.2134&rep=rep1&type=pdf
8. Gilbert K, Forsch N, Hegde S, Mauger C, Omens JH, Perry JC, Pontré B, Suinesiaputra A, Young AA, McCulloch AD. Atlas-based computational analysis of heart shape and function in congenital heart disease. J Cardiovasc Transl Res. 2018;11(2):123–32. https://www.ncbi.nlm.nih.gov/pmc/articles/PMC5910190/
9. McLeod K, Mansi T, Sermesant M, Pongiglione G, Pennec X. Statistical shape analysis of surfaces in medical images applied to the Tetralogy of Fallot heart. In: Modeling in computational biology and biomedicine. Berlin, Heidelberg: Springer; 2013. p. 165–91. https://hal.inria.fr/docs/00/81/38/50/PDF/McLeod_Pennec_CBB.pdf.
10. Zhou Z, Geva T, Rathod RH, Tang A, Yang C, Billiar KL, Tang D, del Nido P. Combining Smaller Patch, RV Remodeling and Tissue Regeneration in Pulmonary Valve Replacement Surgery Design May Lead to Better Post-Surgery RV Cardiac Function for Patients with Tetralogy of Fallot. Mol Cell Biomech. 2018;15(2):99–115. https://doi.org/10.3970/mcb.2018.000.558.pdf.
11. Tang D, Yang C, Pedro J, Zuo H, Rathod RH, Huang X, Gooty V, Tang A, Billiar KL, Wu Z, Geva

T. Mechanical stress is associated with right ventricular response to pulmonary valve replacement in patients with repaired tetralogy of Fallot. J Thorac Cardiovasc Surg. 2016;151(3):687–94. https://www.sciencedirect.com/science/article/pii/S0022522315018152
12. Watson SR, Dormer JD, Fei B. Imaging technologies for cardiac fiber and heart failure: a review. Heart Fail Rev. 2018;23(2):273–89. https://www.ncbi.nlm.nih.gov/pmc/articles/PMC6121700/
13. Crozier A, Augustin CM, Neic A, Prassl AJ, Holler M, Fastl TE, Hennemuth A, Bredies K, Kuehne T, Bishop MJ, Niederer SA. Image-based personalization of cardiac anatomy for coupled electromechanical modeling. Ann Biomed Eng. 2016;44(1):58–70. https://doi.org/10.1007/s10439-015-1474-5.
14. Kimura Y, Fukuda K, Nakano M, Hasebe Y, Fukasawa K, Chiba T, Miki K, Tatebe S, Miyata S, Ota H, Kimura M. Prognostic significance of PR interval prolongation in adult patients with total correction of tetralogy of Fallot. Circ Arrhythm Electrophysiol. 2018;11(11):e006234. https://doi.org/10.1161/CIRCEP.118.006234.
15. Pazzano V, Di Mambro C, Russo MS, Silvetti MS, Palmieri R, Fabrizio GF, Giannico S, Leonardi B, Amodeo A, Di Ciommo VM, Drago F. Role of right ventricular three-dimensional electroanatomic voltage mapping for the arrhythmic risk stratification of patients with corrected tetralogy of Fallot or other congenital heart disease involving the right ventricle outflow tract. Thorac Cardiovasc Surg. 2016;64(S 02):OPIT1. https://www.thieme-connect.com/products/ejournals/abstract/10.1055/s-0036-1571900
16. Shade JK, Cartoski MJ, Nikolov P, Prakosa A, Doshi A, Binka E, Olivieri L, Boyle PM, Spevak PJ, Trayanova NA. Ventricular arrhythmia risk prediction in repaired Tetralogy of Fallot using personalized computational cardiac models. Heart Rhythm. 2019; https://www.sciencedirect.com/science/article/abs/pii/S1547527119309087
17. Behdadfar S, Navarro L, Sundnes J, Maleckar M, Odland HH, Avril S. Abnormal tissue zone detection and average active stress estimation in patients with LV dysfunction. In: Medical and biological image analysis. IntechOpen; 2018. p. 69.

Complete Transposition of Great Arteries Post-Arterial Switch Operation: An Integrated Approach of Imaging and Modelling to Assess Long-Term Outcomes

7

Claudio Capelli, Teodora Popa, Andrei-Mihai Iacob, and Hopewell Ntsinjana

7.1 Introduction

Complete transposition of the great arteries (TGA) is a singular congenital lesion in which the aorta arises from the right ventricle and the pulmonary artery from the left ventricle. TGA accounts for 5–7% of all congenital heart defects (CHDs), with a prevalence rate of 0.3 per 1000 live births [1, 2]. TGA affects preponderantly males with a ratio to females of 2:1 and is the commonest cyanotic CHD presenting in the neonatal period [3]. In TGA patients, the pulmonary and systemic circulations are in parallel, a situation not compatible with life. Therefore, there has to be an obligatory communication between the two, either with an atrial septal defect (ASD), a ventricular septal defect (VSD), or at the great arterial level (patent ductus arteriosus) to support postnatal life. Thus, a palliative or definitive surgery is the ultimate goal to ensure long-term survival of these patients.

Morphologically, the most common form of TGA is the dextro-transposition of great arteries (referred to as d-TGA), in which there is dextroposition of the bulboventricular loop (i.e. the right ventricle is positioned to the right of the left ventricle). In this condition, the aorta also tends to be on the right and anterior, and the great arteries are parallel rather than crossing as they do in the normal heart. The d-TGA is characterised by discordant ventriculoarterial alignment such that the aorta arises entirely or in large part from the right ventricle and the pulmonary artery arises entirely or in large part from the left ventricle. Concordant atrioventricular arrangement is nearly always recognised in these hearts (Fig. 7.1).

Diagnosis of TGA can be made in utero or postnatally on two-dimensional echocardiography and surgical correction or palliation can be instituted based on the echocardiographic findings alone. A surgical procedure to switch the position of aorta and pulmonary arteries, the

C. Capelli (✉) · T. Popa
UCL Institute of Cardiovascular Science & Great Ormond Street Hospital for Children, London, UK
e-mail: c.capelli@ucl.ac.uk; teodora.popa.15@ucl.ac.uk

A.-M. Iacob
Carol Davila University of Medicine and Pharmacy, Bucureşti, Romania
e-mail: andrei.iacob@stud.umfcd.ro

H. Ntsinjana
Division of Paediatric Cardiology, Department of Paediatrics and Child Health, School of Clinical Medicine, Nelson Mandela Children's Hospital, University of the Witwatersrand, Parktown, Johannesburg, South Africa
e-mail: hopewell.ntsinjana@wits.ac.za

G. Butera et al. (eds.), *Modelling Congenital Heart Disease*,
https://doi.org/10.1007/978-3-030-88892-3_7

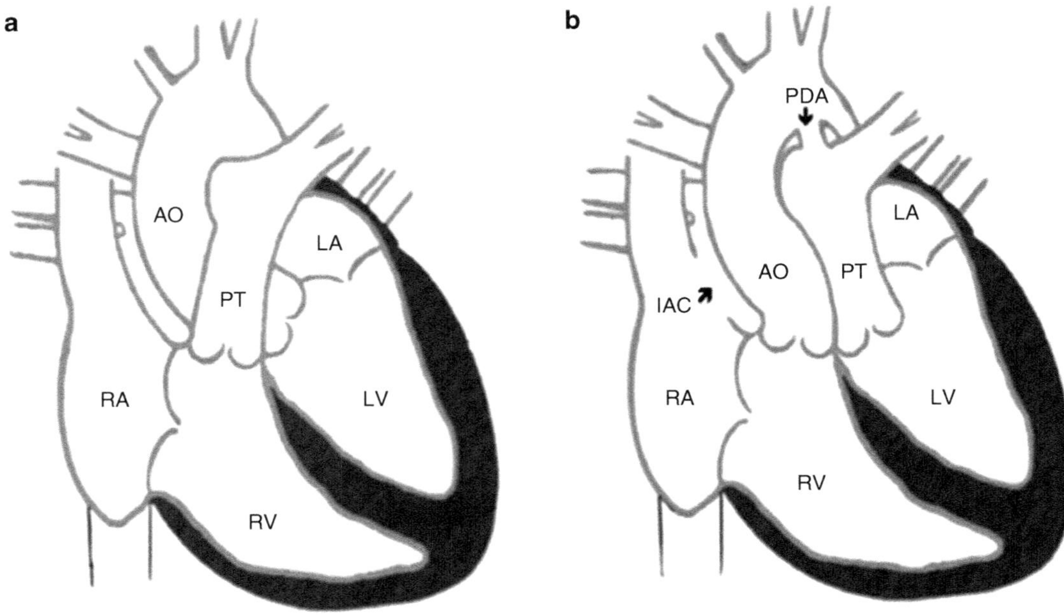

Fig. 7.1 Schematic drawing of the normal heart (**a**), and heart with transposition of great arteries (**b**). Labels denote the following: RA (right atrium), RV (right ventricle), PT (pulmonary trunk), LA (left atrium), LV (left ventricle), AO (aorta), IAC (intra-atrial communication) and PDA (patent ductus ateriosus)

arterial switch operation (ASO), is a life-saving intervention which is typically performed within the first 2 weeks of life. Follow-up of TGA patients post-ASO is crucial in order to estimate prognosis and monitor the integrity of the vessels including re-implanted coronary arteries and the translocated great arteries. Various complimentary imaging modalities are fundamental to address all the relevant clinical questions posed by the complexities of lesions occurring after surgery. In addition, modelling and simulations, typically used in the field of biomedical engineering, recently brought new perspectives to provide further insights to understand physiological and pathological processes on both specific patient and the whole TGA population.

In this chapter, we aim to comprehensively assess the consequences of ASO in TGA patients, through the use of non-invasive multi-modality imaging modalities, which can be complemented by 3D modelling and computational simulations in order to understand the long-term sequelae of the ASO, with a potential to impact follow-up strategies.

7.2 Surgical Treatments

Historically, definitive surgery for d-TGA has come a long way. To date, there are still adult patients who were operated on with intra-atrial repair surgeries such as Senning and Mustard [4], but these have since been replaced with ventricular level (Rastelli) operation [5] and great artery level surgery [6]. The atrial baffle procedures work by re-routing pulmonary and systemic venous return at the atrial level with resulting physiological correction (Fig. 7.2). However, they suffer from a number of long-term problems such as arrhythmias, Tricuspid valve insufficiency, RV dysfunction, baffle obstructions or leaks and sudden death [7].

The arterial switch operation (ASO) is currently the standard for surgical repair of d-TGA [8]. The procedure involves transecting the aorta and the pulmonary trunk 5 to 10 mm below its bifurcation, harvesting the coronary arteries together with two coronary buttons and re-implanting them in the neo-aorta. After this, the Lecompte manoeuvre is performed to relocate

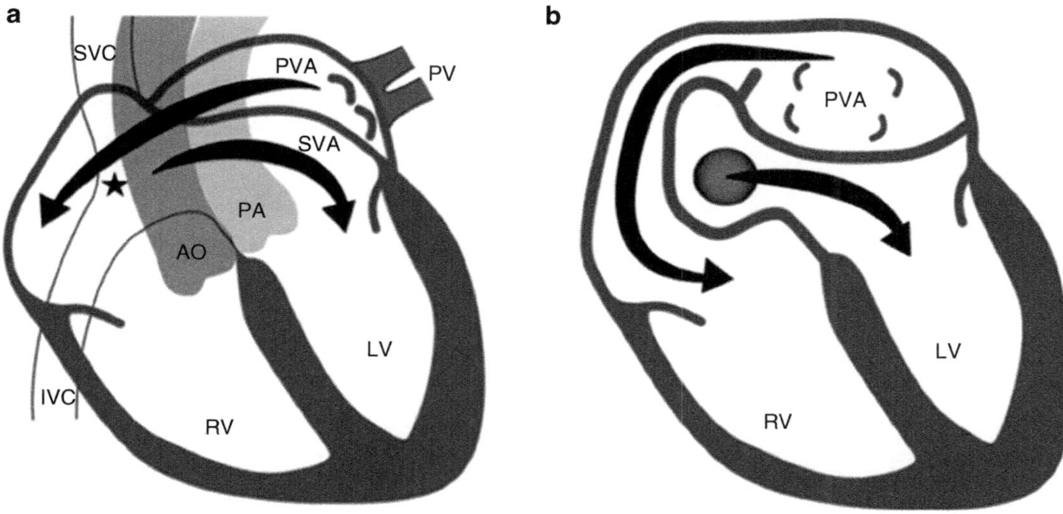

Fig. 7.2 Cartoon diagram (**a**) and (**b**) is an illustration of Senning/Mustard procedure, as performed by means of right atrial incision. Shown are SVC (superior vena cava) and IVC (inferior vena cava) baffle (*) channelling blood through the systemic venous atrium (SVA) for systemic venous return, to the LV then to lungs via PA (pulmonary artery). PV (pulmonary veins) returns oxygenated blood to the (PVA) pulmonary venous atrium, then to the systemic RV (right ventricle) to be pumped into systemic circulation for metabolic demands. Black thick arrows denote direction of flow in both diagrams (**a**) and (**b**)

the neo-pulmonary trunk anterior to the aorta (Fig. 7.3). Prior to the arterial switch, the intracardiac defect is repaired in patients presenting with TGA and an associated VSD. The timing of repair must be tailored to the infant's medical status and to technical considerations of the centre's surgical team. However, it is typically performed within the first 2 weeks of life. In most centres, the optimal age to perform the arterial switch operation is day 3 of life, but it is still possible up to 10 weeks postnatally, if adequate post-operative ventilator and circulatory support are available.

Mortality associated with d-TGA has dramatically improved from approximately 90% for unoperated patients to rates of <5% following corrective surgery with the ASO [9, 10]. Coronary artery transfer is the most intricate part of the surgical treatment of TGA and unusual coronary arrangement is associated with considerable mortality, therefore understanding the particular coronary anatomy through pre-operative imaging is a very important step.

7.3 Imaging a Child or Young Adult with ASO

Technological advances in medical diagnostics have enabled the medical field to leap into a new level, where imaging has become the integral part of patient evaluation prior to surgical- or catheter-based intervention. Ultrasound assessment is adequate for pre-operative work-up prior to the ASO, providing adequate anatomical and functional information, including coronary artery anatomy according to the Leiden classification [11]. However, for patients presenting beyond the neonatal period, diagnostic cardiac catheterisation may be warranted. Intra-operative echocardiography during ASO can guide the surgeon during the procedure in order to ensure a suitable outcome, by assessing LV function and colour plus spectral doppler flows through the coronary arteries [12]. Also, echocardiography can diagnose majority of post-operative sequelae following ASO. However, as child grows cross-sectional imaging is recommended for surveillance

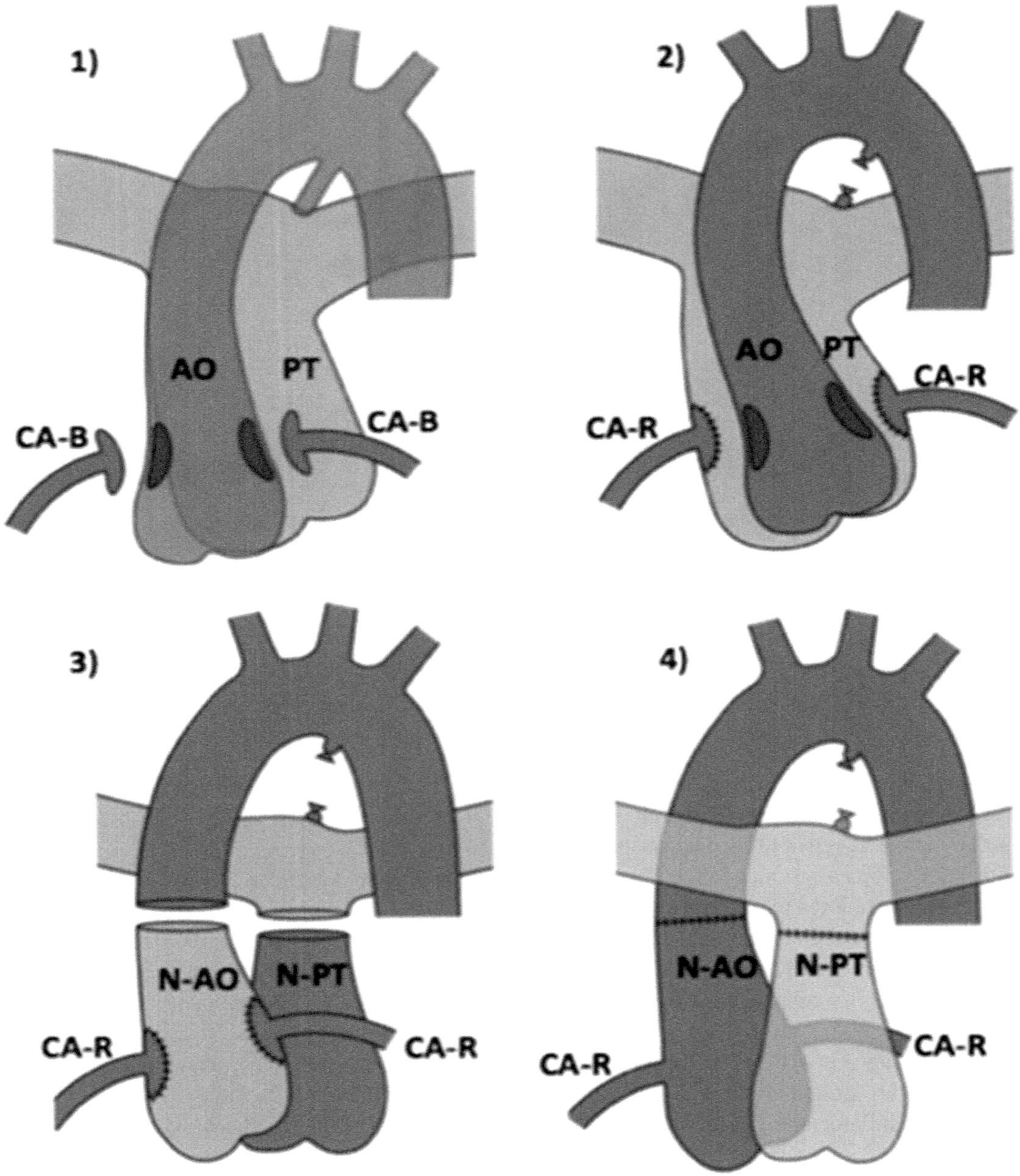

Fig. 7.3 Cartoon diagrams showing stages of Jatene's arterial switch operation stages 1–4. CA-B (coronary artery with button), AO (aorta), PT (pulmonary trunk), CA-R (coronary artery re-implanted), N-AO (neo-aorta), N-PT (neo-pulmonary trunk)

purposes. Currently, computed tomography (CT) and cardiac magnetic resonance (CMR) imaging are modalities used for cross-sectional imaging in TGA patients following ASO. CT and CMR are often used in a complementary fashion to echocardiography, while exploring strength and weaknesses of each modality [13]. More recently, computer models and simulations from the field of biomedical engineering have shown capabilities to enrich further the imaging modalities. These are based on advanced image post-processing strategies to create patient-specific models that are then subjected to computational mathematical solvers (in-silico simulations) or benchtop (in-vitro) experiments with an intention to understand physiological and pathological processes.

7.4 Late Complications of ASO and Monitoring with Imaging and Modelling

Peri-operative mortality and re-operation rate in ASO patients have dropped to less than 6% [14, 15]. Over time, the techniques of coronary artery transfer and re-implantation have improved making even the most complex anatomical variants feasible to successful transfer. However, there are still lingering clinical and subclinical late post-operative residua. Both left and right heart together with coronary arteries post-ASO pose a difficult challenge to the clinicians because, to date, there are no standardised protocols on how to investigate and follow-up these patients. Moreover, the impacts these lesions will pose as patients enter adulthood are yet to be fully understood. In the next sections, we analyse how the main potential late complications can be monitored and tackled with up-to-date techniques of imaging and engineering modelling.

7.4.1 Late Complications of Neo-aortic, Root Valve and Arch

Following ASO and the Lecompte manuevre, the neo-aortic root is known to be prone to dilation. The aetiology for this lesion is multifactorial, including structural intrinsic abnormality of the pulmonary root which possesses histopathological differences in the systemic circulation [16]. The extrinsic risk factors for neo-aortic root dilatation include previous pulmonary artery band, ASO at age greater than 1 year and presence of VSD [17]. Interestingly, dilatations warrant a lifelong surveillance for these patients. Monitoring the neo-aortic valve for durability is a major part of follow-up assessment in ASO survivors. Neo-aortic valve incompetence has been recorded in up to 40% of patients [18–21]. Another notable lesion which can lead to haemodynamic instability is the acute angulation of the neo-aortic arch. This feature is a direct consequence of the Lecompte manoeuvre, which causes the neo-pulmonary trunk to push against the aorta, nudging it posteriorly. The aortic arch then assumes a more acutely angulated geometry resembling a 'gothic' arch, with implications on the aetiology of neo-aortic valve incompetence and root dilatation. One of the most worrying findings is the reduced aortic bioelastic properties, suggesting abnormal ventricular arterial (V-A) coupling. This raises concerns about how the neo-aorta will behave in the context of systemic hypertension at a later stage in life and the risk for aortic dissection under increased wall shear stress (WSS), despite no reported cases up to date.

Anatomical and functional assessment of neo-aortic root, valve and arch can be achieved with multi-modality imaging assessment including ultrasounds, CT and CMR. Flow and area changes over time are used to address vessel stiffness. In addition, patient-specific 3D models have been used to perform a morphometric analysis to quantify geometric differences between TGA patients repaired with ASO and healthy control subjects [22, 23]. These studies showed how the relative position of the LV was not different between TGA and control subject. Instead, a significant difference in the angle between the aortic valve plane and sino-tubular junction plane was detected. This is likely due to the indentation of the aortic wall as a result of the anterior position of the pulmonary trunk following the Lecompte manoeuvre. Aortic root volume was found to be twice as large in TGA post-ASO patients compared to healthy controls with comparable age and body size. This may lead to surgical treatment of a dilated aortic root [24]. Importantly, the 'gothic' arch curvature, often described in the TGA cohort, is clearly

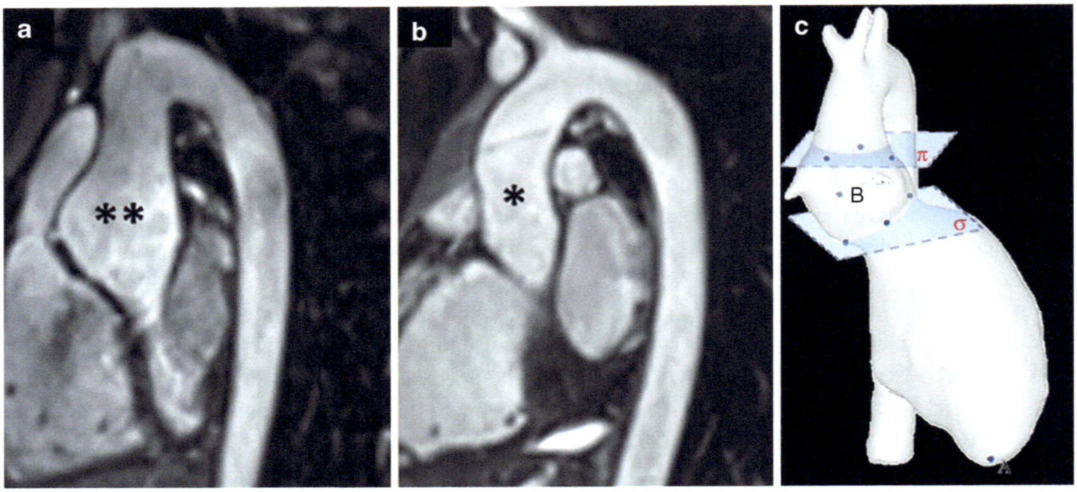

Fig. 7.4 MRI images and 3D model of TGA and control subject. (**a**) Cardiac MRI sagittal view of arterial switch operated patient, (**b**) cardiac MRI sagittal view of normal healthy control. (**) shows dilated aortic root for the ASO patient compared to control (*). (**c**) Three-dimensional model of LV, LVOT, aorta up to the level of diaphragm highlighting the points used as anatomical landmarks (A, apex of the left ventricle; and B, centre of coaptation of aortic valve leaflets); the annular plane σ; and the sinutubular plane π (adapted with permission from Ref. [23])

identified and quantified with patient-specific models (Fig. 7.4).

The relation between geometry and function of the aortic arch can also be assessed combining images and modelling. The flow within gothic aortic arch was investigated on experiments on TGA patient-specific models aimed at detecting the impact of different geometries on the haemodynamic profile of a patient. 4D flow-sensitive in vitro analysis of CMR of patient-specific ASO model [25] revealed a systolic high velocity jet directed at the enlarged segment of the aortic root wall, with swirling, lower velocity vortices accompanying it as shown in Fig. 7.5. This behaviour was confirmed by CFD analyses and clinical studies [26] which consistently showed WSS increase, justifying a concern related to likely aneurysm formation later in life.

Finally, the root distensibility in TGA post-ASO patients was also assessed by wave intensity analysis (WIA). WIA is a non-invasive, numerical technique which can be used to assess vessels' stiffness in the paediatric population. WIA is not operator-dependent compared to the established diagnostic tests for diastolic dysfunction. The use of WIA showed a lower aortic distensibility in TGA patients compared to control subjects, overriding the physiological difference due to increasing age. This could potentially impact negatively on their aortic properties and these patients may be at higher risk of aortic dissection and premature atherosclerosis. Albeit this can only be speculated at present, aortic stiffness is known to have independent predictive value for cardiovascular events in adults [27].

7.4.2 Late Complications of Coronary Arteries

Coronary artery issues after ASO are reported in 3–11% of patients [28]. Coronary artery issues exhibit a bimodal nature suggesting an early peak related to initial peri-operative issues and a second peak mostly due to growth-related stretch, compressions and progressive fibrocellular intimal thickening [29] and some patients remain asymptomatic with negative perfusion imaging despite presence of significant coronary lesions.

CMR is a feasible screening test for assessing fibrotic lesions in paediatric patients following ASO. A pharmacological stress test is performed during CMR, using an adenosine perfusion to induce wall motion abnormalities. This procedure has a 100% negative predictive value in identifying paediatric coronary artery disease,

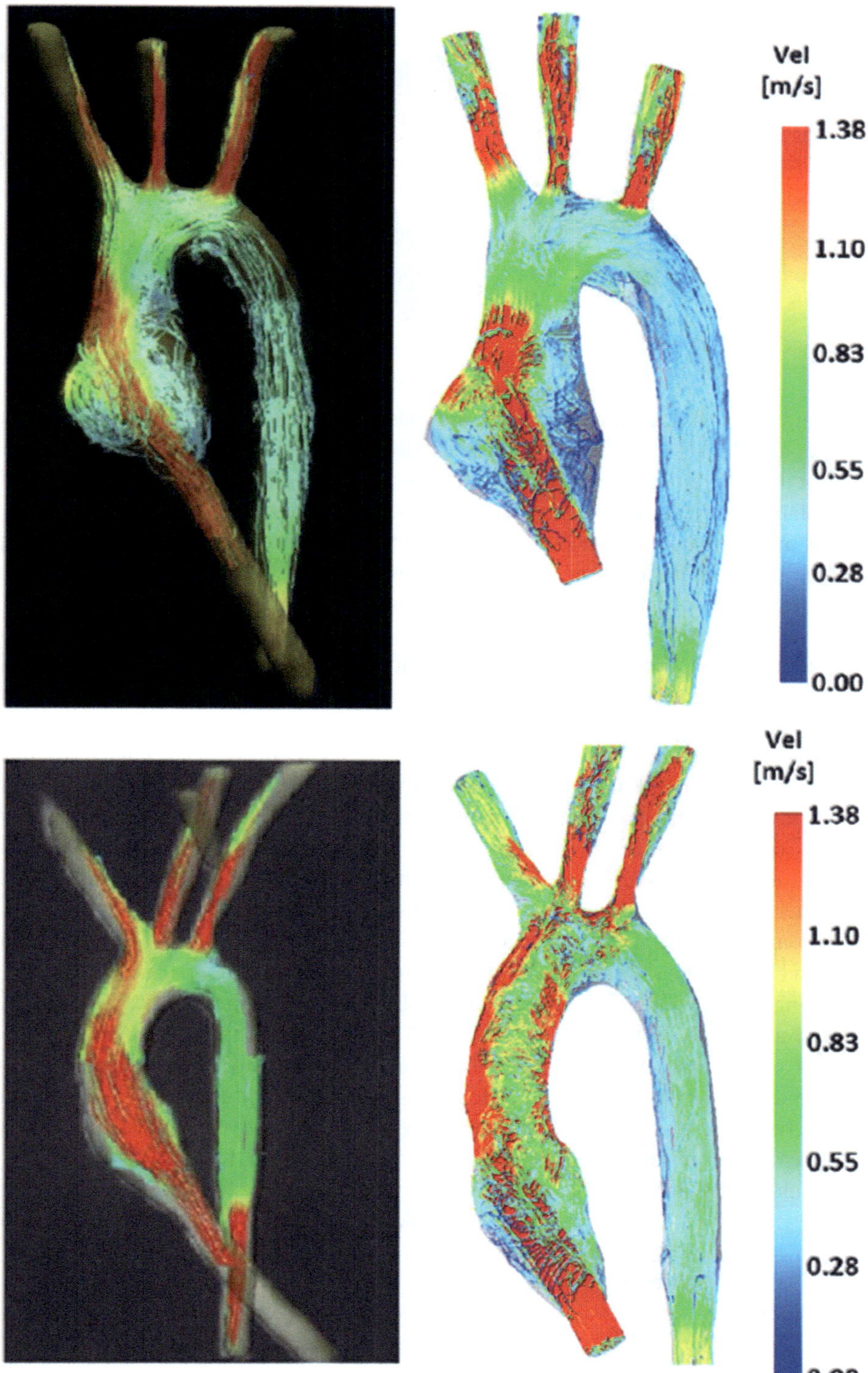

Fig. 7.5 Top row: 4D flow streamlines (left) compared with CFD streamlines (right) at t2 = 0.2 s (peak systole), in the TGA model. Bottom row: 4D flow streamlines (left) compared with CFD streamlines (right) at t2 = 0.2 s, (peak systole), in the Control model (reproduced from ref. [25] which is an open-access article distributed under the terms of the Creative Commons Attribution License (CC BY))

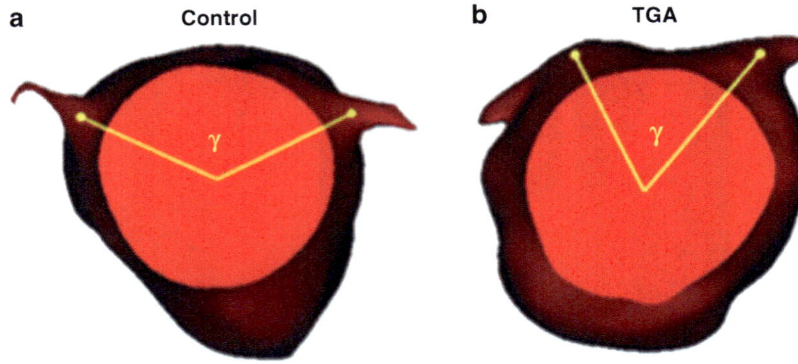

Fig. 7.6 Difference in the angle between the coronary ostia (γ) in controls and TGA post-ASO patients (reproduced with permission from ref. [23])

proving it is suitable as a screening biomarker in this patient population.

Patient-specific 3D models can be used to explore the relation between morphology and its long-term risk of obstruction. Geometric features such as height, position and relation to the pulmonary trunk can be immediately assessed visually and precisely measured. This allows a comparison with control subjects. Such analysis showed how the repositioning of the coronaries results in significant morphological differences with respect to the normal arrangement (Fig. 7.6). Coronary arteries appear more likely to be asymmetrical with respect to the annular plane, with one of the two ostia situated more superiorly in the sinus. This morphological difference could be a reason for later torsion and stretching of the re-implanted coronary arteries, as they essentially have to travel a longer distance to reach the myocardium. During surgery, coronary buttons are also positioned more closely to each other on the anterior aspect of the aorta, as this is the easily accessible surgical view during the operation. This could increase the possibility of compression due to the antero-posterior position of the pulmonary trunk and the aorta.

7.4.3 Late Complications of the Left Ventricle

Majority of ASO survivors demonstrate normal global systolic function on both echocardiography and invasive haemodynamic data. However, patients who were repaired late and those who had abnormal coronary arteries demonstrated stiffer left ventricles with decreased contractile reserve [30–32]. Moreover, CMR studies have shown reduced aortic distensibility with effect on ventricular-arterial (VA) coupling. These patients have dilated LV with increased ventricular mass and borderline systolic function [33, 34].

There is reason to suspect a difference in LV morphology in ASO patients compared to the general population, as suggested by an increase in parameters such as ventricular mass and indexed end diastolic volume, despite normal LVEF at late follow-up. A way to quantify the ventricular shape of patients with repaired TGA is by analysing 3D anatomical models reconstructed from CMR patient-specific data in a statistical shape modelling framework (SSM). By matching controls for age and BSA, a recent study [35] was able to identify significant morphological differences in the LV of patients with ASO. Compared to controls, TGA patients had an overall dilated LV and presented a bulged septum. Interestingly, these changes do not seem to be associated with differences in LVEF.

7.4.4 Late Complications of the Pulmonary Arteries

During the Lecompte procedure, when the neo-pulmonary trunk is moved anteriorly, the branched pulmonary arteries are stretched to aid the anastomosis of the neo-pulmonary trunk to the RVOT. Overall, this procedure leads abnormal pulmonary trunk morphology and a higher

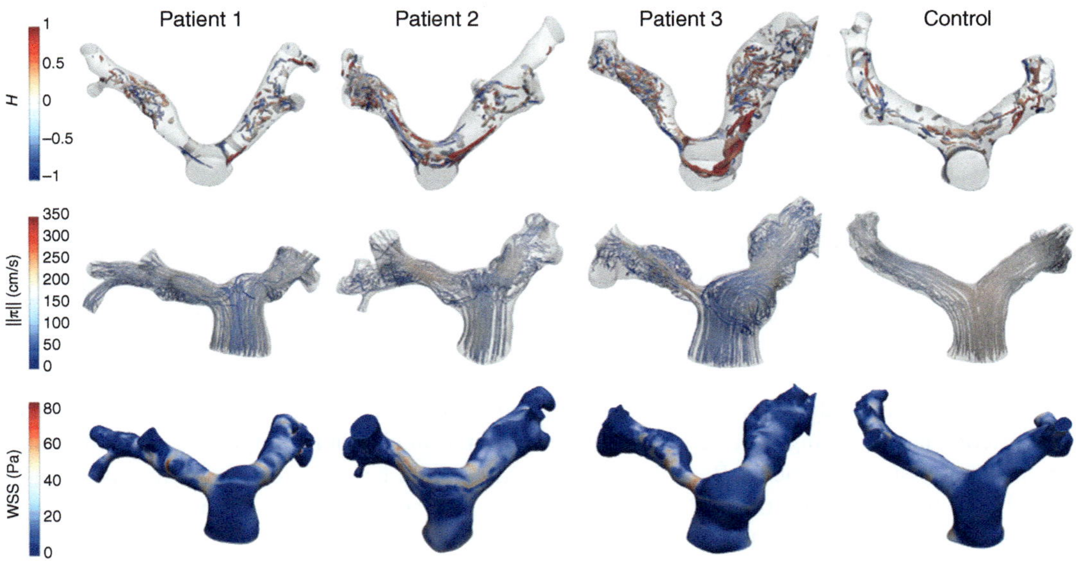

Fig. 7.7 Snapshot of the flow structure for the four subjects at peak systole. Top row: Vortical structures identified by isosurfaces of $Q = 0.05$ Qmax coloured by normalized helicity; middle row: streamlines coloured by velocity magnitude; bottom row: wall shear stress (reproduced with permission from ref. [36])

incidence of branched pulmonary artery (BPA) stenosis. As a consequence, pulmonary artery branches exhibit abnormal bending angles and increased radius of the curvature of the BPA take off, correlating with increased right ventricular afterload independently of PA stenosis [36].

BPA stenosis results in reduced flow to the affected lung and abnormal differential pulmonary blood flow. Balloon angioplasty, performed with or without stenting, is the preferred therapeutic option. It reduces the resistance to flow caused by the narrowed artery lumen, thus improving pulmonary blood flow. However, some patients have little to no benefit on clinical symptoms after relieving the stenosis by this procedure. This might be explained by the fact that the link between reduced pulmonary blood flow and stenosis does not hold true in the case of peripheral pulmonary vascular remodelling. It is not yet clear if this event is due to vasoconstriction of the contralateral lung vasculature, vasodilatation in the stenotic side or a combination of the previous two.

The relationship between expected abnormal differential pulmonary blood flow due to stenosis and the effective measured flow distribution has not yet been fully explained. One possible explanation for this phenomenon is the remodelling of the peripheral pulmonary vascular resistances (PVR), which counteract the presence of the stenosis, thus readjusting the flow split. CFD analyses can provide more information about the possibility of PVR remodelling in a non-invasive manner. Patient-specific CFD models were employed to assess the flow fields within the pulmonary arteries after correction of TGA with ASO [37] as shown in Fig. 7.7. CFD allows identifying the onset mechanisms of abnormal flow patterns and links them to certain shape features of the post-operative morphology (Fig. 7.7). The narrow branch angulation and small radius of curvature of the LPA/RPA bends were found to contribute to flow separation and shear layer instability.

A zero-dimensional lumped parameter network (LPN)—an electrical equivalent representation of the pulmonary artery segments—was set up to simulate the pulmonary circulation of each patient. Models of patients with no stenosis showed little influence of the proximal geometry on the flow split for any total unknown PVR. In the stenotic patients, increased total PVR resulted in a decreased effect of the stenosis. Even though computational analysis cannot fully explain the phenomenon behind PVR, it has the potential to enrich the information gathered with only non-invasive CMR measurements.

Given the complexity of ASO, the geometry of PAs following the procedure widely varies. Numerous computational studies looked at patient-specific 3D anatomical models obtained from CMR image reconstructions of ASO patients to obtain a 3D morphometric analysis of the PA. The 3D analysis revealed significant differences in regards to the position of the pulmonary arteries and ascending aorta between healthy controls and post-ASO patients. Regarding shape and size of the pulmonary trunk, TGA patients presented larger PT area and also higher values of eccentricity of the PT cross-sectional area compared to controls. These findings suggest that PT is rather elliptical in TGA rather than round in the control group. The assessment of the 3D volumes also showed a size mismatch between RPA and LPA, with smaller LPA in the TGA group. This might suggest that unilateral branch stenosis is more likely in TGA, a fact which is also described in the literature as post-ASO complication.

7.5 Future Perspective and Conclusion

This chapter explores the different events occurring late after the ASO operation using different imaging and modelling tools in an attempt to assess both anatomical and physiological consequences of ASO. Although ASO has transformed the surgical management of d-TGA, its long-term sequelae remain a concern despite current excellent survival. Complications described above are all related to the loss of spiral nature of the great vessels secondary to ASO and Lecompte manoeuvre. Morphometric analysis on 3D models demonstrates that altered geometry of the great arteries modifies flow dynamics leading to haemodynamic changes that may lead to abnormal clinical ramifications. The impact of onset of non-communicable diseases such as hypertension and coronary artery disease in these patients as the enter adulthood is worrying in the background of these vessel abnormalities albeit some being subclinical. Interestingly, in 1998, Chiu and colleagues from the National University of Taiwan invented an ASO which maintains the spiral nature of the great arteries with great success even though there is no literature that shows it being taken up by other centres [38]. Computational fluid dynamics comparing flow patterns in Lecompte and spiral models has demonstrated favourable streamlines in the latter as expected [39]. Therefore, the combined use of advanced imaging and tools computational models and simulation for surveillance of ASO survivors should be exploited further in order to help us understand this condition more. Although ASO with Lecompte is the gold standard for repair of dTGA, it may be the time to think out of the box about modifying this procedure into a more favourable alternative.

References

1. Hoffman JI, Kaplan S. The incidence of congenital heart disease. J Am Coll Cardiol. 2002;39(12):1890–900.
2. van der Linde D, Konings EE, Slager MA, Witsenburg M, Helbing WA, Takkenberg JJ, Roos-Hesselink JW. Birth prevalence of congenital heart disease worldwide: a systematic review and meta-analysis. J Am Coll Cardiol. 2011;58(21):2241–7.
3. Ferencz C, Rubin JD, Mccarter RJ, Brenner JI, Neill CA, Perry LW, Hepner SI, Downing JW. Congenital heart disease: prevalence at livebirth: the Baltimore-Washington Infant Study. Am J Epidemiol. 1985;121(1):31–6.
4. Mustard WT, Chute AL, Keith JD, Sirek A, Rowe RD, Vlad P. A surgical approach to transposition of the great vessels with extracorporeal circuit. Surgery. 1954;36(1):31–51.
5. Kreutzer C, De Vive J, Oppido G, Kreutzer J, Gauvreau K, Freed M, Mayer JE Jr, Jonas R, Pedro J. Twenty-five–year experience with Rastelli repair for transposition of the great arteries. J Thorac Cardiovasc Surg. 2000;120(2):211–23.
6. Villafañe J, Lantin-Hermoso MR, Bhatt AB, Tweddell JS, Geva T, Nathan M, Elliott MJ, Vetter VL, Paridon SM, Kochilas L, Jenkins KJ. D-transposition of the great arteries: the current era of the arterial switch operation. J Am College Cardiol. 2014;64(5):498–511.
7. Dos L, Teruel L, Ferreira IJ, Rodriguez-Larrea J, Miro L, Girona J, Albert DC, Gonçalves A, Murtra M, Casaldaliga J. Late outcome of Senning and mustard

procedures for correction of transposition of the great arteries. Heart. 2005;91(5):652–6.
8. Jatene AD, Fontes VF, Paulista PP, Souza LC, Neger F, Galantier M, Sousa JE, Zerbini EJ. Anatomic correction of transposition of the great vessels. J Thorac Cardiovasc Surg. 1976;72(3):364–70.
9. Daebritz SH, Tiete AR, Sachweh JS, Engelhardt W, von Bernuth G, Messmer BJ. Systemic right ventricular failure after atrial switch operation: midterm results of conversion into an arterial switch. Ann Thorac Surg. 2001;71(4):1255–9.
10. Jaggers JJ, Cameron DE, Herlong JR, Ungerleider RM. Congenital Heart Surgery Nomenclature and Database Project: transposition of the great arteries. Ann Thorac Surg. 2000;69(Suppl. 4):S205–35.
11. Sithamparanathan S, Padley SP, Rubens MB, Gatzoulis MA, Ho SY, Nicol ED. Great vessel and coronary artery anatomy in transposition and other coronary anomalies: a universal descriptive and alphanumerical sequential classification. JACC: Cardiovasc Imaging. 2013;6(5):624–30.
12. Wong D, Golding F, Hess L, Caldarone CA, Van Arsdell G, Manlhiot C, McCrindle BW, Miner SE, Nield LE. Intraoperative coronary artery pulse Doppler patterns in patients with complete transposition of the great arteries undergoing the arterial switch operation. Am Heart J. 2008;156(3):466–72.
13. Cohen MS, Eidem BW, Cetta F, Fogel MA, Frommelt PC, Ganame J, Han BK, Kimball TR, Johnson RK, Mertens L, Paridon SM. Multimodality imaging guidelines of patients with transposition of the great arteries: a report from the American Society of Echocardiography developed in collaboration with the Society for Cardiovascular Magnetic Resonance and the Society of Cardiovascular Computed Tomography. J Am Soc Echocardiogr. 2016;29(7):571–621.
14. Brown JW, Park HJ, Turrentine MW. Arterial switch operation: factors impacting survival in the current era. Ann Thorac Surg. 2001;71(6):1978–84.
15. Planche C, Lacour-Gayet F, Serraf A. Arterial switch. Pediatr Cardiol. 1998;19(4):297–307.
16. Lalezari S, Hazekamp MG, Bartelings MM, Schoof PH, Gittenberger-de Groot AC. Pulmonary artery remodeling in transposition of the great arteries: relevance for neoaortic root dilatation. J Thorac Cardiovasc Surg. 2003;126(4):1053–60.
17. Michalak KW, Moll JA, Moll M, Dryżek P, Moszura T, Kopala M, Młudzik K, Moll JJ. The neoaortic root in children with transposition of the great arteries after an arterial switch operation. Eur J Cardiothorac Surg. 2013;43(6):1101–8.
18. Haas F, Wottke M, Poppert H, Meisner H. Long-term survival and functional follow-up in patients after the arterial switch operation. Ann Thorac Surg. 1999;68(5):1692–7.
19. Losay J, Touchot A, Serraf A, Litvinova A, Lambert V, Piot JD, Lacour-Gayet F, Capderou A, Planche C. Late outcome after arterial switch operation for transposition of the great arteries. Circulation. 2001;104(12 Suppl. 1):I121–6.
20. Bove T, De Meulder F, Vandenplas G, De Groote K, Panzer J, Suys B, DeWolf D, Francois K. Midterm assessment of the reconstructed arteries after the arterial switch operation. Ann Thorac Surg. 2008;85(3):823–30.
21. Lim HG, Kim WH, Lee JR, Kim YJ. Long-term results of the arterial switch operation for ventriculo-arterial discordance. Eur J Cardiothorac Surg. 2013;43(2):325–34.
22. Vignon-Clementel IE, Marsden AL, Feinstein JA. A primer on computational simulation in congenital heart disease for the clinician. Prog Pediatric Cardiol. 2010;30(1–2):3–13.
23. Ntsinjana HN, Capelli C, Biglino G, Cook AC, Tann O, Derrick G, Taylor AM, Schievano S. 3D morphometric analysis of the arterial switch operation using in vivo MRI data. Clin Anat. 2014;27(8):1212–22.
24. Fricke TA, Brizard CP, d'Udekem Y, Konstantinov IE. Aortic root and valve surgery after arterial switch operation. J Thorac Cardiovasc Surg. 2012;144(5):1269–71.
25. Biglino G, Cosentino D, Steeden JA, De Nova L, Castelli M, Ntsinjana H, Pennati G, Taylor AM, Schievano S. Using 4D cardiovascular magnetic resonance imaging to validate computational fluid dynamics: a case study. Front Pediatr. 2015;3:107.
26. Jarvis K, Vonder M, Barker AJ, Schnell S, Rose M, Carr J, Robinson JD, Markl M, Rigsby CK. Hemodynamic evaluation in patients with transposition of the great arteries after the arterial switch operation: 4D flow and 2D phase contrast cardiovascular magnetic resonance compared with Doppler echocardiography. J Cardiovasc Magn Reson. 2016;18(1):59.
27. Antonini-Canterin F, Carerj S, Di Bello V, Di Salvo G, La Carrubba S, Vriz O, Pavan D, Balbarini A, Nicolosi GL. Arterial stiffness and ventricular stiffness: a couple of diseases or a coupling disease? A review from the cardiologist point of view. Eur Heart J: Cardiovasc Imaging. 2009;10(1):36–43.
28. Agnoletti G, Bajolle F, Bonnet D, Sidi D, Vouhe P. Late coronary complications after arterial switch operation for transposition of great arteries. Clinical and therapeutic implications. Images Paediatr Cardiol. 2005;7(3):1–11.
29. Bonnet D, Bonhoeffer P, Piechaud JF, Aggoun Y, Sidi D, Planche C, Kachaner J. Long-term fate of the coronary arteries after the arterial switch operation in newborns with transposition of the great arteries. Heart. 1996;76(3):274–9.
30. Colan SD, Boutin C, Castaneda AR, Wernovsky G. Status of the left ventricle after arterial switch operation for transposition of the great arteries. Hemodynamic and echocardiographic evaluation. J Thorac Cardiovasc Surg. 1995;109(2):311–21.
31. Borow KM, Arensman FW, Webb C, Radley-Smith R, Yacoub MH. Assessment of left ventricular contractile state after anatomic correction of transposition of the great arteries. Circulation. 1984;69(1):106–12.
32. Hausdorf G, Grävinghoff L, Sieg K, Keck EW, Raddley-Smith R, Yacoub MH. Left ventricular per-

formance after anatomic correction of D-transposition of the great arteries. In: Doyle E, Engle M, Gersony W, Rashkind W, Talner N, editors. Pediatric cardiology. Springer: New York; 1986. p. 717–20.

33. Chen RH, Wong SJ, Wong WH, Cheung YF. Arterial mechanics at rest and during exercise in adolescents and young adults after arterial switch operation for complete transposition of the great arteries. Am J Cardiol. 2014;113(4):713–8.
34. Biglino G, Ntsinjana H, Plymen C, Tann O, Giardini A, Derrick G, Parker KH, Tsang V, Schievano S, Taylor AM. Ventriculovascular interactions late after atrial and arterial repair of transposition of the great arteries. J Thorac Cardiovasc Surg. 2014;148(6):2627–33.
35. Bruse JL, Ntsinjana H, Capelli C, Biglino G, McLeod K, Sermesant M, Pennec X, Hsia TY, Schievano S, Taylor A. CMR-based 3D statistical shape modelling reveals left ventricular morphological differences between healthy controls and arterial switch operation survivors. J Cardiovasc Magn Reson. 2016;18(S1):Q2.
36. Loke YH, Capuano F, Mandell J, Cross RR, Cronin I, Mass P, Balaras E, Olivieri LJ. Abnormal pulmonary artery bending correlates with increased right ventricular afterload following the arterial switch operation. World J Pediatr Congenit Heart Surg. 2019;10(5):572–81.
37. Capuano F, Loke YH, Cronin I, Olivieri LJ, Balaras E. Computational study of pulmonary flow patterns after repair of transposition of great arteries. J Biomech Eng. 2019;141(5)
38. Chiu IS, Huang SC, Chen YS, Chang CI, Lee ML, Chen SJ, Chen MR, Wu MH. Restoring the spiral flow of nature in transposed great arteries. Eur J Cardiothorac Surg. 2010;37(6):1239–45.
39. Tang T, Chiu S, Chen HC, Cheng KY, Chen SJ. Comparison of pulmonary arterial flow phenomena in spiral and Lecompte models by computational fluid dynamics. J Thorac Cardiovasc Surg. 2001;122(3):529–34.

8 Double Outlet Right Ventricle: Introductory Concepts and Applications

Justin T. Tretter, Claudio Capelli, and Puneet Bhatla

8.1 Introduction

Double outlet right ventricle (DORV) describes a group of lesions in which both arterial trunks are predominately supported by the morphological right ventricle [1, 2]. Various classifications have been proposed, although none of them are able to incorporate all the possible variations in anatomy, including the position of the commonly present ventricular septal defect (VSD), the relationship of the great arteries to each other or to the VSD, the common presence of outflow tract obstruction, nor the presence of other associated lesions [1]. The complex 3D anatomy, including the potential pathways to appropriately baffle and septate into a successful biventricular repair varies in each case, and is often poorly understood solely by two-dimensional imaging, making it difficult to generalize management algorithms. 3D patient-specific models, either virtual or 3D-printed, allow for visualization and physical maneuvering of the patient-specific heart, and have appropriately been increasingly utilized in this lesion, including attempts toward surgical simulation [3], potentially resulting in improvement of presurgical planning [4]. In this chapter, we review the current concepts and application for 3D modeling in the wide array of lesions classified to have DORV.

Supplementary Information The online version of this chapter (https://doi.org/10.1007/978-3-030-88892-3_8) contains supplementary material, which is available to authorized users.

J. T. Tretter (✉)
Heart Institute, Cincinnati Children's Hospital Medical Center, Cincinnati, OH, USA
e-mail: justin.tretter@cchmc.org

C. Capelli
Centre for Cardiovascular Imaging, UCL Institute of Cardiovascular Science & Great Ormond Street Hospital for Children, London, UK
e-mail: c.capelli@ucl.ac.uk

P. Bhatla
Division of Pediatric Cardiology, Hassenfeld Children Hospital, New York Langone Health, New York, NY, USA
e-mail: puneet.bhatla@nyumc.org

8.1.1 Morphology

Although DORV comprises only approximately 1% of all forms of congenital heart disease [5], due to its wide spectrum of morphological variation it presents one of the most clinically complex cardiac lesions encountered in regards to the possibilities for both medical and surgical management [6]. Despite the historical debates, it is now clear that the defining feature of this lesion is that the greater parts of both arterial trunks are supported by the right ventricle, and that the presence of bilateral infundibulum, or "conus", is not an essential feature to the defect [7]. In fact, a recent morphological study demonstrated that

G. Butera et al. (eds.), *Modelling Congenital Heart Disease*, https://doi.org/10.1007/978-3-030-88892-3_8

bilateral infundibulum was only present in approximately one quarter of lesions clearly defined to have DORV [2].

As initially proposed by Lev et al. [8], the lesion can broadly be categorized based on the location and commitment of the VSD to one or both great arteries. Using this classification, the most commonly encountered subtype is one with a subaortic VSD (anatomically close to the aorta), followed by the subpulmonary VSD (anatomically close to the pulmonary artery (PA)) (Fig. 8.1), followed by the doubly committed VSD type (DCVSD) (anatomically committed to both the aorta and PA) and finally the remote type (noncommitted) (Fig. 8.2). The size, location, and morphology of the VSD are certainly the most important aspect of this lesion since it serves as the outflow to the left ventricle. The interventricular communication in the subaortic, subpulmonary, and doubly committed phenotypes is consistently found within the limbs of the septomarginal trabeculation, with the variability in morphology and resulting physiology largely dependent on the location of the muscular or fibrous outlet septum relative to the limbs of the septomarginal trabeculation. The caudal

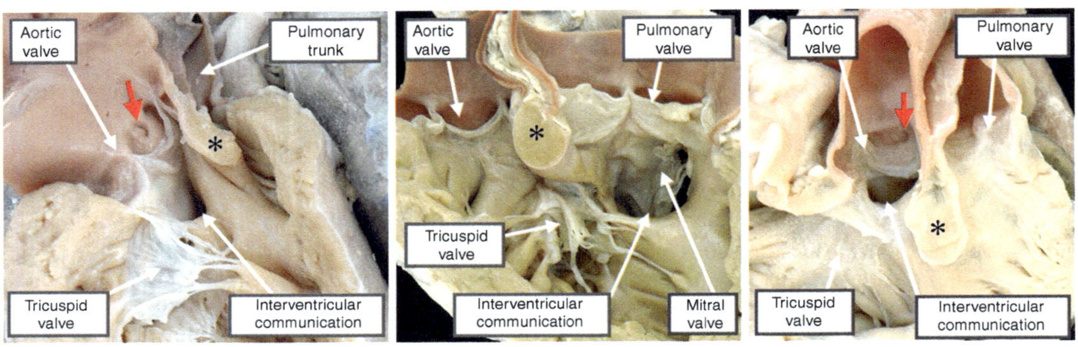

Fig. 8.1 Common forms of DORV. (**a**) DORV with subaortic VSD and subpulmonary stenosis. The outlet septum (black asterisk) is deviated anteriorly under the pulmonary valve causing subpulmonary obstruction. (**b**) DORV with subpulmonary VSD. (**c**) DORV with subaortic VSD without pulmonary stenosis. In contrast to Panel A, the outlet septum is not deviated and there is no subarterial obstruction. The visible coronary arteries are marked with a red arrow

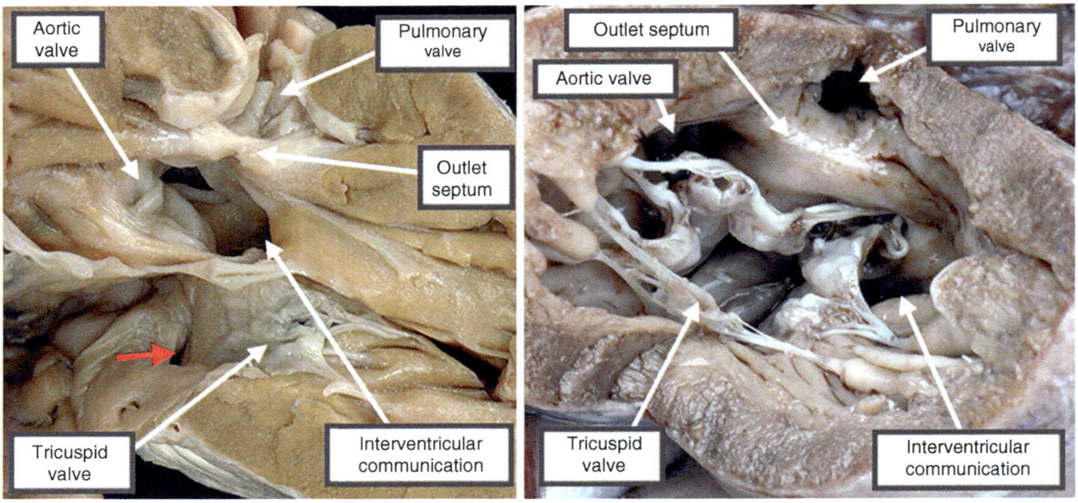

Fig. 8.2 Less common forms of DORV. (**a**) DORV with doubly committed VSD. There is fibrous continuity between the aortic and pulmonary valves at the roof of the VSD through the fibrous remnant of the outlet septum. (**b**) DORV with noncommitted VSD. The VSD is remote from the aortic and pulmonary valves, found within the inferior part of the muscular ventricular septum. The red arrow marks the coronary sinus as it opens to the right atrium

margin of the VSD is formed by the crest of the muscular interventricular septum, while its cranial margin is now a completely right ventricular structure, formed by the muscular or fibrous outlet septum which resides between the bases of the arterial roots. The VSD in most cases is perimembranous outlet with a fibrous posteroinferior rim (which must be avoided by the surgeon to avoid conduction damage), and less commonly a muscular outlet defect with completely muscular borders (Fig. 8.3). In contrast, the noncommitted defect either represents the ventricular component of an atrioventricular septal defect, typically with a common atrioventricular valve, or an isolated muscular defect opening to the inlet or apical part of the right ventricle [1, 2]. Due to the marked phenotypic variability and its ramifications on the surgical repair, despite the desire to easily subcategorize those with DORV, each patient requires individualized morphological categorization.

8.1.2 Pathophysiology and Clinical Presentation

As wide the possibilities in morphological variation seen in DORV, so is the resulting pathophysiology and clinical presentation encountered. The general clinical presentations include that of pulmonary overcirculation seen in an isolated VSD, to the varying degrees of cyanosis encountered in tetralogy of Fallot with pulmonary outflow obstruction or those with transposition of the great arteries, to single ventricle physiology in patients with semilunar valvar atresia. Pulmonary

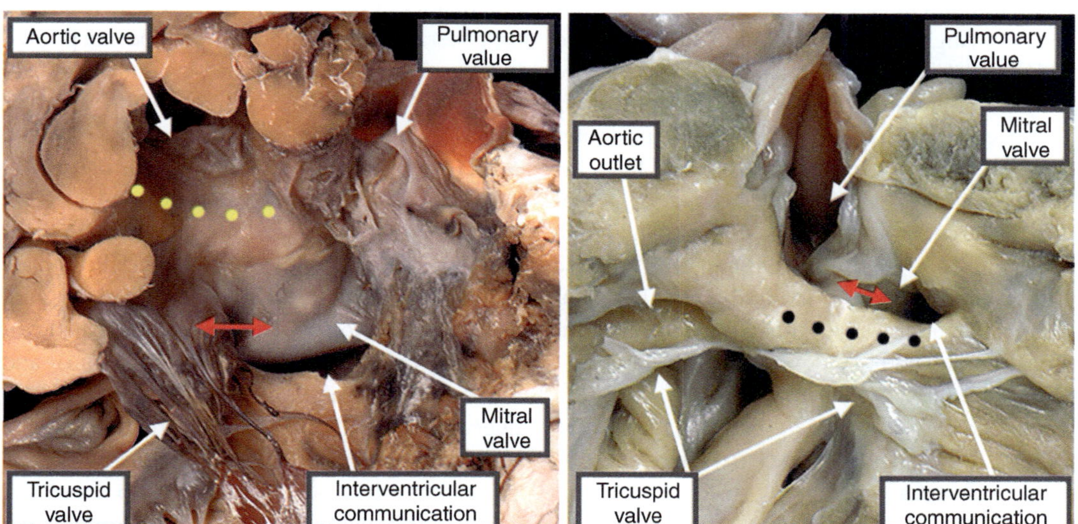

Fig. 8.3 The VSD observed in the more common forms of DORV (common forms displayed in Fig. 8.1) opens to the outlet of the RV, located within the limbs of the septomarginal trabeculation. The borders of the defect are most commonly either perimembranous (**a**) or muscular (**b**). Less commonly the borders are doubly committed (see Fig. 8.2a). (**a**) In this DORV with subaortic VSD, the VSD is within the limbs of the septomarginal trabeculation with a fibrous posteroinferior rim created by fibrous continuity between the tricuspid and mitral valves (red double-headed arrow) (**perimembranous outlet VSD**). The atrioventricular conduction axis courses along this posteroinferior fibrous border and is vulnerable to surgical insult. (**b**) In this DORV with subpulmonary VSD, the VSD is again within the limbs of the septomarginal trabeculation; however, its borders are now muscular (**muscular outlet VSD**) with a muscular bar (black dotted line) forming the posteroinferior border and protecting the atrioventricular conduction axis from surgical insult. While there is fibrous continuity between the pulmonary and mitral valves (red double-headed arrow), this area represents the inner heart curvature (this structure is highlighted in Panel A with the yellow dotted line) which forms the cranial plane of the interventricular communication but not the cranial plane of the VSD. Technically speaking, the cranial border of the VSD is formed by the outlet septum which is now an exclusively RV structure

stenosis is the most common associated defect, resulting in cyanosis, while aortic coarctation and other forms of aortic arch obstruction are present in up to 25% of patients and can lead to poor systemic perfusion following birth [9].

8.1.3 Possible Therapies

Surgical palliation or repair is necessary for all forms of DORV. Both the timing and specific type of surgery performed are dependent on the morphological variation described above and the resulting clinical presentation. If surgical palliation or repair is indicated, appropriate imaging as described below will help guide those involved in the decision-making process on the appropriate surgical approach. A primary biventricular repair is possible in the majority [9], and often involves baffling the left ventricle through the interventricular communication to one of the great arteries, relieving any outflow tract obstruction and addressing associated lesions. If the baffle is created to the pulmonary root, an arterial switch is also necessary. In the setting of pulmonary outflow obstruction, the Rastelli procedure [10], the REV (réparation à l'étage ventriculaire) procedure [11], and aortic translocation or Naikaidoh procedure [12, 13] are in common use. In those with subaortic stenosis and aortic arch hypoplasia or interruption, the Yasui procedure is commonly employed [14].

8.2 Potential Impact on Clinical Arena

8.2.1 Image-Based Assessment and Planning

Although echocardiography remains the mainstay for preoperative diagnosis [15], it remains challenging to interpret the often complex 3D anatomy and structural relationships present in DORV. Cross-sectional imaging, including both computed tomography (CT) and magnetic resonance imaging (MRI), have become essential tools for presurgical planning in DORV [1]. The initial question is whether biventricular repair is feasible and appropriate versus functionally univentricular palliation. Once the patient is deemed appropriate for biventricular repair, the question then becomes which repair strategy is best. In order to answer these questions, those involved in the decision-making process must have a thorough 3D spatial understanding of the intracardiac relationships, including knowledge of the relationships between the interventricular communication, the outflow tracts and great arteries, the atrioventricular valves, and any other associated lesion.

Both CT and MRI allow for excellent 3D imaging of these structures, both with their pros and cons, and both with varied experience and expertise across institutions [16]. Prior to acquiring the 3D data, it is important to consider the requirements for each patient including the size of the cardiovascular structures and the heart rate. The quality of the raw 2D imaging data is critical for generating an accurate patient-specific printed heart model. The data must show good endoluminal contrast in order to highlight the different cardiovascular structures. There must also be sufficient spatial and temporal resolution. ECG-gated CT or MR images performed under breath holding/apnea or with respiration navigation are best suited for 3D printing. Both modalities have various strengths and weaknesses. CT angiogram has a much shorter scan time and provides a higher spatial resolution (~0.3–0.7 mm) and hence superior ability to accurately delineate finer structures like trabeculae around complex VSD's and other complex intracardiac anatomy. To maintain accuracy, it is critical to avoid any cardiac- or breathing-related motion artifact, and hence it is recommended to perform the scan with retrospective ECG gating (throughout the cardiac cycle), ideally with some heart rate control, especially in neonates and infants. The breathing-related motion artifact can be offset by the scan being performed under anesthesia for smaller children and images acquired with apnea or with breath holding for older children. A single heart beat scan may be good enough on newer generation scanners (with rapid data acquisition) even at relatively high heart rates. The scan timing in relation to the contrast administration is also critical, with the goal of scanning when all the cardiac chambers are homogeneously enhanced. While CT allows for better spatial resolution with improved anatomical detail, this comes with exposure to radiation. With current technology and improved scanning techniques, the radiation dose can often be minimized to less than 1 millisievert [17]. MRI has an inherent advantage of no radiation exposure, along with the ability to obtain accurate functional and hemodynamic information. However, this comes at a much longer scan time and hence a longer time under general anesthesia for smaller children. Though we can push the MRI spatial resolution to a voxel size of ~1 mm, it is obtained at the expense of decreased signal-to-noise ratio. In conventional MRI angiography using an extracellular contrast agent, ECG-gating is not possible and a degree of artifact from cardiac motion is unavoidable [18]. Contrast-enhanced 3D whole-heart MRI sequences allow the ability to obtain free-breathing, ECG-gated and respiratory-navigated images with higher signal-to-noise ratio and isotropic voxel resolution, ideal for 3D printing [18, 19].

8.2.2 Engineer's Perspective

3D imaging modalities can be further postprocessed to create 3D models of DORV and the common associated multiple malformations. These reconstructions are instrumental in relaying this complex anatomy, guiding presurgical decision-making and planning (Fig. 8.4). 3D models can be handled with ease, displayed from any viewpoint, and once printed, allow tangible manipulation [3, 4].

The approach of transforming medical images into 3D virtual and physical objects has been used to enhance the imaging of both extracardiac and intracardiac anatomy for several congenital heart defects [20]. In the context of DORV, 3D models are mainly used to improve the evaluation of the complex anatomic features prior to the surgical treatment. In particular, the feasibility of an unobstructed baffle from the left ventricle to the aorta through the interventricular communication can be explored directly on replica of the patient's anatomy together with the mapping of VSD(s), the atrioventricular valvar-supporting apparatus, the relationship of the great vessels, and coronary arterial anatomy [21]. While experts in cardiovascular imaging have a great ability to appreciate dimensions and spatial relationships from traditional 2D cross-sectional images, navigations into 3D models can be more helpful for those with less imaging expertise, for instance cardiac surgeons and interventional cardiologists. Specifically, models can be used to test the feasibility of creation of an unobstructed baffle from the left ventricle to aorta through the interventricular communication, or to the PA necessitating an arterial switch.

The models can help in medical education and research to improve insights into both patient- and also disease-specific anatomies either through viewing of the anatomy or by direct interaction with the models and prints [22, 23].

Finally, the prints and models are also a valuable tool to improve the quality of counselling of parents and their families. Through direct interaction with the 3D models and prints, the parent may reach a better understanding of their child's cardiac disease and the potential risk and advantages of advised treatment routes [24].

3D models are normally generated by means of whole-heart segmentation. This process is performed to label cardiovascular structures and to allow their volume rendering. In the case of DORV, generally the blood pool is segmented corresponding to the four chambers, systemic and pulmonary veins, and great arteries. Ideally, valvar structures should also be included to allow a full visualization of intracardiac structures and the relation of the VSDs with the great arteries and valvar structures. A wide range of software and tools are available for segmentation of cardiac images. The significant anatomical variability seen in DORV often precludes a fully automatic atlas-based segmentation. However, the increasing quality of medical images together with advances in machine learning techniques have shown promising results to reduce time and user inputs [25].

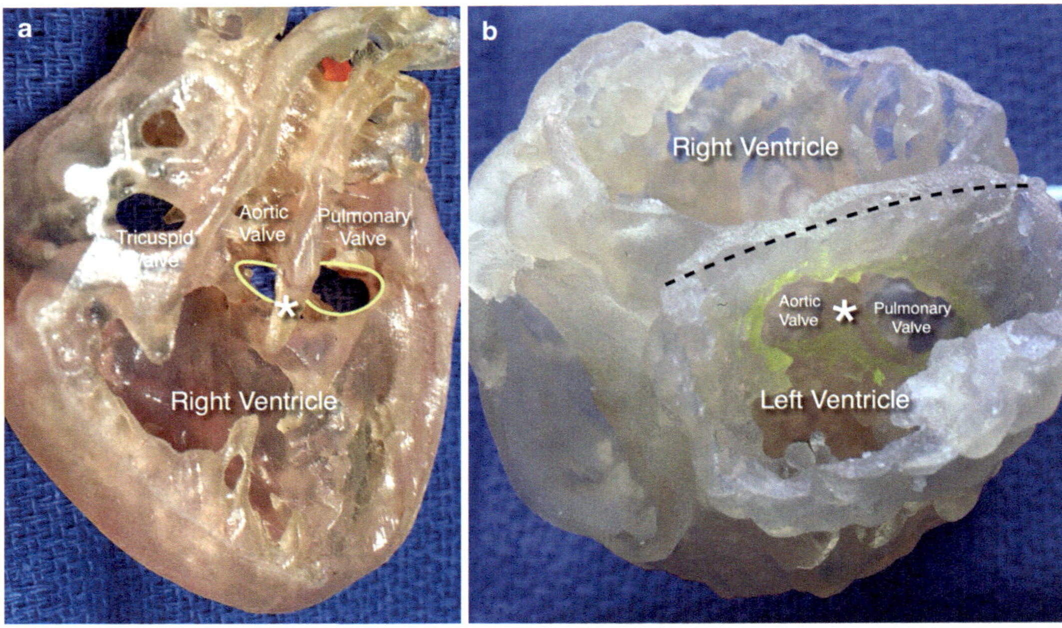

Fig. 8.4 This example highlights the anatomical clarification provided by 3D printing in a newborn with DORV. In this patient with DORV, the aortic root was positioned slightly rightward and posterior of the pulmonary root, with the large oblong-shaped doubly committed VSD intersected by a large outlet septum. The details of this anatomy and what would be the ideal reparative strategy were ambiguous on 2D echocardiography and CT. Multiple 3D-printed models were created, both a whole model and models cut in various planes. (**a**) In this first model cut in a coronal plane, the right ventricular free wall is removed and the rightward aspect of the interventricular septum is visualized. The unusually long outlet septum (white asterisk) bisects the rightward plane of the interventricular communication (the borders of which are highlighted with a yellow line). It is clear from this model that the pathway from the left ventricle through the interventricular communication to the pulmonary valve is larger than that to the aortic valve. (**b**) An additional model was created cutting the apices of both the right and left ventricles. The model is positioned so that we are looking from the inferior aspect of the ventricles toward the base. The interventricular septum is highlighted with a black dashed line. The leftward aspect of the margins of the interventricular communication is highlighted in yellow with the outlet septum (white asterisk) seen through the communication bisecting it between the aortic and pulmonary valves. Again, the pathway from the left ventricle through the interventricular communication to the pulmonary valve is much larger than that to the aortic valve. Additionally, a whole-heart model allowed surgical simulation of the approach. This provided the surgeon with important information, that the initial plan to approach through a right ventriculotomy would not be ideal, and instead guided the surgeon to take an approach through a right atriotomy. The patient underwent successful arterial switch with baffling the left ventricle through the interventricular communication to what was the native pulmonary valve (and now the neoaortic valve) with excellent surgical outcome

Following segmentation, 3D models can then be handled in three main modalities: on the screen, as 3D-printed objects and, more recently, within augmented and virtual reality (Fig. 8.5). On the screen, the user can interact with the rendered volume by means of operations such as handling, zooming, and cutting. In particular, common file formats such as 3D PDF have enhanced portability of patient-specific models making them more accessible for more users and institutions. 3D-printed models are generated from the same output from the image reconstruction, typically as .stl format. The manufacturing of 3D-printed heart models can be achieved with multiple technologies and materials. For replica of patient-specific DORV, normally a cut which allows a selected intracardiac view is decided before the printing (Videos 8.1 and 8.2). Then, a material is selected amongst the ones compatible with the 3D-printing technologies. For DORV

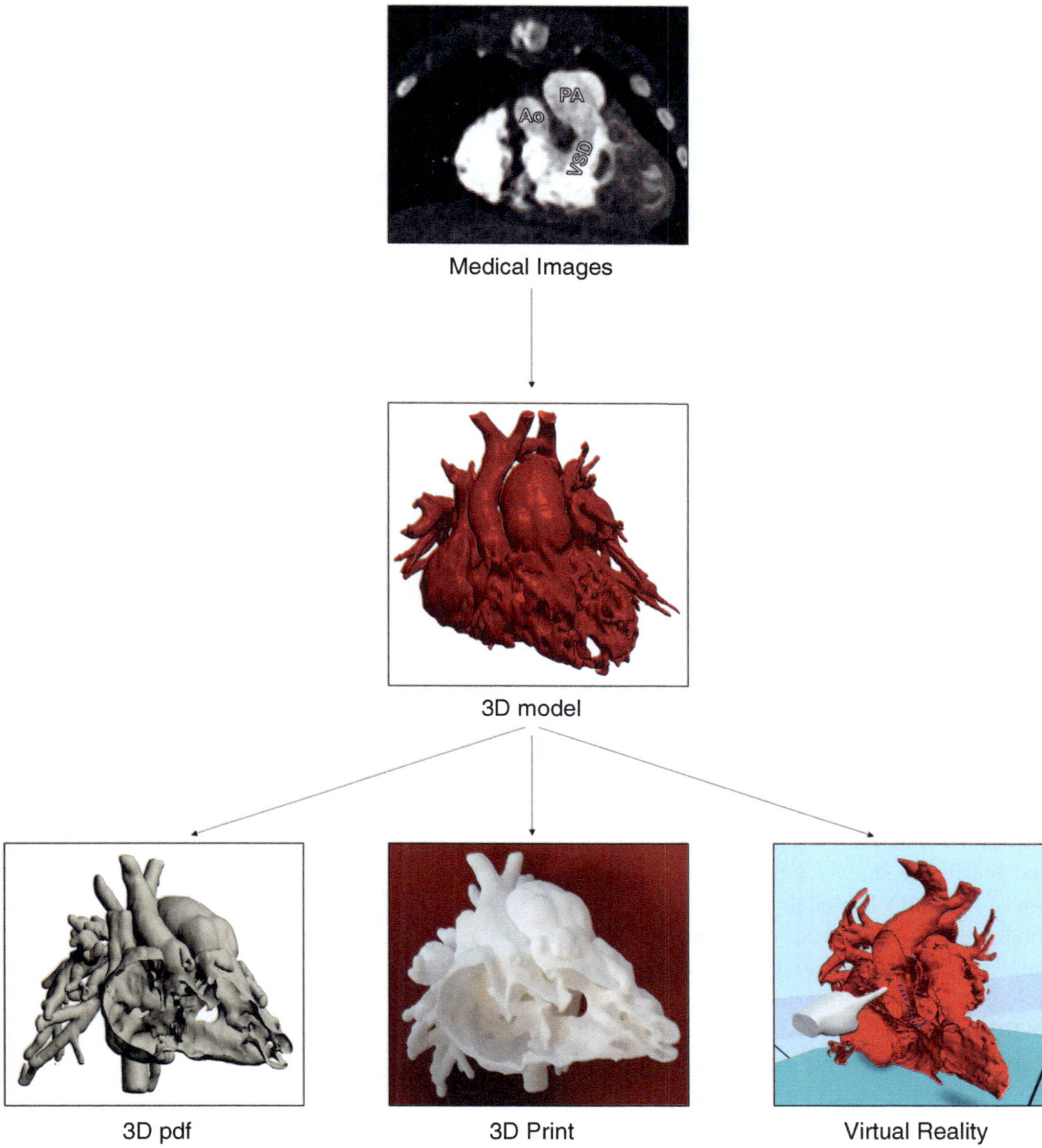

Fig. 8.5 Possibilities of transforming DORV medical images into a 3D model which can be subsequently used as 3D pdf, 3D-printed model, and within virtual reality. *Ao* aorta, *PA* pulmonary artery, *VSD* ventricular septal defect

replicas, rigid materials are normally chosen for visualizing the defects and their spatial relations while softer materials are selected for manipulating the anatomical models and simulating operations (Fig. 8.6). Time and costs of a print are strongly dependent on the size of the model and chosen materials. Resolution of state-of-the-art 3D-printing machines allows printing structures with details below 1 mm. Clinical studies on DORV patients reported benefit of using 3D-printed models of several patients with additional information gathered with models which can enhance confidence of the surgical team to predicting suitability for biventricular repair [3, 26–28]. More recently, 3D models of DORVs can be handled in digital environments

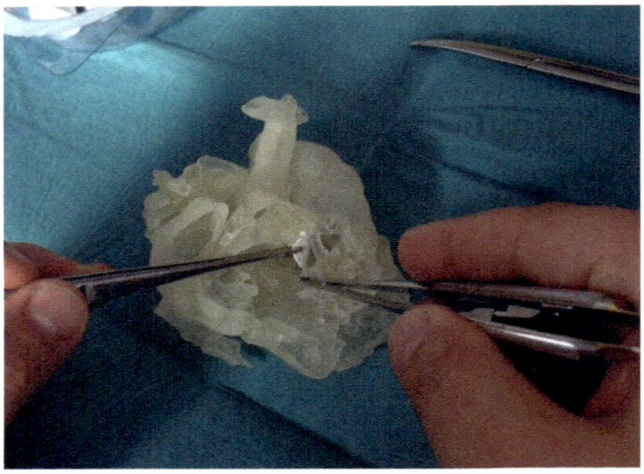

Fig. 8.6 A DORV 3D model printed with a soft material to practice the suturing of a patch to baffle the left ventricle through the VSD to the adjacent great artery

of extended reality such as augmented reality (AR) and virtual reality (VR). AR allows the enrichment of the real world by the projection of interactive digital objects (i.e., holograms). VR can be described as an experience in which the user is completely immersed in a simulated environment. This is most commonly achieved through the use of a VR headset, which contains a stereoscopic display and some tracking capabilities via external sensors. A proof of concept and feasibility study for preoperative diagnostic use of AR and VR models of individual 3D heart models showed potential enhancement of surgical planning with potential of improvement of the hologram manipulation tools. A patient-specific DORV VR model showed the benefit of an interactive experience which can provide unlimited views of cardiac structures with the ability to virtually navigate into the heart and, potentially, interact with it by simulating the surgical intervention [29, 30].

Beyond the visualization purposes, these models can be further processed for numerical analyses such as computational fluid dynamics (CFD). CFD can be used to compute hemodynamics before and after the surgical procedure. Results of the analysis can potentially identify which combinations of angles and positions produce the best geometry and also the sensitivity of that measure to local variations of the parameters. In the future, the surgeons can use these analysis results when planning the details of the surgery [31, 32].

8.3 Conclusions

The understanding of the complex DORV anatomy is crucial to personalize the planning of repair. Advances in images and modeling can help address the feasibility of a biventricular repair in the preoperative planning of this group of patients. Novel physical and digital tools can show whether a newly created left ventricle to arterial valvar baffle, through the interventricular communication, will be achievable without significant obstruction.

Acknowledgment We are thankful to Diane E. Spicer (Department of Pediatric Cardiology University of Florida, Gainesville, FL) for providing the specimen images displayed in Figs. 8.1–8.3.

References

1. Bharucha T, Hlavacek AM, Spicer DE, Theocharis P, Anderson RH. How should we diagnose and differentiate hearts with double-outlet right ventricle? Cardiol Young. 2017;27:1–15.
2. Ebadi A, Spicer DE, Backer CL, Fricker FJ, Anderson RH. Double-outlet right ventricle revisited. J Thorac Cardiovasc Surg. 2017;154:598–604.
3. Bhatla P, Tretter JT, Chikkabyrappa S, Chakravarti S, Mosca RS. Surgical planning for a complex double-outlet right ventricle using 3D printing. Echocardiography. 2017;34:802–4.
4. Bhatla P, Tretter JT, Ludomirsky A, et al. Utility and scope of rapid prototyping in patients with complex muscular ventricular septal defects or double-outlet right ventricle: does it alter management decisions? Pediatr Cardiol. 2017;38:103–14.
5. Schwedler G, Lindinger A, Lange PE, et al. Frequency and spectrum of congenital heart defects among live births in Germany: a study of the competence network for congenital heart defects. Clin Res Cardiol. 2011;100:1111–7.
6. Lagopoulos ME, Manlhiot C, McCrindle BW, Jaeggi ET, Friedberg MK, Nield LE. Impact of prenatal diagnosis and anatomical subtype on outcome in double outlet right ventricle. Am Heart J. 2010;160:692–700.

7. Walters HL 3rd, Mavroudis C, Tchervenkov CI, Jacobs JP, Lacour-Gayet F, Jacobs ML. Congenital heart surgery nomenclature and database project: double outlet right ventricle. Ann Thorac Surg. 2000;69:S249–63.
8. Lev M, Bharati S, Meng CC, Liberthson RR, Paul MH, Idriss F. A concept of double-outlet right ventricle. J Thorac Cardiovasc Surg. 1972;64:271–81.
9. Bradley TJ, Karamlou T, Kulik A, et al. Determinants of repair type, reintervention, and mortality in 393 children with double-outlet right ventricle. J Thorac Cardiovasc Surg. 2007;134:967–973.e6.
10. Kreutzer C, De Vive J, Oppido G, et al. Twenty-five-year experience with rastelli repair for transposition of the great arteries. J Thorac Cardiovasc Surg. 2000;120:211–23.
11. Lecompte Y. Reparation a l'Etage Ventriculaire—the REV procedure: technique and clinical results. Cardiol Young. 1991;1:63–70.
12. Yeh T Jr, Ramaciotti C, Leonard SR, Roy L, Nikaidoh H. The aortic translocation (Nikaidoh) procedure: midterm results superior to the Rastelli procedure. J Thorac Cardiovasc Surg. 2007;133:461–9.
13. Hu S, Xie Y, Li S, et al. Double-root translocation for double-outlet right ventricle with noncommitted ventricular septal defect or double-outlet right ventricle with subpulmonary ventricular septal defect associated with pulmonary stenosis: an optimized solution. Ann Thorac Surg. 2010;89:1360–5.
14. Yasui H, Kado H, Nakano E, et al. Primary repair of interrupted aortic arch and severe aortic stenosis in neonates. J Thorac Cardiovasc Surg. 1987;93:539–45.
15. Mahle WT, Martinez R, Silverman N, Cohen MS, Anderson RH. Anatomy, echocardiography, and surgical approach to double outlet right ventricle. Cardiol Young. 2008;18(Suppl. 3):39–51.
16. Han BK, Hlavacek AM, Kay WA, et al. Multi-institutional evaluation of the indications and radiation dose of functional cardiovascular computed tomography (CCT) imaging in congenital heart disease. Int J Cardiovasc Imaging. 2016;32:339–46.
17. Han BK, Rigsby CK, Leipsic J, et al. Computed tomography imaging in patients with congenital heart disease, Part 2: Technical recommendations. An expert consensus document of the Society of Cardiovascular Computed Tomography (SCCT): endorsed by the Society of Pediatric Radiology (SPR) and the North American Society of Cardiac Imaging (NASCI). J Cardiovasc Comput Tomogr. 2015;9:493–513.
18. Yoo SJ, Thabit O, Kim EK, et al. 3D printing in medicine of congenital heart diseases. 3D Print Med. 2015;2:3.
19. Greil G, Tandon AA, Silva Vieira M, Hussain T. 3D Whole-heart imaging for congenital heart disease. Front Pediatr. 2017;5:36.
20. Olivieri LJ, Krieger A, Loke YH, Nath DS, Kim PC, Sable CA. Three-dimensional printing of intracardiac defects from three-dimensional echocardiographic images: feasibility and relative accuracy. J Am Soc Echocardiogr. 2015;28:392–7.
21. Farooqi KM, Uppu SC, Nguyen K, et al. Application of virtual three-dimensional models for simultaneous visualization of intracardiac anatomic relationships in double outlet right ventricle. Pediatr Cardiol. 2016;37:90–8.
22. Giannopoulos AA, Chepelev L, Sheikh A, et al. 3D-printed ventricular septal defect patch: a primer for the 2015 Radiological Society of North America (RSNA) hands-on course in 3D printing. 3D Print Med. 2015;1:3.
23. Biglino G, Milano EG, Capelli C, et al. Three-dimensional printing in congenital heart disease: considerations on training and clinical implementation from a teaching session. Int J Artif Organs. 2019;42(10):595–59. https://doi.org/10.1177/0391398819849074.
24. Biglino G, Capelli C, Wray J, et al. 3D-manufactured patient-specific models of congenital heart defects for communication in clinical practice: feasibility and acceptability. BMJ Open. 2015;5:e007165.
25. Pace DF, Dalca AV, Geva T, Powell AJ, Moghari MH, Golland P. Interactive whole-heart segmentation in congenital heart disease. Med Image Comput Comput Assist Interv. 2015;9351:80–8.
26. Garekar S, Bharati A, Chokhandre M, et al. Clinical application and multidisciplinary assessment of three dimensional printing in double outlet right ventricle with remote ventricular septal defect. World J Pediatr Congenit Heart Surg. 2016;7:344–50.
27. Farooqi KM, Nielsen JC, Uppu SC, et al. Use of 3-dimensional printing to demonstrate complex intracardiac relationships in double-outlet right ventricle for surgical planning. Circ Cardiovasc Imaging. 2015;8
28. Yoo SJ, van Arsdell GS. 3D Printing in surgical management of double outlet right ventricle. Front Pediatr. 2017;5:289.
29. Brun H, Bugge RAB, Suther LKR, et al. Mixed reality holograms for heart surgery planning: first user experience in congenital heart disease. Eur Heart J Cardiovasc Imaging. 2019;20:883–8.
30. Milano EG, Pajaziti E, Sauvage E, et al. Taking surgery out of reality. Circ Cardiovasc Imaging. 2019;12:e009297.
31. Luffel M, Sati M, Rossignac J, et al. SURGEM: a solid modeling tool for planning and optimizing pediatric heart surgeries. Computer-Aided Des. 2016;70:3–12.
32. Capelli C, Sauvage E, Giusti G, et al. Patient-specific simulations for planning treatment in congenital heart disease. Interface Focus. 2018;8:20170021.

9 Hypoplastic Left Heart Syndrome: Introductory Concepts

Lorenzo Galletti and Nicola Uricchio

9.1 Introduction

Hypoplastic Left Heart Syndrome (HLHS) describes a spectrum of congenital cardiac anomalies characterized by the presence of a hypoplastic left ventricle (LV), associated to hypoplasia, stenosis or atresia of both aortic and/or mitral valve. The first description of HLHS was made by Lev in 1952 while Noonan and Nadas first employed its name in 1958 [1, 2]. According to the International Congenital Nomenclature [3], different subtypes of HLHS have been identified:

1. Mitral atresia/Aortic atresia (MA/AA).
2. Mitral atresia/Aortic stenosis (MA/AS).
3. Mitral stenosis/Aortic atresia (MS/AA).
4. Mitral stenosis/Aortic stenosis (MS/AS).
5. MA/AS + ventricular septal defect (VSD).
6. Hypoplastic left heart complex (HLHC).

The term Hypoplastic Left Heart Complex (HLHC) applies to patients at the favorable end of HLHS spectrum, in which the hypoplasia of left structures consists in aortic and mitral valve hypoplasia without intrinsic stenosis, hypoplastic LV, and left ventricular outflow tract, frequently associated with hypoplasia of ascending aorta and arch with or without coarctation. Other complex diseases, like unbalanced double outlet right ventricle (uDORV) and unbalanced atrioventricular canal (uAVC), could present with significative LV and mitral valve (MV) hypoplasia and usually are treated similarly to a classic HLHS. Associated anomalies frequently affect the pulmonary venous drainage, the tricuspid valve, and the coronary circulation, and are critical for the management strategy [4].

9.2 Epidemiology

Among critical heart defects, HLHS represents 8% to 12% of cases [5] and, if untreated, causes approximately 25% of deaths in the first few weeks of life. Its prevalence has more recently dropped due to the increased rate of pregnancy termination caused by the unclear long-term results of the surgical palliation (viz. Fontan circulation) in patients with HLHS [5]. In recent years, there has been a particular interest in researching the genetic mechanisms causing the development of HLHS. Genetic syndromes (like Turner, Noonan, Trisomy, etc.) are present in around 25% of cases, but studies involving non-syndromic family members have suggested that heritability is complex and not related to a single pathway anomaly [6].

L. Galletti (✉)
Bambino Gesù Hospital, Rome, Italy
e-mail: lorenzo.galletti@opbg.net

N. Uricchio
Congenital Cardiac Unit, Papa Giovanni XXIII Hospital, Bergamo, Italy
e-mail: nuricchio@asst-pg23.it

G. Butera et al. (eds.), *Modelling Congenital Heart Disease*,
https://doi.org/10.1007/978-3-030-88892-3_9

9.3 Pathophysiological Models and Clinical Picture

Despite the spectrum of anatomical subtype is variable, the associate pathophysiological pictures are similar:

1. There is an obligatory Left to Right shunt at atrial level.
2. The right ventricle (RV) supplies both the pulmonary circulation and the systemic one, through ductus arteriosus.

Thus, maintenance of ductal patency and adequate mixing at the atrial level are essential for postnatal survival. A restrictive or closed foramen ovale (FO) may result in pulmonary edema and hypoxia and fatal lung parenchymal disease whereas a premature closure of ductus will result in low cardiac output, cardiogenic shock, and death [7]. Also, the postnatal drop in pulmonary vascular resistances (PVR) will lead to a progressive increase in pulmonary flow in a few days after birth. In milder form of HLHS and HLHC, systemic circulation could be only partially dependent by RV and LV support flow into ascending aorta and proximal arch branches.

Currently, the majority of babies with HLHS are prenatally diagnosed and their delivery is planned in a tertiary-care center where timing and course of any patient's treatment depend on the initial presentation [7].

Once ductal patency is ensured by PGE infusion, the only potential restriction to pulmonary flow is at atrial level. According to the size of interatrial communication, there are three possible clinical scenarios:

1. Around 30–40% of patients show only moderately restrictive ASD (gradient 4–8 mm/Hg). In this case, there is a limited increase in pulmonary overcirculation usually well tolerated without signs of systemic hypoperfusion. Patients remain stable on PGE infusion and spontaneous ventilation on air and do not need any treatment before surgery.
2. Neonates with mild or no restriction at atrial level show a rapid increase in pulmonary flow, which is poorly compensated by systemic RV. These patients rapidly show progression of heart failure with signs of poor systemic perfusion, tachycardia, tachypnea metabolic acidosis, and oliguria; this clinical picture could further progress in cardio circulatory collapse and sudden death. Initial treatment should focus on stabilization by balancing pulmonary and systemic flow to optimize systemic oxygen delivery. Patients often require emergency measures recommended in neonatal advanced life-support algorithms, including use of vasoactive agents (mainly systemic vasodilators), intubation, and ventilation [8]. In extreme case, emergent bilateral pulmonary artery banding should be considered to limit pulmonary run-off. Newborns admitted in shock with metabolic acidosis because of an obstructed DA, unresponsive to high doses of PGE1, with catecholamine (e.g., norepinephrine) infusion, should be considered for a percutaneous or trans-PA ductal stent [9].
3. A third subgroup, 6–15% of patients, presents with severely restricted or intact atrial septum, or obstructed total anomalous pulmonary vein defect (TAPVD). Clinical presentation is characterized by immediate postnatal deep cyanosis, acidosis, and respiratory distress requiring urgent intubation and resuscitation. These groups of patients have a worse prognosis, both for their critical clinical status and for the associated abnormal pulmonary vascular and parenchymal development [10]. Due to the poor results in this subset of patients, early left atrial decompression with balloon septostomy and bilateral PA banding, or ECMO support before the surgical palliation, is advocated by some authors. Recently, fetal intervention has been offered to fetuses with HLHS with restrictive or intact atrial septum.

9.4 Diagnosis

Screening by pulse oximetry is universally renowned to be effective in postnatal detection of congenital heart disease (CHD) and may allow the early uncovering of HLHS, should it be unrecognized in utero. On the physical examination, the neonate is tachypneic and in case of ductal constriction exhibits signs of poor peripheral perfusion, with tachycardia and low-volume pulses. The chest x-ray may show signs of congestion, or a "white" lung usually associated with cardiomegaly. The ECG shows typical neonatal findings of right-axis deviation and right ventricular hypertrophy.

Echocardiography is the gold standard for the diagnosis of HLHS [7, 11]. It permits to recognize the essential anatomical and functional aspects useful for the future treatment:

- Definition of HLHS subtypes and LV structures.
- RV function.
- TV anatomy and function.
- Morphology of interatrial septum and pulmonary venous return.
- Morphology of ascending aorta and aortic arch.
- Pulmonary artery size.
- Associated anomalies (i.e., coronary arteries).

Preoperative diagnostic cardiac catheterization should be reserved for diagnosis of complex additional lesions. Advanced techniques like CT scan or cardiac MRI are useful if the echocardiography leaves doubts about the anatomy of the pulmonary veins and/or the aortic arch morphology.

9.5 Treatment

The natural history of HLHS is almost universally fatal, and the only treatment option is surgical. Following the pioneering work started by Dr. Norwood in the 80s [12], the treatment protocol of HLHS is based on 3-stages surgical procedure to achieve, a Fontan circulation [13]. The first stage, known as Norwood Stage I, is performed in the first week of life, ideally once reached stable hemodynamic conditions. Goals of Stage I palliation are [13]:

1. Creation of an unobstructed pathway from RV to descending aorta in order to suppress ductal-dependent systemic circulation. This is technically achieved with the amalgamation of the ascending aorta with the pulmonary trunk obtaining a systemic root, along with a patch enlargement of the ascending aorta, arch, and descending aorta beyond ductal insertion.
2. Prevention of pulmonary venous hypertension by a wide atrial septectomy.
3. Modulation of pulmonary blood flow, in order to achieve a pulmonary to systemic flow ratio (QP/QS) around 1, provides an arterial oxygen saturation of 75–85%. In this way, the development pulmonary vascular tree is allowed, while its architecture is preserved, and the systemic ventricular is not overloaded. There are two ways to achieve this target: creating an systemic to pulmonary shunt, modified Blalock-Taussing shunt, (mBT shunt) using a gore-tex conduit (3–3.5 mm) between the innominate artery and right pulmonary artery (Fig. 9.1); the alternative way, Sano modification, is to connect the RV to the pulmonary artery confluence inserting a 5–6 mm conduit (Fig. 9.2) [14, 15]. Both procedures have advantages and disadvantages represented by diastolic steal (cerebral and coronary) in case of mBT shunt and need for a ventriculotomy that could have an impact on the long-term RV function. In order to establish the optimal pulmonary flow source, an important multicentric randomized trial has been started in 2005 [16].

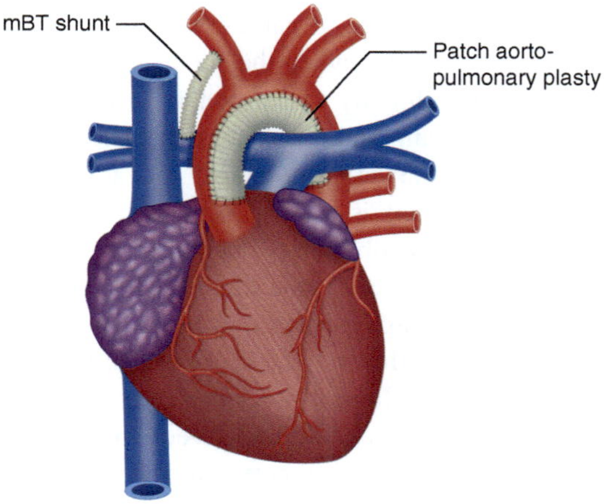

Fig. 9.1 Classic Norwood operation

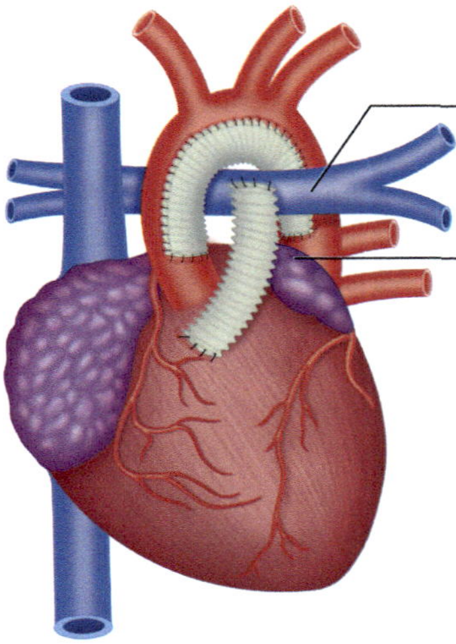

Fig. 9.2 Sano modification of Norwood operation

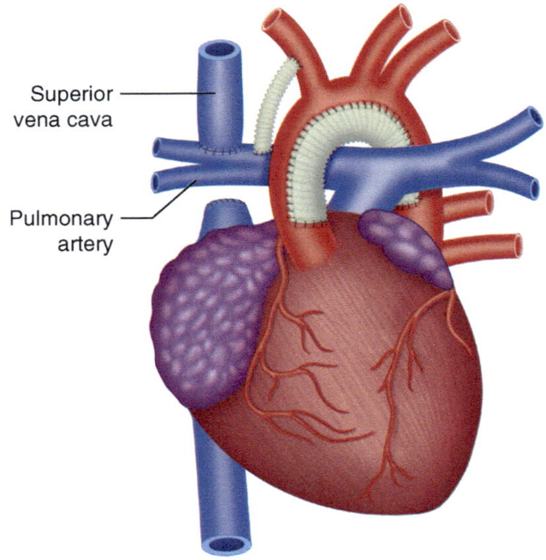

Fig. 9.3 Glenn shunt

The Norwood Stage I remains one of higher-risk procedure in congenital heart surgery due to its nature of complex although palliative procedure, sometimes performed on neonates in critical clinical conditions [17]. Additionally, the operative risk is particularly high in presence of additional factors like: concomitant genetical syndrome or extracardiac disease, prematurity and/or low birth weight, clinical status, and presence of preoperative tricuspid regurgitation or severe restriction to pulmonary venous return [4, 18, 19]. Another significant factors related to mortality are individual surgeon and center experience with the Norwood Stage I procedure [16, 20].

The second stage of palliation is achieved, at 3–6 months of age, when the superior vena cava is connected directly to pulmonary artery (bidirectional Glenn shunt or hemi-Fontan procedure), (Fig. 9.3). In this stage, the arterial or ventricular shunt is usually removed, allowing suppression of obligatory ventricular volume overload consequent to the stage I palliation.

The third stage is the completion of Fontan circulation: in this stage, the systemic and the pulmonary circulations are separated, by inserting an extracardiac conduit between the inferior vena cava and the pulmonary artery (Fig. 9.4) or by a lateral tunnel procedure, as described by De Leval and coll. The Fontan completion allows the separation of pulmonary and systemic circulation obtaining a normal or nearly normal of arterial saturation at the expense of a hypertension of inferior venous district. The completion Fontan is usually performed around 3 years of age.

In order to decrease the in-hospital mortality of stage I, particularly in face of high-risk patients, a less invasive procedure, the so-called Hybrid Stage I has been developed [21, 22]. This procedure consists in (1) a bilateral pulmonary artery banding to regulate the pulmonary flow, (2) positioning of a stent in the arterial duct to maintain its patency, and if necessary (3) a balloon septostomy or stenting in the atrial septal defect (Fig. 9.5). The hybrid procedure is followed at 3–6 months of age by a comprehensive second stage that includes the same steps of stage 1 (reconstruction of aortic arch and aortopulmonary anastomosis, pulmonary confluence reconstruction, and septostomy) in association with a cavopulmonary shunt as pulmonary blood flow source [22, 23]. Some centers have adopted widely the hybrid stage I, but long-term results

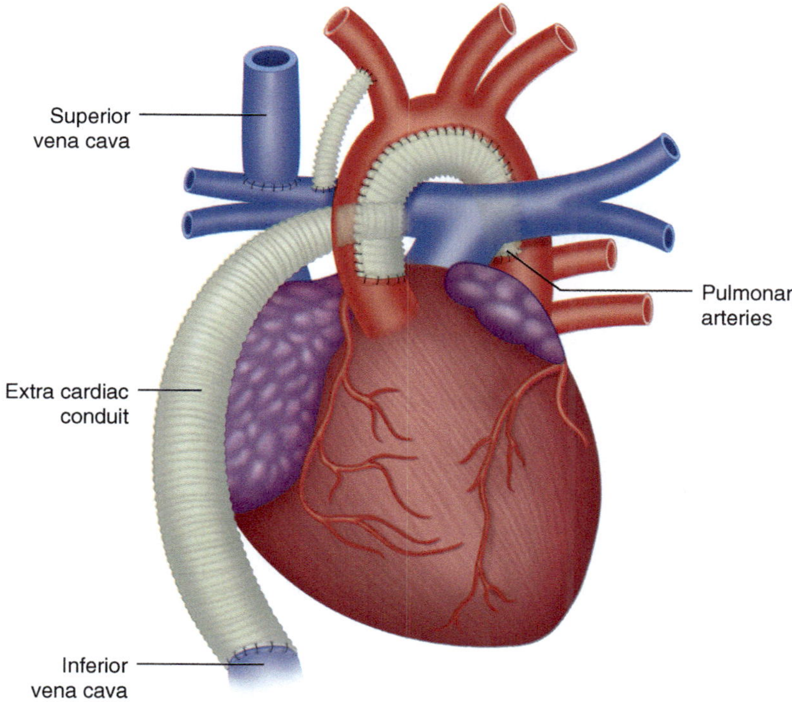

Fig. 9.4 Fontan completion

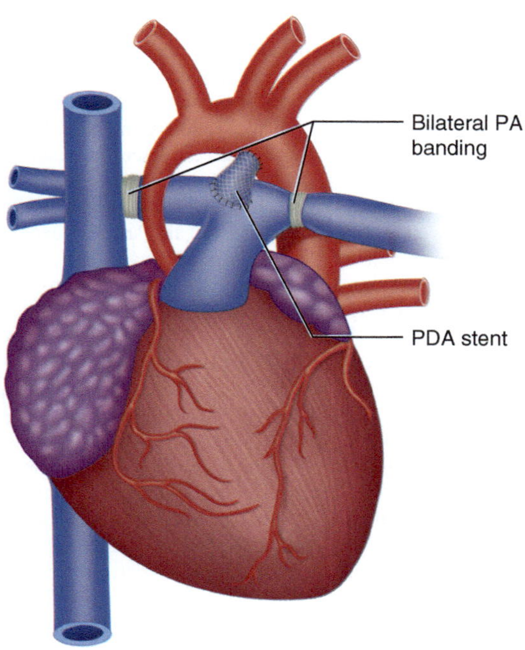

Fig. 9.5 Hybrid approach

did not demonstrate superiority of the hybrid strategy over the surgical stage I.

Neonatal heart transplantation (HTx) as primary treatment of HLHS was popularized particularly by Dr Bailey in Loma Linda during the 80s [24]. The main limiting factor for this strategy is shortage of donors; hence, HTX is reserved for babies in which reconstructive surgery is not opportune [25] or more commonly, in patients with a failure of the single ventricle reconstruction pathway.

9.6 Results

Mortality for treatment of HLHS dropped drastically over the last 25 years [17] and many leading institutions reported survival close to 85% after stage I procedure [25]. One of the most significant studies conducted on HLHS is the Single Ventricle Reconstruction (SVR) trial [16]. This is a randomized, prospective study, which compared the outcomes of the modified Blalock Taussig (mBT) shunt with the RVPA conduit modification. The RVPA shunt had greater transplantation-free survival 12 months after randomization compared to the mBT shunt (74% vs. 64%, $P = 0.01$). Conversely, the RVPA shunt group had more unintentional interventions ($P = 0.003$) and complications ($P = 0.002$). Right ventricular volume and function at the age of 14 months and the rate of nonfatal serious adverse events at the age of 12 months were similar in the two groups [16]. After a mean follow-up of 32 ± 11 months [26, 27], there was a nonsignificant difference in transplantation-free survival between the two groups ($P = 0.06$). On nonproportional hazards analysis, the size of the treatment effect differed before and after 12 months ($P = 0.02$).

Starting from the original cohort, transplant-free survival at 6 years and other interventions were investigated [28]. The 6-years transplant-free survival was unaffected by the shunt type (64% RVPA group vs. 59% mBT shunt group, $P = 0.13$). Interestingly, the treatment effect on the transplant-free survival varied across the three intervals. The RVPAS group had a lower risk of death or transplant before stage II (hazard ratio HR 0.66; $P = 0.003$), with a trend toward increased risk of death during the second interval (HR, 1.36; $P = 0.17$). In the third interval, shunt type was not significantly associated with transplant-free survival, although the HR was lower for the RVPAS group (HR, 0.76; $P = 0.52$). The incidence of catheter interventions was greater among the RVPA group. Lastly, serious adverse clinical events were widespread among all survivors, with 20% having a neurologic event (13% seizure and 7% stroke), and nearly 16% having a thrombotic event.

It can be concluded that RVPA modification brings a survival advantage before stage II palliation but a greater risk of catheter interventions until the third stage of palliation procedure is achieved. After the Fontan procedure, there is no persistent advantage of the initial systemic to pulmonary artery shunt on transplant-free survival or catheter intervention. Morbidity begins early in life and gradually increases for children in both shunt groups. These data stress the importance of continued and close follow-up of this cohort in order to find new strategies to improve the long-term outlook for those with single ventricle anomalies.

References

1. Lev M. Pathologic anatomy and interrelationship of hypoplasia of the aortic tract complexes. Lab Investig. 1952;1(1):61–70.
2. Noonan JA, Nadas AS. The hypoplastic left heart syndrome: an analysis of 101 cases. Pediatr Clin N Am. 1958;5(4):1029–56.
3. Tchervenkov CI, Jacobs JP, Weinberg PM, Aiello VD, Béland MJ, Colan SD, Elliott MJ, Franklin RC, Gaynor JW, Krogmann ON, Kurosawa H, Maruszewski B, Stellin G. The nomenclature, definition and classification of hypoplastic left heart syndrome. Cardiol Young. 2006;16(4):339–68.
4. Bove EL, Lloyd TR. Staged reconstruction for hypoplastic left heart syndrome. Contemporary results. Ann Surg. 1996;224(3):387–94. discussion 394–395
5. van der Linde D, Konings EEM, Slager MA, Witsenburg M, Helbing WA, Takkenberg JJM, et al. Birth prevalence of congenital heart disease worldwide: a systematic review and meta-analysis. J Am Coll Cardiol. 2011;58(21):2241–7.
6. Yagi H, Liu X, Gabriel GC, Wu Y, Peterson K, Murray SA, et al. The genetic landscape of hypoplastic left heart syndrome. Pediatr Cardiol. 2018;39(6):1069–81.
7. Guidelines for the Management of Congenital Heart Diseases in childhood and adolescence. Cardiol Young. 2017;27(S3):S1–105.
8. Wyckoff MH, Aziz K, Escobedo MB, Kapadia VS, Kattwinkel J, Perlman JM, et al. Part 13: neonatal resuscitation: 2015 American Heart Association guidelines update for cardiopulmonary resuscitation and emergency cardiovascular care. Circulation. 2015;132(18 Suppl 2):S543–60.
9. Alphonso N, Angelini A, Barron DJ, Bellsham-Revell H, Blom NA, Brown K, Davis D, Duncan D, Fedrigo M, Galletti L, Hehir D, Herberg U, Jacobs JP, Januszewska K, Karl TR, Chaiman HLHS Guidelines Task Force, Malec E, Maruszewski B, Montgomerie J, Pizzaro C, Schranz D, Shillingford AJ, Simpson JM. Guidelines for the management of neonates and infants with hypoplastic left heart syndrome: The European Association for Cardio-Thoracic Surgery (EACTS) and the Association for European Paediatric and Congenital Cardiology (AEPC) Hypoplastic Left Heart Syndrome Guidelines Task Force. Eur J Cardiothorac Surg. 2020;58(3):416–99. https://doi.org/10.1093/ejcts/ezaa188.
10. Rychik J, Rome JJ, Collins MH, DeCampli WM, Spray TL. The hypoplastic left heart syndrome with

intact atrial septum: atrial morphology, pulmonary vascular histopathology and outcome. J Am Coll Cardiol. 1999;34(2):554–60.
11. Nguyen T, Miller M, Gonzalez J, Nardell K, Galas J, John JB, et al. Echocardiography of hypoplastic left heart syndrome. Cardiol Young. 2011;21(Suppl. 2):28–37.
12. Norwood WI, Lang P, Hansen DD. Physiologic repair of aortic atresia-hypoplastic left heart syndrome. N Engl J Med. 1983;308(1):23–6.
13. Norwood WI. Hypoplastic left heart syndrome. Ann Thorac Surg. 1991;52(3):688–95.
14. Sano S, Huang SC, Kasahara S, Yoshizumi K, Kotani Y, Ishino K. Risk factors for mortality after the Norwood procedure using right ventricle to pulmonary artery shunt. Ann Thorac Surg. 2009;87(1):178–85. discussion 185-6
15. Maher KO, Pizarro C, Gidding SS, Januszewska K, Malec E, Norwood WI, et al. Hemodynamic profile after the Norwood procedure with right ventricle to pulmonary artery conduit. Circulation. 2003;108(7):782–4.
16. Ohye RG, Sleeper LA, Mahony L, Newburger JW, Pearson GD, Lu M, et al. Comparison of shunt types in the Norwood procedure for single-ventricle lesions. N Engl J Med. 2010;362(21):1980–92.
17. Mahle WT, Spray TL, Wernovsky G, Gaynor JW, Clark BJ. Survival after reconstructive surgery for hypoplastic left heart syndrome: a 15-year experience from a single institution. Circulation. 2000, 102;(19 Suppl. 3):III136–41.
18. McGuirk SP, Stickley J, Griselli M, Stumper OF, Laker SJ, Barron DJ, et al. Risk assessment and early outcome following the Norwood procedure for hypoplastic left heart syndrome. Eur J Cardiothorac Surg. 2006;29(5):675–81.
19. Shamszad P, Gospin TA, Hong BJ, McKenzie ED, Petit CJ. Impact of preoperative risk factors on outcomes after Norwood palliation for hypoplastic left heart syndrome. J Thorac Cardiovasc Surg. 2014;147(3):897–901.
20. Hirsch JC, Gurney JG, Donohue JE, Gebremariam A, Bove EL, Ohye RG. Hospital mortality for Norwood and arterial switch operations as a function of institutional volume. Pediatr Cardiol. 2008;29(4):713–7.
21. Schranz D, Akintuerk H. Hypoplastic Left Heart Syndrome: The Giessen approach; history, technique, and results. In: Butera G, Cheatham J, Pedra C, Schranz D, Tulzer G, editors. Fetal and hybrid procedures in congenital heart diseases. Springer; 2016. p. 209–20.
22. Pizarro C, Derby CD, Baffa JM, Murdison KA, Radtke WA. Improving the outcome of high-risk neonates with hypoplastic left heart syndrome: hybrid procedure or conventional surgical palliation? Eur J Cardiothorac Surg. 2008;33(4):613–8.
23. Galantowicz M, Yates AR. Improved outcomes with the comprehensive stage 2 procedure after an initial hybrid stage 1. J Thorac Cardiovasc Surg. 2016;151(2):424–9.
24. Bailey L, Concepcion W, Shattuck H, Huang L. Method of heart transplantation for treatment of hypoplastic left heart syndrome. J Thorac Cardiovasc Surg. 1986;92(1):1–5.
25. Ohye RG, Schranz D, D'Udekem Y. Current therapy for hypoplastic left heart syndrome and related single ventricle lesions. Circulation. 2016;134(17):1265–79.
26. Newburger JW, Sleeper LA, Frommelt PC, Pearson GD, Mahle WT, Chen S, Dunbar-Masterson C, Mital S, Williams IA, Ghanayem NS, Goldberg CS, Jacobs JP, Krawczeski CD, Lewis AB, Pasquali SK, Pizarro C, Gruber PJ, Atz AM, Khaikin S, Gaynor JW, Ohye RG. Pediatric heart network investigators. Transplantation-free survival and interventions at 3 years in the single ventricle reconstruction trial. Circulation. 2014;129(20):2013–20.
27. Ghanayem NS, Allen KR, Tabbutt S, Atz AM, Clabby ML, Cooper DS, et al. Interstage mortality after the Norwood procedure: results of the multicenter single ventricle reconstruction trial. J Thorac Cardiovasc Surg. 2012;144(4):896–906.
28. Newburger JW, Sleeper LA, Gaynor JW, Hollenbeck-Pringle D, Frommelt PC, Li JS, et al. Transplant-free survival and interventions at 6 years in the SVR trial. Circulation. 2018;137(21):2246–53.

10 Hypoplastic Left Heart Syndrome, Norwood and Variants: Clinical Concepts, Engineering Applications and Impact of an Integrated Medico-Engineering Approach

Adelaide de Vecchi

Since Norwood's first successful experience of staged palliation in a child [1], significant advances in this complex procedure have contributed to increase 1-year and 5-year survival statistics to approximately 60–80% overall [2, 3]. Despite this progress, morbidity and mortality rates are still high, especially interstage mortality between Stage I and II, which nears 20% [4]. This wide range in survival statistics is partly due to the variability of the experience in different centres. The choice of repair strategy (e.g. modified Norwood vs. hybrid procedure) or even of the surgical procedure within a single stage of repair (e.g. Sano vs. Blalock-Taussig shunt at Stage I) is often dictated by the expertise of the surgeon performing the operation. Similarly, there is no definitive evidence on what type of repair strategy produces the best outcomes, not only in relation to short-term survival but also in the long-term as Fontan patients enter adulthood. Alongside in vivo cohort-based research, recent years have seen a steady rise in bioengineering technology to shed light on the best course of action to treat this complex disease. In particular, due to the heterogeneity of manifestation in HLHS and the variety of treatment options, personalised modelling has emerged as a unique tool to understand the disease physiology and predict the effects of surgical palliation in relation to the individual characteristics [5]. Most of these applications focus on cardiac haemodynamics, which is a key area for insight into HLHS mechanisms. It is now widely recognised that early indication of pathophysiological processes can be inferred from abnormal blood flow dynamics, since alterations in cardiac function are immediately reflected in flow patterns before pathological structural remodelling takes place [6]. The intraventricular fluid dynamics is dominated by the formation (during diastole) and unravelling (during systole) of vortices of blood, which are in turn associated with changes in pressure magnitude and gradients [7]. In normal conditions, ventricular filling and ejection follow a well-defined pattern that optimises flow transport and, consequently, haemodynamic efficiency [8, 9]. In pathological conditions, abnormal values in haemodynamic parameters can indicate compensatory mechanisms heralding maladaptive changes in both anatomy and function. In HLHS, the single ventricle operates in precarious conditions, especially in the first stage

A. de Vecchi (✉)
School of Biomedical Engineering and Imaging Sciences, King's College London, St Thomas' Hospital, London, UK
e-mail: adelaide.de_vecchi@kcl.ac.uk

G. Butera et al. (eds.), *Modelling Congenital Heart Disease*,
https://doi.org/10.1007/978-3-030-88892-3_10

where it has to support both the pulmonary and the systemic circulation, and thus haemodynamic analysis can timely reveal the onset of maladaptive mechanisms potentially leading to interstage heart failure. From a structural point of view, abnormal flow patterns result in a shift in the momentum associated with the blood flow to different regions of the myocardium, which in the long-term triggers self-sustaining adaptive mechanisms in the muscle. Adverse ventricular remodelling can lead to mechanical dyssynchrony, ventricular stiffening, cavity dilation, and power loss in the blood flow – all factors that are associated with negative outcomes in HLHS [10, 11]. As structural remodelling can lead to irreversible deterioration in these fragile patients, even seemingly minor changes in the flow energy efficiency can have a drastic impact on treatment outcomes.

The synergy between bioengineering technology, clinical data and imaging sciences has the potential to address these issues by providing means to quantify the HLHS haemodynamics using computational modelling alongside in vitro experiments [5]. The information provided by these models can be used to characterise and understand the HLHS pathophysiology at different stages of repair, as well as to predict the impact of variations in surgical repair on ventricular function by performing simulations of virtual treatment. The main in silico approaches for blood flow modelling can be broadly divided into blood flow modelling via Computational Fluid Dynamics (CFD) and Fluid-Structure Interaction (FSI) modelling for the coupled simulation of blood flow and myocardial dynamics and structural mechanics. Typical workflows for the generation of personalised CFD and FSI models are illustrated in Fig. 10.1a, b, respectively.

Historically, CFD models based on finite-element or finite-volume methods have been extensively used in HLHS to simulate the 3D

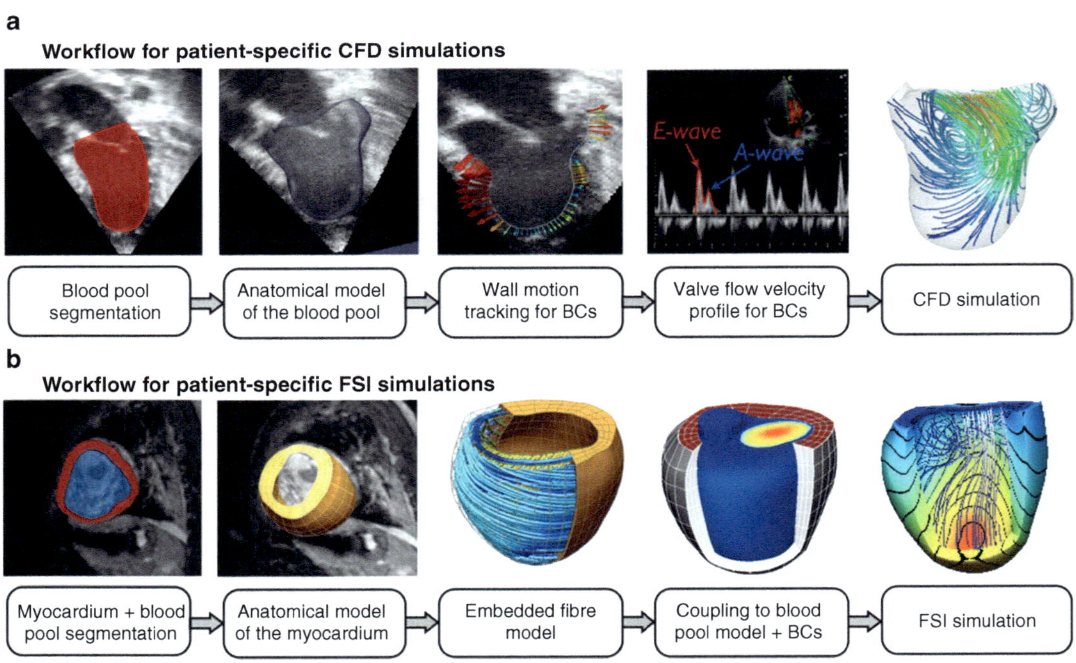

Fig. 10.1 Examples of workflow for the generation of image-based personalised models for Computational Fluid Dynamics simulations (**a**) and Fluid-Structure Interaction simulations (**b**)

blood flow dynamics in patient-specific anatomies derived from imaging data. These patient-specific anatomical models can be coupled to lumped parameters models of the downstream and upstream boundary conditions, which are also derived from clinical data [12, 13]. Such approaches usually rely on the rigid wall assumption, which holds when the transient structural deformation is not significant.

The most common CFD modelling applications in HLHS aim at evaluating the flow efficiency in the systemic-to-pulmonary shunts and neo-aortic reconstruction at Stage I, and in the total cavopulmonary connection (TCPC) of the completed Fontan circulation at Stage III.

In Stage I, a shunt is inserted to ensure blood flow to the pulmonary circulation. Several variations in the position of the shunt have been proposed, i.e. between the innominate artery and the right pulmonary artery (modified Blalock-Taussig shunt), between the ascending aorta and the pulmonary trunk (central shunt) or between the right ventricle and the pulmonary artery (Sano shunt). Each technique has been modelled computationally and experimentally to investigate the comparative advantages and disadvantages, as well as the effect of varying sizes and types of shunting on oxygenation and haemodynamics [14, 15]. Different types of shunt have been tested using models to analyse the hydraulic performance of the systemic-to-pulmonary connection, with the Sano shunt generally performing better than the Blalock-Taussig shunt [16–18], although an alternative design with a flare at the distal end has been recently proposed to improve the pulmonary circulation with the Blalock-Taussig shunt [19]. The effect of different types of shunt on the pressure gradients has also been widely investigated in vitro using pulsatile flow rigs to test the accuracy of Doppler-derived measurements [20, 21]. Mock circulatory loops with 3D-printed patient-specific anatomies and lumped parameter networks have been used to test the effect of the Sano modification and the Fontan circulation [22–24].

Another important area of investigation, where modelling provides valuable quantitative insight into the optimisation of surgical techniques, is the neo-aortic reconstruction. The creation of the neo-aortic tract is arguably the most challenging part of the Norwood procedures and even minor increases in afterload might lead to a critical reduction of the cardiac output, which in HLHS patients is already severely compromised by the chronic preload starvation that characterises the pulmonary circulation. The changes in ventricular afterload, pulmonary pressure and flow energy efficiency associated with the shape of the neo-aortic reconstruction and Damus-Kaye-Stansel (DSK) anastomosis have been investigated by means of both statistical shape analysis [25] and CFD modelling on patient-specific geometries [26–28], showing that a smooth artic arch and a larger anastomosis result in higher energy efficiency in the blood flow. Wave intensity analysis combined with MRI has also been performed to characterise the changes in the neo-aortic distensibility and shape, and their impact on ventriculo-arterial coupling, highlighting that reduced distensibility and neo-aortic size mismatch lead to a deterioration in ventricular mechanics [29, 30]. Information from patient-specific fluid dynamic modelling and statistical shape analysis can also provide key insight into the performance of the single ventricle in relation to a specific morphology of the diminutive left ventricle [31, 32]—an area of investigation which has been often overlooked in modelling studies despite the evident impact of the remnant left ventricle on right ventricular function.

Using models to determine the effect of various degrees of pulmonary banding in the hybrid procedure is another example of a bioengineering application to HLHS, whereby different levels of banding can be simulated to determine the optimal systemic to pulmonary flow ratio and oxygen delivery [33, 34]. In the future, these analyses can play a role in providing guidance to optimising the bandage level to promote adequate growth of the pulmonary arteries in patients who received hybrid palliation.

Finally, perhaps the most widely investigated problem in HLHS is the optimisation of the Fontan system which completes the passive pulmonary circulation in the final stage of repair by creating the total cavopulmonary connection (TCPC), a 4-way junction of vessels. Over the past decade, computational modelling has provided extensive insight into this final stage of repair [35]. The TCPC optimisation mainly focusses on whether an intra-atrial or extra-cardial tunnel should be performed and on where to position the inferior vena cava. Personalised finite-volume models lent support to the hypothesis that the extra-cardiac tunnel with left pulmonary artery anastomosis exhibits the lowest power loss in the blood flow [36]. A comparison using computational models between the Hemi-Fontan and the bidirectional Glenn procedures showed that the Glenn connection achieves a superior hydraulic efficiency than the Hemi-Fontan, contradicting results from previous modelling studies [37, 38]. Furthermore, image-based models of the TCPC showed that power loss could be minimised, and lung perfusion optimised, by positioning the inferior vena cava between the two superior venae cavae [39], while another computational study concluded that the adoption of a Y-shaped graft could further improve both energy efficiency and flow distribution [40]. Similarly, enlarging the inferior vena cava anastomosis was shown to lead to a better balanced flow distribution to the lungs with minimal power loss [41]. The efficiency of the Fontan circulation was also studied in vitro to assess the impact of patient-specific anatomies as well as respiration and exercise in HLHS patients [24, 42], a fundamental aspect that has been scarcely investigated with computational models.

Such examples demonstrate how CFD models are a versatile tool to simulate the patient-specific heamodynamics in the great vessels; however, they cannot provide information on the structural mechanics of the myocardial wall, which has to be modelled either separately or in conjunction with the blood flow. In this latter approach, FSI modelling gives insight into the coupled dynamics of blood and muscle, providing a more comprehensive way to model cardiac function that can also include elements of electrophysiology to model myocardial contraction. This approach is thus inherently more complex that CFD alone. FSI models is well suited to represent the pathophysiology of diseases with complex substrates such as HLHS, since they allow the simulation of complex feedback mechanisms between haemodynamics and tissue mechanics. Unlike CFD models, where only pressure differences can be quantified unless a reference pressure value is provided from clinical measurements (e.g. invasive catheterisation), FSI models also have the ability to provide a full non-invasive quantification of absolute pressures. This is a crucial information in HLHS patients, where invasive pressure measurements would pose additional risks for the patient. Pressure quantification from FSI models is however dependent on the accuracy of the myocardial modelling, which involves estimation of fibre orientation and constitutive parameters for the myocardial muscle. Variation in fibre architecture and material parameters can result in significantly different local mechanical structural behaviour of the myocardium [43]. This information can be derived from state-of-the-art clinical imaging, such as Diffusion Tensor Magnetic Resonance Imaging, dissection analysis from excise hearts and experimental studies to characterise the non-linear, anisotropic nature of the myocardial behaviour ex-vivo [44, 45]. A wealth of advanced imaging and clinical data is therefore required for FSI models to be correctly calibrated and tailored to each individual physiology, meaning this approach, although potentially capable of providing the most complete representation of the HLHS pathophysiology, often relies on simplifying assumptions that limit its accuracy and clinical applicability. Techniques of data assimilations have been proposed to infer the actual mechanical behaviour of the tissue from the measured kinematics. Despite the promise of this approach, no unique set of constitutive parameters can be derived from single direct measurements of myocardial displacements, meaning additional constraints are required to fully characterise the material law governing myocardial behaviour [46]. Nevertheless some studies attempted to analyse the energy efficiency of the single ventricle using FSI, showing how the timing and magnitude of the diastolic inflow and the cavity shape can affect the filling capacity of the ventricle by altering the transfer of energy between blood flow and the myocardium [47, 48]. Despite the promise of the approach, FSI modelling of HLHS remains an area of research significantly less active than CFD modelling.

This body of work shows the significant impact of computational modelling on surgical planning in HLHS. The integration of state-of-the-art imaging and computational modelling is at the forefront of bioengineering technologies for a realistic representation of the complexity of HLHS. With recent progress in mathematical methods and imaging techniques, increasingly accurate boundary conditions can be derived from imaging data and incorporated in computer models, while solution times decrease significantly thanks to the ever-growing capacity of high performance computing. With this progress comes the ability to use bioengineering tools to assist and improve patient selection, tailor intervention to the specific individual needs and personalise patient management [49, 50]. These advances have been made possible by the shift in focus from a population-based approach to a patient-specific one, which represented a key step to enable a better understanding of the pathophysiology in HLHS patients. An excellent example of how the synergy between computational modelling and imaging can help improve pre- and peri-procedural planning is given by non-invasive measurement of blood pressure in the single ventricle and Fontan circulation. Obtaining invasive haemodynamic measurements poses significant challenges and risks in HLHS patients due to the very small size of the ventricles and their critical condition. Advanced imaging techniques such as Phase-Contrast MRI now allow us to acquire flow information non-invasively, but this type of image acquisition is challenging in HLHS patients due to the long acquisition times required to achieve images of the single ventricle with sufficient spatiotemporal resolution to allow a detailed reconstruction of the intraventricular flow field. These haemodynamic measurements are generally not available as part of the normal clinical routine in HLHS cohorts, who would potentially benefit the most from this type of information. Image-based computational models allow quantifying haemodynamic metrics such as pressure gradients in the blood flow at key locations in the single ventricle and Fontan circulation to provide a detailed haemodynamic characterisation that would not be otherwise possible without invasive catheterisation. As mentioned above, pressure quantification can also be obtained from FSI modelling with myocardial models derived from Diffusion Tensor MRI. Going forward, image-based modelling has also potential to predict surgical outcomes by simulating the effect of a virtual intervention on the individual patient – another key area of improvement in HLHS, which is characterised by a high variability amongst individuals in both anatomy and function. Achieving this impact, however, relies on a seamless integration and validation of these state-of-the-art techniques. The lack of effective and robust validation techniques is currently hindering a direct clinical translation of these bioengineering applications in the day-to-day clinical routine. Experimental models can provide reproducible, controlled data for validation and benchmarking of computational data, in addition to a wealth of information on the pathophysiology of HLHS, but they require time-consuming set-up and calibration, and are not readily available. Similarly, advanced imaging can provide accurate validation data in vivo but these techniques are not routinely available in standard clinical settings and would require separate patient studies. It is now accepted that the cost of acquiring extra data for cross-validation of in vitro, in silico and in vivo data is a crucial step towards the demonstration that modelling can achieve clinical impact and will also contribute to increase the use and acceptance of computer modelling by regulatory bodies to provide regulatory-grade scientific evidence. In the case of HLHS, where the success of surgical palliation relies on the multifactorial interplay between individual patient characteristics and choice of repair strategy, validated in silico models would have the capability to predict the outcome of a specific procedure via personalised simulations of 'virtual' treatment. The Norwood and hybrid procedures pose significant surgical challenges, which make outcome prediction and personalisation of surgery a vital element to reduce the heterogeneity of outcomes. A robust derivation of boundary conditions and parameter estimation is however essential to achieve predictive modelling of clinical value. Finally, it should also be added that haemodynamics and structural mechanics are not the only elements at play in determining HLHS outcomes: the interaction between mechanics and multi-scale processes at cellular level is equally important, especially if the focus of the modelling contribution shifts to analysing longer-term biological processes. An example of this application is modelling of blood coagulation dynamics in the completed Fontan circulation to predict the long-term likelihood of thrombus formation, expanding on previous findings on pro-thrombotic factors such as low wall shear stress in the TCPC [51]. Ultimately, the value of integrating modelling into the clinics depends on how it can contribute to improving quality and quantity of life for the patient. Advanced modelling and imaging techniques have already contributed to tangible improvements in understanding the disease physiology and the effects of surgical techniques in HLHS but increasing the complexity of the models is not a guarantee that this goal will be achieved. The path leading to an effective clinical translation of these new technologies is arduous and its success relies on a constructive collaboration between surgeons, radiologists, imaging scientists and bioengineers.

References

1. Norwood WI, Lang P, Hansen DD. Physiologic repair of aortic atresia–hypoplastic left heart syndrome. N Engl J Med. 1983;308:23–6.
2. McGuirk SP, et al. Staged surgical management of hypoplastic left heart syndrome: a single institution 12-year experience. Heart. 2006;92:364–70.
3. DiBardino DJ. Long-term progression and survival following Norwood single ventricle reconstruction. Curr Opin Cardiol. 2015;30:95–9.
4. Ghanayem NS, et al. Interstage mortality after the Norwood procedure: results of the multicenter Single Ventricle Reconstruction trial. J Thorac Cardiovasc Surg. 2012;144:896–906.
5. Biglino G, et al. Modeling single ventricle physiology: review of engineering tools to study first stage palliation of hypoplastic left heart syndrome. Front Pediatr. 2013;1:31.
6. Pedrizzetti G, et al. Cardiac fluid dynamics anticipates heart adaptation. J Biomech. 2015;48:388–91.
7. Pedrizzetti G, La Canna G, Alfieri O, Tonti G. The vortex—an early predictor of cardiovascular outcome? Nat Rev Cardiol. 2014;11:545–53.
8. Gharib M, Rambod E, Kheradvar A, Sahn DJ, Dabiri JO. Optimal vortex formation as an index of cardiac health. Proc Natl Acad Sci U S A. 2006;103:6305–8.
9. Kilner PJ, et al. Asymmetric redirection of flow through the heart. Nature. 2000;404:759–61.
10. Ho P-K, Lai CTM, Wong SJ, Cheung Y-F. Three-dimensional mechanical dyssynchrony and myocardial deformation of the left ventricle in patients with tricuspid atresia after Fontan procedure. J Am Soc Echocardiogr. 2012;25:393–400.
11. Feinstein JA, et al. Hypoplastic left heart syndrome: current considerations and expectations. J Am Coll Cardiol. 2012;59:S1–42.
12. de Zélicourt DA, Marsden A, Fogel MA, Yoganathan AP. Imaging and patient-specific simulations for the Fontan surgery: Current methodologies and clinical applications. Prog Pediatr Cardiol. 2010;30:31–44.
13. Marsden AL, et al. A new multiparameter approach to computational simulation for Fontan assessment and redesign. Congenit Heart Dis. 2010;5:104–17.
14. Migliavacca F, et al. Modeling of the Norwood circulation: effects of shunt size, vascular resistances, and heart rate. Am J Physiol Circ Physiol. 2001;280:H2076–86.
15. Migliavacca F, et al. Multiscale modelling in biofluidynamics: application to reconstructive paediatric cardiac surgery. J Biomech. 2006;39:1010–20.
16. Bove EL, et al. Use of mathematic modeling to compare and predict hemodynamic effects of the modified Blalock–Taussig and right ventricle–pulmonary artery shunts for hypoplastic left heart syndrome. J Thorac Cardiovasc Surg. 2008;136:312–320.e2.
17. Pennati G, Migliavacca F, Dubini G, Bove EL. Modeling of systemic-to-pulmonary shunts in newborns with a univentricular circulation: State of the art and future directions. Prog Pediatr Cardiol. 2010;30:23–9.
18. Mroczek T, Małota Z, Wójcik E, Nawrat Z, Skalski J. Norwood with right ventricle-to-pulmonary artery conduit is more effective than Norwood with Blalock–Taussig shunt for hypoplastic left heart syndrome: mathematic modeling of hemodynamics. Eur J Cardio-Thoracic Surg. 2011;40:1412–8.
19. Arthurs CJ, et al. Reproducing patient-specific hemodynamics in the Blalock–Taussig circulation using a flexible multi-domain simulation framework: applications for optimal shunt design. Front Pediatr. 2017;5:78.
20. DeGroff CG, Shandas R, Kwon J, Valdes-Cruz L. Accuracy of the Bernoulli equation for estimation of pressure gradient across stenotic Blalock–Taussig Shunts: an in vitro and numerical study. Pediatr Cardiol. 2000;21:439–47.
21. Tacy TA, Whitehead KK, Cape EG. In vitro Doppler assessment of pressure gradients across

modified Blalock-Taussig shunts. Am J Cardiol. 1998;81:1219–23.
22. Biglino G, et al. In vitro study of the Norwood palliation. ASAIO J. 2012;58:25–31.
23. Biglino G, et al. Implementing the sano modification in an experimental model of first-stage palliation of hypoplastic left heart syndrome. ASAIO J. 2013;59:86–9.
24. Vukicevic M, et al. Mock circulatory system of the Fontan circulation to study respiration effects on venous flow behavior. ASAIO J. 2013;59:253–60.
25. Bruse JL, et al. Looks do matter! aortic arch shape after hypoplastic left heart syndrome palliation correlates with cavopulmonary outcomes. Ann Thorac Surg. 2017;103:645–54.
26. Itatani K, et al. Influence of surgical arch reconstruction methods on single ventricle workload in the Norwood procedure. J Thorac Cardiovasc Surg. 2012;144:130–8.
27. Miyazaki S, et al. Surgical strategy for aortic arch reconstruction after the Norwood procedure based on numerical flow analysis†. Interact Cardiovasc Thorac Surg. 2018;26:460–7.
28. Qian Y, Liu JL, Itatani K, Miyaji K, Umezu M. Computational hemodynamic analysis in congenital heart disease: simulation of the norwood procedure. Ann Biomed Eng. 2010;38:2302–13.
29. Biglino G, et al. Reduced ascending aorta distensibility relates to adverse ventricular mechanics in patients with hypoplastic left heart syndrome: noninvasive study using wave intensity analysis. J Thorac Cardiovasc Surg. 2012;144:1307–14.
30. Biglino G, et al. Ventriculoarterial coupling in palliated hypoplastic left heart syndrome: Noninvasive assessment of the effects of surgical arch reconstruction and shunt type. J Thorac Cardiovasc Surg. 2014; https://doi.org/10.1016/j.jtcvs.2014.02.012.
31. Walsh MA, et al. Left ventricular morphology influences mortality after the Norwood operation. Heart. 2009;95:1238–44.
32. Bellsham-Revell HR, et al. Serial magnetic resonance imaging in hypoplastic left heart syndrome gives valuable insight into ventricular and vascular adaptation. J Am Coll Cardiol. 2013;61:561–70.
33. Corsini C, et al. Multiscale models of the hybrid palliation for hypoplastic left heart syndrome. J Biomech. 2011;44:767–70.
34. Baker CE, et al. Effects of pulmonary artery banding and retrograde aortic arch obstruction on the hybrid palliation of hypoplastic left heart syndrome. J Thorac Cardiovasc Surg. 2013;146:1341–8.
35. Wang C, et al. Progress in the CFD modeling of flow instabilities in anatomical total cavopulmonary connections. Ann Biomed Eng. 2007;35:1840–56.
36. Hsia T-Y, et al. Computational fluid dynamic study of flow optimization in realistic models of the total cavopulmonary connections. J Surg Res. 2004;116:305–13.
37. Bove EL, de Leval MR, Migliavacca F, Guadagni G, Dubini G. Computational fluid dynamics in the evaluation of hemodynamic performance of cavopulmonary connections after the norwood procedure for hypoplastic left heart syndrome. J Thorac Cardiovasc Surg. 2003;126:1040–7.
38. Pekkan K, et al. Hemodynamic performance of stage-2 univentricular reconstruction: glenn vs hemi-Fontan templates. Ann Biomed Eng. 2009;37:50–63.
39. de Zélicourt DA, et al. Flow study of an extracardiac connection with persistent left superior vena cava. J Thorac Cardiovasc Surg. 2006;131:785–91.
40. Marsden AL, et al. Evaluation of a novel Y-shaped extracardiac Fontan baffle using computational fluid dynamics. J Thorac Cardiovasc Surg. 2009;137:394–403.e2.
41. de Leval MR, et al. Use of computational fluid dynamics in the design of surgical procedures: application to the study of competitive flows in cavopulmonary connections. J Thorac Cardiovasc Surg. 1996;111:502–13.
42. Tree M, et al. Using a novel in vitro fontan model and condition-specific real-time MRI data to examine hemodynamic effects of respiration and exercise. Ann Biomed Eng. 2018;46:135–47.
43. Cutrì E, et al. Patient-specific biomechanical model of hypoplastic left heart to predict post-operative cardio-circulatory behaviour. Med Eng Phys. 2017;47: 85–92.
44. Krishnamurthy A, et al. Patient-specific models of cardiac biomechanics. J Comput Phys. 2013;244:4–21.
45. Walmsley J, et al. Combining computer modelling and cardiac imaging to understand right ventricular pump function. Cardiovasc Res. 2017;113:1486–98.
46. Xi J, et al. An automatic data assimilation framework for patient-specific myocardial mechanical parameter estimation. Berlin, Heidelberg: Springer; 2011. p. 392–400. https://doi.org/10.1007/978-3-642-21028-0_50.
47. de Vecchi A, et al. Inflow typology and ventricular geometry determine efficiency of filling in the hypoplastic left heart. Ann Thorac Surg. 2012;94:1562–9.
48. de Vecchi A, Nordsletten D, Razavi R, Greil G, Smith NP. Patient specific fluid-structure ventricular modelling for integrated cardiac care. Med Biol Eng Comput. 2013;51:1261–70.
49. Conover T, et al. An interactive simulation tool for patient-specific clinical decision support in single-ventricle physiology. J Thorac Cardiovasc Surg. 2018;155:712–21.
50. Capelli C, et al. Patient-specific simulations for planning treatment in congenital heart disease. Interface Focus. 2018;8:20170021.
51. Yang W, Chan FP, Reddy VM, Marsden AL, Feinstein JA. Flow simulations and validation for the first cohort of patients undergoing the Y-graft Fontan procedure. J Thorac Cardiovasc Surg. 2015;149:247–55.

Univentricular Heart: Clinical Concepts and Impact of an Integrated Medico-Engineering Approach

11

Mario Giordano, Gianpiero Gaio, Maria Giovanna Russo, and Gianfranco Butera

11.1 Univentricular Heart: A Congenital Heart Disease with Multiple Anatomies and Physiologies

Univentricular Heart (UVH) is a complex congenital heart disease with different anatomies and physiologies. A ventricular chamber normally is characterized by three different elements: an inlet tract, surrounding the atrioventricular valve; an outlet tract, surrounding the ventriculoarterial valve; and an apical trabecular part, in between the last ones. Right ventricle shows a tricuspid atrioventricular valve with the presence of a septal papillary muscle, coarse apical trabeculations, and a muscular infundibulum supporting the ventriculoarterial valve with consequent lack of fibrous continuity between the atrioventricular and the ventriculoarterial valve. Left ventricle shows a bileaflet atrioventricular valve with the absence of a septal papillary muscle, fine apical trabeculations, and the lack of a muscular infundibulum supporting the ventriculoarterial valve with consequent fibrous continuity between the atrioventricular and the ventriculoarterial valve. The assessment of the apical trabeculation is the main element to distinguish the right ventricle from the left one, because the ventricular apical trabeculation part is always present, contrary to the inlet and outlet tract. The inlet tract is considered the discriminant to define a chamber (within the ventricular mass) as "ventricle" (presence of the inlet) or "rudimentary accessory chamber" (lack of the inlet).

"Anatomically", the UVH is characterized by a dominant large and well-developed ventricle, eventually accompanied by one accessory chamber, which lacks an inlet portion. The absence of the inlet tract differentiates an accessory chamber by a true ventricle (even if hypoplastic) [1]. Morphologically, the UVH may be characterized by either a true unique ventricle (without accessory chamber) or one large and functional ventricular chamber with a rudimentary accessory chamber. The communication between the dominant ventricle and the accessory chamber is called bulboventricular foramen (BVF).

Concerning the anatomic definition of UVH, numerous congenital heart diseases (CHDs) requiring univentricular surgical palliation (Glenn shunt and Fontan circuit) were not included in the definition of anatomic UVH. For this reason, the STS-Congenital Heart Surgery Database Committee [2] and Robert H. Anderson [3] introduced the concept of "functionally" UVH to identify every CHD in which one ven-

M. Giordano (✉) · G. Gaio · M. G. Russo
Department Paediatric Cardiology, Monaldi Hospital, University of Campania "L. Vanvitelli", Naples, Italy

G. Butera
Department of Pediatric Cardiology, Cardiac Surgery and Heart Lung Transplantation - ERN GUARD HEART, Bambin Gesù Hospital and Research Institute, Rome, Italy

G. Butera et al. (eds.), *Modelling Congenital Heart Disease*,
https://doi.org/10.1007/978-3-030-88892-3_11

tricle (right or left) is unable to sustain the pulmonary or the systemic circulation. This definition included into the UVH and also the CHDs with a right or left ventricle so much undeveloped to be unable to support the pulmonary or systemic circulation, such as pulmonary and aortic atresia. This definition widened the concept of UVH, although some CHDs unsuitable for biventricular correction (large ventricular septal defect, straddling anomaly of atrioventricular valve) rested excluded.

Segmental analysis is crucial for a comprehensive assessment of UVH pathophysiology. Step-by-step analysis requires the evaluation of atrial arrangement, atrioventricular connection, singleventricle morphology, and ventriculoarterial connection. The atria may be arranged in solitus, inversus or ambiguous (right and left isomerism) situs. The atrioventricular connections may show a tricuspid atresia, a mitral atresia, a double-inlet ventricle or a single-inlet ventricle (in the context of an unbalanced complete atrioventricular septal defect with a common atrioventricular valve) (Fig. 11.1). The dominant ventricle may be a right ventricle, a left ventricle or an indeterminate ventricle when it is not possible to detect features of neither a right nor a left ventricle. The great vessels may be normally related, transposed or they may be arranged in a context of double-outlet right ventricle or double-outlet left ventricle. The presence of concomitant congenital heart lesions (pulmonary or aortic stenosis/atresia, ventricular septal defect, restrictive bulboventricular foramen) characterizes the pathophysiology of the UVH: pulmonary blood flow overload and systemic hypoperfusion, or pulmonary blood flow hypoperfusion (with cyanosis) or well-balanced pulmonary-systemic blood flow. In the first-step management of UVH, the aim is to achieve a well-balanced pulmonary and systemic circulation (QP:QS $\approx$ 1.00). Systemic saturation is a "mirror" of pulmonary-systemic blood flow ratio. A systemic saturation between 75 and 85% suggests an adequate balancing between systemic and pulmonary blood flows [4, 5]. When systemic saturation is over 90%, the systemic blood flow may be reduced (with tissue hypoperfusion, metabolic acidosis, and a low cardiac output state), whereas the pulmonary blood flow may be increased with signs of congestive heart failure. This is a typical setting in patients affected by a functionally UVH without pulmonary obstructive lesions. In these cases, a well-balanced circulation may be detected in the first days of life; however, when the pulmonary vascular resistances (PVRs) fall, a pulmonary overload appears. Reversely, associated critic pulmonary obstructive lesions may reduce the pulmonary blood flow with a significant displacement toward the systemic circulation, with an oxygen saturation $\leq$ 70% and cyanosis. In this setting, additional sources of pulmonary blood flow are required to increase oxygen saturation as well as to achieve an adequate growth-up of pulmonary arteries over time.

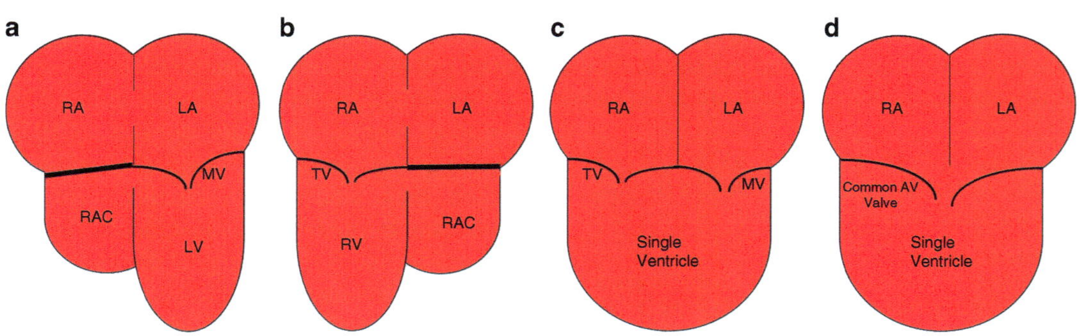

Fig. 11.1 Different patterns of atrioventricular valve arrangement: tricuspid atresia (**a**), mitral atresia (**b**), double-inlet ventricle (**c**), and common atrioventricular valve (**d**). Abbreviations: left atrium = LA; left ventricle = LV; mitral valve = MV; right atrium = RA; rudimentary accessory chamber = RCA; tricuspid valve = TV

11.2 First-Step Palliation of Univentricular Heart

The first surgical palliation of UVH is usually performed in neonatal age (or within the first year of life) to achieve both the normalization of the pressure and volume load on the systemic ventricle as well as an adequate balance of pulmonary and systemic blood flows. A comprehensive knowledge of the anatomic and hemodynamic findings is necessary to address the surgical management. The presence of pulmonary and/or systemic blood flow obstructions is the main element to choose the surgical strategy [6] (Fig. 11.2).

The detection of a critical pulmonary stenosis (or pulmonary atresia) identifies a cyanotic univentricular heart. At the birth, the pulmonary circulation is duct-dependent. The first surgical palliation is a modified Blalock-Taussig-Thomas (mBTT) shunt: a conduit (usually 3–3.5 mm) anastomosed between an epiaortic vessel (innominate artery, carotid artery, and subclavian artery) and the pulmonary artery [7]. Currently, arterial duct stenting is an effective percutaneous alternative strategy to mBTT shunt in selected cases [8] (Fig. 11.3).

A concomitant presence of moderate-severe pulmonary or subpulmonary stenosis without systemic obstruction determines a well-balanced pulmonary-systemic blood flow (QP:QS ≈ 1.00). Arterial oxygen saturation is ≈ 80–85%. In these cases, the patient may rest in natural history up to the "second step" palliation: Glenn shunt. Exceptionally, undiagnosed cases of UVH, with

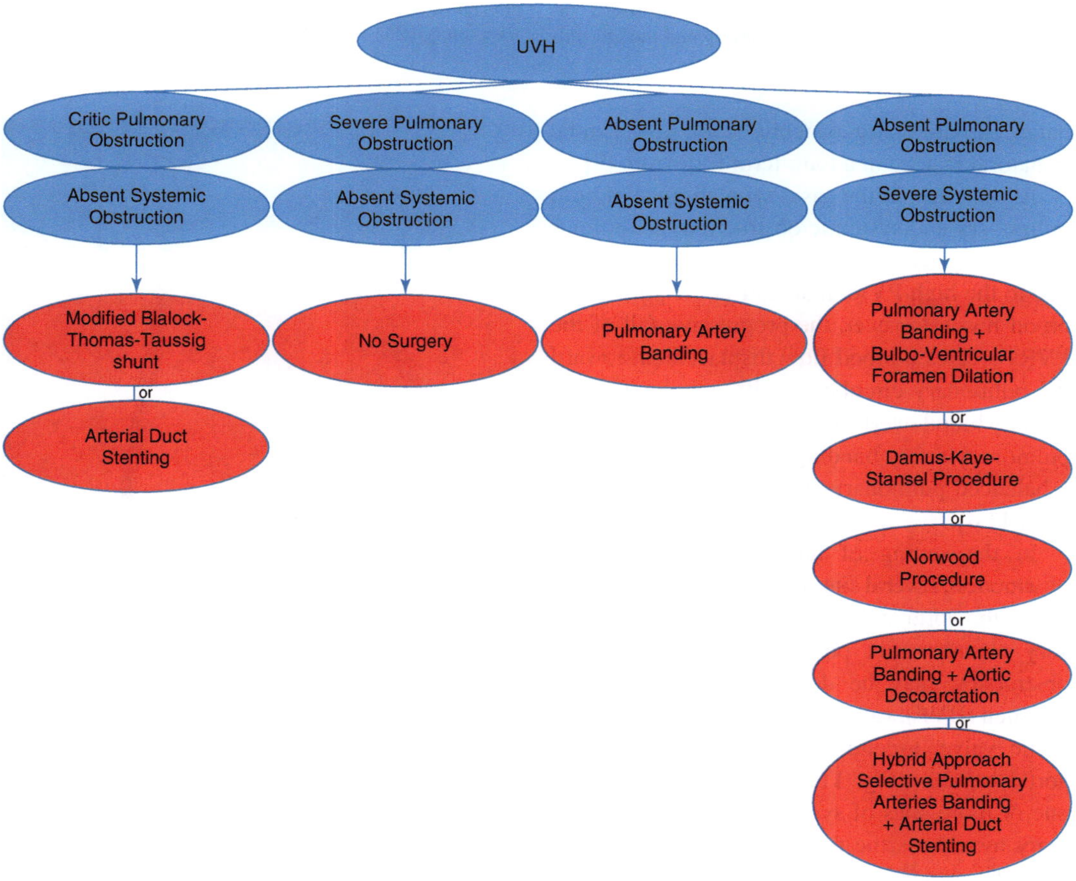

Fig. 11.2 Schematic picture of first-step surgical management of a functional UVH in relation to pulmonary and/or systemic obstruction

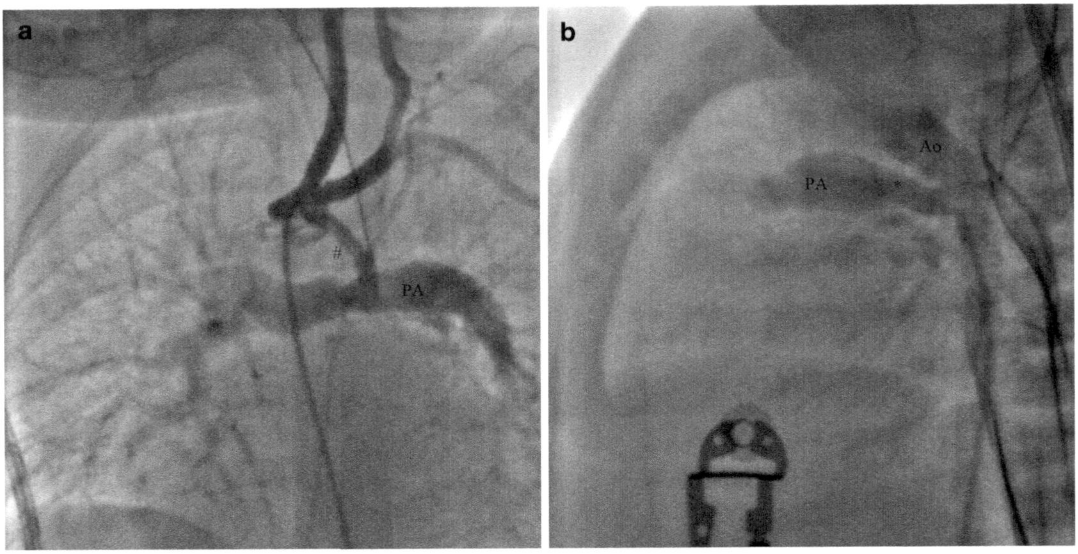

Fig. 11.3 First-step palliation strategy in two cases of cyanotic UVH. (**a**) Fluoroscopic view of an mBTT (#) shunt between the left innominate artery and the left pulmonary artery (PA). (**b**) Fluoroscopic view of a stented arterial duct (*), between the aorta (Ao) and the pulmonary artery (PA)

balanced pulmonary-systemic blood flow, can be in natural history in a long-term follow-up [9].

At the birth, in the cases with absence of both pulmonary and systemic obstruction, the pulmonary and systemic blood flows are well-balanced due to the high PVRs that restrict the pulmonary blood flow. However, the progressive fall of the PVRs creates a blood flow displacement toward the pulmonary circulation with consequent pulmonary blood overload. In these circumstances, a pulmonary artery banding is necessary to restore a balanced pulmonary-systemic blood flow [10] (Fig. 11.4).

In the setting of an UVH with systemic obstruction, several sites of stenosis can be found. When the aorta arises from a rudimentary accessory chamber, six potential sites of systemic obstruction may be detected: bulboventricular foramen (BVF), rudimentary accessory chamber, aortic valve, ascending aorta, aortic arch, and aortic isthmus (Fig. 11.5). The identification of one or more sites of systemic obstruction is primary to address the first surgical palliation [11, 12]. The presence of a restrictive BVF foramen

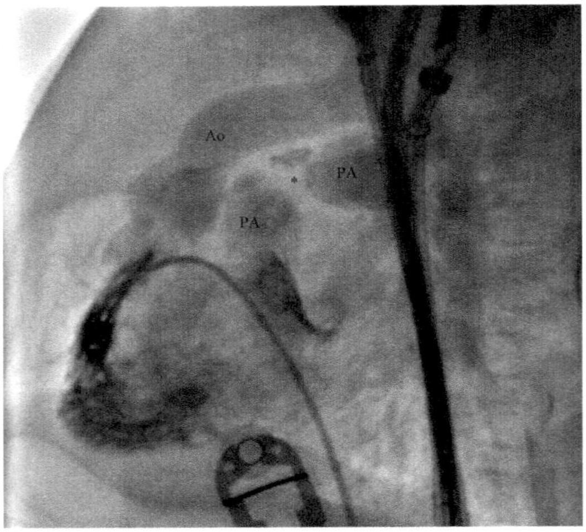

Fig. 11.4 Fluoroscopic lateral view of a banded pulmonary artery (PA) in a case of UVH (double-inlet left ventricle) with transposed great vessels: aorta (Ao) anterior and PA posterior. The banding (*) is adequate and well-placed

and/or an aortic valvular/subvalvular stenosis addresses to a Damus-Kaye-Stansel (DKS) operation. In DKS procedure, the systemic outflow is

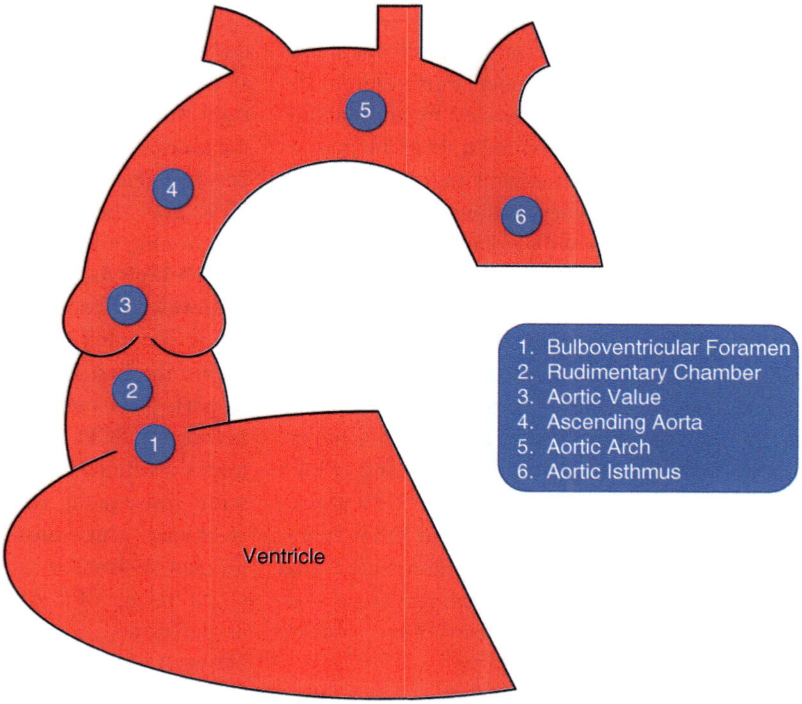

Fig. 11.5 Multiple sites of systemic obstruction in a setting of functional UVH with aorta arising by a rudimentary accessory chamber

restored through the pulmonary arterial trunk which is anastomosed to the ascending aorta. The pulmonary blood flow is supported by an mBTT shunt. The resection of BVF may be performed in cases with isolated restrictive BVF, even if this surgical technique is not applicable during neonatal period. Moreover, the resection of BVF is associated with a high-risk of recurrences and atrioventricular block due to conduction system injuries [13]. DKS procedure shows more advantages than the BVF resection: better outflow gradient relief and lower risk of postoperative heart block. A systemic obstruction extended up to the aortic arch (with hypoplasia of ascending aorta) forces to perform the Norwood procedure. In this surgical technique, the pulmonary arterial trunk is used to rebuild the ascending aorta and the aortic arch (and is named neoaorta). The pulmonary blood flow is restored by an mBTT shunt or a conduit between right ventricle and pulmonary artery (the Sano modification of the Norwood procedure). The right ventricle – pulmonary artery conduit has the advantage to supply a pulsatile blood flow to the pulmonary arteries, deleting the systemic diastolic run-off into the pulmonary circulation that might be cause of diastolic coronary blood stealing with consequent myocardial ischemia. Furthermore, the pulsatile flow of right ventricle – pulmonary artery conduit should allow a better growth of the pulmonary artery [14]. In the patients with high surgical risk, a hybrid approach may be an alternative to the Norwood procedure. The hybrid strategy is characterized by a combined surgical and percutaneous approach in which the surgeon bands selectively the pulmonary arteries while the interventional cardiologist performs a stenting of the arterial duct. The surgical and percutaneous acts may be performed in two steps (Giessen approach) [15] or in one step (Columbus approach) [16]. However, the placement of selective pulmonary arterial bands in newborns <2 kg is very challenge, since small changes in the diameter of the band determine significant alterations of intraluminal diameter. Furthermore, the hybrid approach is characterized by an unavoidable distortion of the proximal tract of the pulmonary arteries with consequent increased risk of pulmonary arterial interventions during follow-up [17]. In the cases with isolated aortic coarctation without pulmonary obstruction, a simpler surgical palliation is performed: pulmonary artery banding with surgical decoarctation. This surgical intervention is characterized by lower surgical risks. However, during follow-up, the single ventricle may develop a progressive hypertrophy due to pulmonary artery banding, which might produce a restriction of BVF with consequent need of BVF enlargement or DKS procedure. Matitiau et al. [18] demonstrated that neonates with a BVF area < 2 cm^2/m^2 are at high risk to develop BVF obstruction, during follow-up.

Pulmonary and systemic obstructions address significantly the first step of UVH palliation. However, both pulmonary and systemic obstructions are not fix entities, but they may develop progressively, requiring a careful follow-up of infants with a functional UVH to perform early the appropriate palliation.

11.3 Cavopulmonary Connections: Glenn and Fontan Shunts

In the management of UVH, the final aim is to achieve a complete separation of the pulmonary and systemic circulation so as the single ventricle supports only the systemic circulation. This result is achieved through other two palliative steps: firstly, a superior cavopulmonary connection (SCPC) (or bidirectional Glenn shunt or hemi-Fontan operation) and then a total cavopulmonary connection (TCPC) (or Fontan operation).

The Glenn shunt is performed with a direct end-to-side anastomosis between the superior vena cava and the pulmonary artery (usually the right pulmonary artery). The SCPC (considered the second-stage palliative procedure of UVH) is performed at 6–12 months of life. At this age, the PVR gets the maximum fall allowing a good functioning of the cavopulmonary anastomosis [19]. A high risk of Glenn shunt dysfunction (due to elevated cavopulmonary pressures) and cyanosis is recorded when the SCPC is performed within the first 2 months of life. The advantages to perform a partial cavopulmonary connection before the TCPC are multiple. The SCPC determines a volume unloading of the single ventricle with consequent reduction of the ventricular workload. This volume unloading allows a significant ventricular remodeling prior to TCPC.

At the age of 5–6 years, the Fontan procedure is performed to allow a complete separation between the pulmonary and systemic circulation through an anastomosis between the inferior vena cava and the pulmonary artery. In this way, the systemic venous return drains passively toward the pulmonary arteriolar system without the support of a ventricular pump. The TCPC allows a "normalization" of the systemic oxygen saturation and a reduction of the risk of paradoxical embolization. However, these "successes" are burdened by some disadvantages: chronic systemic venous congestion (with the risk of liver dysfunction), low cardiac output reserve, and high ventricular afterload [20]. For these reasons, the Fontan circuit is considered a paradox circuit characterized by high systemic venous pressure and low pulmonary arterial pressure. Concerning these physiopathological features, a pre-Fontan accurate evaluation is mandatory. In 1977, Choussat et al. [21] published the "Ten Commandments" of the ideal Fontan:

- Age > 4 years.
- Sinus rhythm.
- Normal systemic venous return.
- Normal right atrial volume.

- Mean pulmonary arterial pressure < 15 mmHg.
- PVR < 4 WU*m^2.
- Pulmonary artery–aorta ratio more than 0.75.
- Ventricular ejection fraction >60%.
- Competent atrioventricular valve.
- Absence of pulmonary artery distortion.

Over time, these rules have become more flexible. Currently, a preserved ventricular ejection fraction and a normal pulmonary arterial pressure are the most important "commandments". Before performing a TCPC, the preoperative cardiac catheterization is often fundamental allowing to detect: Glenn shunt status, inferior vena cava position and drainage, distortion or stenosis of the pulmonary arteries, and crucial hemodynamic findings (pulmonary arterial pressure, PVR, and ventricular end-diastolic pressure) [22].

The first Fontan operation was performed by Fontan and Baudet [23] in 1971 to repair a patient with tricuspid atresia. The TCPC, called atriopulmonary connection, was executed closing the interatrial septal defect and connecting the right atrium directly to the right pulmonary artery. The idea of this surgical technique was that the right atrium might work like a "pump" giving a pulsatile flow to pulmonary circulation. This idea was promptly disowned. The right atrium did not have enough contractile force to work like a pulsatile pump and it went to a significant enlargement with consequent higher risk of supraventricular tachyarrhythmias over time. In 1987, de Leval et al. [24] proposed a variation of the previous surgical technique: the intracardiac total cavopulmonary connection or lateral tunnel procedure. In this technique, the superior vena cava is directly anastomosed to the right pulmonary artery and the inferior vena cava blood flow is drained into the superior vena cava through a conduit created inside the atrium. This kind of TCPC excludes the right atrium by the Fontan circuit, reducing the risk of atrial enlargement and arrhythmias over time. Moreover, the coronary sinus remains to drain in a low-pressure atrial chamber allowing a physiological myocardial venous drainage. In 1990, Marcelletti et al. [25] developed a novel surgical technique: the extracardiac Fontan. The superior vena cava is directly connected to the right pulmonary artery while the inferior vena cava is connected to the right pulmonary artery with the interposition of an extracardiac prosthesis. This surgical technique shows several advantages: myocardial ischemia not required during the procedure, fewer suture lines in the right atrium, no foreign material in the right atrium, lower risk of sinus node dysfunction, and postoperative arrhythmias. Nowadays, the extracardiac conduit is the most common Fontan procedure, followed by the lateral tunnel technique [26]. The atriopulmonary connection is almost abandoned. Lateral tunnel and extracardiac conduit techniques provide a more physiological laminar blood flow (with low turbulence) inside the Fontan circuit.

A peculiar alternative to Fontan operation is represented by the Kawashima operation [27]. This procedure is performed in patients with left isomerism characterized by an interrupted inferior vena cava with azygos/hemiazygos continuation (and consequent drainage to the superior vena cava) and direct connection of the suprahepatic veins to the atrium. In these patients, the anastomosis of the superior vena cava to the pulmonary artery allows an immediate redirection of both the upper- and lower-body venous drainage. However, the direct connection of the suprahepatic veins to the right atrium leads only a partial "total cavopulmonary connection" with consequent lower systemic arterial oxygen saturation than the patient with "true" TCPC. These patients may develop a progressive arterial oxygen desaturation during follow-up. If a significant cyanosis is established, a surgical redirection of the hepatic veins toward the pulmonary circulation allows restoring an acceptable systemic arterial saturation [28].

The various cavopulmonary connections are summarized in Fig. 11.6.

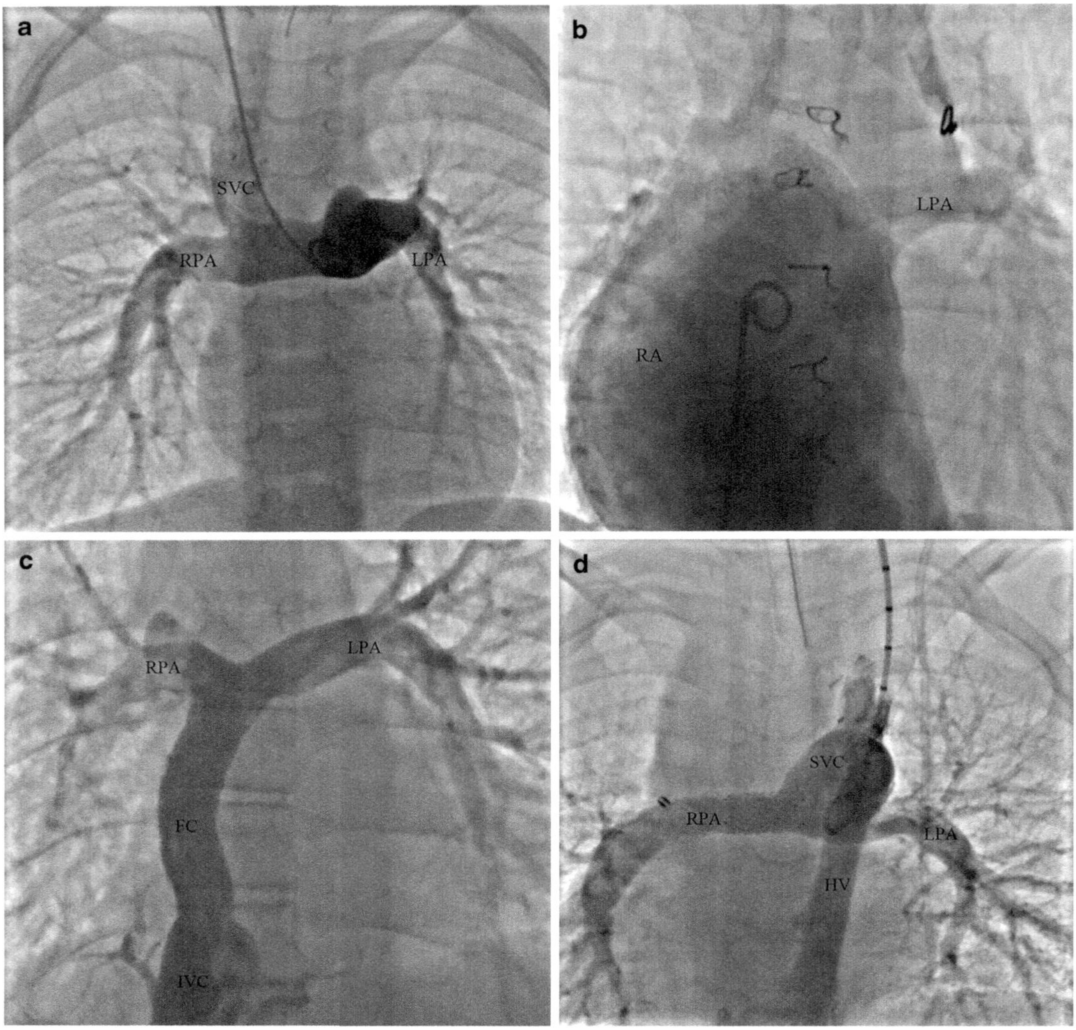

Fig. 11.6 Various types of cavopulmonary connection: superior cavopulmonary connection or bidirectional Glenn shunt or hemi-Fontan operation (**a**); Fontan operation with atriopulmonary connection technique (**b**); Fontan operation with extracardiac conduit (**c**); and Kawashima operation (**d**). Abbreviations: hemiazygos vein = HV; extracardiac Fontan conduit = FC; inferior vena cava = IVC; left pulmonary artery = LPA; right atrium = RA; right pulmonary artery = RPA; superior vena cava = SVC

11.4 Fontan Physiology

Normal heart is characterized by an in-series circulation wherein both the pulmonary and the systemic blood flows are surrounded by a ventricular pump. Otherwise, UVH is a congenital heart disease characterized by a single ventricular chamber supporting in-parallel both the pulmonary and the systemic circulation. The various palliative stages allow a progressive separation of the two circulations up to restore an in-series circulation. However, this "new" in-series circulation is not comparable to a normal heart's one due to the absence of a ventricular pump supporting the pulmonary vascular bed (Fig. 11.7). The absence of a subpulmonary ventricle leads to the paradox of a circulation characterized by high central venous pressure (CVP) and low pulmonary arterial pressure. This hemodynamic condition carries out a reduced preload for the single ventricle with con-

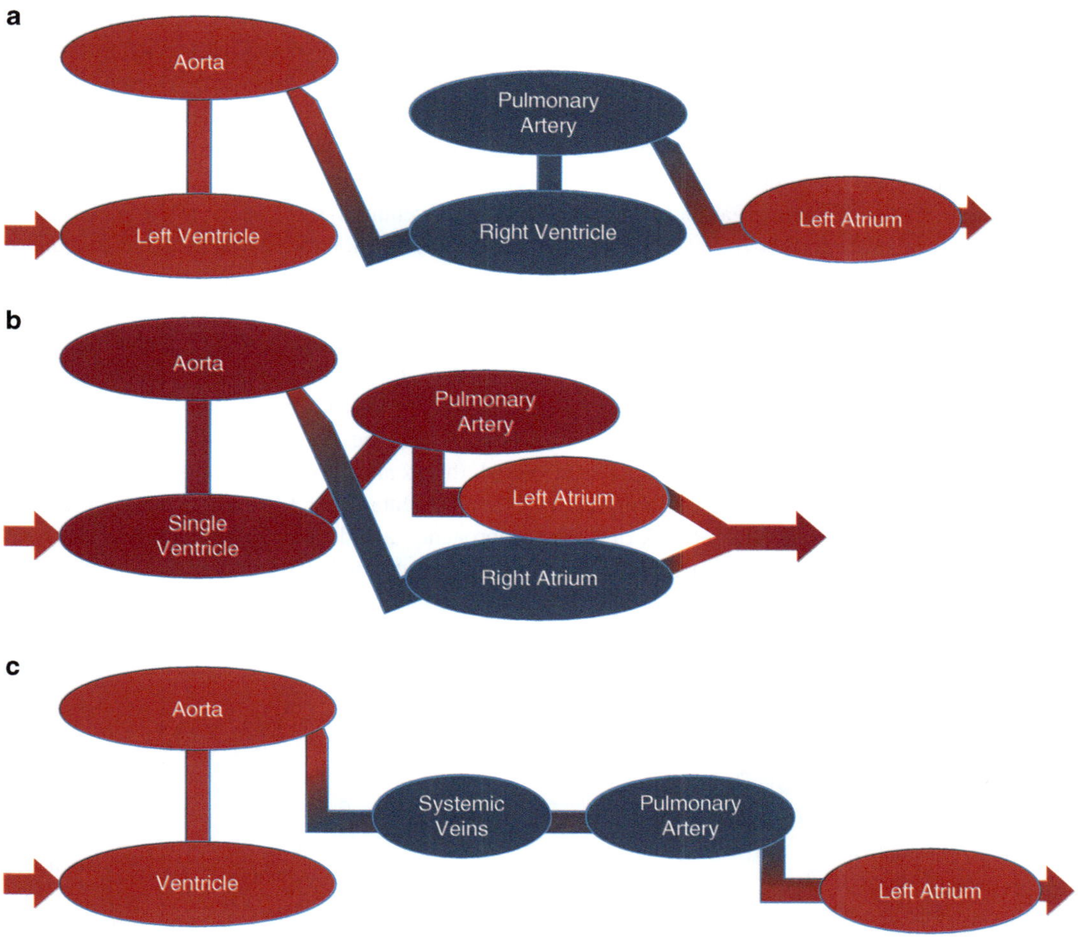

Fig. 11.7 Schematic picture of a normal in-series circulation (**a**), a univentricular heart in-parallel circulation (**b**), and a Fontan circuit in-series circulation (**c**). The colormap highlights the blood oxygen saturation: red (high oxygen saturation), purple (mixed oxygen saturation), and blue (low oxygen saturation)

sequent chronic low cardiac output. Thus, the pulmonary and systemic blood flows are mainly influenced by vascular resistance. Pulmonary blood flow has a direct inverse relationship with CVP. A high CVP is associated with lower pulmonary blood flow. Both precapillary (Fontan conduit and pulmonary artery stenosis or high arteriolar pulmonary vascular resistances) and postcapillary (diastolic/systolic ventricular dysfunction or atrioventricular valve regurgitation) factors influence pulmonary arterial pressure and blood flow. A reduced pulmonary blood flow leads to pulmonary ventilation-perfusion mismatch, lower systemic cardiac output, and systemic venous congestion. A low systemic perfusion pressure is the consequence of the low cardiac output and venous congestion. The "excellent Fontan" is characterized by: CVP $\leq$10 mmHg, arterial oxygen saturation $\geq$ 94%, cardiac index $\geq$2.6 l/min/m^2, end-diastolic ventricular volume index $\leq$70 ml/m^2, and single ventricle ejection fraction $\geq$55% [29]. Moreover, PVR index plays a crucial role for the optimal functioning of Fontan circuit. Egbe et al. [30] highlight that patients with elevated PVR index (>2 WU*m^2) showed an increased risk of Fontan failure, and this risk was even higher when a low cardiac index (<2.5 l/min/m^2) was associated. High mean pulmonary arterial pressure (>15 mmHg), high PVR index (>2 WU*m^2), pulmonary artery distortion, high pulmonary arterial wedge pressure (≥12 mmHg), significant atrioventricular valve regurgitation, and nonsinus rhythm are considered preoperative risk factors for poor prognosis after the Fontan procedure. In these cases, the creation of a Fontan conduit fenestration improves the cardiac output through a right-to-left shunt at the atrial level. However, the cost to pay is a lower systemic arterial oxygenation and a theoretical risk of paradoxical embolism due to the presence of the right-to-left shunt [31].

The lymphatic circulation is often a “forgotten” system, which undergoes several significant changes in patients with Fontan circuit. Lymphatic system is characterized by a fine network of lymphatic vessels that drain to the thoracic duct. The thoracic duct releases the lymph into the systemic venous system (left innominate vein). In the normal heart, both the pulmonary arterial pressure and pulmonary arterial wedge pressures are higher than the CVP, allowing an adequate pleural/bronchial lymphatic production and flow toward the systemic venous system. In the cases of SCPC and TCPC, the pulmonary arterial pressure is equal or lower (in presence of circuit stenosis) than the CVP, hindering the lymphatic drainage. There is a “paradox circuit” in which the lymph is forced to drain without or versus a positive pressure gradient [32, 33] (Fig. 11.8). This anomalous lymphatic circulation may be the physiopathological cause of chylothorax and plastic bronchitis in these patients. Novel percutaneous [34] and surgical [35, 36] approaches are developing to improve the lymphatic circulation in TCPC.

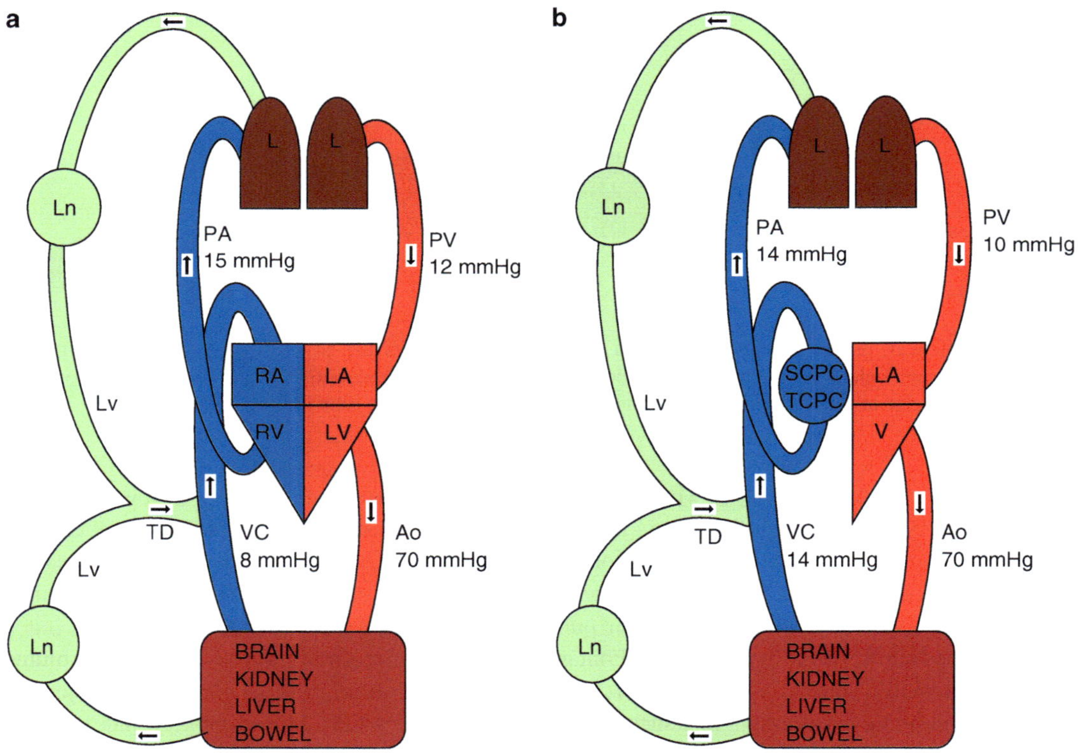

Fig. 11.8 Schematic picture of a blood and lymphatic circulation in a normal heart (**a**) and in a Fontan circuit (**b**). Abbreviations: aorta = Ao; left atrium = LA; left ventricle = LV; lymph node = Ln; lymphatic vessels: = Lv; lung = L; pulmonary artery = PA; pulmonary veins = PV; right atrium = RA; right ventricle = RV; superior cavopulmonary connection = SCPC; total cavopulmonary connection = TCPC; thoracic duct = TD; ventricle = V; vena cava = VC. The color-map highlights the blood oxygen saturation: red (high oxygen saturation), blue (low oxygen saturation). The values of pressure indicate the mean pressure of the single districts

11.5 Conclusion

UVH is a complex congenital heart disease that shows a broad spectrum of different anatomy and physiology. The knowledge of the single heart lesion that constitutes the "cardiopathy" allows to understand the below pathophysiology. The first-step palliation is usually surgical with the purpose to achieve an acceptable balance between the pulmonary and systemic blood flows. Arterial saturation is a useful noninvasive tool that allows having a description of pulmonary-systemic blood flow ratio. The final management of UVH palliation consists of a complete separation between the pulmonary and the systemic circulation. This result is achieved through other two palliation stages: the SCPC (6–12 months of life) and TCPC (5–6 years of life). The final TCPC allows dividing the pulmonary and the systemic circulation, at the price of a nonphysiological hemodynamic. The absence of ventricular pump surrounding the pulmonary blood flow exposes the patient to various risks of circuit dysfunction and thus a strict follow-up is always necessary.

References

1. Van Praagh R, Ongley PA, Swan HJ. Anatomic types of single or common ventricle in man: Morphologic and geometric aspects of 60 necropsied cases. Am J Cardiol. 1964;13:367–86.
2. Jacobs ML, Mayer JE Jr. Congenital heart surgery nomenclature and database project: single ventricle. Ann Thorac Surg. 2000;69(Suppl. 4):S197–204.
3. Anderson RH, Cook AC. Morphology of the functionally univentricular heart. Cardiol Young. 2004;14(Suppl 1):3–12.
4. Graham EM, Bradley SM, Atz AM. Preoperative management of hypoplastic left heart syndrome. Expert Opin Pharmacother. 2005;6(5):687–93.
5. Diller GP, Uebing A, Willson K, Davies LC, Dimopoulos K, Thorne SA, Gatzoulis MA, Francis DP. Analytical identification of ideal pulmonary systemic flow balance in patients with bidirectional cavo-pulmonary shunt and univentricular circulation: oxygen delivery or tissue oxygenation? Circulation. 2006;114:1243–50.
6. Davies RR, Pizarro C. Decision-making for surgery in the management of patients with univentricular heart. Front Pediatr. 2015;3:61.
7. Sasaki T, Takeda Y, Ohnakatomi Y, Asou T. Surgical approach for systemic-pulmonary shunt in neonates with functionally univentricular heart: comparison between sternotomy and thoracotomy. Gen Thorac Cardiovasc Surg. 2016;64(9):529–36.
8. Celebi A, Yucel IK, Bulut MO, Kucuk M, Balli S. Stenting of the ductus arteriosus in infants with functionally univentricular heart disease and ductal-dependent pulmonary blood flow: a single-center experience. Catheter Cardiovasc Interv. 2017;89(4):699–708.
9. Gesuete V, Fabi M, Bonvicini M. A 74-year-old unoperated univentricular heart: the oldest reported survival. Cardiol Young. 2016;26(4):805–7.
10. Alsoufi B, Manlhiot C, Ehrlich A, Oster M, Kogon B, Mahle WT, Maher K, McCrindle BW, Kanter K. Results of palliation with an initial pulmonary artery band in patients with single ventricle associated with unrestricted pulmonary blood flow. J Thorac Cardiovasc Surg. 2015;149(1):213–20.
11. Serraf A, Conte S, Lacour-Gayet F, Bruniaux J, Sousa-Uva M, Roussin R, Planché C. Systemic obstruction in univentricular hearts: surgical options for neonates. Ann Thorac Surg. 1995;60(4):970–6. discussion 976–7
12. Alsoufi B. Management of the single ventricle and potentially obstructive systemic ventricular outflow tract. J Saudi Heart Assoc. 2013;25(3):191–202.
13. Pass RH, Solowiejczyk DE, Quaegebeur JM, Liberman L, Altmann K, Gersony WM, Hordof AJ. Bulboventricular Foramen resection: hemodynamic and electrophysiologic results. Ann Thorac Surg. 2001;71(4):1251–4.
14. Ohye RG, Ludomirsky A, Devaney EJ, Bove EL. Comparison of right ventricle to pulmonary artery conduit and modified Blalock-Taussig shunt hemodynamics after the Norwood operation. Ann Thorac Surg. 2004;78:1090–3.
15. Michel-Behnke I, Akintürk H, Schranz D. Fate of the stented arterial duct. Circulation. 2000;102:E178.
16. Galantowicz M, Cheatham JP, Phillips A, Cua CL, Hoffman TM, Hill SL, Rodeman R. Hybrid approach for hypoplastic left heart syndrome: intermediate results after the learning curve. Ann Thorac Surg. 2008;85:2063–70.
17. Dave H, Rosser B, Knirsch W, Hubler M, Prêtre R, Kretschmar O. Hybrid approach for hypoplastic left heart syndrome and its variants: the fate of the pulmonary arteries. Eur J Cardiothorac Surg. 2014;46(1):14–9.
18. Matitiau A, Geva T, ColanSD ST, Parness IA, Spevak PJ, Van Der Velde M, Mayer JE Jr, Sanders SP. Bulboventricular Foramen size in infants with double-inlet left ventricle or tricuspid atresia with transposed great arteries: influence on initial palliative operation and rate of growth. J Am Coll Cardiol. 1992;19(1):142–8.
19. Teitel DF, Iwamoto HS, Rudolph AM. Changes in the pulmonary circulation during birth-related events. Pediatr Res. 1990;27:372–8.
20. Senzaki H, Masutani S, Kobayashi J, Kobayashi T, Sasaki N, Asano H, Kyo S, Yokote Y, Ishizawa

A. Ventricular afterload and ventricular work in Fontan circulation: comparison with normal two-ventricle circulation and single-ventricle circulation with Blalock-Taussig shunts. Circulation. 2002;105:2885–92.
21. Choussat A, Fontan F, Besse P. Patient selection for the Fontan procedure. In: Anderson R, Shinebourne E, editors. Paediatric cardiology. Edinburgh, New York: Churchill Livingstone; 1978. p. 559–66. New York: distributed by Longman.
22. Nakanishi T. Cardiac catheterization is necessary before bidirectional Glenn and Fontan procedures in single ventricle physiology. Pediatr Cardiol. 2005;26(2):159–61.
23. Fontan F, Baudet E. Surgical repair of tricuspid atresia. Thorax. 1971;26(3):240–8.
24. de Leval MR, Kilner P, Gewillig M, Bull C. Total cavopulmonary connection: a logical alternative to atriopulmonary connection for complex Fontan operations. Experimental studies and early clinical experience. J Thorac Cardiovasc Surg. 1988;96(5):682–95.
25. Marcelletti C, Corno A, Giannico S, Marino B. Inferior vena cava-pulmonary artery extracardiac conduit. A new form of right heart bypass. J Thorac Cardiovasc Surg. 1990;100(2):228–32.
26. de Leval MR, Deanfield JE. Four decades of Fontan palliation. Nat Rev Cardiol. 2010;7(9):520–7.
27. Kawashima Y, Kitamura S, Matsuda H, Shimazaki Y, Nakano S, Hirose H. Total cavopulmonary shunt operation in complex cardiac anomalies. A new operation. J Thorac Cardiovasc Surg. 1984;87(1):74–81.
28. Larsen SH, Emmertsen K, Bjerre J, Hjortdal VE. Progressive cyanosis following Kawashima operation: slow resolution after redirection of hepatic veins. J Cardiothorac Surg. 2013;8:67.
29. Ohuchi H. Where Is the "Optimal" Fontan hemodynamics? Korean Circ J. 2017;47(6):842–57.
30. Egbe AC, Connolly HM, Miranda WR, Ammash NM, Hagler DJ, Veldtman GR, Borlaug BA. Hemodynamics of Fontan failure: the role of pulmonary vascular disease. Circ Heart Fail. 2017;10(12):e004515.
31. Lemler MS, Scott WA, Leonard SR, Stromberg D, Ramaciotti C. Fenestration improves clinical outcome of the fontan procedure: a prospective, randomized study. Circulation. 2002;105(2):207–12.
32. Kreutzer C, Kreutzer G. The lymphatic system: the achilles heel of the Fontan-Kreutzer circulation. World J Pediatr Congenit Heart Surg. 2017;8(5):613–23.
33. Kreutzer J, Kreutzer C. Lymphodynamics in congenital heart disease: the forgotten circulation. J Am Coll Cardiol. 2017;69(19):2423–7.
34. Dori Y, Keller MS, Rome JJ, Gillespie MJ, Glatz AC, Dodds K, Goldberg DJ, Goldfarb S, Rychik J, Itkin M. Percutaneous lymphatic embolization of abnormal pulmonary lymphatic flow as treatment of plastic bronchitis in patients with congenital heart disease. Circulation. 2016;133(12):1160–70.
35. Hraska V. Decompression of thoracic duct: new approach for the treatment of failing Fontan. Ann Thorac Surg. 2013;96(2):709–11.
36. Antonio M, Gordo A, Pereira C, Pinto F, Fragata I, FragataJ. Thoracic duct decompression for protein-losing enteropathy in failing Fontan circulation. Ann Thorac Surg. 2016;101(6):2370–3.

Fontan Surgery and Fluid Dynamics

12

Ethan Kung and Alison Marsden

12.1 Engineering Applications

12.1.1 Relevant Methods

12.1.1.1 Lumped-Parameter Physiology Modeling

Low-order models are simplified representations of the cardiovascular system often used to capture the systemic behaviors of the entire or a large portion of the circulation. Such an approach offers a "bird's eye view" of the system where the approximated behavior of each subcomponent can interact and produce the overall systemic behavior. A common type of low-order model for the cardiovascular system is the 0-dimensional lumped-parameter network (LPN) typically represented in the form of an electrical circuit [1, 2]. This type of model contains no spatial information, but only the structure of the vascular network represented as the subcomponent organization in the circuit.

The passive components in this type of model typically consist of the resistance element which describes the viscous dissipation of pressure energy as blood flows through vasculature, the capacitance element which captures the effects of blood vessel compliance, the inductance element which describes how the momentum of blood resists changes in flow, flow-dependent resistances which describe viscous dissipation in non-laminar flows, and diodes which describe the function of valves in the body.

The active components of the LPN generate pressures to drive blood through the passive components. The main active components are the heart chambers which are often modeled using time-varying elastance functions [3, 4] or passive and active pressure-volume relationships and activation functions [5]. These approaches allow the simulation to capture effects of preload on cardiac output due to the Frank-Starling mechanism. Pressures generated by other active components are selectively included in each specific model depending on their importance to the topic being investigated. These can include respiration, muscle pumps, and medical devices such as blood pumps.

A closed-loop LPN model serves the purpose of describing the entire circulation and often includes sections that describe the heart, upper and

E. Kung (✉)
Dept. of Mechanical Engineering, Clemson University, Clemson, SC, USA

Dept. of Bioengineering, Clemson University, Clemson, SC, USA
e-mail: ekung@clemson.edu

A. Marsden
Department of Pediatrics—Cardiology, Stanford University, Stanford, CA, USA
e-mail: amarsden@stanford.edu

G. Butera et al. (eds.), *Modelling Congenital Heart Disease*,
https://doi.org/10.1007/978-3-030-88892-3_12

lower body vasculatures with venous valves, pulmonary vasculature, and vascular beds in important abdominal organs such as kidneys, liver, and intestine (Fig. 12.1a). To create a patient-specific model, the LPN component values are iteratively tuned such that the circuit reproduces the patient-specific clinical data, which may include time-varying blood flow and pressure measurements at various locations as well as heart chamber volumes (Fig. 12.1a) [5–8].

12.1.1.2 Multiscale Computational Fluid Dynamics Modeling

In scenarios where detailed patient-specific anatomy plays an important factor in the resulting hemodynamics, three-dimensional models of the patient anatomy can be reconstructed from 3D imaging data (Fig. 12.1b) using open-source [9] or commercially available (i.e., Materialise Mimics) software via a process of region-growing and segmentation [9–11]. Such models are useful for performing computational "virtual surgeries" under the collaborative guidance of cardiac surgeons and engineers to obtain reconstructed geometries for testing different potential options of the surgical procedure. This approach has been used to study patient-specific single ventricle physiology and predict hemodynamic changes resulting from various surgical approaches [5, 11–14].

The geometry to be tested can be discretized into a large number of entities (elements or control volumes) [15, 16] via meshing software (i.e., TetGen [17], MESHSIM, Simmetrix Inc., New York). The resulting mesh then serves as the domain for conducting finite volume or finite element analyses using the mass conservation and incompressible Navier-Stokes equations to compute blood flow velocity and pressure inside the domain; this is referred to as computational fluid dynamics (CFD) modeling. The CFD domain can then be coupled to a closed-loop LPN via boundary conditions by exchanging information of flow and pressure between the domains at every simulation timestep, often employing implicit coupling and outflow stabilization [18]. This forms a multiscale simulation (Fig. 12.2a) capable of capturing both the detailed 3-dimensional local hemodynamics as well as systemic physiologic responses.

12.1.1.3 In Vitro and Hybrid Modeling

In Vitro mock circulation loops offer the advantage of directly testing medical devices and investigating physical phenomena that are not

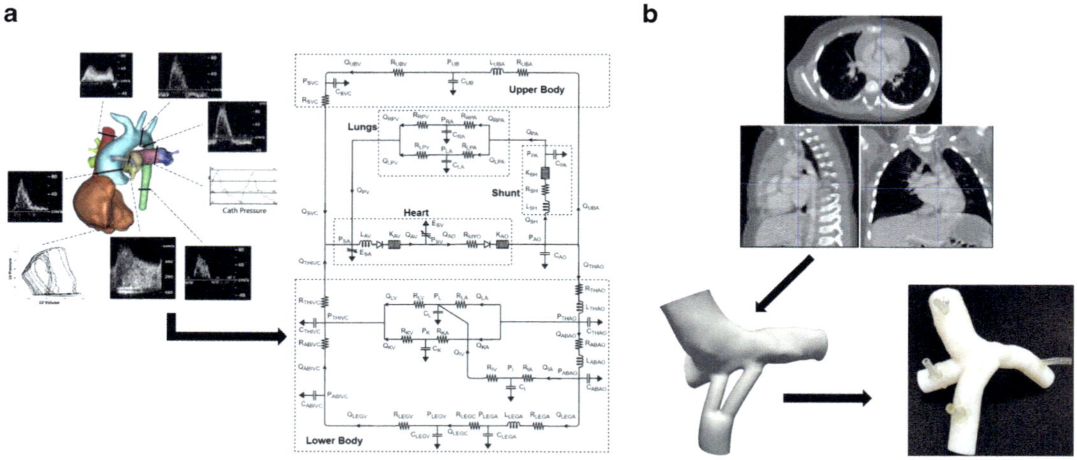

Fig. 12.1 (**a**) Clinical measurements of pressure, flow, and heart volume provide useful data for tuning a patient-specific lump-parameter network physiology model; (**b**) Clinical 3D imaging data can be used to construct a patient-specific 3D anatomical model, and a corresponding physical phantom for flow experiments

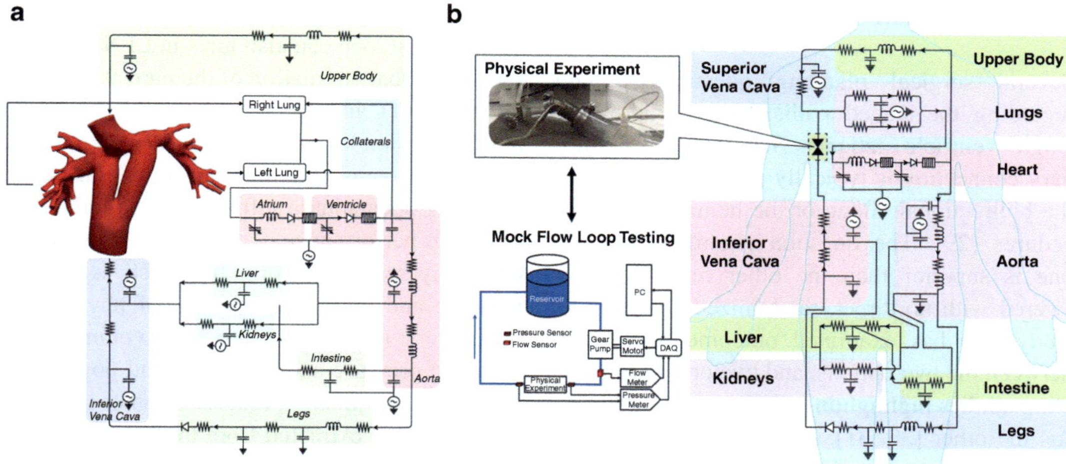

Fig. 12.2 (**a**) A multiscale computational fluid dynamics simulation serves to test a 12 mm Y-graft in a virtual surgery for the stage 3 Fontan procedure; (**b**) A Physiology Simulation Coupled Experiment [27] serves to test a Jarvik 2000 blood pump implemented for Fontan cavopulmonary support; such a hybrid model directly couples a medical device bench test with a computational simulation of patient physiology in closed-loop

fully captured by computational models. In vitro setups need to mimic in vivo conditions in various aspects depending on the relevant dynamics to be investigated. An accurate physical reconstruction of the patient-specific anatomy (Fig. 12.1b) with which flow experiments can be performed is referred to as a "phantom;" this can be created using 3D printing or other methods such as casting [19, 20]. Precisely controlled programmable actuators can replicate realistic cardiovascular flows [19–21] and accurate in vivo pressures can then be achieved by well-designed impedance modules [19, 20, 22]. A working fluid consisting of 40% glycerol solution possesses density and viscosity that are similar to those of blood and hence exhibits realistic hemodynamic behaviors. Such physiologically realistic benchtop experimental environments are valuable for testing the performance of medical devices and to predict the results of surgical procedures. Physical testing in these setups can provide important information toward the design of devices and translation of novel surgical concepts before placing real patients at risk during clinical trials. For example, similar experimental techniques have been applied to study medical devices such as prosthetic heart valves [23, 24] and blood pumps [25, 26].

A hybrid model can directly combine an in vitro experiment with an LPN physiology simulation using the hardware-in-the-loop concept (Fig. 12.2b), capitalizing on the strengths of both in vitro and numerical platforms in a "Physiology Simulation Coupled Experiment" [27]. This can be achieved with high fidelity using an iterative coupling method to exchange information between the two domains [27, 28]. The main strength of the hybrid model is the ability to experimentally investigate a scenario that is challenging for numerical simulations (i.e., fluid-structure interactions), and fully couple the experimental results with a physiology simulation (which is challenging to produce experimentally). Since the bench top physical experiment is coupled directly in-the-loop with a computational physiology simulation, it operates in a dynamically changing feedback environment with the physiology simulation. This allows one to obtain relevant information such as device operation and physiologic impacts as if performing an *in-vivo* human experiment. Both computational and experimental platforms are within a single hybrid framework, enabling the physical testing of medical devices or surgical procedures with simulated physiologic feedback.

12.1.2 Surgical Planning

Several surgical planning questions remain regarding the staged palliation procedures for single ventricle patients. Superior cavopulmonary connection is typically performed as either the bidirectional Glenn or the hemi-Fontan procedures [29]. The question regarding whether one is superior than the other remains unanswered with no direct, randomized comparative study that has examined outcome differences between the two options, and two opposing modeling studies highlighting superiority of one versus the other [30, 31]. Also, various degrees of left pulmonary stenosis are common among these patients; surgical planning for the stage 2 procedure needs to determine whether to perform the additional surgical steps to remove any existing stenosis. For the stage 3 total cavopulmonary connection procedure, several grafts types and surgical configurations have been considered, including the traditional extracardiac conduit and the 9 mm [32] and 12 mm [33] Y-graft at various anastomosis locations and angles. The surgical reconstruction aims to reduce power loss and improve distribution of inferior venous flow to the right and left pulmonary circulations, which would balance hepatic factors to both lungs [32–34]. It remains unclear to what degree these outcomes are achieved with the various surgical approaches and whether power loss of the surgical junction equates to improved performance of the cavopulmonary circulation [35].

Using multiscale CFD modeling and virtual surgeries, investigators can examine the predicted hemodynamic differences between different surgical options as well as the physiologic impacts of left pulmonary stenoses for a specific patient. For example, from 3D imaging data of a patient, one can construct a 3D computational model that is a representation of the patient's preoperative anatomy, then perform "virtual" Glenn or Hemi-Fontan surgeries (Fig. 12.3a), "virtual" Fontan surgery with different types of grafts (Fig. 12.3b), or adjust a "virtual" left pulmonary stenosis of various severities (Fig. 12.3c) by computationally modifying the preoperative anatomic geometry. Using the preoperative clinical data from the same patient, one can also tune an LPN to represent the global circulation of the patient and couple the CFD and LPN models to complete a multiscale simulation setup (Fig. 12.2a) [5, 7, 11, 36]. Such a coupled multiscale simulation can be run using open-source software Simvascular [37] on high-performance computing equipment. Clinically relevant information such as the 3D pressure and flow, time tracings of physiologic parameters including ventricular volumes and systemic pressure and flowrates, and power loss in the surgical junction relative to other systemic powers can be extracted from the CFD and LPN parts of the multiscale simulation results. These results can reveal insights regarding the local and global hemodynamic differences between surgical options [5, 11, 38] or the physiological effects of various degrees of pulmonary stenosis [39] on a patient-specific basis. For example, a virtual stenosis study revealed that for two specific patients, a 50% or 85% area stenosis does not produce a significant impact to the global physiology due to the fact that the power loss in the surgical junction is relatively small compared to the systemic power losses [5].

12.1.3 New Device Testing and Evaluation

The "Fontan paradox" characterized by pulmonary arterial hypotension concurrent with central venous hypertension persists in single ventricle patients even after the staged palliation procedures [40–42]. Chronic high-caval pressure is a main determinant of many Fontan-related diseases and circulatory failure [43]. A significant number of Fontan patients will eventually develop circulatory failure as a result of the poor hemodynamics following the Fontan procedure [40, 44]. Most failing Fontan patients are unable to receive heart transplantation either due to being poor candidates as a result of several previous surgeries and multiple organ dysfunctions, or due to the unavailability of matching donor organs. Therefore, the use of a blood pump for

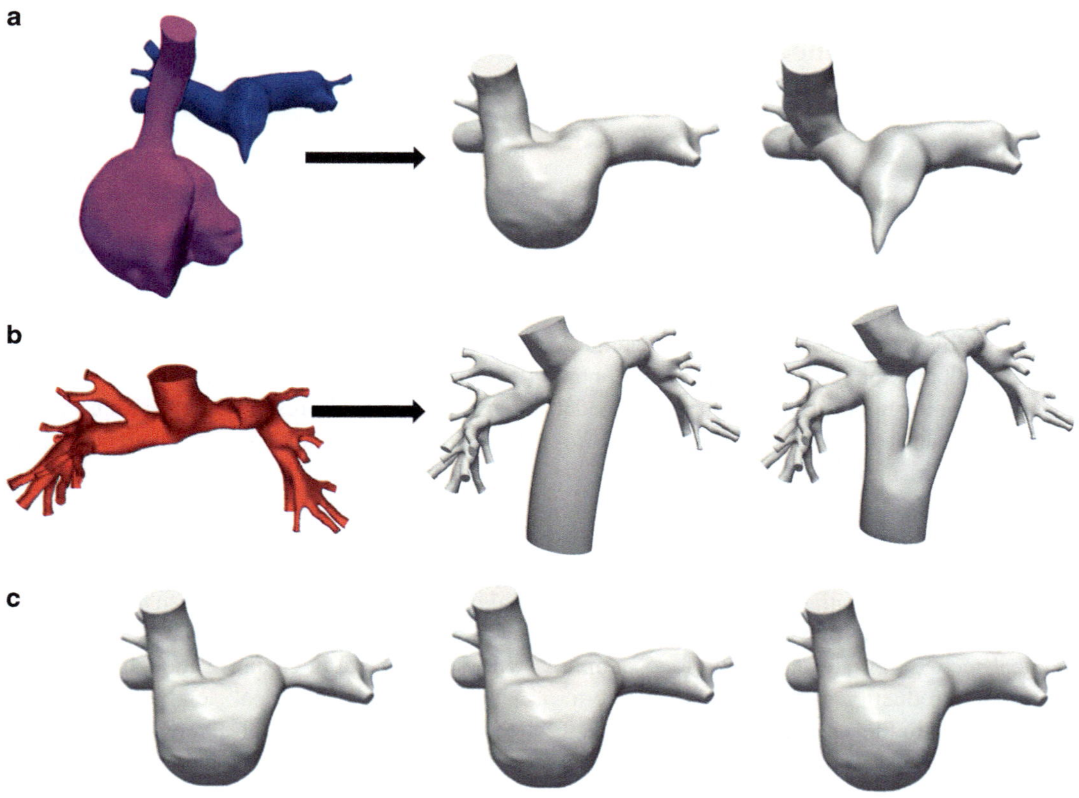

Fig. 12.3 Virtual surgery can investigate different surgical options based on preoperative anatomy: (**a**) the hemi-Fontan and bidirectional Glenn surgical options for the stage 2 superior cavopulmonary connection, (**b**) a traditional extracardiac conduit versus a Y-graft for the stage 3 Fontan surgery, and (**c**) virtual construction of a left pulmonary stenosis of various severities

cavopulmonary support [45–47] has been proposed for treatment of Fontan failure. However, currently there is no existing commercially available device designed for this clinical application.

Combinations of in-vitro, LPN, and hybrid modeling can offer methods to investigate new device prototypes or the off-label use of existing devices for cavopulmonary support. For example, one can use in-vitro experiments to identify the pressure-flow characteristics of a device, then substitute a corresponding mathematical surrogate into an LPN physiology model to examine the physiologic impacts of device implantation [28, 48, 49]. Alternately, a hybrid model can directly test a blood pump in a physical setup while simultaneously fully coupling the device interactions with a LPN simulation of physiology to study the relevant physiologic responses (Fig. 12.2b) [27]. This would also allow the investigator to examine how device operation is affected by physiologic changes (i.e., by measuring device power consumption fluctuations) and the interactions between device settings and patient physiology. For example, a hybrid modeling study revealed that for the cavopulmonary support application, an off-the-shelf Jarvik 2000 blood pump would create dangerously high pulmonary pressures and potential venous collapse, while one with a modified controller to achieve a 5000RPM pump speed can improve central venous pressure, ventricular preload, and cardiac output [27].

12.2 Potential Impact on Clinical Arena (Engineer's Perspective)

12.2.1 Now

Current state of the art modeling methods is capable of estimating the relative hemodynamic and physiologic outcomes between different clinical scenarios of interest, for example, the overall trends of physiologic changes from implementing a blood pump of various strengths for Fontan cavopulmonary support [48]. This type of model application can help provide concrete, quantitative reasons for supporting and establishing clinical protocols as well as decision-making. Furthermore, the development of novel medical devices and exploration of new surgical concepts can be investigated with zero risk to patients. In contrast with the traditional trial-and-error-based medical technology advancement (for example, the initial introduction of the stage 1 systemic-to-pulmonary shunting procedure had very high mortality rate due to pulmonary over circulation), the use of modeling tools can now test new palliation approaches and improve them to safer levels before they are performed in humans. Noninvasive clinical measurement techniques enabled by modeling [50] also reduce the risk and cost of patient assessment, increasing patient screening, and the early detection/prevention of diseases. As the modeling of heart disease continues to demonstrate its clinical value, the synergy between clinicians and engineers is fortified, encouraging increased collaboration and the important clinical data collection necessary for accurate model construction [36]. The training of next generation physicians can also make use of modeling tools [51] to increase exposure to realistic clinical scenarios and develop clinical intuition via hands-on learning under controlled, simulated scenarios. As clinicians become aware of the modeling tools available, more and more clinical problems that can be solved or improved via powerful modeling approaches are increasingly being identified.

12.2.2 Going Forward

Going forward, the next steps in technological advancements of modeling methods will incorporate physiologic changes during surgery [52] as well as postsurgical remodeling [53]. This will be necessary to drastically improve the accuracy of model predictions and therefore its utility for patient-specific treatment planning. The ability to accurately predict long-term therapeutic outcomes will not only enable the optimization of treatments, but also of the comprehensive patient management and resource allocation plans. Furthermore, the analysis of patient-specific models can help push fundamental understanding regarding the etiology and progression of pathologic conditions [54]. The development of a multiscale modeling method capable of accurately linking biomechanical factors with thrombus formation is an important research direction that will enable the application of modeling tools in answering many critical clinical questions (Fig. 12.4) [55]. There will likely be a paradigm shift regarding the process of developing new medical devices and therapies, as much of the design and testing can be based on modeling (Fig. 12.5), resulting in accelerated turnaround times as well as reduced animal and clinical trials necessary.

In order to realize a wide clinical translation of engineering modeling tools, the biggest need of the immediate future is the validation of modeling predictions directly against clinical data. This is the crucial step necessary to establish confidence in the models and subsequently buy-in from clinicians to use them in clinical practice. Next, logistical studies as well as improvements in human-machine interface are necessary to facilitate incorporating the use of modeling tools into the often complex clinical workflow. Lastly, a coherent, synergized clinician-engineer operation will mean that the standard-of-care will evolve to include the collection of clinical data needed for high-quality model construction, enabling a modeling-based healthcare approach.

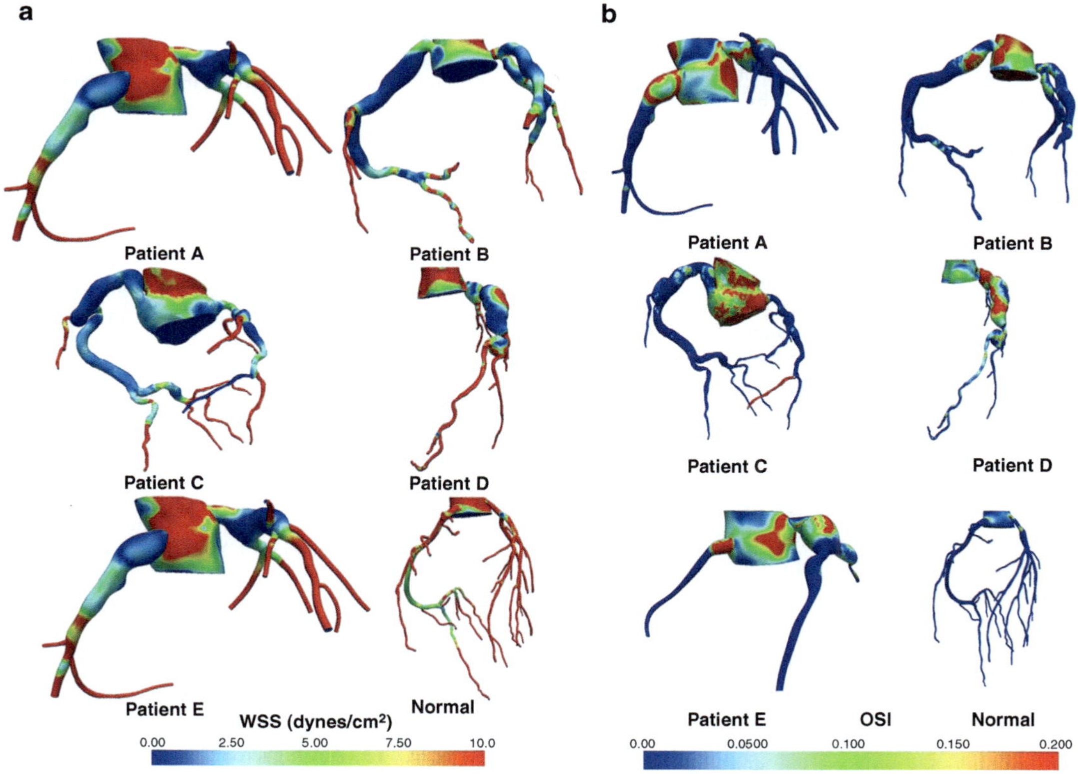

Fig. 12.4 Coronary aneurysm hemodynamics in five patients compared to a normal reference (modified with permission from [55]). Current assessment of thrombotic risk using modeling results typically examines (**a**) wall shear stress and (**b**) oscillatory shear index as surrogate quantities relating to thrombosis. An accurate multiscale method that can directly couple biomechanical factors with thrombus formation is a critical research direction forward

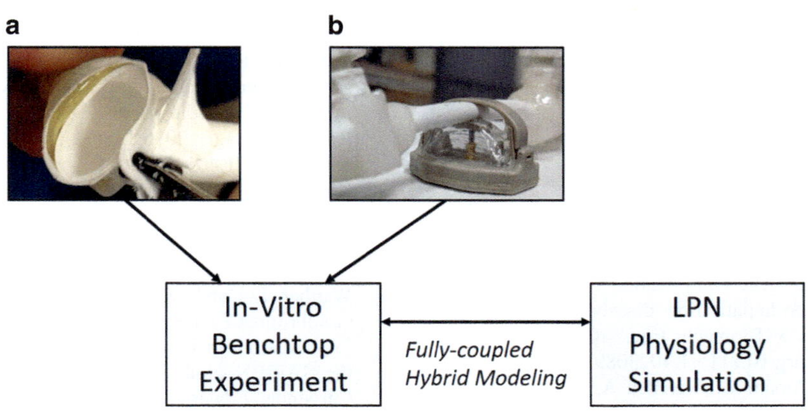

Fig. 12.5 (**a**) The development of a novel external-compression-based device prototype for Fontan cavopulmonary support [56], (**b**) the redesign of a wireless banding device (FloWatch© [57–60]) for single ventricle systemic-to-pulmonary shunt banding, can both make use of (**c**) in-vitro benchtop testing for device characterization, and (**d**) fully coupled hybrid modeling to investigate the closed-loop interactions between the device and patient physiologic responses

References

1. Snyder MF, Rideout VC. Computer simulation studies of the venous circulation. IEEE Trans Biomed Eng. 1969;16(4):325–34.
2. Kung E, Pennati G, Migliavacca F, Hsia T-Y, Figliola R, Marsden A, et al. A simulation protocol for exercise physiology in Fontan patients using a closed loop lumped-parameter model. J Biomech Eng. 2014;136(8):081007. https://doi.org/10.1115/1.4027271.
3. Suga H, Sagawa K, Shoukas AA. Load independence of the instantaneous pressure-volume ratio of the canine left ventricle and effects of epinephrine and heart rate on the ratio. Circ Res. 1973;32(3):314–22.
4. Senzaki H, Chen CH, Kass DA. Single-beat estimation of end-systolic pressure-volume relation in humans: a new method with the potential for noninvasive application. Circulation. 1996;94(10):2497–506.
5. Kung E, Baretta A, Baker C, Arbia G, Biglino G, Corsini C, et al. Predictive modeling of the virtual Hemi-Fontan operation for second stage single ventricle palliation: two patient-specific cases. J Biomech. 2013;46(2):423–9.
6. Arbia G, Corsini C, Baker C, Pennati G, Hsia TY, Vignon-Clementel IE. Pulmonary hemodynamics simulations before stage 2 single ventricle surgery: patient-specific parameter identification and clinical data assessment. Cardiovasc Eng Technol. 2015;6(3):268–80. https://doi.org/10.1007/s13239-015-0212-3.
7. Corsini C, Baker C, Kung E, Schievano S, Arbia G, Baretta A, et al. An integrated approach to patient-specific predictive modeling for single ventricle heart palliation. Comput Methods Biomech Biomed Eng. 2014;17(14):1572–89. https://doi.org/10.1080/10255842.2012.758254.
8. Tran JS, Schiavazzi DE, Ramachandra AB, Kahn AM, Marsden AL. Automated tuning for parameter identification and uncertainty quantification in multiscale coronary simulations. Comput Fluids. 2017 Jan;5(142):128–38.
9. Yushkevich PA, Piven J, Hazlett HC, Smith RG, Ho S, Gee JC, et al. User-guided 3D active contour segmentation of anatomical structures: significantly improved efficiency and reliability. NeuroImage. 2006;31(3):1116–28.
10. Yang W, Feinstein JA, Shadden SC, Vignon-Clementel IE, Marsden AL. Optimization of a Y-graft design for improved hepatic flow distribution in the Fontan circulation. J Biomech Eng. 2012;135(1):011002. https://doi.org/10.1115/1.4023089.
11. Kung E, Corsini C, Marsden A, Vignon-Clementel I, Pennati G, Figliola R, et al. Multiscale modeling of superior cavopulmonary circulation: hemi-Fontan and bidirectional Glenn are equivalent. Semin Thorac Cardiovasc Surg. 2019; Available from: https://linkinghub.elsevier.com/retrieve/pii/S104306791930276X. [cited 2019 Sep 12]
12. Pekkan K, Whited B, Kanter K, Sharma S, Zelicourt D, Sundareswaran K, et al. Patient-specific surgical planning and hemodynamic computational fluid dynamics optimization through free-form haptic anatomy editing tool (SURGEM). Med Biol Eng Comput. 2008;46(11):1139–52.
13. Sundareswaran KS, de Zélicourt D, Sharma S, Kanter KR, Spray TL, Rossignac J, et al. Correction of pulmonary arteriovenous malformation using image-based surgical planning. JACC Cardiovasc Imaging. 2009;2:1024–30.
14. Vignon-Clementel IE, Marsden AL, Feinstein JA. A primer on computational simulation in congenital heart disease for the clinician. Prog Pediatr Cardiol. 2010 Dec;30(1–2):3–13.
15. Krause E. Computational fluid dynamics: its present status and future direction. Comput Fluids. 1985;13(3):239–69.
16. Peskin CS. Flow patterns around heart valves: a numerical method. J Comput Phys. 1972;10(2):252–71.
17. Si H. TetGen, a delaunay-based quality tetrahedral mesh generator. ACM Trans Math Softw. 2015;41(2):1–36.
18. Esmaily Moghadam M, Bazilevs Y, Hsia TY, Vignon-Clementel IE, Marsden AL. A comparison of outlet boundary treatments for prevention of backflow divergence with relevance to blood flow simulations. In: Computational mechanics. Springer-Verlag; 2011. p. 277–91.
19. Kung EO, Les AS, Figueroa CA, Medina F, Arcaute K, Wicker RB, et al. In vitro validation of finite element analysis of blood flow in deformable models. Ann Biomed Eng. 2011;39(7):1947–60.
20. Kung EO, Les AS, Medina F, Wicker RB, McConnell MV, Taylor CA. In vitro validation of finite-element model of AAA hemodynamics incorporating realistic outlet boundary conditions. J Biomech Eng. 2011;133(4):041003. Available from: http://biomechanical.asmedigitalcollection.asme.org/article.aspx?articleid=1421454
21. Mechoor RR, Schmidt T, Kung E. A real-time programmable pulsatile flow pump for in vitro cardiovascular experimentation. J Biomech Eng. 2016;138(11):111002. https://doi.org/10.1115/1.4034561.
22. Kung EO, Taylor CA. Development of a physical windkessel module to re-create in vivo vascular flow impedance for in vitro experiments. Cardiovasc Eng Technol. 2011;2(1):2–14.
23. Walther T, Dehdashtian MM, Khanna R, Young E, Goldbrunner PJ, Lee W. Trans-catheter valve-in-valve implantation: in vitro hydrodynamic performance of the SAPIEN+cloth trans-catheter heart valve in the Carpentier-Edwards Perimount valves. Eur J Cardio-Thoracic Surg. 2011;40(5):1120–6. https://doi.org/10.1016/j.ejcts.2011.02.056.
24. Barbara V, Daniele C, Grigioni M, Palombo A, Sargentini A. In-vitro evaluation of prosthetic heart valves: towards comparable testing. J

Med Eng Technol. 1992;16(1):10–4. https://doi.org/10.3109/03091909209021951.
25. Spurlock DJ, Raney DN, Fracz EM, Mazur DE, Bartlet RH, Haft JW. In vitro testing of a novel blood pump designed for temporary extracorporeal support. ASAIO J. 2012;58(2):1. Available from: http://www.ncbi.nlm.nih.gov/pubmed/22236624
26. Mizuguchi K, Kitamura S, Damm G, Benkowsky R, Glueck J, Aber G, et al. In vitro hemolysis test method for developing an axial flow ventricular assist device. In: Heart replacement. Tokyo, Japan: Springer; 1996. p. 237–43. https://doi.org/10.1007/978-4-431-67020-9_32.
27. Kung E, Farahmand M, Gupta A. A hybrid experimental-computational modeling framework for cardiovascular device testing. J Biomech Eng. 2019;141:1–8.
28. Mirzaei E, Farahmand M, Kung E. An algorithm for coupling multi-branch in-vitro experiment to numerical physiology simulation for a hybrid cardiovascular model. Int J Numer Method Biomed Eng. 2019:e3289. https://doi.org/10.1002/cnm.3289.
29. Norwood WI. Hypoplastic left heart syndrome. Ann Thorac Surg. 1991;52(3):688–95.
30. Bove EL, de Leval MR, Migliavacca F, Guadagni G, Dubini G. Computational fluid dynamics in the evaluation of hemodynamic performance of cavopulmonary connections after the Norwood procedure for hypoplastic left heart syndrome. J Thorac Cardiovasc Surg. 2003;126(4):1040–7. [cited 2019 Nov 28]. Available from: http://www.ncbi.nlm.nih.gov/pubmed/14566244
31. Pekkan K, Dasi LP, De Zélicourt D, Sundareswaran KS, Fogel MA, Kanter KR, et al. Hemodynamic performance of stage-2 univentricular reconstruction: Glenn vs. Hemi-fontan templates. Ann Biomed Eng. 2009 Jan;37(1):50–63.
32. Haggerty CM, Kanter KR, Restrepo M, De Zélicourt DA, Parks WJ, Rossignac J, et al. Simulating hemodynamics of the Fontan Y-graft based on patient-specific in vivo connections. J Thorac Cardiovasc Surg. 2013 Mar;145(3):663–70.
33. Marsden AL, Bernstein AJ, Reddy VM, Shadden SC, Spilker RL, Chan FP, et al. Evaluation of a novel Y-shaped extracardiac Fontan baffle using computational fluid dynamics. J Thorac Cardiovasc Surg. 2009;137(2)
34. Yang W, Feinstein JA, Shadden SC, Vignon-Clementel IE, Marsden AL. Optimization of a Y-Graft design for improved hepatic flow distribution in the fontan circulation. J Biomech Eng. 2013;135(1):011002.
35. Hsia TY. Taming the Fontan with the Y-graft: a nod and a wink to the great Yu. J Thorac Cardiovasc Surg. 2016;151(6):1537–9. https://doi.org/10.1016/j.jtcvs.2016.03.027.
36. Corsini C, Baker C, Baretta A, Biglino G, Hlavacek AM, Hsia TY, et al. Integration of clinical data collected at different times for virtual surgery in single ventricle patients: a case study. Ann Biomed Eng. 2015;43(6):1310–20.
37. Updegrove A, Wilson NM, Merkow J, Lan H, Marsden AL, Shadden SC. SimVascular: an open-source pipeline for cardiovascular simulation. Ann Biomed Eng. 2017;45(3):525–41. Springer, New York LLC
38. Haggerty CM, de Zélicourt DA, Restrepo M, Rossignac J, Spray TL, Kanter KR, et al. Comparing pre- and post-operative Fontan hemodynamic simulations: implications for the reliability of surgical planning. Ann Biomed Eng. 2012;40(12):2639–51.
39. Schiavazzi DE, Kung EO, Marsden AL, Baker C, Pennati G, Hsia T-Y, et al. Hemodynamic effects of left pulmonary artery stenosis after superior cavopulmonary connection: a patient-specific multiscale modeling study. J Thorac Cardiovasc Surg. 2015;149(3):689–96. e3. Available from: http://linkinghub.elsevier.com/retrieve/pii/S0022522314021849
40. Khairy P, Fernandes SM, Mayer JE, Triedman JK, Walsh EP, Lock JE, et al. Long-term survival, modes of death, and predictors of mortality in patients with Fontan surgery. Circulation. 2008;117(1):85–92. Available from: http://www.ncbi.nlm.nih.gov/pubmed/18071068
41. Hebson CL, McCabe NM, Elder RW, Mahle WT, McConnell M, Kogon BE, et al. Hemodynamic phenotype of the failing Fontan in an adult population. Am J Cardiol. 2013;112(12):1943–7.
42. Egbe AC, Connolly HM, Miranda WR, Ammash NM, Hagler DJ, Veldtman GR, et al. Hemodynamics of Fontan failure: the role of pulmonary vascular disease. Circ Heart Fail. 2017;10(12):e004515. [cited 2019 Nov 28]. Available from: http://www.ncbi.nlm.nih.gov/pubmed/29246897
43. Gewillig M, Brown SC. The Fontan circulation after 45 years: update in physiology. Heart. 2016;102(14):1081–6. https://doi.org/10.1136/heartjnl-2015-307467.
44. Rossano JW, Woods RK, Berger S, Gaynor JW, Ghanayem N, Morales DLS, et al. Mechanical Support as Failure Intervention in Patients with Cavopulmonary Shunts (MFICS): rationale and aims of a new registry of mechanical circulatory support in single ventricle patients. Congenit Heart Dis. 2013;8(3):182–6. https://doi.org/10.1111/chd.12053. [cited 2019 Nov 28]
45. Prêtre R, Häussler A, Bettex D, Genoni M. Right-sided univentricular cardiac assistance in a failing fontan circulation. Ann Thorac Surg [Internet]. 2008;86(3):1018–20. Available from: http://www.ncbi.nlm.nih.gov/pubmed/18721610
46. Riemer RK, Amir G, Reichenbach SH, Reinhartz O. Mechanical support of total cavopulmonary connection with an axial flow pump. J Thorac Cardiovasc Surg. 2005;130(2):351–4. Available from: http://linkinghub.elsevier.com/retrieve/pii/S0022522305001066
47. Myers CD, Boyd JH, Presson RG, Vijay P, Coats AC, Brown JW, et al. Neonatal cavopulmonary assist: pulsatile versus steady-flow pulmonary perfusion. Ann Thorac Surg. 2006;81(1):257–63.
48. Farahmand M, Kavarana MN, Kung E. Risks and benefits of using a commercially available ventricular

assist device for failing fontan cavopulmonary support: a modeling investigation. IEEE Trans Biomed Eng. 2020;67(1):213–9.

49. Schmidt T, Rosenthal D, Reinhartz O, Riemer K, He F, Hsia TY, et al. Superior performance of continuous over pulsatile flow ventricular assist devices in the single ventricle circulation: a computational study. J Biomech. 2017;52:48–54.
50. Chinnaiyan KM, Akasaka T, Amano T, Bax JJ, Blanke P, De Bruyne B, et al. Rationale, design and goals of the HeartFlow assessing diagnostic value of non-invasive FFRCT in Coronary Care (ADVANCE) registry. J Cardiovasc Comput Tomogr. 2017;11(1):62–7.
51. Conover T, Hlavacek AM, Migliavacca F, Kung E, Dorfman A, Figliola RS, et al. An interactive simulation tool for patient-specific clinical decision support in single-ventricle physiology. J Thorac Cardiovasc Surg. 2018;155(2):712–21.
52. Kaufmann J, Kung E. Factors affecting cardiovascular physiology in cardiothoracic surgery: implications for lumped-parameter modeling. Front Surg. 2019;6:62. https://doi.org/10.3389/fsurg.2019.00062.
53. Ramachandra AB, Humphrey JD, Marsden AL. Gradual loading ameliorates maladaptation in computational simulations of vein graft growth and remodelling. J R Soc Interface. 2017;14(130):20160995.
54. Goergen CJ, Barr KN, Huynh DT, Eastham-Anderson JR, Choi G, Hedehus M, et al. In vivo quantification of murine aortic cyclic strain, motion, and curvature: implications for abdominal aortic aneurysm growth. J Magn Reson Imaging. 2010;32(4):847–58. https://doi.org/10.1002/jmri.22331. Sep 29 [cited 2019 Nov 28]
55. Sengupta D, Kahn AM, Kung E, Esmaily Moghadam M, Shirinsky O, Lyskina GA, et al. Thrombotic risk stratification using computational modeling in patients with coronary artery aneurysms following Kawasaki disease. Biomech Model Mechanobiol. 2014;13(6):1261–76.
56. Farahmand M, Mirzaei E, Shabanisamghabady M, Kung E. A Novel right-side assist device for univentricular fontan patients. In: 8th World Congress of Biomechanics. 2018.
57. Corno AF, Prosi M, Fridez P, Zunino P, Quarteroni A, Von Segesser LK. The non-circular shape of FloWatch 1 -PAB prevents the need for pulmonary artery reconstruction after banding. Computational fluid dynamics and clinical correlations. Eur J Cardiothorac Surg. 2006;29(1):93–9.
58. Bonnet D, Corno AF, Sidi D, Sekarski N, Beghetti M, Schulze-Neick I, et al. Early clinical results of the telemetric adjustable pulmonary artery banding FloWatch-PAB. Circulation. 2004;110(11 Suppl. 1):II-158—163. [cited 2018 Nov 21]; Available from: http://www.ncbi.nlm.nih.gov/pubmed/15364856
59. Corno AF, Ladusans EJ, Pozzi M, Kerr S. FloWatch versus conventional pulmonary artery banding. J Thorac Cardiovasc Surg. 2007;134(6):4–7. [cited 2018 Nov 21]. Available from: http://www.ncbi.nlm.nih.gov/pubmed/18023654
60. Corno AF, Bonnet D, Sekarski N, Sidi D, Vouhé P, von Segesser LK. Remote control of pulmonary blood flow: initial clinical experience. J Thorac Cardiovasc Surg. 2003;126(6):1775–80. 1 [cited 2018 Nov 21]. Available from: http://linkinghub.elsevier.com/retrieve/pii/S0022522303014995

Ventriculo-arterial Coupling in Fontan Patients

13

Giovanni Biglino, Maria Victoria Ordonez, and Andrew M. Taylor

13.1 A Misalliance

The concept of ventriculoarterial (VA) or ventriculovascular coupling is an established concept in cardiovascular physiology, recognized as a central determinant of cardiovascular energetics [1]. Fundamentally, the heart does not pump in isolation from the remainder of the vasculature and changes in the pumping efficiency of the heart are not only due to intrinsic properties of the myocardium, but also to properties of the aorta and other vessels that represent the downstream impedance for the heart. This problem, from an engineering perspective, can be simplified with a hydraulic or an electrical model [2] whereby the heart is modeled by a pump or a source of energy connected to a set of resistive and compliant elements, which recapitulate key variables such as changes in vessel caliber, local obstructions (e.g., stenosis, aortic coarctation, etc.), variations in arterial stiffness or vascular tone, or alterations in the arterial Windkessel [2]. The relationship between the different elements can be synthetized by Ohm's law, whereby current equals the ratio between voltage and resistance (or flow equals the ratio of pressure and resistance).

Normal VA coupling settings, in the absence of disease, include a systemic left ventricle (LV) pumping into the aorta. In the presence of congenital heart disease (CHD) however, both the "V" and the "A" component of the coupling equation can be compromised. An extreme example of such a "mis-matched" VA coupling is found in patients with a single ventricle physiology undergoing Fontan palliation, and particularly those patients with a primary diagnosis of hypoplastic left heart syndrome (HLHS). In this case, on the "V" side of the equation, the systemic single ventricle is a morphologic right ventricle (RV), which is not designed to pump against aortic pressure levels, and there may also be a residual underdeveloped LV connected via a ventricular septal defect. Other factors inherent to the Fontan palliation and the complex-staged surgical strategy can further affect the "V" side, particularly the takedown of the Sano shunt at the time of the Glenn operation, as this can result in scarring at the site of ventriculotomy, thus potentially negatively affecting the mechanical

G. Biglino (✉)
University of Bristol, Bristol Medical School, Bristol, UK
e-mail: g.biglino@bristol.ac.uk

M. V. Ordonez
University Hospitals Bristol and Weston, NHS Foundation Trust, Bristol, UK
e-mail: victoria.ordonez@uhbw.nhs.uk

A. M. Taylor
Great Ormond Street Hospital for Children, NHS Foundation Trust, London, UK
e-mail: a.taylor76@ucl.ac.uk

G. Butera et al. (eds.), *Modelling Congenital Heart Disease*,
https://doi.org/10.1007/978-3-030-88892-3_13

properties of the single ventricle. Moreover, on the "A" side of the equation, the originally hypoplastic aorta has been surgically enlarged and reconstructed with a homograft patch in the early days of life, when the patient underwent the Norwood procedure [3] with the Damus-Kaye-Stansel [4] operation, resulting in a neoaorta composed of pulmonary artery tissue and homograft patch, the latter known to calcify over time. Comorbidities known to occur in HLHS patients such as coarctation of the aorta, residual arterial hypertension, or substantial degrees of valve incompetence, can further negatively affect VA coupling in these patients. We can therefore refer to an unfavorable coupling scenario, or mismatched VA coupling, in patients with a Fontan circulation.

The Fontan circulation itself is a unique scenario that, since its first introduction in 1971 [5] and subsequent evolutions in surgical techniques [6], whilst representing a viable way to palliate the circulation in these patients, also represents a "paradox" from a hemodynamic/functional perspective [7]. By connecting the systemic and the pulmonary circulations in series rather than in parallel, the Fontan procedure creates a condition whereby passive caval flows from the superior and inferior venae cavae (SVC, IVC) drive pulmonary flow, the pulmonary arteries being joined to the SVC and IVC in the total cavopulmonary connection (TCPC). As a pressure gradient is required to drive flow, caval pressures need to exceed pulmonary wedge pressure even just by 1–2 mmHg, resulting in a circulation where IVC/SVC pressures are high and leading long-term to venous hypertension and consequent Fontan complications such as protein losing enteropathy, liver congestion/fibrosis, venous congestion, and plastic bronchitis [8]. The Fontan circulation is a complex arrangement that has fascinated surgeons and modelers alike over the years [9–11].

Overtime, the Fontan circulation is not sustainable, not just in relation to the aforementioned venous hypertension, but also due to other key factors such as limited growth of the pulmonary arteries [12] resulting in changes both in preload and afterload in these patients. Therefore, assessment and quantification of VA coupling can be particularly relevant and physiologically insightful in these patients.

13.2 Invasive In Vivo Measurements

Measurement of VA coupling is complex and relies on the use of conductance catheters [13]. The principle of these catheters is that they measure the electrical conductance of the blood in the LV by means of electrodes that generate an electric field inside the ventricular cavity and record the resulting voltage. Typically, the catheter is mounted with twelve electrodes and is positioned along the LV long axis. The equipment thus generates continuous readings of segmental ventricular volumes and pressures. From the pressure-volume data it is possible to evaluate systolic and diastolic function, including measures of elastance. Early work showed that the method can be applied also to the right ventricle and assessment of right ventricular function [14]. Measurements with conductance catheters are also feasible and reliable in single ventricle patients [15]. In the context of the Fontan circulation, pressure-volume measurements derived from conductance catheters have been used to compare indices of systolic and diastolic function versus echocardiographic measures [16]. These measurements have also been used to assess potentially clinically meaningful interventions, such as the acute administration of Sildenafil in Fontan patients which, despite reported improvements in exercise capacity in these patients, was not found to improve ventricular function according to conductance catheter measures [17]. These measurements can also highlight differences in ventricular mechanics in patients with hypoplastic left heart syndrome palliated with a modified Blalock-Taussig shunt compared to Sano shunt palliation, at Stage 1, based on substantially different pressure-volume loop contours [18]. Invasive measurements can also be acquired during dobutamine stress in the context of hybrid cardiovascular magnetic resonance imaging with catheterization ("XMR") also for single ventricle patients [19].

13.3 Noninvasive In Vivo Measurements

One technique that lends itself to estimating VA coupling is wave intensity analysis [20]. Introduced in 1990, this analysis looks at changes in wave intensity (defined as the product of simultaneous pressure and velocity differentials at one location, or $dI = dPdU$) and as such provides a measure of the working relationship between the heart and the remainder of the vasculature. The clinical meaning of wave intensity measures, discussed more in detail elsewhere [21, 22], essentially allows us to consider key properties of ventricular contractility and ventricular filling as well as wave reflections at different sites. Wave reflections have been discussed as holding prognostic value [23] and a typical wave intensity pattern is shown in Fig. 13.1.

Traditional wave intensity measurements are also invasive, relying on pressure catheters and flow probes [24] or catheters mounted with both a pressure and a flow-velocity sensor [25]. More recently, wave intensity and particularly its formulation employing either vessel diameter [26] or area [27] as surrogates for blood pressure have rendered the analysis compatible with medical imaging techniques such as echocardiography or cardiovascular magnetic resonance (CMR) imaging, thus rendering it noninvasive.

Noninvasive wave intensity measurements have been used in different scenarios, including the assessment of carotid hemodynamics [28] or left ventricular systolic dysfunction [29]. In the context of CHD, CMR-derived wave intensity has been employed in scenarios such as repaired aortic coarctation [30] and tetralogy of Fallot and other conotruncal anomalies [31] or assessing ventricular function in pediatric patients [32]. In the case of the Fontan circulation, noninvasive wave intensity analysis has provided some quantitative insight into the mis-matched nature of VA coupling. Considering patients with a single ventricle physiology and comparing those with HLHS and other single ventricle scenarios (e.g., pulmonary atresia), a small study identified significant differences in wave intensity magnitudes and wave energy between the two groups, with lower values in the HLHS cohort, suggesting that the effect of aortic reconstruction and consequent homograft patch calcification can reflect on coupling efficiency [33, 34]. Indeed, aortic distensi-

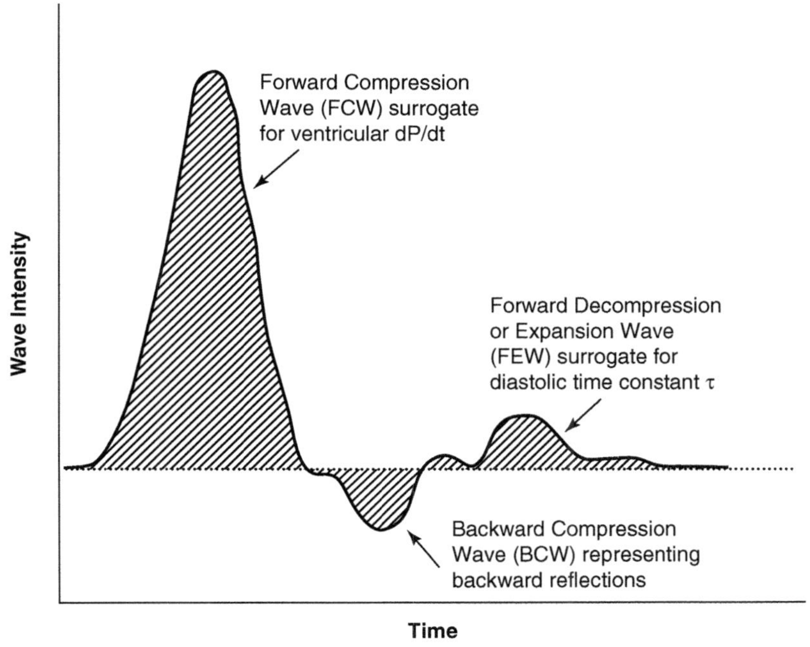

Fig. 13.1 Example of wave intensity pattern showing the main three waves (early systolic forward compression wave and late systolic forward decompression or expansion wave, as well as the backward compression wave representing reflected waves traveling back toward the heart)

bility in the HLHS patients was significantly reduced compared to other single ventricle patients [33]. These observations were expanded in a follow-up study, exploring two additional effects: firstly the type of shunt employed during Stage 1 palliation (modified Blalock-Taussig vs. Sano), and secondly not just the presence but the size of aortic reconstruction (more bulbous transverse arch vs. more tubular). Results revealed that shunt type did not affect the major wave intensity peak measures (taken as surrogate for contractility and filling), but that the size of the transverse arch did have an impact on ventricular function, identifying an inverse correlation between increasing size of the aortic reconstruction (i.e., more bulbous transverse arch setting) and reduced amplitude of the peak of the systolic forward compression wave on the wave intensity pattern.

13.4 Estimating PV Loops from Computational Simulations

As mentioned, the Fontan circulation has been the focus of modeling studies for a long time, and indeed early applications of mathematical modeling to the cardiovascular system include pioneering simulations of the hemodynamics in different TCPC configurations [35]. Detailed simulations in the form of multiscale models can account for global as well as local changes in different parameters in the Fontan circulation. As part of the outputs from multiscale simulations, the inclusion of a ventricular model connected to a three-dimensional anatomical domain and a lumped parameter network [36] allows extracting also information on ventricular function in relation to changes in the circulation. By performing parametric studies, this can reveal insightful information on the impact of different variables on the function of the single ventricle. For instance, a combination of different mBT shunt diameters together with varying severity of aortic coarctation in Norwoord patients was shown to result in substantially different pressure-volume loops, as indicative of different (and detrimental, in some cases) impacts on ventricular function [37]. By varying the boundary conditions, computational simulations allow in an easily reproducible way to assess different pathophysiological scenarios and/or vascular adaptations, outputting information on cardiac workload [38]. A family of pressure-volume loops can thus be simulated for scenarios such as the hybrid Norwood, incrementally varying the size of the pulmonary arteries banding and the diameter of the ductal stent [39]. Interestingly, such principles of mathematical modeling have also been translated into an interactive Web-based simulation tool for providing clinical decision support for all three stages of palliation and including simulation of pressure-volume loops in the simulation outputs [40].

13.5 Back to the Hydraulic Model: Experimental Studies

Bench experiments have also been shown to provide information on the Fontan circulation, including studies detailing the construction and verification of mock circulatory systems suitable to replicate different stages of Fontan palliation, including Stage 1 with mBT and Sano shunts, Stage 2 with a Glenn connection and the full Fontan circulation with different TCPC configurations including the effect of respiration [41–44]. Of interest, experimental setups such as mock circulatory loops can also output pressure-volume loops and as such provide information on ventriculoarterial coupling under a range of different conditions and/or for the purpose of parametric studies. This has been demonstrated in the context of a mock circulatory system for testing ventricular assist devices, showing that the setup was indeed able to mimic the Frank-Starling response to a range of scenarios (i.e., normal circulation, heart failure, and cardiac recovery) over a range of preload and afterload conditions [45]. This also stresses the importance of being able to extract such information from experimental setups when evaluating novel devices, also in the context of the Fontan circulation. Indeed, experimental setups represent a valuable resource for conducting parametric studies and are an essential tool for validating computational models, yet their role in specifically measuring VA coupling is probably somewhat limited at present, compared for example to other applications such as testing novel devices for Fontan patients [46]. The ability of an experimental system to provide this kind of data is nonetheless very appealing, particularly in the context of device testing, but the implications on the complexity of the experimental setup should not be underestimated.

13.6 Translational Outlook: Unraveling a Complex Physiology

A good functioning univentricular physiology circulation depends on a fine interaction between the systemic venous system (SVC/IVC), pulmonary vascular resistances, diastolic and systolic function, and afterload and systemic vascular resistances. The lack of a subpulmonary ventricle means that the only mechanism that the univentricular system has to drive the pulmonary flow toward the lungs is by expenses of a high pressure in the venous system (above the values of the pulmonary wedge pressure), low PVR, normal ventricular diastolic deformation, efficient contractility, and low afterload. Therefore, any alteration in each part of the system can lead to a well-known state of "failing" Fontan circulation.

Ventricular and arterial efficiency assessments are highly challenging in this population as no standardized values have been established, leading to animperfect risk stratification and therefore fortuneless outcomes in the absence of identifiable risk factors. As Fontan patients may have altered ventricular loading conditions, ejection fraction may not accurately reflect systolic ventricular function. The analysis of VA properties has been identified as a key element for a normal cardiovascular function in biventricular hearts. VA coupling includes the analysis of Ea (arterial elastance) and Ees (end-systolic elastance, a load-insensitive measure of ventricular contractility). This has been broadly used for risk stratification and outcomes prediction in biventricular circulation. Hence, the extrapolation of this action has been recently span toward univentricular physiology bearing in mind that ventricular and arterial dysfunction may impact the homeostasis of the Fontan circulation and predispose to heart failure.

In the normal circulation, arterial and end-systolic ventricular elastances are coupled to optimize cardiovascular performance and efficiency. It has been shown that normal coupling is perturbed in adult Fontan patients owing to disproportionate impairment in systolic function [47]. Impaired contractile function was worse in the systemic right ventricle than systemic left ventricle. Surprisingly, contractile dysfunction and ventricular inefficiency were apparent even in Fontan patients with preserved (>50%) EF. The marked reduction in ventricular efficiency (expressed by a low VA ratio), may suggest a novel therapeutic target.

The analysis of diastolic function properties is also of paramount importance when evaluating the performance of the univentricular physiology and the VA interaction. Ventricular deformation during diastole is reflective of ventricular filling (preload) and CI, suggesting that ventricular diastolic function is a critical factor influencing circulatory success in Fontan patients. Studies have begun to gather evidence of the inefficient ventriculoarterial coupling in Fontan patients [48], with patients with low VA coupling at higher risk of death and heart transplantation listing. Moreover, a low VA ratio was an independent risk factor for death and heart transplantation in patients without ventricular dilation, in whom traditional risk factors were not associated with ventricular inefficiency or outcomes.

The understanding of VA coupling at rest is fundamental in order to extend the same analysis to the higher physiological demands of states such as exercise or pregnancy. Fontan patients struggle to cope against increasing hemodynamic demand, as there is a lack of HR and SV augmentation. Chronotropic insufficiency is thought to be secondary to (a) myocardial scarring and conduction system damage (after surgery), (b) insufficient coronary perfusion (secondary to low CI), and (c) low SV, in patients with curtailed preload due to either high PVR or diastolic myocardial dysfunction.

The Fontan system should be analyzed both globally and in its individual components, ultimately aiming to unravel this complex and delicate physiology, and tools such as computational models, imaging-derived indices, and in-vitro studies could all add to such an analysis. Further investigations are needed in order to fully understand the meaning and clinical implications of changes in VA coupling in this population and to plan a solid and uniform stratification score.

Acknowledgments The authors gratefully acknowledge the contribution by Dr. Sandra Neumann (Bristol Medical School, University of Bristol) who helped producing Fig. 13.1.

References

1. Chantler PD, Lakatta EG, Najjar SS. Arterial-ventricular coupling: mechanistic insights into cardiovascular performance at rest and during exercise. J Appl Physiol (1985). 2008;105(4):1342–51.
2. Westerhof N, Lankhaar JW, Westerhof BE. The arterial Windkessel. Med Biol Eng Comput. 2009;47(2):131–41.
3. Norwood WI, Lang P, Hansen DD. Physiologic repair of aortic atresia-hypoplastic left heart syndrome. N Engl J Med. 1983;308(1):23–6.
4. Gates RN, Laks H, Elami A, Drinkwater DC Jr, Pearl JM, George BL, Jarmakani JM, Williams RG. Damus-Stansel-Kaye procedure: current indications and results. Ann Thorac Surg. 1993;56(1):111–9.
5. Fontan F, Baudet E. Surgical repair of tricuspid atresia. Thorax. 1971;26(3):240–8.
6. Feinstein JA, Benson DW, Dubin AM, Cohen MS, Maxey DM, Mahle WT, Pahl E, Villafañe J, Bhatt AB, Peng LF, Johnson BA, Marsden AL, Daniels CJ, Rudd NA, Caldarone CA, Mussatto KA, Morales DL, Ivy DD, Gaynor JW, Tweddell JS, Deal BJ, Furck AK, Rosenthal GL, Ohye RG, Ghanayem NS, Cheatham JP, Tworetzky W, Martin GR. Hypoplastic left heart syndrome: current considerations and expectations. J Am Coll Cardiol. 2012;59(Suppl. 1):S1–42.
7. de Leval MR. The Fontan circulation: a challenge to William Harvey? Nat Clin Pract Cardiovasc Med. 2005;2(4):202–8.
8. Gewillig M, Brown SC. The Fontan circulation after 45 years: update in physiology. Heart. 2016;102(14):1081–6.
9. Migliavacca F, de Leval MR, Dubini G, Pietrabissa R, Fumero R. Computational fluid dynamic simulations of cavopulmonary connections with an extracardiac lateral conduit. Med Eng Phys. 1999;21(3):187–93.
10. Biglino G, Giardini A, Hsia TY, Figliola R, Taylor AM, Schievano S. Modeling single ventricle physiology: review of engineering tools to study first-stage palliation of hypoplastic left heart syndrome. Front Pediatr. 2013;1:31.
11. Bove EL, de Leval MR, Migliavacca F, Balossino R, Dubini G. Toward optimal hemodynamics: computer modeling of the Fontan circuit. Pediatr Cardiol. 2007;28(6):477–81.
12. De Rita F, Crossland D, Griselli M, Hasan A. Management of the failing Fontan. Semin Thorac Cardiovasc Surg Pediatr Card Surg Annu. 2015;18(1):2–6.
13. Baan J, van der Velde ET, Steendijk P. Ventricular pressure-volume relations in vivo. Eur Heart J. 1992;13(Suppl. E):2–6.
14. Bishop A, White P, Oldershaw P, Chaturvedi R, Brookes C, Redington A. Clinical application of the conductance catheter technique in the adult human right ventricle. Int J Cardiol. 1997;58(3):211–21.
15. Butts RJ, Chowdhury SM, Buckley J, Hlavacek AM, Hsia TY, Khambadkone S, Baker GH, MOCHA investigators. Comparison of echocardiographic and pressure-volume loop indices of systolic function in patients with single ventricle physiology: a preliminary report. Congenit Heart Dis. 2015;10(1):E17–24.
16. Chowdhury SM, Butts RJ, Buckley J, Hlavacek AM, Hsia TY, Khambadkone S, Baker GH, MOCHA Investigators. Comparison of pressure-volume loop and echocardiographic measures of diastolic function in patients with a single-ventricle physiology. Pediatr Cardiol. 2014;35(6):998–1006.
17. Butts RJ, Chowdhury SM, Baker GH, Bandisode V, Savage AJ, Atz AM. Effect of sildenafil on pressure-volume loop measures of ventricular function in Fontan patients. Pediatr Cardiol. 2016;37(1):184–91.
18. Butts RJ, Hsia TY, Hamilton Baker G, MOCHA Investigators. Feasibility of conductance catheter-derived pressure-volume loops to investigate ventricular mechanics in shunted single ventricles. Cardiol Young. 2013;23(5):776–9.
19. Wong J, Pushparajah K, de Vecchi A, Ruijsink B, Greil GF, Hussain T, Razavi R. Pressure-volume loop-derived cardiac indices during dobutamine stress: a step towards understanding limitations in cardiac output in children with hypoplastic left heart syndrome. Int J Cardiol. 2017;230:439–46.
20. Parker KH, Jones CJ. Forward and backward running waves in the arteries: analysis using the method of characteristics. J Biomech Eng. 1990;112(3):322–6.
21. Kolyva K, Khir AW. Wave intensity analysis in the great arteries – what has been learning during the last 25 years? Part 1. Int Cardiovasc Forum J. 2013;1(2)
22. Kolyva K, Khir AW. Wave intensity analysis in the ventricles, carotid and coronary arteries – what has been learning during the last 25 years? Part 2. Int Cardiovasc Forum J. 2013;1(3)

23. Cheng HM, Chuang SY, Wang JJ, Shih YT, Wang HN, Huang CJ, Huang JT, Sung SH, Lakatta EG, Yin FC, Chou P, Yeh CJ, Bai CH, Pan WH, Chen CH. Prognostic significance of mechanical biomarkers derived from pulse wave analysis for predicting long-term cardiovascular mortality in two population-based cohorts. Int J Cardiol. 2016;215:388–95.
24. Penny DJ, Mynard JP, Smolich JJ. Aortic wave intensity analysis of ventricular-vascular interaction during incremental dobutamine infusion in adult sheep. Am J Physiol Heart Circ Physiol. 2008;294(1):H481–9.
25. Khir AW, Parker KH. Wave intensity in the ascending aorta: effects of arterial occlusion. J Biomech. 2005;38(4):647–55.
26. Borlotti A, Khir AW, Rietzschel ER, De Buyzere ML, Vermeersch S, Segers P. Noninvasive determination of local pulse wave velocity and wave intensity: changes with age and gender in the carotid and femoral arteries of healthy human. J Appl Physiol (1985). 2012;113(5):727-735.
27. Biglino G, Steeden JA, Baker C, Schievano S, Taylor AM, Parker KH, Muthurangu V. A non-invasive clinical application of wave intensity analysis based on ultrahigh temporal resolution phase-contrast cardiovascular magnetic resonance. J Cardiovasc Magn Reson. 2012;14(1):57.
28. Neumann S, Sophocleous F, Kobetic MD, Hart EC, Nightingale AK, Parker KH, Hamilton MK, Biglino G. Wave intensity analysis in the internal carotid artery of hypertensive subjects using phase-contrast MR angiography and preliminary assessment of the effect of vessel morphology on wave dynamics. Physiol Meas. 2018;39(10):104003.
29. Hametner B, Parragh S, Weber T, Wassertheurer S. Wave intensity of aortic root pressure as diagnostic marker of left ventricular systolic dysfunction. PLoS One. 2017;12(6):e0179938.
30. Quail MA, Short R, Pandya B, Steeden JA, Khushnood A, Taylor AM, Segers P, Muthurangu V. Abnormal wave reflections and left ventricular hypertrophy late after coarctation repair. Hypertension. 2017;69(3):501–9.
31. Ordonez MV, Neumann S, Caputo M, Curtis S, Biglino G. Feasibility of wave intensity analysis in patients with conotruncal anomalies before and after pregnancy: new physiological insights? Front Pediatr. 2021;8:557407.
32. Ntsinjana HN, Chung R, Ciliberti P, Muthurangu V, Schievano S, Marek J, Parker KH, Taylor AM, Biglino G. Utility of cardiovascular magnetic resonance-derived wave intensity analysis as a marker of ventricular function in children with heart failure and normal ejection fraction. Front Pediatr. 2017;5:65.
33. Biglino G, Schievano S, Steeden JA, Ntsinjana H, Baker C, Khambadkone S, de Leval MR, Hsia TY, Taylor AM, Giardini A, Modeling of Congenital Hearts Alliance (MOCHA) Collaborative Group. Reduced ascending aorta distensibility relates to adverse ventricular mechanics in patients with hypoplastic left heart syndrome: noninvasive study using wave intensity analysis. J Thorac Cardiovasc Surg. 2012;144(6):1307–13.
34. Biglino G, Giardini A, Ntsinjana HN, Schievano S, Hsia TY, Taylor AM, Modeling of Congenital Hearts Alliance Collaborative Group. Ventriculoarterial coupling in palliated hypoplastic left heart syndrome: noninvasive assessment of the effects of surgical arch reconstruction and shunt type. J Thorac Cardiovasc Surg. 2014;148(4):1526–33.
35. de Zélicourt DA, Marsden A, Fogel MA, Yoganathan AP. Imaging and patient-specific simulations for the Fontan surgery: current methodologies and clinical applications. Prog Pediatr Cardiol. 2010;30(1-2):31–44.
36. Meoli A, Cutrì E, Krishnamurthy A, Dubini G, Migliavacca F, Hsia TY, Pennati G. A multiscale model for the study of cardiac biomechanics in single-ventricle surgeries: a clinical case. Interface Focus. 2015;5(2):20140079.
37. Corsini C, Biglino G, Schievano S, Hsia TY, Migliavacca F, Pennati G, Taylor AM, MOCHA Collaborative Group. The effect of modified Blalock-Taussig shunt size and coarctation severity on coronary perfusion after the Norwood operation. Ann Thorac Surg. 2014;98(2):648–54.
38. Coogan JS, Chan FP, Ladisa JF Jr, Taylor CA, Hanley FL, Feinstein JA. Computational fluid dynamic simulations for determination of ventricular workload in aortic arch obstructions. J Thorac Cardiovasc Surg. 2013;145(2):489–95.
39. Young A, Gourlay T, McKee S, Danton MH. Computational modelling to optimize the hybrid configuration for hypoplastic left heart syndrome. Eur J Cardiothorac Surg. 2013;44(4):664–72.
40. Conover T, Hlavacek AM, Migliavacca F, Kung E, Dorfman A, Figliola RS, Hsia TY. An interactive simulation tool for patient-specific clinical decision support in single-ventricle physiology. J Thorac Cardiovasc Surg. 2018;155(2):712–21.
41. Biglino G, Giardini A, Baker C, Figliola RS, Hsia TY, Taylor AM, Schievano S, MOCHA Collaborative Group. In vitro study of the Norwood palliation: a patient-specific mock circulatory system. ASAIO J. 2012;58(1):25–31.
42. Biglino G, Giardini A, Baker C, Figliola RS, Hsia TY, Taylor AM, Schievano S, MOCHA Collaborative Group. Implementing the Sano modification in an experimental model of first-stage palliation of hypoplastic left heart syndrome. ASAIO J. 2013;59(1):86–9.
43. Figliola RS, Giardini A, Conover T, Camp TA, Biglino G, Chiulli J, Hsia TY. In vitro simulation and validation of the circulation with congenital heart defects. Prog Pediatr Cardiol. 2010;30(1-2):71–80.
44. Vukicevic M, Chiulli JA, Conover T, Pennati G, Hsia TY, Figliola RS, MOCHA Network. Mock circulatory system of the Fontan circulation to study respiration effects on venous flow behavior. ASAIO J. 2013;59(3):253–60.

45. Pantalos GM, Koenig SC, Gillars KJ, Giridharan GA, Ewert DL. Characterization of an adult mock circulation for testing cardiac support devices. ASAIO J. 2004;50(1):37–46.
46. Rodefeld MD, Marsden A, Figliola R, Jonas T, Neary M, Giridharan GA. Cavopulmonary assist: long-term reversal of the Fontan paradox. J Thorac Cardiovasc Surg. 2019;158(6):1627–36.
47. Saiki H, Eidem BW, Ohtani T, Grogan MA, Redfield MM. Ventricular-arterial function and coupling in the adult Fontan circulation. J Am Heart Assoc. 2016;5(9):e003887.
48. Godfrey ME, Rathod RH, Keenan E, Gauvreau K, Powell AJ, Geva T, Prakash A. Inefficient ventriculo-arterial coupling in Fontan patients: a cardiac magnetic resonance study. Pediatr Cardiol. 2018;39(4):763–73.

Modeling the Pulmonary Circulation in CHD: Clinical Concepts, Engineering Applications, and an Integrated Medico-Engineering Approach

14

Weiguang Yang, Jeffrey A. Feinstein, and Irene E. Vignon-Clementel

14.1 Introduction to the Pulmonary Circulation

The cardiovascular system is comprised of the heart and blood vessels, which function to circulate the blood and deliver oxygen and nutrients to each organ in the body. The pulmonary circulation is the portion of the cardiovascular system, which serves to oxygenate the blood and remove carbon dioxide. From the right ventricle (RV), deoxygenated blood is pumped into the main pulmonary artery (MPA) and then distributed to the right and left pulmonary arteries (RPA and LPA). In general, the central RPA and LPA branch into 10 and 8 lung segments in the right and left lungs, respectively, and within each segment, subsegmental PAs, and multiple levels of smaller pulmonary arteries divide again before terminating in the pulmonary arterioles, which are connected to pulmonary capillaries. The adult MPA measures 26.0 ± 2.7 mm and 24.2 ± 2.7 mm in diameter for men and women, respectively [1]. In each lung, there are about 15 orders of pulmonary arteries (PAs) ranging from ~15 mm to 0.02 mm [2]. The pulmonary capillary networks surround millions of alveoli (tiny air sacs). The alveolar–capillary membrane allows for passive diffusion of oxygen from the alveoli to the capillaries and of carbon dioxide from the capillaries to the alveoli; the uptake of oxygen and the release of carbon dioxide are carried out through gas diffusion across the alveolar–capillary membrane. The pulmonary capillaries measure 5–10 μm forcing red blood cells to travel single file, while the alveolar surface area by comparison is remarkably large, ~80 cm^2 for human lungs, promoting exceptionally efficient gas exchange [3]. The alveolar–capillary bed can simply be considered as two sheets held by cells and connective tissues that act as columns inside [4]. Oxygenated blood returns to the left atrium via the pulmonary venules and veins, the course of which, in general, parallels the arterial vessels.

The pulmonary arteries consist of three layers: the tunica intima, media, and adventitia (also known as tunica externa). The innermost layer, the intima, is primarily comprised of endothelial cells and the internal elastic lamina. The media, middle layer, consists primarily of smooth muscle cells, with some elastin and collagen, while the adventitia, the outermost layer, contains collagen and elastin. Despite being of similar

W. Yang · J. A. Feinstein
Department of Pediatrics (Cardiology), Stanford University, Stanford, USA
e-mail: wgyang@stanford.edu; jeff.feinstein@stanford.edu

I. E. Vignon-Clementel (✉)
Inria Saclay-Île-de-France Research Centre, Inria, Palaiseau, France
e-mail: Irene.vignon-clementel@inria.fr

G. Butera et al. (eds.), *Modelling Congenital Heart Disease*,
https://doi.org/10.1007/978-3-030-88892-3_14

structures, the composition and contribution of each layer vary greatly from the largest main pulmonary artery to the smallest arteriolar vessels. The "larger" proximal PAs (down to ~2 mm in diameter) are quite compliant due to the presence of multiple elastic laminae in the media. With decreasing vessel size, smooth muscle cell concentration in the media increases, and elastic laminae decrease. The PAs are muscular when the diameter is between 0.1 mm and 1 mm. As the size further decreases to less than 0.1 mm, however, the PAs become partially muscular and finally non-muscular when adjacent to the bronchioles and alveolar ducts. In the human lung, at the level of the capillaries, the blood–gas barrier thickness is approximately 0.3 μm. The structure of the pulmonary veins at each branch segmentation generally mirrors that of the pulmonary arteries, though the venous walls are thinner compared to similar-sized arteries, and the large pulmonary veins in human lungs are more muscular [5]. The transition from muscular vessels to non-muscular vessels in the distal veins tends to occur at larger sizes when compared to the arteries [5].

The pulmonary circulation displays dramatically different hemodynamics when compared to the systemic circulation. Specifically, the pulmonary circulation is designed to be a low-pressure system with a normal mean pressure of ~10–15 mmHg. The basic hemodynamic parameters, pressure, blood flow (rate), and resistance are related through Ohm's law:

$$\mathrm{PVR} = \frac{\Delta P}{Q} \qquad (14.1)$$

where PVR is the pulmonary vascular resistance (in dynes-s/cm^5 or Wood units), ΔP is the pressure difference between the mean pressures of the MPA and left atrium (in mmHg), and Q is the pulmonary flow rate (in L/min). PVR is normally very low, approximately 1–3 Wood units (mmHg min/L), as compared to systemic vascular resistance (~10–30 Wood Units). The longitudinal distribution of PVR characterizes the contribution from the arteries, capillaries, and veins to the total PVR. Brody et al. [6] found that the pulmonary arteries, capillaries, and veins are responsible for 46%, 34%, and 20%, respectively, of total lobar vascular resistance in a canine lung. Siegel et al. [7] showed that the pressure drop from the pulmonary artery to the capillaries is ~6.7 mmHg and, by analyzing the decay of PA occlusion pressure curves, determined the pre-capillary pulmonary vascular resistance accounts for ~70% of the total PVR.

Relative blood flow distribution to the two lungs is felt to be preserved within a narrow range, and a ratio of right 55%/left 45% is generally reported as the normal value for pulmonary flow split in clinical practice. Cheng et al., [8] using MRI to measure pulmonary flow at rest and during exercise in healthy children (n = 10) and adults (n = 7), found a right/left pulmonary artery flow ratio of 1.3:1 ± 0.4 (unchanged with exercise) and 1.1:1 ± 0.2 (exercise: 1.0:1 ± 0.2) for children and adults, respectively. They subsequently analyzed 206 normal adult scintigraphy-based lung perfusion scans and described a slightly more balanced distribution to the right and left lungs of 52.5%/47.5% ± 2.1%, respectively. Blood flow is not distributed evenly within each lung, as more blood is directed to the lower lobes due to a gravitational effect and the structure of the pulmonary vasculature [9–11].

14.2 Pulmonary Vascular Abnormalities in Congenital Heart Diseases

Structural forms of pulmonary vascular abnormalities may affect all levels of the pulmonary vascular tree, "from the valve to the veins," range from mild to severe, and present immediately after birth, during adulthood, or never at all [12–14]. The pulmonary valve may be dysplastic and/or stenosed, atretic, or absent [12]. Pulmonary arteries may (a) be discontinuous from each other or from the right ventricle, (b) display stenoses of one or more vessels, which may be isolated or concomitant with other components of CHD, and/or (c) display normal branching patterns but suffer from underdevelopment secondary to other disease processes [13]. Additionally, even in the setting of a structurally normal pulmonary vascular tree, various forms of congenital heart disease, and the associated abnormal flow volumes and pressure levels can lead to alterations in the biomechanical properties and 3-dimensional hemodynamics in these vessels.

Pulmonary stenosis is characterized by a narrowing below (subvalvular), at (valvular), or above (supravalvular) the pulmonary valve resulting in RV hypertension and the potential for decreased pulmonary flow and/or cyanosis. Valvular stenosis accounts for 80–90% of cases of RV outflow tract obstruction. For example, the infundibular (right ventricular outflow tract) narrowing and ventricular septal defect in tetralogy of Fallot combine to shunt deoxygenated blood to the systemic circulation resulting in variable degrees of desaturation depending on the severity of the obstruction. Peripheral pulmonary artery stenosis, characterized by a single or multiple focal stenoses or diffuse hypoplasia in the branch PAs, may be found in isolation but is more commonly associated with genetic syndromes such as Alagille [15], Keutel [16], and Williams syndromes [17]. These stenoses generally lead to pulmonary blood flow disparity [18] with or without RV hypertension.

When a pulmonary artery is congenitally discontinuous, a patent ductus arteriosus or aortopulmonary collateral artery supplies blood from the aorta (or one of its branches) to the lungs [19]. In patients with tetralogy of Fallot and major aortopulmonary collateral arteries (MAPCAs), the degree to which the collaterals restrict pulmonary blood flow determines whether the patients present with cyanosis due to inadequate pulmonary flow, or with pulmonary congestion and heart failure due to pulmonary overcirculation. Structural pulmonary artery abnormalities can also result in cyanosis as a result of the admixing of oxygenated blood with deoxygenated blood as seen in truncus arteriosus, where the pulmonary artery and aorta arise from the heart as a single blood vessel rather than separately, exposing the pulmonary artery to exaggerated flow and pressure. Lastly, anomalous venous return of one or more pulmonary veins may be accompanied by lung (and vascular) underdevelopment as seen in Scimitar syndrome [20].

Isolated pulmonary vein stenosis may be found in isolation or in association with other disease processes such as prematurity [21]. Pulmonary vein stenosis may involve a single vein or multiple veins, be discrete or long segment, and progress slowly or rapidly. The obstruction to the egress of blood from the lung often leads to elevated pulmonary pressures [22].

Altered hemodynamic environments due to congenital heart defects may lead to abnormal flow patterns, biomechanical properties, and/or pulmonary vascular maladaptation. Septal defects cause increased pulmonary blood flow and elevated wall shear stress. The increased flow may be accompanied by a significant increase in right ventricular and pulmonary artery pressures and may lead to pulmonary arterial hypertension (PAH, see Chap. 18) [23–25]. Pulmonary vascular disease has been documented in patients with dextro-transposition of the great arteries (D-TGA) both pre- and postrepair [26–28] and abnormalities in flow persist after complete repair with the arterial switch operation. Patients with various single ventricle forms of congenital heart disease will be palliated to non-pulsatile repairs, leading to altered flow patterns and shear, and altered vascular growth [8, 29, 30]. Single ventricle defects are discussed in detail in Chaps. 11–13.

Figure 14.1 presents a gross schematic representation of the pulmonary circulation in the setting of CHD. Structural pulmonary artery abnormalities generally increase the right ventricular workload. Therefore, surgical and/or catheter-based treatments are employed in an attempt to "normalize" the pulmonary circulation. Treatments, however, are not always curative; issues of persistent obstruction, re-stenosis, RV hypertension, ventricular volume loading secondary to pulmonary regurgitation, and ventricular dysfunction may persist or develop, particularly for complex defects [13, 31, 32].

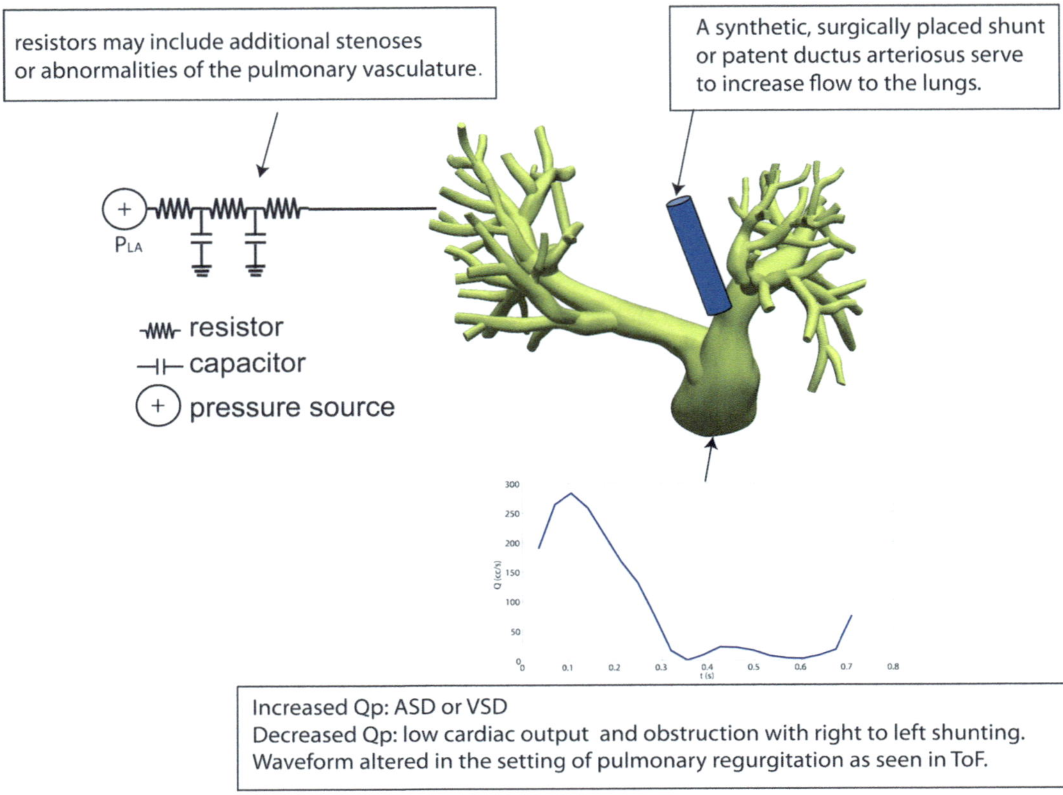

Fig. 14.1 Pulmonary circulation schematic diagram. 3D model portions created from CT/MRI imaging data. Pulmonary vasculature distal to 3D model represented by electrical circuits

Decision making in the complex scenarios is often based on anecdotes, institutional experience, and/or interpolation from other forms of congenital heart disease, and as such, surgical planning and patient management may benefit from the development and use of computational modeling.

14.3 Numerical Simulations of the Pulmonary Circulation: Healthy and Disease Modeling and Integration of Patient Data

Computational simulations of blood flow in the PAs have the potential to answer both simple and complex clinical and research-based questions, but require careful attention to, and understanding of, the unknowns and assumptions applied in the engineering pipeline [33, 34].

14.3.1 Modeling the Normal Pulmonary Circulation: Multidimensional Models and Parameterization from Data

The process of patient-specific modeling of the pulmonary circulation is generally comprised of (1) vessel centerline generation, (2) lumen segmentation, (3) lofting segmentations, (4) mesh generation, (5) prescribing boundary conditions, (6) numerical computation, and (7) post-processing. Figure 14.2 illustrates the first few steps for a patient-specific PA model [35, 36].

The integration of clinical data to build and simulate realistic models is critical: Geometry is segmented from magnetic resonance angiography or CT angiograms, flow information comes from phase-contrast MRI, echo-Doppler ultrasound, or cardiac catheterization, and PA, PV, or intracar-

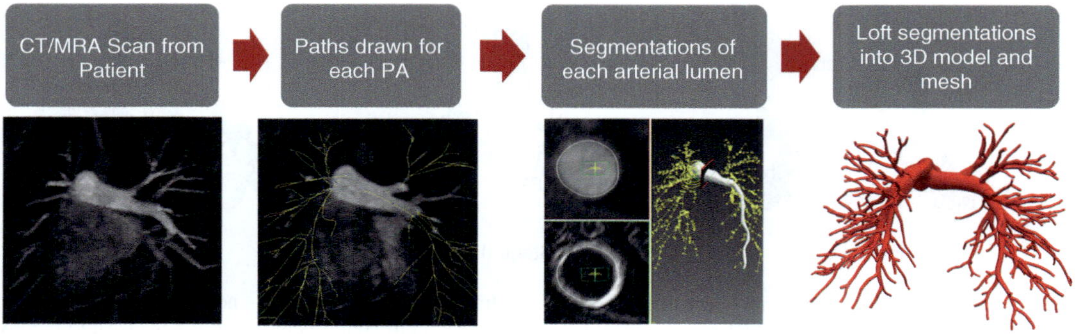

Fig. 14.2 Patient-specific modeling construction using SimVascular

diac pressures are obtained via catheterization [33]. The application of these measures combined with (when available) literature-based assumptions defines reduced models of the pulmonary circulation, though with the obvious inherent flaws (see, e.g., [37–39]). The specific question asked (e.g., "How will removal of multiple proximal stenoses impact post-operative PA pressure?"), and level of detail required to answer it will define the region of interest (ROI). The complexity of the flow and biomechanics in the region, and the fidelity of the associated available clinical data set the level of refinement needed and available to be mathematically modeled. 3D Navier–Stokes equations relating blood velocity and pressure in a few PA branches along with consideration of vessel wall elasticity through fluid–solid interaction (FSI) models are used to solve complex three-dimensional flows [40]. In general, blood enters the ROI through one inlet and leaves it through multiple outlet branches (pulmonary vein modeling would be a mirror situation of multiple inlets and a single outlet). The mathematical model requires parameterization for both the inlet and outlet boundary conditions. Predictive modeling, e.g., as with surgical planning, requires careful selection of baseline and post-treatment values [33]. If, for example, the simulation goal is to predict a pressure change, flow waveforms or a relationship between pressure and flow, such as a reduced model of the right heart, would be imposed at the inlet. Since surgical intervention typically affects pressure and flow distribution among the multiple branches, outlet boundary conditions need to incorporate the downstream pulmonary circulation not incorporated into the 3D model. This is typically done as a reduced model, which provides a relationship between pressure and flow at the coupling boundary [41]. Reduced models are either 1D models, which take into account wave propagation along the pulmonary vessel main axis, or 0D electric analog models. The simplest model incorporates a resistance boundary condition [27] similar to Eq. (14.1). In pulsatile simulations, this can be extended to (a) Windkessel models, or as a better representation of the pulmonary circulation, (b) *RCRCR* representing the large PA resistance and capacitance, the small vessels resistance, and venous capacitance and resistance [38], (c) structured tree impedances [42], or (d) morphometric Womersley-based impedances [43]. The ROI itself can be modeled with reduced models [44], and the 3D ROI can be embedded within a closed-loop model of the entire circulation [45].

14.3.2 Modeling the Pulmonary Circulation in the Setting of Congenital Heart Disease

As noted above, CHD can affect various portions of the pulmonary circulation, influencing hemodynamics and/or disease progression. Pulmonary circulation models may be of value in numerous forms of CHD including those of septal or valvular defects, tetralogy of Fallot, and single ventricle (patho)physiology (see the relevant chapters) as a means to study their influence on pulmonary vascular biomechanics. Shunts that are either palliative (Norwood case) or developing due disease (pulmonary–arteriovenous collaterals) have been

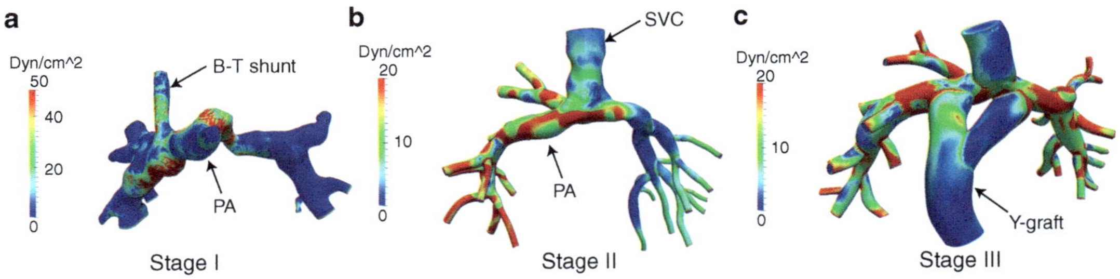

Fig. 14.3 Examples of pulmonary artery time-averaged wall shear stress (WSS) at each palliative surgical stage in a single ventricle defects: (**a**) In stage I: Blalock–Taussig (B-T) shunt connects the aorta to the PA [37], (**b**) stage II (Glenn shunt): B-T shunt removed, superior vena cava connected to the PA [47], (**c**) stage III (Fontan, total cavo-pulmonary connection): A non-traditional Gore-Tex Y-shaped graft connects the inferior vena cava to the PA [59, 63]. Shear varies greatly between the first and last two stages, and areas of increased shear are seen at the bifurcations even under the low pressure, low flow, and non-pulsatile final two stages

demonstrated, using numerical simulations, to show altered pulmonary circulation hemodynamics in the large PAs and distally [37, 46]. Patient-specific PA simulations demonstrate complex flow, pressure levels, and wall shear stress alterations unique to each single ventricle palliation stage (Norwood [37], Glenn [47], Fontan (Chap. 12)) and have been used extensively in Fontan surgical planning. Figure 14.3 shows time-averaged wall shear stress in the PAs through staged surgical palliation. The application of typical pathophysiology conditions via modeling can also be useful to evaluate PA pressure and flow during the device design process as reported for a percutaneous valve for TOF patients [48].

Other CHDs directly modify the pulmonary circulation itself at different levels. The geometry (and often the wall elasticity) needs to be modified in case of stenosis (Sect. 14.4) and pulmonary hypertension [49] in the large arteries or more distally. Moreover, boundary conditions that account for vessel adaptation to increased flow due to treatment [50] or to increase pressure due to disease progression [49, 51] are important components for clinically relevant modeling.

14.4 A Case Study of Integrated Medico-engineering Approach

Peripheral pulmonary artery stenosis (PPS) is characterized by multiple and often diffusely distributed narrowed branch pulmonary artery ostia and is commonly associated with Alagille and Williams syndromes [31, 52, 53]. PPS can result in grossly abnormal pulmonary blood flow distribution, significant right ventricular (RV) hypertension, abnormal pulmonary vascular development, and sequelae including right heart failure. The RV hypertension and right heart failure also impose significant operative risks for Alagille patients who may require liver transplantation. Current PPS treatment options include catheter-based [31] and surgical interventions [13]. Despite recent advances in balloon angioplasty and stenting, due to the risk of incomplete relief of the obstruction, vessel dissection, and/or rupture, surgical reconstruction when technically feasible is performed to treat complex PPS. The surgical approach is technically exceptionally challenging, and results are therefore quite mixed. Regardless of the treatment choice, postoperative results are highly variable and often unpredictable.

A case study serves to demonstrate the application of numerical simulations to predict postoperative pulmonary flow distribution in the setting of PPS. Three male pediatric patients (age: 1.8–2.9 years) undergoing surgical repair for PPS with near systemic RV hypertension and pulmonary flow disparity with 65–85% of flow to the right pulmonary artery were included. 3D CT-based preoperative PA models were constructed using SimVascular (simvascular.org) [35, 36]. To perform a virtual surgical repair, the initial, diseased model of the PAs with PPS was modified as per the surgeon's specifications. The resulting patient-specific models were discretized into meshes of ~1.5 million elements. Blood flow was computed with a 3D FSI model [54] under incompressibility and Newtonian assumptions. Pulmonary mean flow rates were determined from catheterization-derived values. Inflow waveforms were taken from Tang et al. [55] and scaled iteratively with Young's modulus for the wall, and outflow boundary conditions to match preoperative PA systolic, diastolic, and mean pressures obtained by catheterization. For outflow boundary conditions, Windkessel models were employed to model the distal PAs [47]. In contrast to previous surgical designs, outflow boundary conditions were adapted in addition to the geometric modification in order to account for distal vascular adaptation (vasodilation) in response to increased blood flow [56, 57]. A structured tree model representing the distal vascular bed in conjunction with the constant wall shear stress (WSS) assumption was used to estimate the postoperative resistance [44]. For each outlet, a baseline structured tree was generated such that the total resistance was equal to the resistance of the Windkessel model used for the preoperative simulations (Fig. 14.4a). A simulation was then performed on the virtual postoperative model using the same preoperative boundary conditions. By comparing flow changes, the segments in the baseline structured trees could be dilated or constricted to restore the WSS level of the preoperative stage. A new set of outlet resistances was defined by the adapted structured trees and applied to the postoperative model. Figure 14.4b summarizes the major steps for modeling surgical repair of PPS.

Figure 14.5 shows the mean pressure distribution and velocity magnitude between pre- and postoperative stages. Table 14.1 compares the predicted postoperative flow distribution using adaptive and constant outflow boundary conditions to in vivo lung perfusion scans, suggesting that satisfactory agreement with in vivo data could be achieved by incorporating the downstream vascular adaptation in response to the surgical repair into the simulation. The "acceptable" level of flow maldistribution or whether trading uneven distribution for slightly lower pressure is currently unknown; pre-op surgical planning,

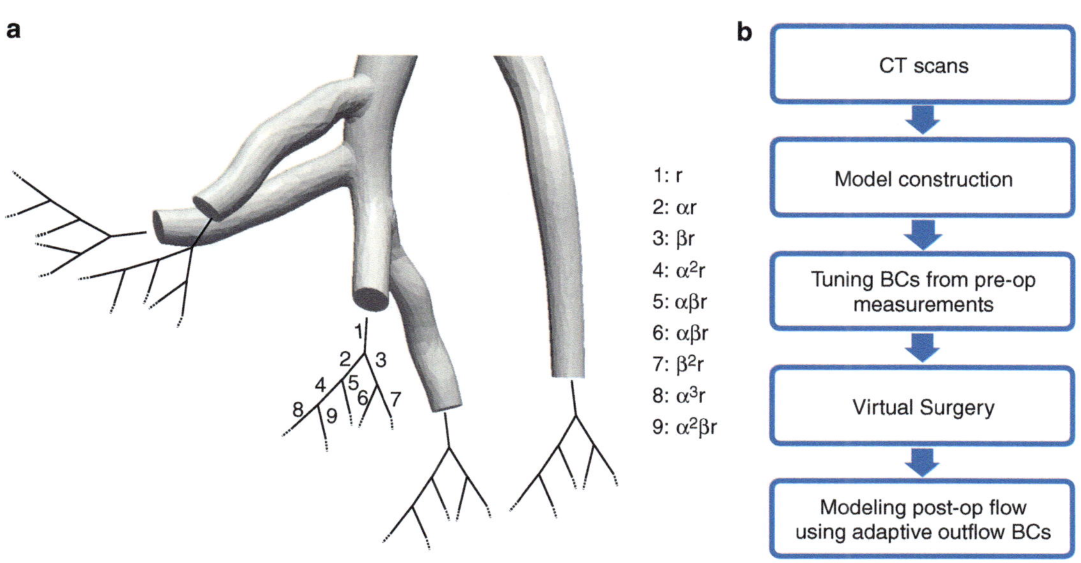

Fig. 14.4 (**a**) An illustration of the structured tree model used to estimate the downstream vascular resistance. In the structured tree, the child segment size is scaled by α and β, starting from the root segment with a radius of r. (**b**) Flow chart for modeling postoperative pulmonary flow distribution in patients undergoing PPS repair (modified and reproduced with permission from [50, 58])

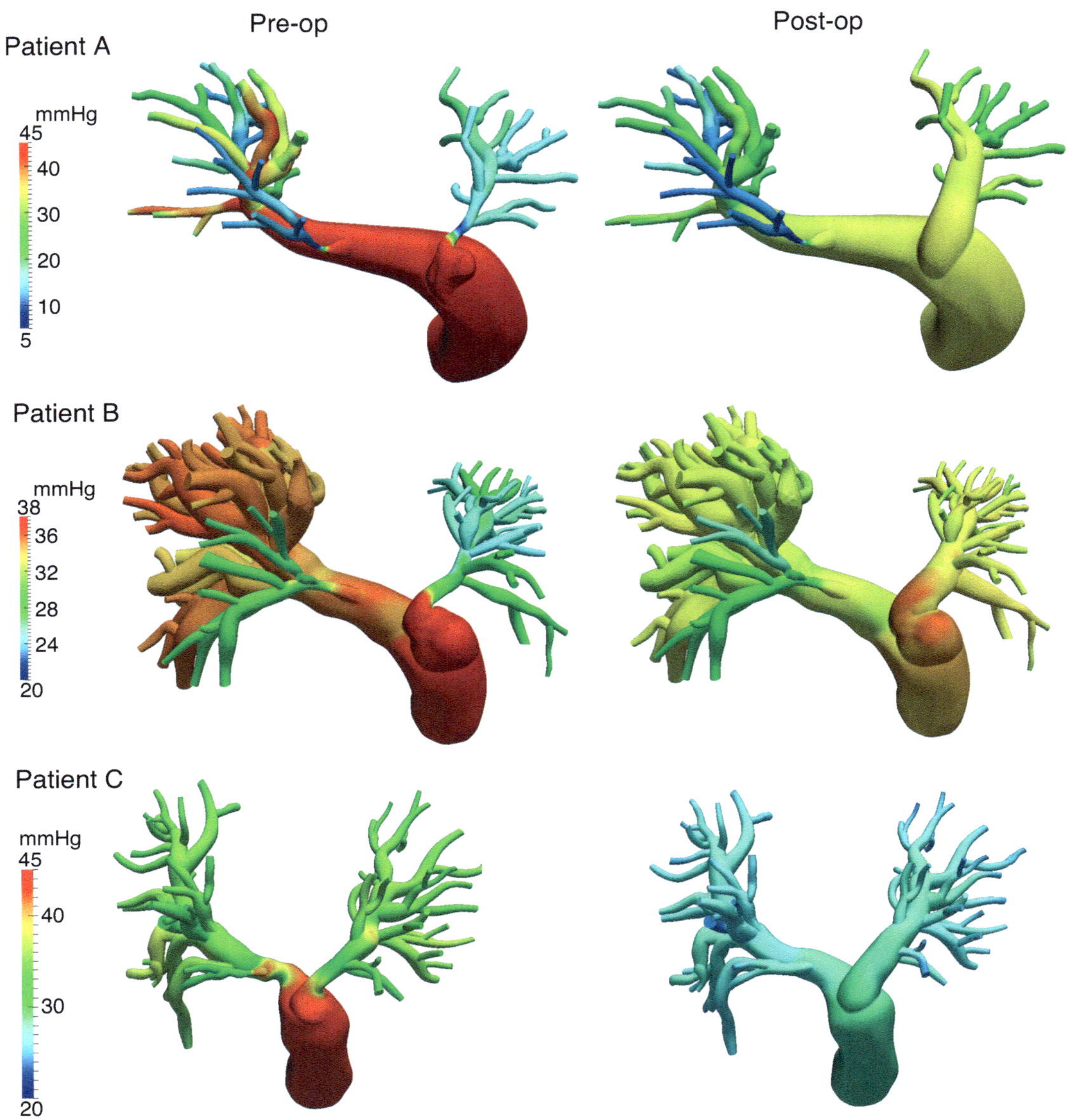

Fig. 14.5 Pre- and post-operative mean pressure distribution for three Alagille patients with PPS. The preoperative baseline boundary conditions are tuned to match the clinical measurements. Postoperative results are obtained with adaptive outflow boundary conditions (modified and reproduced with permission from [58])

Table 14.1 Validation of simulation-based surgical treatment of PPS

Patient	Measured: post-op LPS	Simulation: constant BCs	Simulation: adaptive BCs	Error (constant → adaptive)
1	53%/47%	65%/35%	52%/48%	23% →2%
2	63%/37%	74%/26%	67%/33%	17% →6%
3	49%/51%	59%/41%	54%/46%	20% →10%

Results listed as right%/left%; Simulations based on surgeon-directed virtual postoperative models. Prediction errors relative to measured RPA percent flow. *LPS* lung perfusion scan; *BC* boundary conditions however, allows for assessment of multiple scenarios and selection of the clinician's preferred choice. For more information, we refer the reader to Yang et al. [50, 58].

14.5 Challenges and Future Work

The pulmonary circulation has received far less modeling attention when compared with the systemic circulation or the left heart, yet it carries a crucial function. More recently, it has been the focus of a growing number of research groups. Validation of predictive models remains a challenge, due to the need for ongoing assumptions in the absence of actual clinical or basic science data [33, 58, 59]. Both somatic growth and congenital and acquired disease-related pulmonary vascular remodeling remain significant challenges [46], and while animal models, when available, may provide insights into pulmonary disease model characterization [60] and predictive model validation [61, 62], the number available and applicability to human disease suggest the need for a greater effort toward understanding the basic pathophysiology of human pulmonary vascular disease to allow for accurate representation in computational models.

References

1. Truong QA, Massaro JM, Rogers IS, Mahabadi AA, Kriegel MF, Fox CS, et al. Reference values for normal pulmonary artery dimensions by noncontrast cardiac computed tomography the Framingham Heart Study. Circ Cardiovasc Imaging. 2012;5(1):147–54.
2. Huang W, Yen RT, McLaurine M, Bledsoe G. Morphometry of the human pulmonary vasculature. J Appl Physiol Bethesda Md 1985. 1996;81(5):2123–2133.
3. Wiebe BM, Laursen H. Human lung volume, alveolar surface area, and capillary length. Microsc Res Tech. 1995;32(3):255–62.
4. Sobin SS, Tremer HM, Fung YC. Morphometric basis of the sheet-flow concept of the pulmonary alveolar microcirculation in the cat. Circ Res. 1970;26(3):397–414.
5. Townsley MI. Structure and composition of pulmonary arteries, capillaries and veins. Compr Physiol. 2012;2:675–709.
6. Brody JS, Stemmler EJ, DuBois AB. Longitudinal distribution of vascular resistance in the pulmonary arteries, capillaries, and veins. J Clin Invest. 1968;47(4):783–99.
7. Siegel LC, Pearl RG, Shafer SL, Ream AK, Prielipp RC. The longitudinal distribution of pulmonary vascular resistance during unilateral hypoxia. Anesthesiology. 1989;70(3):527–32.
8. Cheng CP, Herfkens RJ, Taylor CA, Feinstein JA. Proximal pulmonary artery blood flow characteristics in healthy subjects measured in an upright posture using MRI: the effects of exercise and age. J Magn Reson Imaging. 2005;33(3):257–65.
9. West JB, Dollery CT. Distribution of blood flow and ventilation-perfusion ratio in the lung, measured with radioactive carbon dioxide. J Appl Physiol. 1960;15:405–10.
10. Glenny RW, Bernard S, Robertson HT, Hlastala MP. Gravity is an important but secondary determinant of regional pulmonary blood flow in upright primates. J Appl Physiol Bethesda Md 1985. 1999;86(2):623–632.
11. Glenny RW, Robertson HT. Determinants of pulmonary blood flow distribution. Compr Physiol. 2011;1(1):39–59.
12. Hill G. Repair and follow-up of Tetralogy of Fallot with pulmonary stenosis. Congenit Heart Dis. 2013;8(2):174–7.
13. Monge MC, Mainwaring RD, Sheikh AY, Punn R, Reddy VM, Hanley FL. Surgical reconstruction of peripheral pulmonary artery stenosis in Williams and Alagille syndromes. J Thorac Cardiovasc Surg. 2013;145(2):476–81.
14. Drubach LA, Jenkins KJ, Stamoulis C, Palmer EL, Lee EY. Evaluation of primary pulmonary vein stenosis in children: comparison of radionuclide perfusion lung scan and angiography. Am J Roentgenol. 2015;205(4):873–7.
15. Rodriguez RM, Feinstein JA, Chan FP. CT-defined phenotype of pulmonary artery stenoses in Alagille syndrome. Pediatr Radiol. 2016 Jul;46(8):1120–7.
16. Khosroshahi HE, Uluoğlu O, Olguntürk R, Başaklar C. Keutel syndrome: a report of four cases. Eur J Pediatr. 1989;149(3):188–91.
17. Collins RT. Cardiovascular disease in Williams syndrome. Circulation. 2013;127(21):2125–34.
18. Fathallah M, Krasuski RA. Pulmonic valve disease: review of pathology and current treatment options. Curr Cardiol Rep. 2017;19(11):108.
19. Eltayeb O, Mongé MC, Popescu AR, Sarwark AE, Harris T, Backer CL. Discontinuous pulmonary artery. World J Pediatr Congenit Heart Surg. 2017;8(1):106–10.
20. Vida VL, Guariento A, Milanesi O, Gregori D, Stellin G, Scimitar Syndrome Study Group. The natural history and surgical outcome of patients with scimitar syndrome: a multi-centre European study. Eur Heart J. 2018;39(12):1002–11.
21. DiLorenzo MP, Santo A, Rome JJ, Zhang H, Faerber JA, Mercer-Rosa L, et al. Pulmonary vein stenosis:

outcomes in children with congenital heart disease and prematurity. Semin Thorac Cardiovasc Surg. 2019;31(2):266–73.
22. Latson LA, Prieto LR. Congenital and acquired pulmonary vein stenosis. Circulation. 2007;115(1):103–8.
23. D'Alto M, Mahadevan VS. Pulmonary arterial hypertension associated with congenital heart disease. Eur Respir Rev Off J Eur Respir Soc. 2012;21(126):328–37.
24. Abman SH, Ivy DD. Recent progress in understanding pediatric pulmonary hypertension. Curr Opin Pediatr. 2011;23(3):298–304.
25. Abman SH, Hansmann G, Archer SL, Ivy DD, Adatia I, Chung WK, et al. Pediatric pulmonary hypertension: guidelines from the American Heart Association and American Thoracic Society. Circulation. 2015;132(21):2037–99.
26. Roofthooft MTR, Bergman KA, Waterbolk TW, Ebels T, Bartelds B, Berger RMF. Persistent pulmonary hypertension of the newborn with transposition of the great arteries. Ann Thorac Surg. 2007;83(4):1446–50.
27. Zijlstra WM, Elmasry O, Pepplinkhuizen S, Ivy DD, Bonnet D, Luijendijk P, et al. Pulmonary arterial hypertension in children after neonatal arterial switch operation. Heart Br Card Soc. 2017;103(16):1244–9.
28. Capuano F, Loke Y-H, Cronin I, Olivieri L, Balaras E. Computational study of pulmonary flow patterns after repair of transposition of great arteries. J Biomech Eng. 2019;
29. Bossers SSM, Cibis M, Kapusta L, Potters WV, Snoeren MM, Wentzel JJ, et al. Long-term serial follow-up of pulmonary artery size and wall shear stress in Fontan patients. Pediatr Cardiol. 2016;37:637–45.
30. Morgan VL, Graham TP, Roselli RJ, Lorenz CH. Alterations in pulmonary artery flow patterns and shear stress determined with three-dimensional phase-contrast magnetic resonance imaging in Fontan patients. J Thorac Cardiovasc Surg. 1998;116(2):294–304.
31. Cunningham JW, McElhinney DB, Gauvreau K, Bergersen L, Lacro RV, Marshall AC, et al. Outcomes after primary transcatheter therapy in infants and young children with severe bilateral peripheral pulmonary artery stenosis. Circ Cardiovasc Interv. 2013;6(4):460–7.
32. Reddy VM, Liddicoat JR, Hanley FL. Midline one-stage complete unifocalization and repair of pulmonary atresia with ventricular septal defect and major aortopulmonary collaterals. J Thorac Cardiovasc Surg. 1995;109(5):832–44. discussion 844–845
33. Vignon-Clementel IE, Marsden AL, Feinstein JA. A primer on computational simulation in congenital heart disease for the clinician. Prog Pediatr Cardiol. 2010;30(1–2):3–13.
34. Arbia G, Corsini C, Esmaily Moghadam M, Marsden AL, Migliavacca F, Pennati G, et al. Numerical blood flow simulation in surgical corrections: what do we need for an accurate analysis? J Surg Res. 2014;186(1):44–55.
35. Schmidt JP, Delp SL, Sherman MA, Taylor CA, Pande VS, Altman RB. The Simbios national center: systems biology in motion. Proc IEEE Special Issue Comput Syst Biol. 2008;96(8):1266–80.
36. Lan H, Updegrove A, Wilson NM, Maher GD, Shadden SC, Marsden AL. A re-engineered software interface and workflow for the open-source SimVascular cardiovascular modeling package. J Biomech Eng. 2018;140(2)
37. Arbia G, Corsini C, Baker C, Pennati G, Hsia T-Y, Vignon-Clementel IE. Pulmonary hemodynamics simulations before stage 2 single ventricle surgery: patient-specific parameter identification and clinical data assessment. Cardiovasc Eng Technol. 2015;6(3):268–80.
38. Pant S, Corsini C, Baker C, Hsia T-Y, Pennati G, Vignon-Clementel IE. Data assimilation and modelling of patient-specific single-ventricle physiology with and without valve regurgitation. J Biomech. 2016;49(11):2162–73.
39. Pant S, Corsini C, Baker C, Hsia T-Y, Pennati G, Vignon-Clementel IE. Inverse problems in reduced order models of cardiovascular haemodynamics: aspects of data assimilation and heart rate variability. J R Soc Interface. 2017;14(126)
40. Vignon-Clementel IE, Figueroa CA, Jansen KE, Taylor CA. Outflow boundary conditions for 3D simulations of non-periodic blood flow and pressure fields in deformable arteries. Comput Methods Biomech Biomed Engin. 2010;13(5):625–40.
41. Vignon-Clementel IE, Figueroa CA, Jansen KE, Taylor CA. Outflow boundary conditions for three-dimensional finite element modeling of blood flow and pressure in arteries. Comput Methods Appl Mech Eng. 2006;195(29–32):3776–96.
42. Olufsen MS, Hill NA, Vaughan GDA, Sainsbury C, Johnson M. Rarefaction and blood pressure in systemic and pulmonary arteries. J Fluid Mech. 2012;705:280–305.
43. Spilker RL, Feinstein JA, Parker DW, Reddy VM, Taylor CA. Morphometry-based impedance boundary conditions for patient-specific modeling of blood flow in pulmonary arteries. Ann Biomed Eng. 2007;35(4):546–9.
44. Olufsen MS. Structured tree outflow condition for blood flow in larger systemic arteries. Am J Phys. 1999;276(1 Pt 2):H257–68.
45. Esmaily Moghadam M, Vignon-Clementel IE, Figliola R, Marsden AL. A modular numerical method for implicit 0D/3D coupling in cardiovascular finite element simulations. J Comput Phys. 2013;244:63–79.
46. Corsini C, Baker C, Baretta A, Biglino G, Hlavacek AM, Hsia T-Y, et al. Integration of clinical data collected at different times for virtual surgery in single ventricle patients: a case study. Ann Biomed Eng. 2015;43(6):1310–20.
47. Troianowski G, Taylor CA, Feinstein JA, Vignon-Clementel I. Three-dimensional simulations in Glenn patients: clinically based boundary conditions,

hemodynamic results and sensitivity to input data. J Biomech Eng. 2011;133(11):111006.

48. Caiazzo A, Guibert R, Boudjemline Y, Vignon-Clementel IE. Blood flow simulations for the design of stented valve reducer in enlarged ventricular outflow tracts. Cardiovasc Eng Technol. 2015;6(4):485–500.
49. Yang W, Dong M, Rabinovitch M, Chan FP, Marsden AL, Feinstein JA. Evolution of hemodynamic forces in the pulmonary tree with progressively worsening pulmonary arterial hypertension in pediatric patients. Biomech Model Mechanobiol. 2019;
50. Yang W, Feinstein JA, Vigon-Clementel IE. Adaptive outflow boundary conditions improve post-operative predictions after repair of peripheral pulmonary artery stenosis. Biomech Model Mechanobiol.
51. Acosta S, Puelz C, Rivière B, Penny DJ, Brady KM, Rusin CG. Cardiovascular mechanics in the early stages of pulmonary hypertension: a computational study. Biomech Model Mechanobiol. 2017;16(6):2093–112.
52. Krantz ID, Piccoli DA, Spinner NB. Alagille syndrome. J Med Genet. 1997;34(2):152–7.
53. Turnpenny PD, Ellard S. Alagille syndrome: pathogenesis, diagnosis and management. Eur J Hum Genet EJHG. 2012;20(3):251–7.
54. Figueroa CA, Vignon-Clementel IE, Jansen KE, Hughes TJR, Taylor CA. A coupled momentum method for modeling blood flow in three-dimensional deformable arteries. Comput Methods Appl Mech Eng. 2006;195(41–43):5685–706.
55. Tang BT, Fonte TA, Chan FP, Tsao PS, Feinstein JA, Taylor CA. Three-dimensional hemodynamics in the human pulmonary arteries under resting and exercise conditions. Ann Biomed Eng. 2011;39(1): 347–58.
56. Kamiya A, Togawa T. Adaptive regulation of wall shear stress to flow change in the canine carotid artery. Am J Phys. 1980;239(1):H14–21.
57. Langille BL, O'Donnell F. Reductions in arterial diameter produced by chronic decreases in blood flow are endothelium-dependent. Science. 1986;231(4736):405–7.
58. Yang W, Hanley FL, Chan FP, Marsden AL, Vignon-Clementel IE, Feinstein JA. Computational simulation of postoperative pulmonary flow distribution in Alagille patients with peripheral pulmonary artery stenosis. Congenit Heart Dis. 2018;13(2):241–50.
59. Yang W, Chan FP, Reddy VM, Marsden AL, Feinstein JA. Flow simulations and validation for the first cohort of patients undergoing the Y-graft Fontan procedure. J Thorac Cardiovasc Surg. 2015;149(1):247–55.
60. Qureshi MU, Colebank MJ, Paun LM, Ellwein Fix L, Chesler N, Haider MA, et al. Hemodynamic assessment of pulmonary hypertension in mice: a model-based analysis of the disease mechanism. Biomech Model Mechanobiol. 2019 Feb;18(1):219–43.
61. Audebert C, Bekheit M, Bucur P, Vibert E, Vignon-Clementel IE. Partial hepatectomy hemodynamics changes: experimental data explained by closed-loop lumped modeling. J Biomech. 2017;04(50):202–8.
62. Pewowaruk R, Roldán-Alzate A. 4D Flow MRI estimation of boundary conditions for patient specific cardiovascular simulation. Ann Biomed Eng. 2019;47(8):1786–98.
63. Yang W, Vignon-Clementel IE, Troianowski G, Reddy VM, Feinstein JA, Marsden AL. Hepatic blood flow distribution and performance in conventional and novel Y-graft Fontan geometries: a case series computational fluid dynamics study. J Thorac Cardiovasc Surg. 2012;143(5):1086–97.

Modelling Pulmonary Arterial Hypertension: Clinical Concepts, Engineering Applications and an Integrated Medico-engineering Approach

15

Ryo Torii and Vivek Muthurangu

15.1 Introduction

Historically, pulmonary hypertension (PH) is defined as a mean pulmonary arterial pressure (mPAP) $\geq$ 25 mmHg associated with a raised pulmonary vascular resistance *(PVR)* [1]. More recently, a lower threshold of mPAP >20 mmHg and pulmonary vascular resistance (PVR) $\geq$ 3 Wood units have been proposed [2]. In general, prognosis is poor with increased morbidity and mortality due to progressive right ventricular (RV) failure [3]. Specific disease-modifying pharmaceutical therapies are continually being developed with some success in improving prognosis [4]. However, further improvements still need to be made and novel engineering approaches may hold the key to new device and drug interventions.

15.2 Clinical Features of PH

The most common clinical manifestation of PH is shortness of breath (particularly on exertion), affecting approximately 60% of patients at diagnosis [5]. Other symptoms include fatigue, syncope and, later in the course of the disease, symptoms of right ventricular failure [6]. Unfortunately, patients with concomitant respiratory or cardiac pathologies may already exhibit such symptoms [7]. Thus, the diagnosis of PH may be delayed until symptoms are obviously out of proportion with the existing pathology. Reported symptoms are used in the World Health Organization (WHO) functional classification of the severity of PH shown in Table 15.1.

Specific physical signs of PH are usually only detectable late in the course of the disease. These include a right ventricular heave, a closely split second heart sound with a loud pulmonic component, the murmur of tricuspid regurgitation, raised jugular venous pressure, hepatomegaly and peripheral cyanosis [6–8]. Again, these signs may already be present in patients with existing pathology. Although most patients present in a similar fashion, PH is a heterogeneous disease with multiple aetiologies and clinical classification has important implications for treatment and prediction of prognosis.

15.2.1 Clinical Classification

In 2013, a major reclassification of PH was undertaken at the fifth World Symposium on Pulmonary Hypertension held in Nice [9]. The Nice 2013 classification (Table 15.2) splits PH into 5 groups and is based on pathophysiological mechanisms, clinical presentation and

R. Torii (✉) · V. Muthurangu
University College London, London, UK
e-mail: r.torii@ucl.ac.uk; v.muthurangu@ucl.ac.uk

G. Butera et al. (eds.), *Modelling Congenital Heart Disease*,
https://doi.org/10.1007/978-3-030-88892-3_15

Table 15.1 World Health Organization (WHO) Functional Assessment for Pulmonary Hypertension (PH)

WHO class	Description
Class I	Patients with PH but without resulting limitation of physical activity. Ordinary physical activity does not cause undue dyspnoea or fatigue, chest pain or near syncope.
Class II	Patients with PH resulting in a slight limitation of physical activity. They are comfortable at rest. Ordinary physical activity causes undue dyspnoea or fatigue, chest pain or near syncope.
Class III	Patients with PH resulting in marked limitation of physical activity. They are comfortable at rest. Less than ordinary activity causes undue dyspnoea or fatigue, chest pain or near syncope.
Class IV	Patients with PH with inability to carry out any physical activity without symptoms. These patients manifest signs of right heart failure. Dyspnoea and/or fatigue may even be present at rest. Discomfort is increased by any physical activity.

Table 15.2 Updated classification of pulmonary hypertension

1. Pulmonary arterial hypertension
1.1 Idiopathic PAH
1.2 Heritable PAH
1.2.1 BMPR2
1.2.2 ALK-1, ENG, SMAD9, CAV1, KCNK3
1.2.3 Unknown
1.3 Drug and toxin induced
1.4 Associated with:
1.4.1 Connective tissue disease
1.4.2 HIV infection
1.4.3 Portal hypertension
1.4.4 Congenital heart diseases
1.4.5 Schistosomiasis
1′ Pulmonary veno-occlusive disease and/or pulmonary capillary haemangiomatosis
1″. Persistent pulmonary hypertension of the newborn (PPHN)
2. Pulmonary hypertension due to left heart disease
2.1 Left ventricular systolic dysfunction
2.2 Left ventricular diastolic dysfunction
2.3 Valvular disease
2.4 Congenital/acquired left heart inflow/outflow tract obstruction and congenital cardiomyopathies
3. Pulmonary hypertension due to lung diseases and/or hypoxia
3.1 Chronic obstructive pulmonary disease
3.2 Interstitial lung disease
3.3 Other pulmonary diseases with mixed restrictive and obstructive pattern
3.4 Sleep-disordered breathing
3.5 Alveolar hypoventilation disorders
3.6 Chronic exposure to high altitude
3.7 Developmental lung diseases
4. Chronic thromboembolic pulmonary hypertension (CTEPH)
5. Pulmonary hypertension with unclear multifactorial mechanisms
5.1 Haematologic disorders: chronic haemolytic anaemia, myeloproliferative disorders, splenectomy
5.2 Systemic disorders: sarcoidosis, pulmonary histiocytosis, lymphangioleiomyomatosis
5.3 Metabolic disorders: glycogen storage disease, Gaucher disease, thyroid disorders
5.4 Others: tumoral obstruction, fibrosing mediastinitis, chronic renal failure, segmental PH

therapeutic options. The rest of this chapter will primarily concentrate on group 1 patients who have pulmonary arterial hypertension (PAH), associated with disease of small pulmonary arteries. This group contains important subgroups including idiopathic PAH (IPAH) in which PAH is a primary disease and PAH associated with other diseases such as congenital heart disease and connective tissue disease. Other PH groups are also important, but the benefits of PH-specific therapies in these patients are less well understood.

15.2.2 Natural History

The natural history of PH is dependent on the underlying aetiology and the haemodynamic state of the cardiovascular system [10]. The NIH registry study conducted between 1981 and 1985 involved 165 patients with IPAH. This study was undertaken prior to the advent of specific treatments for PH and thus represents the closest approximation to the true natural history of IPAH [10]. One-year survival was approximately 68% [10], and the median survival time was 2.5 years. In this group, prognosis was closely associated with haemodynamic parameters, i.e. *mPAP*, RV cardiac output and right atrial pressure [11]. Specific treatments for PH have altered the natural history of IPAH, with prostacyclin and endothelin receptor antagonists improving 1-year survival and increasing median survival time [4].

The natural history of other forms of PH is less well described. In patients with PH associated with congenital systemic to pulmonary shunts, although *mPAP* is generally higher than in IPAH mortality that is lower (1-year survival 92% [10]). In PH associated with HIV infection, 1-year survival is 60%, and thus, it has a poorer prognosis than IPAH [10]. Irrespective of underlying aetiology, the majority of patients die of right-sided heart failure, although some die of sudden cardiac death secondary to arrhythmias or pulmonary thromboembolism.

15.2.3 Clinical Investigations

Right heart catheterisation is the 'gold standard' method of assessing PH [12]. However, other clinical investigations are important in the initial diagnostic workup and in subsequent follow-up. Electrocardiography (ECG) often reveals evidence of RV hypertrophy (right axis deviation and right bundle branch block) [7, 10], while on chest X-ray, there is often marked tapering or pruning of the more distal pulmonary arteries. A characteristic finding on pulmonary function testing is a low carbon monoxide transfer factor due to the reduced perfusion of the respiratory gas exchange surfaces.

However, the most important non-invasive test in PH is trans-thoracic echocardiography (ECHO). Two-dimensional (2D) ECHO is used to assess RV hypertrophy and function, while Doppler ECHO is used to measure systolic pulmonary artery pressure through the quantification of the peak tricuspid regurgitant jet velocity and estimation of pressure using the modified Bernoulli equation. Several studies have demonstrated reasonable correlation between ECHO-derived pressure and cardiac catheterisation findings [13–15]. However, this finding is not consistent with some studies demonstrating a discrepancy of more than 30% [16]. In addition, between 40% and 70% of patients with PH do not have analysable tricuspid regurgitant jets [13–15]. Therefore, ECHO is more useful in screening and assessing response to therapy than definitive assessment.

More recently, magnetic resonance imaging (MRI) has become an important part of the diagnosis and follow-up of patients with PH. Cardiovascular MRI is the reference standard method of assessing ventricular function. Pulmonary hypertension is characterised by right ventricular failure, and therefore, MRI enables surveillance of physiological deterioration. Several studies have shown that MRI measures of right ventricular volume and function are prognostic in both adult and paediatric PH [9, 17]. In addition, MRI has been used to estimate pulmonary artery pressure through evaluation of ventricular septal curvature [18].

More importantly from an engineering point of view, MRI is also the reference standard method of measuring blood flow using the velocity phase-contrast technique. This allows measurement of pulmonary blood flow, as well as more complex 4D flow evaluation that has been used as inputs to various computational models. Clinically, measurement of pulmonary blood flow is needed for calculation of PVR during MR right heart catheterisation [19]. This procedure involves combining right heart catheterisation that allows measurement of mPAP with MRI that allows measurement of blood flow.

15.3 Pathology in Pulmonary Hypertension

Pulmonary hypertension is due to increased PVR that results in increased pulmonary artery pressure. The consequence of elevated pressure is increased afterload, which ultimately leads to right ventricular failure. The cause of increased PVR is abnormal vessel tone and vascular remodelling.

15.3.1 Abnormal Vascular Tone

The presence of smooth muscle in the tunica media of pulmonary muscular arteries allows for precise control of vascular diameter. Smooth muscle tone is dependent on intracellular calcium, which activates calmodulin leading to myosin phosphorylation and subsequent smooth muscle contraction. Vascular tone describes the interplay between various pathways that control vessel diameter in order to maintain optimum cardiopulmonary function. In PH, there is evidence of derangement of these pathways, which results in vasoconstriction and increased PVR.

One of the most important pathways involves nitric oxide (NO). Nitric oxide is produced through oxidation of arginine by endothelial nitric oxide synthase (eNOS) [20]. The expression of eNOS is modulated by various stimuli including hypoxia, shear stress, growth factors and cytokines [21]. Endothelially derived NO acts in a paracrine fashion on pulmonary vascular smooth muscle, activating soluble guanylate cyclase and increasing the cytoplasmic cyclic guanosine monophosphate (cGMP) concentration. Increasing the concentration of cGMP results in smooth muscle relaxation and vasodilation by reducing intracellular calcium, through the action of cGMP-dependent protein kinases [22] (Fig. 15.1). Endothelial NOS is reduced in the lungs of patients with PH [23], and there is evidence to suggest that this results in reduced NO production [24]. In addition, eNOS-deficient mice are more prone to the development of PH [25], while eNOS gene transfer is protective against PH [26]. This suggests that reduced NO production is involved in the development of increased tone in PH and vasoconstriction in PH. This hypothesis is supported by the fact that inhaled NO reduces pulmonary arterial pressure in patients with PH.

Prostaglandin levels are also deranged in patients with PH. Prostaglandin is an end product of the arachidonic acid pathway, the last step being catalysed by prostacyclin synthase [27]. Prostaglandin binds to a smooth muscle G-protein-coupled receptor, which activates adenyl cyclase leading to an increase in cyclic adenosine monophosphate (cAMP). This leads to vasodilation through the action of cAMP-dependent protein kinases [7] (Fig. 15.1). Patients with PH have reduced urinary excretion of PGI_2 metabolites [28] as well as reduced levels of PGI_2 synthase in the pulmonary arterial system [29], which contribute to vasoconstriction.

Finally, PH patients have been shown to have abnormalities of their endothelin pathways. Endothelin 1 is a vasoactive peptide produced by endothelin-converting enzyme in the vascular endothelium. Endothelin 1 production is controlled by several stimuli, including hypoxia,

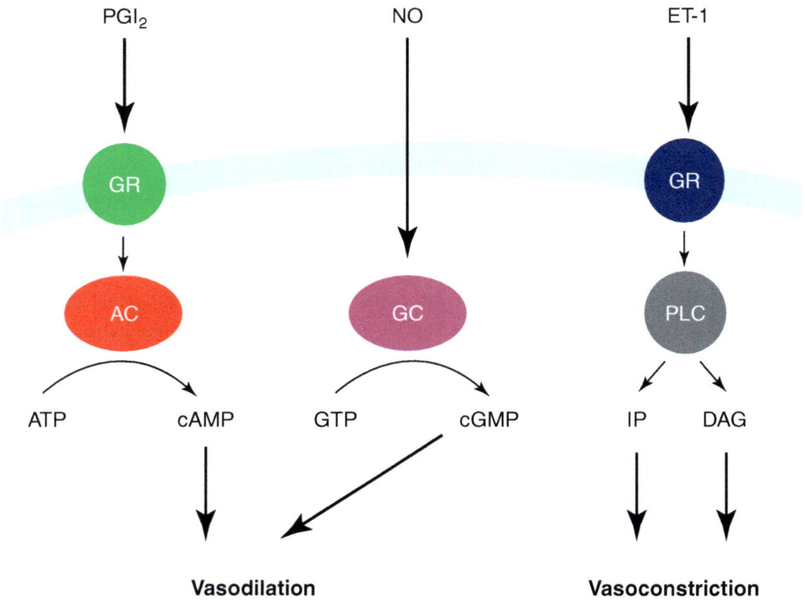

Fig. 15.1 The action of endothelial mediators on smooth muscle function. Prostacyclin (PGI_2), nitric oxide (NO), endothelin 1 (ET-1), G-protein-coupled receptor (GR), adenyl cyclase (AC), guanylate cyclase (GC), phospholipase C (PLC), cyclic adenosine monophosphate (cAMP), cyclic guanosine monophosphate (cAMP), inositol triphosphate (IP) and diacylglycerol (DAG)

shear stress, thrombus and ischaemia (which increase production) and NO (which decreases production [30]). Endothelin 1 acts on two different receptors, the endothelin receptor A (ET_A receptor), which is expressed on vascular smooth muscle cells, and the endothelin receptor B (ET_B receptor), which is expressed predominantly on endothelial cells [31]. Endothelin 1 binding to both ET_A and ET_B receptors leads to activation of phospholipase C, which converts membrane phosphatidylinositol to inositol triphosphate (IP) and diacylglycerol (DAG). In smooth muscle cells, these intracellular messengers cause an increase in intracellular calcium, with consequent smooth muscle contraction and vasoconstriction [32]. In endothelial cells, ET_B receptor activation leads to production of nitric oxide and prostaglandins, causing endothelially mediated vasodilation, and downregulation of ET-1 production [33]. In patients with PH, ET_A receptor concentration is increased, while ET_B receptor concentration is decreased [30]. In addition, there is an increase in plasma levels of ET-1 [34] associated with greater endothelial ET-1 immunoreactivity [23], demonstrating increased endothelial production of ET-1. The overall effect of these changes is increased vasoconstriction.

15.3.2 Vascular Remodelling

Although abnormal vascular tone may be an important aspect of increased PVR early in the disease, irreversible structural changes predominate in established disease. The pulmonary vasculature is a dynamic structure with a low constant turnover of cellular components. In PH, cellular hypertrophy and hyperplasia lead to the structural changes that increase PVR. The characteristic histopathological findings are as follows: intimal expansion, medial enlargement/neo-muscularisation and adventitial thickening. Further features of severe PH are the development of plexiform lesions (cellular conglomerates found within the vessel lumen) and in situ thrombosis. Medial enlargement is predominantly due to hypertrophy and hyperplasia of smooth muscle cells. In non-muscular arteries, neo-muscularisation is due to the surrounding pericytes taking on a smooth muscle phenotype. Increased muscularisation leads to a reduction in luminal diameter due to physical encroachment and increased vascular tone. There is also migration of fibroblasts into the media with increased deposition of extracellular matrix components particularly collagen and elastin. This leads to stiffening of the vessel wall and thus a reduction in compliance. There is also adventitial thickening due to fibroblast proliferation and increased deposition of extracellular matrix components. This is a form of outward remodelling that does not encroach on luminal diameter. However, it does have an effect on wall stiffness and thus vascular compliance. In PH, an imbalance between pro-proliferative and anti-proliferative factors is the mechanism of vascular remodelling.

Endothelin 1 is a potent promoter of remodelling causing proliferation of fibroblasts and smooth muscle cells, and increased production of extracellular matrix components. Both ET_A receptors and ET_B receptors are implicated in the effects through the action of calcium-dependent protein kinases [30, 35]. In addition, ET-1 increases endothelial expression of adhesion molecules, promoting intravascular thrombosis and cellular migration, both of which are implicated in vascular remodelling. The pivotal role the endothelin system plays in vascular remodelling has been demonstrated by the action of endothelin receptor antagonists. Both dual ET_A/ET_B receptor and selective ET_A receptor antagonists have been shown to decrease vascular remodelling in various animal models of PH. This translates into a reduction in *PVR* in patients with PH. Other pro-proliferative factors that are increased in PH are growth factors, angiotensin and serotonin.

Nitric oxide and PGI_2 are inhibitors of vascular remodelling. Nitric oxide has an anti-proliferative effect on smooth muscle and fibroblasts. In vitro studies demonstrate that NO inhibits smooth muscle and fibroblast proliferation through the induction of the cyclin-dependent kinase inhibitor p21 [36]. Furthermore, NO inhibits the release of pro-proliferative factors like ET-1. Prostacyclin has similar effects, reducing cellular proliferation and extracellular matrix deposition while inhibiting release of ET-1. The important anti-proliferative effects of NO and PGI_2 have been demonstrated in animal models of PH. Animals exposed to chronic hypoxia and NO undergo less vascular remodelling than animals exposed to hypoxia alone [37]. In addition, transgenic mice with overexpression of PGI_2 synthase did not undergo vascular remodelling after exposure to chronic hypoxia. The reduction in NO and PGI_2 activity thus promotes remodelling, exacerbating the effect of ET-1 upregulation. The structural sequelae of remodelling are reduced luminal diameter and vessel number which cause an increase in *PVR*.

15.4 Current Treatment Approaches in Pulmonary Hypertension

Treatment strategies in PH revolve around reducing PVR, pulmonary artery pressure and right ventricular afterload. This is achieved through targeting the pathways described above. The first vasodilators that were used in PH were calcium channel blockers. Calcium channels blockers work by reducing calcium influx into the smooth muscle cell, thereby reducing contraction and vascular tone. Calcium channel blockers have been shown to improve survival in IPAH patients in whom PVR or mPAP falls acutely in response to administration of vasodilators such as NO [38, 39]. However, it should be remembered that calcium channel blockers can have a negative inotropic effect. Thus, agents such as nifedipine and diltiazem are more appropriate than more myocardium-specific agents such as verapamil.

Although inhaled NO is used for acute vasodilation (i.e. in the intensive care setting), it is impractical for chronic use. However, the NO/cGMP pathway is targeted by several PH-specific drugs. The most commonly used are the phosphodiesterase 5 (PDE5) inhibitors such as Sildenafil [40]. Phosphodiesterase 5 breakdown cGMP, and therefore, inhibition results in increased intracellular cGMP concentration, smooth muscle relaxation and vasodilation.

The prostaglandin pathway can be targeted through supplementation with synthetic prostaglandins (i.e. epoprostenol) or analogues (i.e. iloprost and treprostinil). Increasing prostaglandin results in increased cAMP, and therefore smooth muscle relaxation and vasodilation. The optimum method of administration is continuously intravenous, although subcutaneous and inhaled preparations are available. Studies demonstrating significant survival benefit with these drugs [41].

Finally, the endothelin pathway can be targeted with endothelin receptor antagonists. Successful agents include both dual ET_A/ET_B (i.e. macitentan) receptor and selective ET_A receptor antagonists (i.e. ambrisentan). The benefit of these agents is that they are orally administered and have been shown to improve symptoms and survival [42]. Their action is complex but involves vasodilation and modification of vascular remodelling.

15.5 Modelling in PH

The application of computational modelling in PH has been sought over decades. In the last decade, computational fluid dynamic (CFD) modelling has come to make a real clinical impact in various areas; a typical example is the non-invasive CT-based fractional flow reserve (FFR) calculation [43]. The success of CT-based FFR calculation is due to the fact that there is a well-established clinical protocol of catheter-based FFR measurement to guide revascularisation, and the computational model is simply a non-invasive alternative of the protocol. Application of computational models to the diagnosis of PH may therefore be more challenging, because the diagnostic method is more versatile and there is no single gold standard, especially for paediatric patients [44].

In principle, computational models can provide unique sets of data that cannot be acquired directly through conventional measurement techniques. Wall shear stress (WSS) on the vascular endothelium, which regulates vascular remodelling, is a typical example of those data. Thanks to the cumulative expertise in translating such technology in clinical setting, and to the steady growth of computing power, it has become feasible to use computationally derived parameters in studies examining a large number of patients; for example, 509 patients' WSS was used as a predictor of atherosclerotic disease progression [45]. Many other studies took similar approach, especially in coronary artery applications, and a summary of good practice in cardiovascular flow modelling has recently been published for the clinical community [46].

One of the other unique capabilities of computational model can be found in prediction of the changes associated with anatomical and/or physiological variations, for example predicting haemodynamic alterations after removal of stenosis [47, 48] and predicting impact of novel surgical technique [49]. In this way, computational model can be used as a pre-operative treatment planning tool and/or testing environment of novel treatment.

In this chapter, state of the art of computational modelling applied in PH context is reviewed, aiming to highlight the most effective use of modelling and the remaining challenges.

15.6 Modelling Pulmonary Vascular Anatomy

An anatomical representation, in other words the spatial extent of the model, needs to be defined in a computational model framework. This can be as simple as one single pulmonary artery resistance representing entire pulmonary circulation or can be an extensive 3D vascular tree model. The former is useful to characterise pulmonary circulation as whole, and details are discussed in the previous section whereas the latter can be utilised to investigate the potential impact of local phenomena to the entire pulmonary circulation, such as the impact of stenosis [48, 50].

For 3D model reconstructions, CT angiograms are in general preferred because it depicts the vessel segments at relatively high resolution (~ 0.4 mm/pixel) with high definition of vessel lumen by the contrast agent. With this, several generations of the branching structure can directly be captured, and 3D anatomical models have been successfully reconstructed for subsequent CFD analyses [48, 50, 51] (Fig. 15.2). Reconstruction of the vascular tree was indeed possible from earlier time, even over larger extent of the tree [52], but the models used in computational flow analysis were rather limited to several upstream branches most likely because obtaining 3D flow characteristics in small vessel segments are not beneficial against the high cost of 3D computation across large tree structure. Additionally, successful attempts have been made to generate the tree structure including small branches that cannot be resolved in CT images, in which generation of branch structure was extended up to 10 based on morphometric law [53] (Fig. 15.2).

Although other imaging modalities can also be used to reconstruct anatomical models, each of them has major limitations and is not used as the mainstream modality for anatomical model reconstruction of pulmonary vessels. For example, optical coherence tomography can provide cross-sectional images of more than 10 times higher resolution (10–20 μm/pixel [54]), but it is an invasive procedure using a catheter and only limited spatial extent in the tree structure can be imaged, normally only a few vessel segments per patient. Cardiovascular magnetic resonance (CMR) is another potential imaging modality. Although the spatial resolution is limited (~1.5 mm/pixel depending on the temporal resolution and scan time [55]) to capture peripheral pulmonary arteries, its non-invasive nature is attractive, especially for longitudinal follow-up studies. CMR could become more attractive option in the near future considering recent advancement of resolution enhancement using machine learning [56].

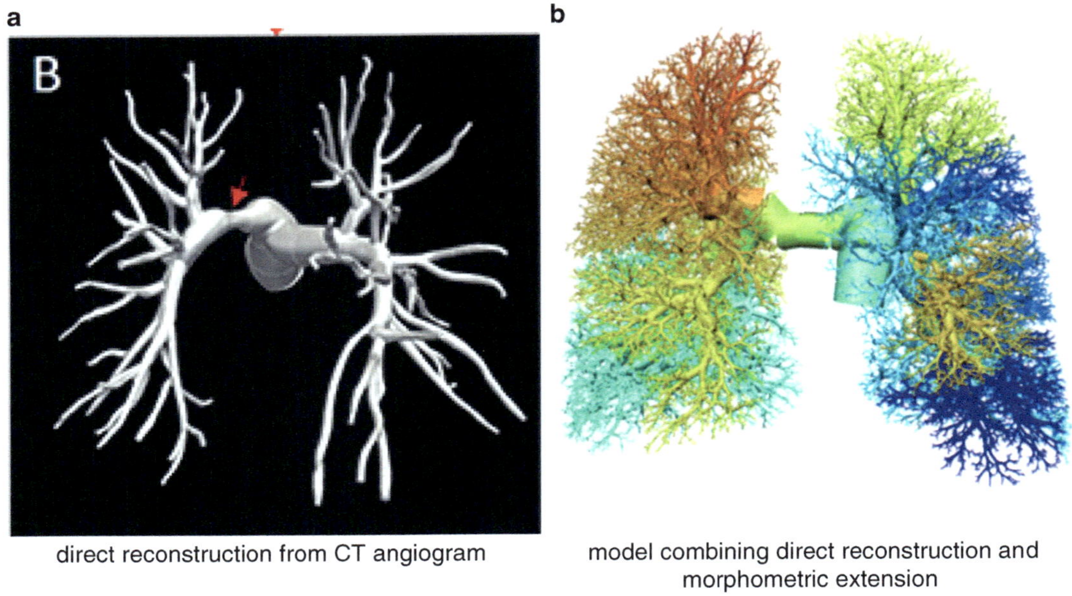

Fig. 15.2 Examples of pulmonary tree reconstruction: (**a**) only from CT images [50] and (**b**) CT-based model with morphometric extension of peripheral branches [53] (Republished with permission of The Royal Society (U.K.), from Burrowes KS, Clark AR, Marcinkowski A, Wilsher ML, Milne DG and Tawhai MH. Pulmonary embolism: predicting disease severity. *Philos Trans A Math Phys Eng Sci*. 369:4255-77 (2011); permission conveyed through Copyright Clearance Center, Inc.)

15.7 Simple Models of Pulmonary Vasculature to Characterise Entire Circulation

15.7.1 Lumped Parameter/0D Model

The simplest form of computational model in pulmonary arterial network is to investigate pressure–flow relationship using single pulmonary vascular resistance. This approach has a long history and is discussed in the previous section and in a review by Chemla et al. [57]. Combination of small number of resistance and compliance has been used as the simplest form to represent the vasculature, sometimes referred as Windkessel model. This can be extended to model a tree of vessels, either approximating the tree as a sheet consisting of many vessels [58] or approximating each vessel segment as a straight tube—typically ignoring tapering—to form an actual tree structure [59]. The latter is more extensively used, often referred as 0D model, lumped parameter model (LPM) or electric circuit analogue model. Here, blood pressure, blood flow and vascular resistance are modelled in analogous to voltage, electric current and resistance, respectively. Compliance of the vessel wall can additionally be included as electric capacitance, and in some advanced models, inertia of the blood and the presence of valve are incorporated with inductance (coil) and diode, respectively.

LPM has been used in studies of PH, for example to better interpret experimental observation of PH in animal models [59, 60]. In Tewari et al. [59], haemodynamics in mice pulmonary circulation with induced hypertension by hypoxic treatment was studied using a simple LPM including ventricles, and systemic and pulmonary circulation represented by only several resistors and capacitors that were calibrated with experimental data using least-squares approach. The model enabled illustration of different modes of systemic response to the hypoxia, e.g. an inward remodelling of the pulmonary artery in the late stage of PH. A more 'predictive' use of model can be found in the prediction of the fate of untreated PH cases [61]. The authors constructed models of both pulmonary and systemic circulations, including all 4 chambers of the heart. Their application of the model in 4 hypothetical PH scenarios—caused by distal arterial stenosis, left ventricular diastolic dysfunction, ventricular septal defect (VSD) and mitral stenosis—indicated that the PH due to VSD leads to the most rapid evolution of PH. However, the predicted evolution of PH appears to be too rapid or too much in long term. The likely reason behind it is that the model does not include adaptive remodelling of the system and the model parameters are all extracted from various literature. This highlights the importance of model calibration, but at the same time, it is not realistic to be able to identify >100 model parameters based on limited experimental data.

In order to incorporate the adaptive nature of the entire system, which plays an important role in the evolution of PH, attempts have been made to couple LPM circulation model with a model of heart including its adaptive nature. The work by Philip et al. on bleomycin-induced PH in mice [62] incorporated such a model to interpret their experimental data set on the development of PH and to predict impact of a cell therapy. The heart model employed there is a biventricular model represented by a combination of electric components, incorporating multiscale contribution of myocyte mechanics [63]. The authors highlighted that the deterioration of myocyte force generation capability plays the most significant role in the development of PH. While their model parameters were based on the literature values, they have included the right set of physiological mechanisms in the model, and the results were discussed comparative manner, which allowed them to draw a reasonable conclusion following their experimental finding. Other types of coupling, especially coupling of LPM and 1D/3D flow models, are common and they are discussed in the later sections.

Although not targeted on pulmonary circulation, an application of lumped parameter network model of cirrhogenesis is an example investigating hypertensive remodelling using modelling approach [64]. Here, the authors used a multi-level LPM of the entire circulatory system of rats with a detailed hepatic circulation whose morphological characteristics were derived from micro-CT scan. They simulated blood flow in various typical cirrhotic conditions, e.g. decreased mean arterial pressure due to vasodilation of peripheral vascular beds. The model illustrated that the morphological change in the vasculature due to cirrhosis causes maladaptive regulation of the blood flow, including increased pressure loss and PH as the result of cirrhosis.

The major advantage of LPM is clearly the possibility of fast computation even with models including wide anatomical territory and complex physiology. At the same time, the level of approximation in each part of the model is inherently high, and model parameters are not directly available from clinical measurements. Thus, the models including a large number of parameters face a challenge in parameter estimation. An intelligent decision-making process is therefore crucial on the complexity of the model, in reference to the research questions and the data available.

15.7.2 1D Model

One-dimensional models are capable to capture wave propagation in the vascular tree, which is considered as an important factor in the diagnosis of PH and not delivered by LPM. Wave intensity analysis (WIA) has been shown as an effective tool to study the age-related change in cardiovascular physiology [65] and to identify patients having proximal pulmonary arterial thrombosis in patients with CTEPH [66].

One-dimensional model involves less assumptions compared to the LPM. The governing equation is the incompressible Navier–Stokes equations in a cylindrical coordinate with axisymmetric approximation, which is often integrated over the vessel cross section to obtain 1D version of the equations [67]. This is still a reasonably accurate representation of fundamental fluid mechanics, especially in the smaller pulmonary arteries downstream to the first few branches

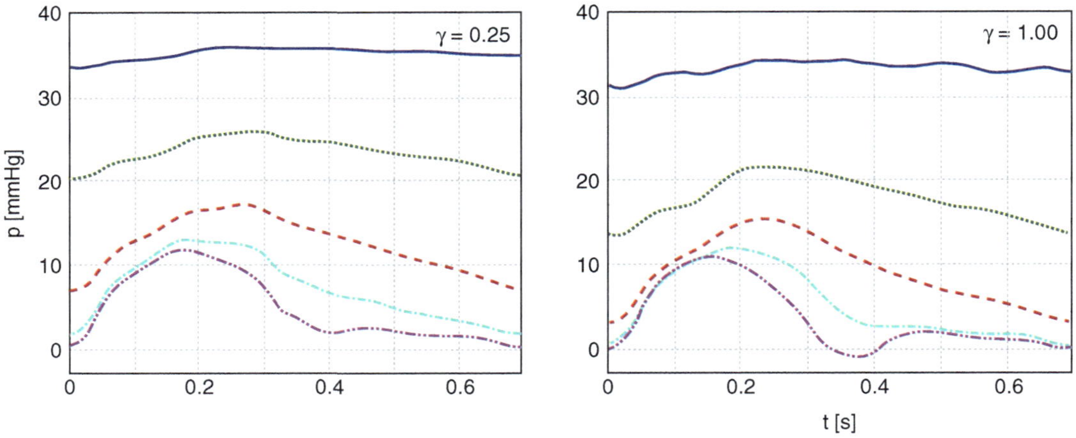

Fig. 15.3 Example results of 1D model haemodynamic computations in pulmonary arterial tree. Effects of vascular rarefaction on flow and pressure waveforms in main pulmonary trunk are examined. Cyan (dash-dot) curve is for normal physiological condition, and the blue curve is the highest degree of rarefaction. While rarefaction significantly affects the blood pressure, the effect of branching asymmetry is moderate (parameter $\gamma = 1$ is symmetry and $\gamma = 0.25$ is 1:4 branching). (Olufsen MS, Hill NA, Vaughan GD, Sainsbury C and Johnson M. Rarefaction and blood pressure in systemic and pulmonary arteries. *J Fluid Mech.* 705:280-305 (2012), reproduced with permission.)

where 3-dimensional nature of the flow is expected to play a less significant role. Even in large arteries such as the aorta where the flow is highly 3-dimensional, it has been shown that 1D model provides similar results as 3D model, especially in diastolic phase [68].

Apart from its simplified anatomy (axisymmetry and circular cross section), the cross-sectional velocity profile—often assumed parabolic—is a major assumption. Additionally, a tube law is often used representing the arterial wall mechanics in response to the blood pressure. A thin-wall approximation of blood vessel wall is typically employed to obtain the pressure–area relationship, assuming the vessel wall material to be linearly elastic [69]. While a more complex model capturing non-linear characteristics of the relationship has been proposed [70], the simple linear model appears to work well and be a preferred choice.

The computation of 1D model is much faster than 3D yet physically more representative than LPM. One-dimensional models have been applied to cover wide range of the circulatory system, e.g. human body circulation including some detailed structure of pulmonary vasculature [71] and even the cardiovascular system of other animals [72]. However, the application of 1D model to PH context is rather scarce. One of the few examples is Olufsen et al. [73] investigating the effect of arterial rarefaction, a reduction in vascular density in hypertensive patients. The study showed that rarefaction in the peripheral microcirculation elevates the mean blood pressure in large pulmonary arteries to >30 mmHg (Fig. 15.3). They also showed that stiffening of vasculature could double the peak pulmonary blood pressure for the same amount of blood flow. Their model was developed further to incorporate venous system [69] and used to study wave reflection in pulmonary circulation [74].

15.8 3D Model and the Value of 3D Data Output

CFD simulations using a 3D anatomical model are a popular method to study cardiovascular system. In many such studies, anatomical models are reconstructed from CT or cardiovascular magnetic resonance CMR images. As the resolution of images has become finer, typically 0.5 mm/pixel and 1 mm/voxel for CT and CMR, respectively, using standard scanners, reconstruction of such models has become relatively straightforward also thanks to many commercial and open-source tools available for anatomical reconstruction. Full 3D incompressible Navier–Stokes equations are solved here, which involves even less assumptions in terms of flow, compared to 0D/1D models. Additionally, flows associated with 3D geometrical features are fully considered. While fluid–structure interaction (FSI) analyses are not uncommon in these days, many of the 3D CFD studies assume the arterial wall to be rigid, which does not allow calculation of wave propagation throughout the vascular tree.

Among the variables derived from 3D CFD analysis, pressure and WSS are most commonly used in analyses of cardiovascular pathophysiology. WSS is often used in the attempt to predict disease progression in association with maladaptive vascular remodelling. In the context of PH, Tang et al. investigated potential value of WSS as a predictive biomarker for the progression of disease [75]. In their study, 3D patient-specific CFD was used to demonstrate significantly low WSS values in proximal pulmonary arteries of 5 PH patients, in comparison with the control group. The authors suggested that the WSS may alter the biomechanics-driven remodelling of vascular wall and the abnormally low WSS could lead to progression of PH. Other study by Kheyfets et al. sought to use WSS as a surrogate marker of known physiological biomarkers (vascular resistance, compliance and wave reflection index) to evaluate right ventricular (RV) afterload towards predicting progression of PH in individual patient [76, 77]. Spatial 3D distribution of WSS is a unique output from 3D CFD; hence, this approach is a strong investigation tool if the 3D feature of biomechanical factor carries a significant meaning in the clinical/pathophysiological problem.

A strength of 3D modelling can also be found in assessment of surgical or intravascular interventions. Spazzapan et al. conducted patient-specific CFD modelling of blood flow in the pulmonary artery of CTEPH patients [50]. They assessed the impact of surgical removal of a proximal pulmonary artery stenosis in terms of the pressure drop across the vascular tree and the flow ratio between left and right pulmonary arterial trees. The study by Yang et al. focused on peripheral pulmonary stenosis, and their 3D CFD model showed a relief of high blood pressure in the main pulmonary trunk, a better left-to-right balance of the flow and WSS after the removal of the stenosis [48]. These studies, though the number of subjects is still small ($n = 5$ [50] and $n = 2$ [48]), demonstrate that such a model could serve as a predictive tool of post-operative haemodynamics.

Biomechanical insight of hypertension has been investigated over decades [78], more in the systemic circulation where the interaction between elevated blood pressure and arterial wall stiffening is proposed as the key mechanism [79]. As an attempt to capture greater degree of the biomechanics involved in remodelling, FSI analysis could be utilised [80, 81], where the interaction between blood flow and arterial wall deformation is considered in the model. One of these studies demonstrated differences in biomechanics—wall deformation and WSS due to different blood pressure and how sensitive they are to hypothetical vascular remodelling (Fig. 15.4). This is a step forward towards more elaborated model including actual vascular remodelling, which can be found in the context of aneurysm growth prediction [82, 83]. Further in this direction, other suspected causes of hypertension, e.g. lung fibrosis [84], can be incorporated by altering the surrounding environment of the vascular system.

In reality, 3D models are often coupled with the lower-order models. Many of the studies mentioned above employ LPM (Windkessel model) as their outflow boundary conditions (BCs) reflecting arteriolar or microvascular networks whose anatomy cannot be captured by in vivo measurement. A more advanced form of outflow BC has been proposed even including the effect of respiration [85]. Since the pulmonary arteries are in the chest cavity and lungs, the effect of respiratory motion could give a considerable impact on the blood flow. However, cardiorespiratory coupling has been considered in pulmonary circulation only as the above-mentioned outflow BC or in simple LPM network model [86]. The impact of respiration was

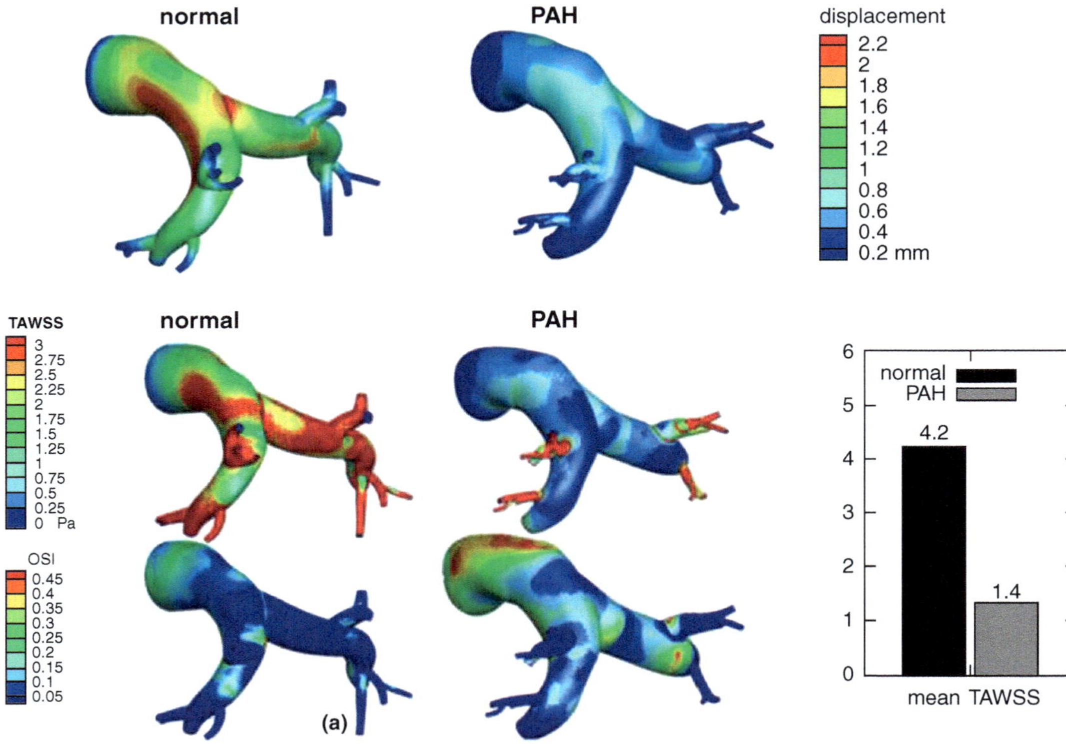

Fig. 15.4 3D FSI models of pulmonary circulation to highlight the difference in haemodynamic indices [80] (time-averaged WSS and oscillatory shear index—in the bottom panels—in this example). (Reprinted from 'Image-based computational assessment of vascular wall mechanics and hemodynamics in pulmonary arterial hypertension patients', Zambrano BA, McLean NA, Zhao X, Tan JL, Zhong L, Figueroa CA, Lee LC and Baek S. *J Biomech*. 68:84–92 (2018), with permission from Elsevier.)

taken into account in 3D model of venous circulation, e.g. in total cavopulmonary connection after Fontan surgery on congenital single ventricle defect [87], and the respiratory coupling may be a future focus since models of respiratory system exist, summarised by Tawhai and Lin [58].

The approach to couple multiple physiological components can be extended in another direction, to include the ventricles as in Shavik et al. [88] and Park et al. [89]. In both studies, the authors aimed to simulate heart failure associated with PH, and the impact of a ventricular assistive device (VAD) to relieve that. They coupled realistic 3D model of the left and right ventricles, simulated by means of finite element solid mechanic analysis, coupled with LPM network of the entire circulation including the VAD. Both successfully captured the unloading of the failed RV by the VAD, demonstrating that computational models can be used as a platform to tune the VAD against the degree of disease progression.

While there are output variables uniquely available from 3D computational models, these models take longer time to run, especially FSI often requiring large computing power. For example, in FSI analysis by Kong et al. [81], maximum of 10,240 processors were utilised. Therefore, similarly to the other types of the models, choice of the most effective method against the research question has a paramount importance. Calculation of WSS distribution on the pulmonary arterial wall can only be done by 3D CFD. Thrombogenic risk estimation (e.g. of aortic dissection [90] or in left atrial appendage [91] could be another attractive topic relevant to PH. Although no computational study in this particular direction seems to exist in pulmonary circulation, a flow-sensitive 4D MRI study showed complex 3D flow patterns in proximal pulmonary artery in relation to Eisenmenger disease [92]. On the other hand, simpler models may be sufficient if estimation of flow and/or pressure distributions across the arterial tree is the primary focus.

15.9 Challenges

As described above, computational models of PH have developed considerably in the last decade. Challenges are, however, still remaining towards making a more significant contribution to the clinical community.

15.9.1 Model Parameters

Among the inputs required to make the models represent real human patients, anatomical and inflow parameters are relatively straightforward to acquire. Anatomy of pulmonary arterial tree can be captured by CT or CMR images down to third or fourth generation of branches as in the examples already introduced. Inflow from the RV can be estimated by phase-contrast MRI [66] or echocardiogram [93].

Parameters representing peripheral microcirculation, which are typically used to define outflow boundary conditions, are more challenging. This is because in vivo flow measurement in the pulmonary microvasculature, such as perfusion imaging by positron emission tomography (PET) [94], is less common compared to the myocardial perfusion measurement [95]. If available, perfusion data can be used to calibrate peripheral resistance [96]. In the absence of such flow data, peripheral vascular resistances derived from the vascular anatomy are commonly used in many different types of modelling, for example outflow boundary condition in 3D models [76, 77]. The resistance can further be calibrated such that experimentally measured flow or pressure is reproduced [97]. Similar calibration approach is also taken in FSI analysis [80] to estimate elasticity of pulmonary arterial wall, which is a parameter difficult to be acquired in vivo for each patient. The use of more patient-specific data naturally leads to more representative computational results [77], but patient-specific in vivo data are not always available. Moreover, there is an important open question about how the physiological parameters would change, should one attempts to predict post-interventional haemodynamics [48]. A potential approach to complement the difficulty to acquire in vivo data is synthetically generated population database [98].

The data acquired in vivo have inherent uncertainty, which is expected to make a substantial impact on the output of the model. Quantification of such uncertainty has been explored. Examples include 1D murine pulmonary arterial model [99] and 3D FSI analysis of human coronary arterial tree [100]. Uncertainty quantification reveals how the variability of the input parameters propagates to the output variables. For instance in Colebank et al. [99], resistance of distal arterial segments has the highest impact on the pressure in the main pulmonary trunk, and the impact varies with the complexity of the model.

15.9.2 Model Validation

Assumptions are inevitable in computational models, and it is extremely important the model is verified and validated. In vitro models have been used for this purpose, especially with realistic vascular anatomy [101] and compliance [102]. Comparison between experimental measurement by particle image velocimetry and CFD showed a good agreement of velocity profiles in proximal pulmonary arteries although the flow predicted by CFD appears to miss some small-scale flow features [101]. Alternatively, flow-sensitive 4D MRI is a good modality to measure in vivo flow patterns and is successfully applied to visualise and quantify pulmonary arterial flows [92, 103], and validate computational model [104]. A new approach to fuse the experimental data and computational model, rather than comparing, has been taken more to complement the experimental uncertainty with the physical principles implemented in the numerical models [105, 106].

15.9.3 Accelerated Computation for Clinical Application

Most of the example studies introduced in this chapter involve less than 10 patients. Typical reasons behind the small number of patients are difficult to acquire large number of high quality and complete data set (e.g. including both CT for anatomy and CMR for inflow velocity) but also the long computational time required especially by 3D computations. Either one aims to carry out population study, or to provide information for clinical diagnostics, and results need to be obtained in relatively short timeframe. The largest population studied using computational model is, as far as the author is aware, PREDICTION study [45] to examine coronary artery WSS to predict atherosclerotic disease progression (n = 509). More complicated, pulsatile flow simulations of coronary arterial tree were performed to derive CT-based FFR (e.g. n = 254 in Norgaard et al. [107]). In PREDICTION study, only straight segment of vessels was examined under assumption of steady inflow, which was sufficient to demonstrate the added value of WSS to predict disease progression. Norgaard et al. invested large resource in their core laboratory, to conduct a more complex CFD simulations per patient. Should one aim to conduct clinically impactful study, therefore, the model complexity needs to be carefully considered in reference to the nature of the problem and the available resource.

15.10 Summary

A considerable number of studies to model and analyse pulmonary circulation have been conducted towards a better understanding of pulmonary hypertension. The wide spectrum of modelling approaches—0D, 1D, 3D, fluid–structure interaction, etc.—indicate that various tools are available, and researchers can select the most suitable one to answer their research question, most effectively in terms of the accuracy and required resources. So far, modelling of physiological adaptation/remodelling is included more in physically simple, 0D, 1D models or coupling those with 3D model in limited anatomical extent. More physically complex models such as fully 3D CFD or FSI are typically used in the form of sensitivity analysis, by alteration of anatomy or boundary conditions. In-depth understanding of both clinical questions and engineering methodology is essential, which can most likely be achieved by close interaction between clinicians and engineers.

References

1. Galie N, Torbicki A, Barst R, Dartevelle P, Haworth S, Higenbottam T, Olschewski H, Peacock A, Pietra G, Rubin LJ, Simonneau G, Priori SG, Garcia MA, Blanc JJ, Budaj A, Cowie M, Dean V, Deckers J, Burgos EF, Lekakis J, Lindahl B, Mazzotta G, McGregor K, Morais J, Oto A, Smiseth OA, Barbera JA, Gibbs S, Hoeper M, Humbert M, Naeije R, Pepke-Zaba J. Guidelines on diagnosis and treatment of pulmonary arterial hypertension. The task force on diagnosis and treatment of pulmonary arterial hypertension of the European Society of Cardiology. Eur Heart J. 2004;25:2243–78.
2. Simonneau G, Montani D, Celermajer DS, Denton CP, Gatzoulis MA, Krowka M, Williams PG, Souza R. Haemodynamic definitions and updated clinical classification of pulmonary hypertension. Eur Respir J. 2019;53
3. Bristow MR, Zisman LS, Lowes BD, Abraham WT, Badesch DB, Groves BM, Voelkel NF, Lynch DM, Quaife RA. The pressure-overloaded right ventricle in pulmonary hypertension. Chest. 1998;114:101S–6S.
4. Woodcock CC, Chan SY. The search for disease-modifying therapies in pulmonary hypertension. J Cardiovasc Pharmacol Ther. 2019;24:334–54.
5. Rich S, Dantzker DR, Ayres SM, Bergofsky EH, Brundage BH, Detre KM, Fishman AP, Goldring RM, Groves BM, Koerner SK, et al. Primary pulmonary hypertension. A national prospective study. Ann Intern Med. 1987;107:216–23.
6. Frost A, Badesch D, Gibbs JSR, Gopalan D, Khanna D, Manes A, Oudiz R, Satoh T, Torres F, Torbicki A. Diagnosis of pulmonary hypertension. Eur Respir J. 2019;53
7. Hughes JMB, Morrell NW. Pulmonary circulation: from basic mechanisms to clinical practice. 1st ed. London: Imperial College Press; 2001.
8. McLaughlin VV, Rich S. Pulmonary hypertension. Curr Probl Cardiol. 2004;29:575–634.
9. Simonneau G, Gatzoulis MA, Adatia I, Celermajer D, Denton C, Ghofrani A, Gomez Sanchez MA, Krishna Kumar R, Landzberg M, Machado RF,

Olschewski H, Robbins IM, Souza R. Updated clinical classification of pulmonary hypertension. J Am Coll Cardiol. 2013;62:D34–41.
10. McLaughlin VV, Presberg KW, Doyle RL, Abman SH, McCrory DC, Fortin T, Ahearn G. Prognosis of pulmonary arterial hypertension: ACCP evidence-based clinical practice guidelines. Chest. 2004;126:78S–92S.
11. Sandoval J, Bauerle O, Palomar A, Gomez A, Martinez-Guerra ML, Beltran M, Guerrero ML. Survival in primary pulmonary hypertension. Validation of a prognostic equation. Circulation. 1994;89:1733–44.
12. Berger RM. Possibilities and impossibilities in the evaluation of pulmonary vascular disease in congenital heart defects. Eur Heart J. 2000;21:17–27.
13. Arcasoy SM, Christie JD, Ferrari VA, Sutton MS, Zisman DA, Blumenthal NP, Pochettino A, Kotloff RM. Echocardiographic assessment of pulmonary hypertension in patients with advanced lung disease. Am J Respir Crit Care Med. 2003;167:735–40.
14. Currie PJ, Seward JB, Chan KL, Fyfe DA, Hagler DJ, Mair DD, Reeder GS, Nishimura RA, Tajik AJ. Continuous wave Doppler determination of right ventricular pressure: a simultaneous Doppler-catheterization study in 127 patients. J Am Coll Cardiol. 1985;6:750–6.
15. Chemla D, Castelain V, Herve P, Lecarpentier Y, Brimioulle S. Haemodynamic evaluation of pulmonary hypertension. Eur Respir J. 2002;20:1314–31.
16. Brecker SJ, Gibbs JS, Fox KM, Yacoub MH, Gibson DG. Comparison of Doppler derived haemodynamic variables and simultaneous high fidelity pressure measurements in severe pulmonary hypertension. Br Heart J. 1994;72:384–9.
17. van Wolferen SA, Marcus JT, Boonstra A, Marques KM, Bronzwaer JG, Spreeuwenberg MD, Postmus PE, Vonk-Noordegraaf A. Prognostic value of right ventricular mass, volume, and function in idiopathic pulmonary arterial hypertension. Eur Heart J. 2007;28:1250–7.
18. Pandya B, Quail MA, Steeden JA, McKee A, Odille F, Taylor AM, Schulze-Neick I, Derrick G, Moledina S, Muthurangu V. Real-time magnetic resonance assessment of septal curvature accurately tracks acute hemodynamic changes in pediatric pulmonary hypertension. Circ Cardiovasc Imaging. 2014;7:706–13.
19. Knight DS, Kotecha T, Martinez-Naharro A, Brown JT, Bertelli M, Fontana M, Muthurangu V, Coghlan JG. Cardiovascular magnetic resonance-guided right heart catheterization in a conventional CMR environment – predictors of procedure success and duration in pulmonary artery hypertension. J Cardiovasc Magn Reson. 2019;21:57.
20. Hampl V, Herget J. Role of nitric oxide in the pathogenesis of chronic pulmonary hypertension. Physiol Rev. 2000;80:1337–72.
21. Forstermann U, Boissel JP, Kleinert H. Expressional control of the 'constitutive' isoforms of nitric oxide synthase (NOS I and NOS III). FASEB J. 1998;12:773–90.
22. Hampl V, Huang JM, Weir EK, Archer SL. Activation of the cGMP-dependent protein kinase mimics the stimulatory effect of nitric oxide and cGMP on calcium-gated potassium channels. Physiol Res. 1995;44:39–44.
23. Giaid A, Yanagisawa M, Langleben D, Michel RP, Levy R, Shennib H, Kimura S, Masaki T, Duguid WP, Stewart DJ. Expression of endothelin-1 in the lungs of patients with pulmonary hypertension. N Engl J Med. 1993;328:1732–9.
24. Kaneko FT, Arroliga AC, Dweik RA, Comhair SA, Laskowski D, Oppedisano R, Thomassen MJ, Erzurum SC. Biochemical reaction products of nitric oxide as quantitative markers of primary pulmonary hypertension. Am J Respir Crit Care Med. 1998;158:917–23.
25. Fagan KA, Fouty BW, Tyler RC, Morris KG Jr, Hepler LK, Sato K, LeCras TD, Abman SH, Weinberger HD, Huang PL, McMurtry IF, Rodman DM. The pulmonary circulation of homozygous or heterozygous eNOS-null mice is hyperresponsive to mild hypoxia. J Clin Invest. 1999;103:291–9.
26. Champion HC, Bivalacqua TJ, Greenberg SS, Giles TD, Hyman AL, Kadowitz PJ. Adenoviral gene transfer of endothelial nitric-oxide synthase (eNOS) partially restores normal pulmonary arterial pressure in eNOS-deficient mice. Proc Natl Acad Sci U S A. 2002;99:13248–53.
27. Budhiraja R, Tuder RM, Hassoun PM. Endothelial dysfunction in pulmonary hypertension. Circulation. 2004;109:159–65.
28. Christman BW, McPherson CD, Newman JH, King GA, Bernard GR, Groves BM, Loyd JE. An imbalance between the excretion of thromboxane and prostacyclin metabolites in pulmonary hypertension. N Engl J Med. 1992;327:70–5.
29. Tuder RM, Cool CD, Geraci MW, Wang J, Abman SH, Wright L, Badesch D, Voelkel NF. Prostacyclin synthase expression is decreased in lungs from patients with severe pulmonary hypertension. Am J Respir Crit Care Med. 1999;159:1925–32.
30. Galie N, Manes A, Branzi A. The endothelin system in pulmonary arterial hypertension. Cardiovasc Res. 2004;61:227–37.
31. Seo B, Luscher TF. ETA and ETB receptors mediate contraction to endothelin-1 in renal artery of aging SHR. Effects of FR139317 and bosentan. Hypertension. 1995;25:501–6.
32. Pollock DM, Keith TL, Highsmith RF. Endothelin receptors and calcium signaling. FASEB J. 1995;9:1196–204.
33. Kelly LK, Wedgwood S, Steinhorn RH, Black SM. Nitric oxide decreases endothelin-1 secretion through the activation of soluble guanylate cyclase. Am J Physiol Lung Cell Mol Physiol. 2004;286:L984–91.
34. Cacoub P, Dorent R, Maistre G, Nataf P, Carayon A, Piette C, Godeau P, Cabrol C, Gandjbakhch

I. Endothelin-1 in primary pulmonary hypertension and the Eisenmenger syndrome. Am J Cardiol. 1993;71:448–50.
35. Zhang YM, Wang KQ, Zhou GM, Zuo J, Ge JB. Endothelin-1 promoted proliferation of vascular smooth muscle cell through pathway of extracellular signal-regulated kinase and cyclin D1. Acta Pharmacol Sin. 2003;24:563–8.
36. Stotz WH, Li D, Johns RA. Exogenous nitric oxide upregulates p21 in pulmonary microvascular smooth muscle cells. J Vasc Res. 2004;41:211–9.
37. Kouyoumdjian C, Adnot S, Levame M, Eddahibi S, Bousbaa H, Raffestin B. Continuous inhalation of nitric oxide protects against development of pulmonary hypertension in chronically hypoxic rats. J Clin Invest. 1994;94:578–84.
38. Rich S, Kaufmann E, Levy PS. The effect of high doses of calcium-channel blockers on survival in primary pulmonary hypertension. N Engl J Med. 1992;327:76–81.
39. Sitbon O, Humbert M, Jais X, Ioos V, Hamid AM, Provencher S, Garcia G, Parent F, Herve P, Simonneau G. Long-term response to calcium channel blockers in idiopathic pulmonary arterial hypertension. Circulation. 2005;111:3105–11.
40. Galie N, Ghofrani HA, Torbicki A, Barst RJ, Rubin LJ, Badesch D, Fleming T, Parpia T, Burgess G, Branzi A, Grimminger F, Kurzyna M, Simonneau G, Sildenafil Use in Pulmonary Arterial Hypertension Study G. Sildenafil citrate therapy for pulmonary arterial hypertension. N Engl J Med. 2005;353:2148–57.
41. McLaughlin VV, Shillington A, Rich S. Survival in primary pulmonary hypertension: the impact of epoprostenol therapy. Circulation. 2002;106:1477–82.
42. Pulido T, Adzerikho I, Channick RN, Delcroix M, Galie N, Ghofrani HA, Jansa P, Jing ZC, Le Brun FO, Mehta S, Mittelholzer CM, Perchenet L, Sastry BK, Sitbon O, Souza R, Torbicki A, Zeng X, Rubin LJ, Simonneau G, Investigators S. Macitentan and morbidity and mortality in pulmonary arterial hypertension. N Engl J Med. 2013;369:809–18.
43. Zarins CK, Taylor CA, Min JK. Computed fractional flow reserve (FFTCT) derived from coronary CT angiography. J Cardiovasc Transl Res. 2013;6:708–14.
44. Hansmann G, Koestenberger M, Alastalo TP, Apitz C, Austin ED, Bonnet D, Budts W, D'Alto M, Gatzoulis MA, Hasan BS, Kozlik-Feldmann R, Kumar RK, Lammers AE, Latus H, Michel-Behnke I, Miera O, Morrell NW, Pieles G, Quandt D, Sallmon H, Schranz D, Tran-Lundmark K, Tulloh RMR, Warnecke G, Wahlander H, Weber SC, Zartner P. 2019 updated consensus statement on the diagnosis and treatment of pediatric pulmonary hypertension: The European Pediatric Pulmonary Vascular Disease Network (EPPVDN), endorsed by AEPC, ESPR and ISHLT. J Heart Lung Transplant. 2019;38:879–901.
45. Stone PH, Saito S, Takahashi S, Makita Y, Nakamura S, Kawasaki T, Takahashi A, Katsuki T, Nakamura S, Namiki A, Hirohata A, Matsumura T, Yamazaki S, Yokoi H, Tanaka S, Otsuji S, Yoshimachi F, Honye J, Harwood D, Reitman M, Coskun AU, Papafaklis MI, Feldman CL, Investigators P. Prediction of progression of coronary artery disease and clinical outcomes using vascular profiling of endothelial shear stress and arterial plaque characteristics: the PREDICTION Study. Circulation. 2012;126:172–81.
46. Gijsen F, Katagiri Y, Barlis P, Bourantas C, Collet C, Coskun U, Daemen J, Dijkstra J, Edelman E, Evans P, van der Heiden K, Hose R, Koo BK, Krams R, Marsden A, Migliavacca F, Onuma Y, Ooi A, Poon E, Samady H, Stone P, Takahashi K, Tang D, Thondapu V, Tenekecioglu E, Timmins L, Torii R, Wentzel J, Serruys P. Expert recommendations on the assessment of wall shear stress in human coronary arteries: existing methodologies, technical considerations, and clinical applications. Eur Heart J. 2019;40:3421–33.
47. Torii R, Wood NB, Hadjiloizou N, Dowsey AW, Wright AR, Hughes AD, Davies J, Francis DP, Mayet J, Yang GZ, Thom SA, Xu XY. Stress phase angle depicts differences in coronary artery hemodynamics due to changes in flow and geometry after percutaneous coronary intervention. Am J Physiol Heart Circ Physiol. 2009;296:H765–76.
48. Yang W, Feinstein JA, Vignon-Clementel IE. Adaptive outflow boundary conditions improve post-operative predictions after repair of peripheral pulmonary artery stenosis. Biomech Model Mechanobiol. 2016;15:1345–53.
49. Yang W, Feinstein JA, Shadden SC, Vignon-Clementel IE, Marsden AL. Optimization of a Y-graft design for improved hepatic flow distribution in the fontan circulation. J Biomech Eng. 2013;135:011002.
50. Spazzapan M, Sastry P, Dunning J, Nordsletten D, de Vecchi A. The use of biophysical flow models in the surgical management of patients affected by chronic thromboembolic pulmonary hypertension. Front Physiol. 2018;9:223.
51. Tang BT, Fonte TA, Chan FP, Tsao PS, Feinstein JA, Taylor CA. Three-dimensional hemodynamics in the human pulmonary arteries under resting and exercise conditions. Ann Biomed Eng. 2011;39:347–58.
52. Bulow T, Wiemker R, Blaffert T, Lorenz C, Renisch S. Automatic extraction of the pulmonary artery tree from multi-slice CT data. Proc SPIE. 2005;5746:730–40.
53. Burrowes KS, Clark AR, Marcinkowski A, Wilsher ML, Milne DG, Tawhai MH. Pulmonary embolism: predicting disease severity. Philos Trans A Math Phys Eng Sci. 2011;369:4255–77.
54. Jorge E, Baptista R, Calisto J, Faria H, Monteiro P, Pan M, Pego M. Optical coherence tomography of the pulmonary arteries: a systematic review. J Cardiol. 2016;67:6–14.

55. Steeden JA, Pandya B, Tann O, Muthurangu V. Free breathing contrast-enhanced time-resolved magnetic resonance angiography in pediatric and adult congenital heart disease. J Cardiovasc Magn Reson. 2015;17:38.
56. Hauptmann A, Arridge S, Lucka F, Muthurangu V, Steeden JA. Real-time cardiovascular MR with spatio-temporal artifact suppression using deep learning-proof of concept in congenital heart disease. Magn Reson Med. 2019;81:1143–56.
57. Chemla D, Lau EM, Papelier Y, Attal P, Herve P. Pulmonary vascular resistance and compliance relationship in pulmonary hypertension. Eur Respir J. 2015;46:1178–89.
58. Clark AR, Burrowes KS, Tawhai MH. Contribution of serial and parallel microperfusion to spatial variability in pulmonary inter- and intra-acinar blood flow. J Appl Physiol. 1985;2010(108):1116–26.
59. Tewari SG, Bugenhagen SM, Wang Z, Schreier DA, Carlson BE, Chesler NC, Beard DA. Analysis of cardiovascular dynamics in pulmonary hypertensive C57BL6/J mice. Front Physiol. 2013;4:355.
60. Gerringer JW, Wagner JC, Velez-Rendon D, Valdez-Jasso D. Lumped-parameter models of the pulmonary vasculature during the progression of pulmonary arterial hypertension. Physiol Rep. 2018;6
61. Tang H, Dai Z, Wang M, Guo B, Wang S, Wen J, Li T. Lumped-parameter circuit platform for simulating typical cases of pulmonary hypertensions from point of hemodynamics. J Cardiovasc Transl Res. 2020;
62. Kalaitzidis P, Orwat S, Kempny A, Robert R, Peters B, Sarikouch S, Beerbaum P, Baumgartner H, Diller GP, Competence Network for Congenital Heart Defects D. Biventricular dyssynchrony on cardiac magnetic resonance imaging and its correlation with myocardial deformation, ventricular function and objective exercise capacity in patients with repaired tetralogy of Fallot. Int J Cardiol. 2018;264:53–7.
63. Tewari SG, Bugenhagen SM, Vinnakota KC, Rice JJ, Janssen PML, Beard DA. Influence of metabolic dysfunction on cardiac mechanics in decompensated hypertrophy and heart failure. J Mol Cell Cardiol. 2016;94:162–75.
64. Audebert C, Peeters G, Segers P, Laleman W, Monbaliu D, Korf H, Trebicka J, Vignon-Clementel IE, Debbaut C. Closed-loop lumped parameter modeling of hemodynamics during cirrhogenesis in rats. IEEE Trans Biomed Eng. 2018;65:2311–22.
65. Borlotti A, Khir AW, Rietzschel ER, De Buyzere ML, Vermeersch S, Segers P. Noninvasive determination of local pulse wave velocity and wave intensity: changes with age and gender in the carotid and femoral arteries of healthy human. J Appl Physiol. 1985;2012(113):727–35.
66. Quail MA, Knight DS, Steeden JA, Taelman L, Moledina S, Taylor AM, Segers P, Coghlan GJ, Muthurangu V. Noninvasive pulmonary artery wave intensity analysis in pulmonary hypertension. Am J Physiol Heart Circ Physiol. 2015;308: H1603–11.
67. Sherwin SJ, Franke V, Peiro J, Parker K. One-dimensional modelling of a vascular network in space-time variables. J Eng Math. 2003;47:217–50.
68. Alastruey J, Xiao N, Fok H, Schaeffter T, Figueroa CA. On the impact of modelling assumptions in multi-scale, subject-specific models of aortic haemodynamics. J R Soc Interface. 2016;13
69. Qureshi MU, Vaughan GD, Sainsbury C, Johnson M, Peskin CS, Olufsen MS, Hill NA. Numerical simulation of blood flow and pressure drop in the pulmonary arterial and venous circulation. Biomech Model Mechanobiol. 2014;13:1137–54.
70. Bassez S, Flaud P, Chauveau M. Modeling of the deformation of flexible tubes using a single law: application to veins of the lower limb in man. J Biomech Eng. 2001;123:58–65.
71. Mynard JP, Smolich JJ. One-dimensional haemodynamic modeling and wave dynamics in the entire adult circulation. Ann Biomed Eng. 2015;43:1443–60.
72. Vera L, Campos Arias D, Muylle S, Stergiopulos N, Segers P, van Loon G. A 1D computer model of the arterial circulation in horses: an important resource for studying global interactions between heart and vessels under normal and pathological conditions. PLoS One. 2019;14:e0221425.
73. Olufsen MS, Hill NA, Vaughan GD, Sainsbury C, Johnson M. Rarefaction and blood pressure in systemic and pulmonary arteries. J Fluid Mech. 2012;705:280–305.
74. Qureshi MU, Hill NA. A computational study of pressure wave reflections in the pulmonary arteries. J Math Biol. 2015;71:1525–49.
75. Tang BT, Pickard SS, Chan FP, Tsao PS, Taylor CA, Feinstein JA. Wall shear stress is decreased in the pulmonary arteries of patients with pulmonary arterial hypertension: an image-based, computational fluid dynamics study. Pulm Circ. 2012;2:470–6.
76. Kheyfets V, Thirugnanasambandam M, Rios L, Evans D, Smith T, Schroeder T, Mueller J, Murali S, Lasorda D, Spotti J, Finol E. The role of wall shear stress in the assessment of right ventricle hydraulic workload. Pulm Circ. 2015;5:90–100.
77. Kheyfets VO, Rios L, Smith T, Schroeder T, Mueller J, Murali S, Lasorda D, Zikos A, Spotti J, Reilly JJ Jr, Finol EA. Patient-specific computational modeling of blood flow in the pulmonary arterial circulation. Comput Methods Prog Biomed. 2015;120:88–101.
78. Hayashi K, Naiki T. Adaptation and remodeling of vascular wall; biomechanical response to hypertension. J Mech Behav Biomed Mater. 2009;2:3–19.
79. Humphrey JD, Harrison DG, Figueroa CA, Lacolley P, Laurent S. Central artery stiffness in hypertension and aging: a problem with cause and consequence. Circ Res. 2016;118:379–81.

80. Zambrano BA, McLean NA, Zhao X, Tan JL, Zhong L, Figueroa CA, Lee LC, Baek S. Image-based computational assessment of vascular wall mechanics and hemodynamics in pulmonary arterial hypertension patients. J Biomech. 2018;68:84–92.
81. Kong F, Kheyfets V, Finol E, Cai XC. Simulation of unsteady blood flows in a patient-specific compliant pulmonary artery with a highly parallel monolithically coupled fluid-structure interaction algorithm. Int J Numer Method Biomed Eng. 2019;35:e3208.
82. Farsad M, Zeinali-Davarani S, Choi J, Baek S. Computational growth and remodeling of abdominal aortic aneurysms constrained by the spine. J Biomech Eng. 2015;137
83. Watton PN, Ventikos Y, Holzapfel GA. Modelling the growth and stabilization of cerebral aneurysms. Math Med Biol. 2009;26:133–64.
84. Shorr AF, Wainright JL, Cors CS, Lettieri CJ, Nathan SD. Pulmonary hypertension in patients with pulmonary fibrosis awaiting lung transplant. Eur Respir J. 2007;30:715–21.
85. Clipp RB, Steele BN. Impedance boundary conditions for the pulmonary vasculature including the effects of geometry, compliance, and respiration. IEEE Trans Biomed Eng. 2009;56:862–70.
86. Fernandes LG, Trenhago PR, Feijoo RA, Blanco PJ. Integrated cardiorespiratory system model with short timescale control mechanisms. Int J Numer Method Biomed Eng. 2020:e3332.
87. Marsden AL, Vignon-Clementel IE, Chan FP, Feinstein JA, Taylor CA. Effects of exercise and respiration on hemodynamic efficiency in CFD simulations of the total cavopulmonary connection. Ann Biomed Eng. 2007;35:250–63.
88. Shavik SM, Zhong L, Zhao X, Lee LC. In-silico assessment of the effects of right ventricular assist device on pulmonary arterial hypertension using an image based biventricular modeling framework. Mech Res Commun. 2019;97:101–11.
89. Park JIK, Heikhmakhtiar AK, Kim CH, Kim YS, Choi SW, Song KS, Lim KM. The effect of heart failure and left ventricular assist device treatment on right ventricular mechanics: a computational study. Biomed Eng Online. 2018;17:62.
90. Menichini C, Cheng Z, Gibbs RG, Xu XY. Predicting false lumen thrombosis in patient-specific models of aortic dissection. J R Soc Interface. 2016;13
91. Bosi GM, Cook A, Rai R, Menezes LJ, Schievano S, Torii R, Burriesci GB. Computational fluid dynamic analysis of the left atrial appendage to predict thrombosis risk. Front Cardiovasc Med. 2018;5:34.
92. Romeih S, Aguib H, Yacoub M. Utility of 4D Flow mapping in Eisenmenger syndrome with pulmonary atresia. Cardiol Young. 2016;26:1581–9.
93. Roberts JD, Forfia PR. Diagnosis and assessment of pulmonary vascular disease by Doppler echocardiography. Pulm Circ. 2011;1:160–81.
94. Matsunaga K, Yanagawa M, Otsuka T, Hirata H, Kijima T, Kumanogoh A, Tomiyama N, Shimosegawa E, Hatazawa J. Quantitative pulmonary blood flow measurement using (15)O-H_2O PET with and without tissue fraction correction: a comparison study. EJNMMI Res. 2017;7:102.
95. Knott KD, Seraphim A, Augusto JB, Xue H, Chacko L, Aung N, Petersen SE, Cooper JA, Manisty C, Bhuva AN, Kotecha T, Bourantas CV, Davies RH, Brown LAE, Plein S, Fontana M, Kellman P, Moon JC. The prognostic significance of quantitative myocardial perfusion: an artificial intelligence-based approach using perfusion mapping. Circulation. 2020;141:1282–91.
96. Lo EW, Menezes LJ, Torii R. On outflow boundary conditions for CT-based computation of FFR: Examination using PET images. Med Eng Phys. 2020;76:79–87.
97. Spilker RL, Feinstein JA, Parker DW, Reddy VM, Taylor CA. Morphometry-based impedance boundary conditions for patient-specific modeling of blood flow in pulmonary arteries. Ann Biomed Eng. 2007;35:546–59.
98. Willemet M, Chowienczyk P, Alastruey J. A database of virtual healthy subjects to assess the accuracy of foot-to-foot pulse wave velocities for estimation of aortic stiffness. Am J Physiol Heart Circ Physiol. 2015;309:H663–75.
99. Colebank MJ, Umar Qureshi M, Olufsen MS. Sensitivity analysis and uncertainty quantification of 1-D models of pulmonary hemodynamics in mice under control and hypertensive conditions. Int J Numer Method Biomed Eng. 2019:e3242.
100. Seo J, Schiavazzi DE, Kahn AM, Marsden AL. The effects of clinically-derived parametric data uncertainty in patient-specific coronary simulations with deformable walls. Int J Numer Method Biomed Eng. 2020;
101. Bordones AD, Leroux M, Kheyfets VO, Wu YA, Chen CY, Finol EA. Computational fluid dynamics modeling of the human pulmonary arteries with experimental validation. Ann Biomed Eng. 2018;46:1309–24.
102. Knoops PGM, Biglino G, Hughes AD, Parker KH, Xu L, Schievano S, Torii R. A Mock circulatory system incorporating a compliant 3D-printed anatomical model to investigate pulmonary hemodynamics. Artif Organs. 2017;41:637–46.
103. Barker AJ, Roldan-Alzate A, Entezari P, Shah SJ, Chesler NC, Wieben O, Markl M, Francois CJ. Four-dimensional flow assessment of pulmonary artery flow and wall shear stress in adult pulmonary arterial hypertension: results from two institutions. Magn Reson Med. 2015;73:1904–13.
104. Biglino G, Cosentino D, Steeden JA, De Nova L, Castelli M, Ntsinjana H, Pennati G, Taylor AM, Schievano S. Using 4D cardiovascular magnetic

resonance imaging to validate computational fluid dynamics: a case study. Front Pediatr. 2015;3:107.
105. Marlevi D, Ha H, Dillon-Murphy D, Fernandes JF, Fovargue D, Colarieti-Tosti M, Larsson M, Lamata P, Figueroa CA, Ebbers T, Nordsletten DA. Non-invasive estimation of relative pressure in turbulent flow using virtual work-energy. Med Image Anal. 2020;60:101627.
106. Annio G, Torii R, Ariff B, O'Regan D, Muthurangu V, Ducci A, Tsang V, Burriesci G. Enhancing magnetic resonance imaging with computational fluid dynamics. ASME J Med Diagn. 2019;2:041010.
107. Norgaard BL, Leipsic J, Gaur S, Seneviratne S, Ko BS, Ito H, Jensen JM, Mauri L, De Bruyne B, Bezerra H, Osawa K, Marwan M, Naber C, Erglis A, Park SJ, Christiansen EH, Kaltoft A, Lassen JF, Botker HE, Achenbach S, Group NXTTS. Diagnostic performance of noninvasive fractional flow reserve derived from coronary computed tomography angiography in suspected coronary artery disease: the NXT trial (Analysis of Coronary Blood Flow Using CT Angiography: Next Steps). J Am Coll Cardiol. 2014;63:1145–55.

Part III

How I Used a Model in Clinical Practice

16 Patient-Specific Numerical Modeling to Predict Coronary Artery Compression in Transcatheter Pulmonary Valve Implantation

Francesca R. Pluchinotta, Alessandro Caimi, Francesco Sturla, and Mario Carminati

16.1 Introduction

The surgical palliation of many complex congenital heart diseases, e.g., the Ross surgery for aortic valve disease dysfunction, involves the implantation of homograft conduits in the right ventricular outflow tract (RVOT) [1, 2]. Long-term postoperative RVOT dysfunctions are frequent, with progressive conduit calcification next to the pulmonary valve leading to RVOT stenosis [3]. Transcatheter pulmonary valve implantation (TPVI) offers a nonsurgical treatment option for failing prosthetic conduits in pulmonary position [3, 4].

When planning a TPVI procedure, clinicians should be aware of the potential life-threatening complications [5]. Specifically, coronary artery (CA) compression is the most worrisome and affects almost 5% of TPVI patients: Some surgically implanted RVOT conduits pass in close apposition to the left CA (LCA), and this proximity places the LCA at risk of distortion or compression depending on how much the RVOT conduit geometry and size change during TPVI [6]. In this scenario, RVOT balloon angioplasty with simultaneous CA angiography is generally performed during cardiac catheterization prior to TPVI to simulate the conduit changes that will occur with valve implantation [6].

Over the last decade, image-based computational, i.e., in silico, strategies, synergically combining with advanced imaging such as cardiac magnetic resonance (CMR) or computed tomography (CT) and finite element (FE) modeling, have become accepted tools to study patient-specific morphologies and simulate interventional procedures in order to predict possible negative outcomes before going "in vivo" [7]. Indeed, to fully exploit FE predictions, a patient-

F. R. Pluchinotta (✉)
Department of Pediatric and Adult Congenital Heart Disease - IRCCS Policlinico San Donato, San Donato Milanese, Italy

Department of Electronics, Information and Bioengineering, Politecnico di Milano, Milan, Italy
e-mail: Francesca.Pluchinotta@grupposandonato.it

A. Caimi
Department of Electronics, Information and Bioengineering, Politecnico di Milano, Milan, Italy
e-mail: alessandro.caimi@polimi.it

F. Sturla
3D and Computer Simulation Laboratory, IRCCS Policlinico San Donato, San Donato Milanese, Italy
e-mail: Francesco.Sturla@grupposandonato.it

M. Carminati
Department of Pediatric and Adult Congenital Heart Disease - IRCCS Policlinico San Donato, San Donato Milanese, Italy
e-mail: Mario.Carminati@grupposandonato.it

G. Butera et al. (eds.), *Modelling Congenital Heart Disease*,
https://doi.org/10.1007/978-3-030-88892-3_16

tailored numerical model is crucial to take both dimension and morphology of the implantation site into account; in addition, adequate numerical reproduction of the employed devices and tuning of their mechanical properties are required to enhance FE model reliability [8, 9].

Herein, we presented a clinical case of a young man with previous RVOT homograft surgery for dysfunctional aortic valve referred for TPVI. CT images showed close proximity of the LCA to the homograft raising the concern for possible compression. Therefore, a patient-specific FE model that included the implantation site anatomy and the aorta and CAs was developed to assess TPVI feasibility and predict procedural complications.

16.2 Clinical History and Imaging

A 40-year-old man with a stenotic bicuspid aortic valve (BAV) underwent Ross procedure [10] at 8 years of age: The dysfunctional aortic valve was replaced with the patient's own native pulmonary valve, and an homograft conduit of 24 mm was used in the RVOT to establish right ventricle to pulmonary artery continuity; in addition, the coronary arteries were detached from the explanted aortic valve and reimplanted onto the *pulmonary autograft*. The patient recently presented with fatigue and dyspnea on mild effort.

Echocardiographic evaluation showed severe homograft stenosis with preserved RV function and aneurysmatic ascending aorta as a possible consequence of the BAV disease. This was confirmed by CMR evaluation, which also highlighted a close proximity of the LCA with the homograft (Fig. 16.1a).

The patient was referred to our center for pulmonary valve replacement. Based on CMR images, the concern for possible LCA compression during a percutaneous procedure was taken into account. However, considering the patient's propensity to avoid surgery, the clinical team decided to deeply investigate TPVI feasibility as a preprocedural planning before going in vivo.

The patient underwent CT angiography on a 64-slice multidetector system with retrospective ECG gating (SOMATOM Definition; Siemens, Munich, Germany) under injection of contrast medium (Iomeprol). The acquisition was performed in diastole, at 70% of the R-R interval (Fig. 16.1b, c), with pixel spacing and slice thickness equal to 0.5 and 0.75 mm, respectively. CT images confirmed the close proximity of the LCA to the homograft. Of note, no visible calcific deposits within the pulmonary homograft were seen.

16.3 FE Model

To numerically investigate the risk of CA compression, we have recently developed a FE protocol [11] able to virtually reproduce the conduit angioplasty within the patient-specific clinical setting. More in detail, DICOM images from the CT acquisition were used to reconstruct the patient-specific 3D geometry using the commercial software Mimics Medical v21 (Materialise, Leuven, Belgium): The reproduced anatomy included the homograft, the two pulmonary branches, the aortic root, and the proximal CAs (Fig. 16.2). 3D reconstruction highlighted the oval shape of the homograft, which was mostly reduced in size in the latero-lateral direction (anteroposterior diameter 23.4 mm, latero-lateral diameter 16.7 mm; Fig. 16.2c).

The reproduced geometry was exported as a triangulated surface mesh (.stl), which was remeshed in Meshmixer (Autodesk; San Rafael, CA, USA) to improve the overall quality of the triangulated grid.

Finally, we simulated homograft balloon angioplasty reproducing the expansion of a 22-mm-diameter balloon, namely the outer balloon of the balloon-in-balloon delivery system (BiB® Numed; Hopkinton, NY USA) [1]. RVOT biomechanics and its interaction with CAs were assessed during the sizing procedure.

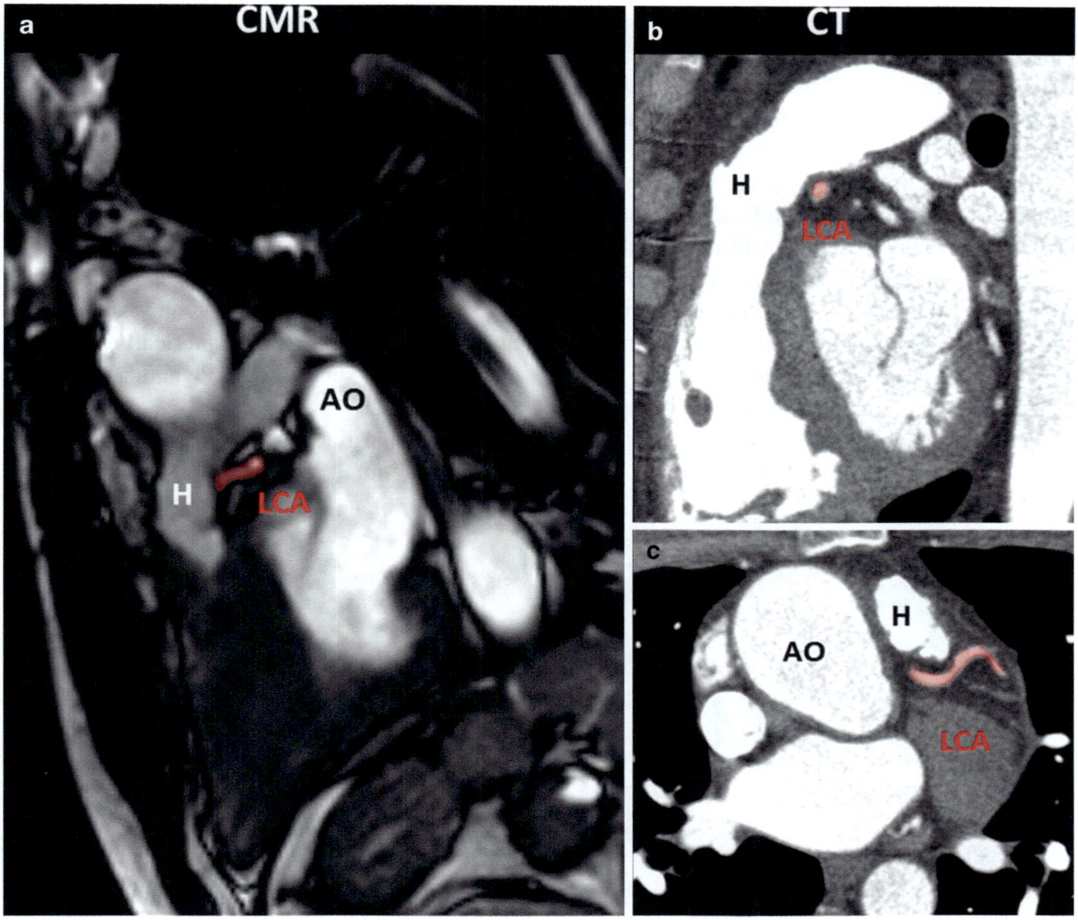

Fig. 16.1 Preoperative imaging: (**a**) CMR sagittal view, (**b**) CT images acquired on a sagittal (**b**) and axial (**c**) plane highlighting the stenotic homograft (H), the dilated aortic root (Ao), and the close proximity of the left coronary artery (LCA) to the homograft

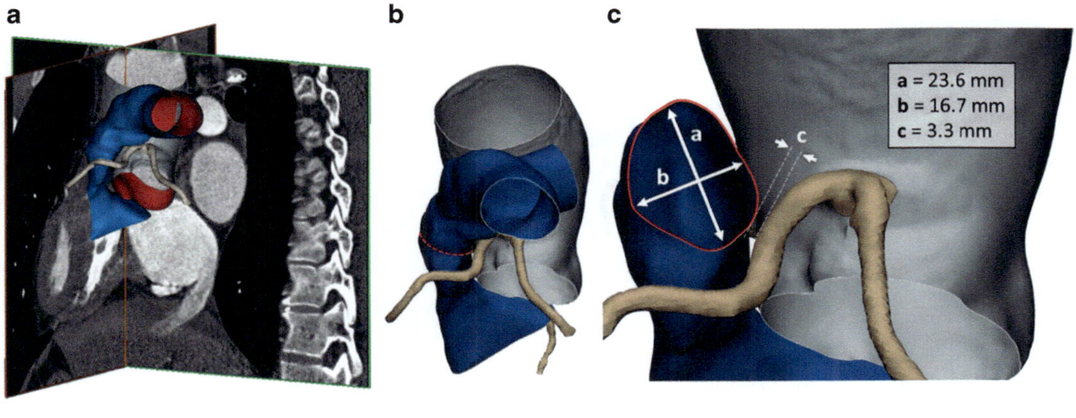

Fig. 16.2 3D reconstruction of the patient-specific anatomy: (**a**) segmentation of CT images to reconstruct the patient-specific 3D geometry (**b**); (**c**) cross-sectional view, at the level of the obstruction, highlighting the maximum and the minimum diameters of the lumen of the homograft and the in-plane distance between the homograft and LCA anatomy

16.4 FE Results

Progressive inflation of the balloon caused deformation of the RVOT conduit and displacement toward the LCA.

The interaction between the homograft and LCA was negligible when inflating the outer BiB balloon to a diameter of about 18 mm; on the contrary, the subsequent balloon expansion up to 22 mm, i.e., the nominal full-expanded diameter of the outer BiB balloon, induced a significant relocation of the LCA, as highlighted in terms of the displacement contour maps (Fig. 16.3a, b), leading to a 10% reduction in LCA lumen cross section (Fig. 16.3c).

Before the planned cardiac catheterization procedure, numerical results were shared within the interventional team to discuss the risk of CA compression related to the potential implantation site anatomy.

16.5 Cardiac Catheterization

To better delineate the CA anatomy and confirm the TPVI unfeasibility, as numerically predicted, the patient was sent for hemodynamic diagnostic evaluation.

During cardiac catheterization, the interventional team performed angiographic

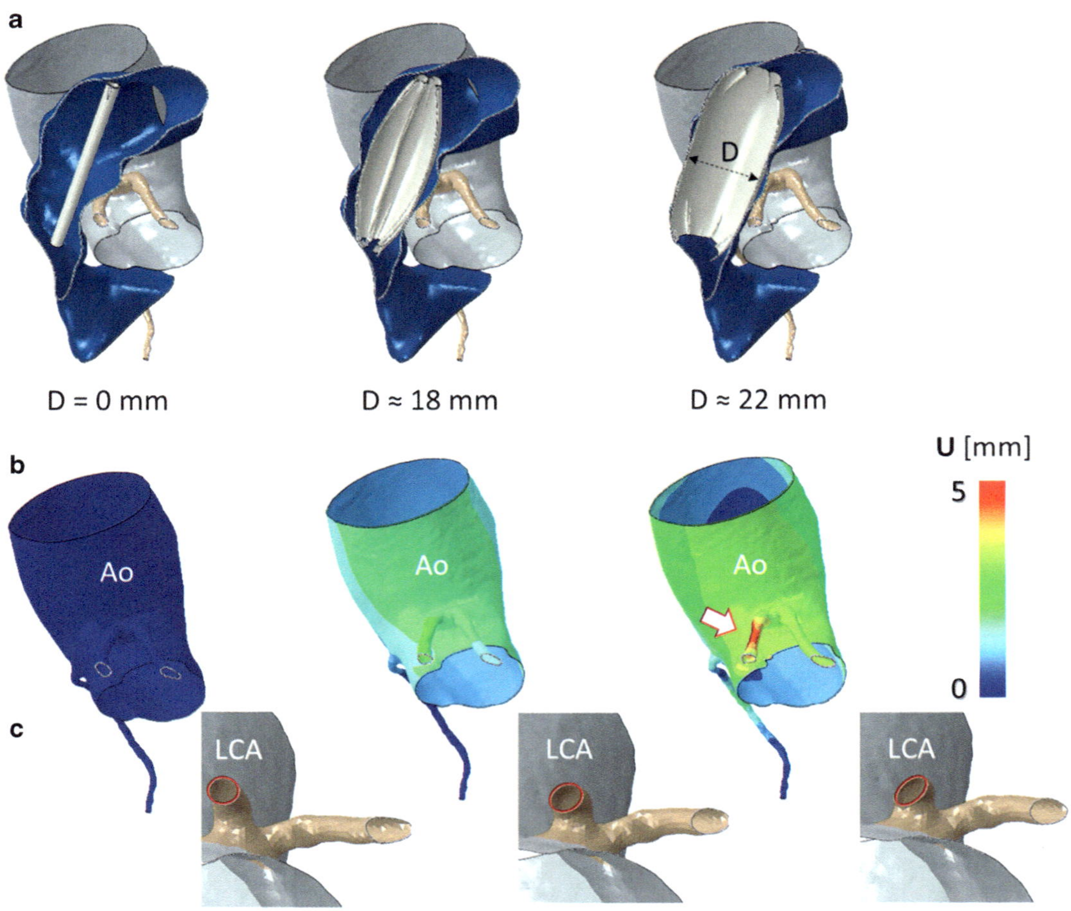

Fig. 16.3 Results of FE analysis: (**a**) main steps of the outer BiB deployment within the homograft on a long-axis cut view of the implantation site; (**b**) contour maps of the displacement magnitude (U) of the aortic root and the ascending aorta (Ao) and the proximal LCA; as visible on the LCA cross section (**c**), the homograft expansion partially relocated and directly impinged the LCA anatomy when the outer BiB diameter was increased from 18 mm up to 22 mm

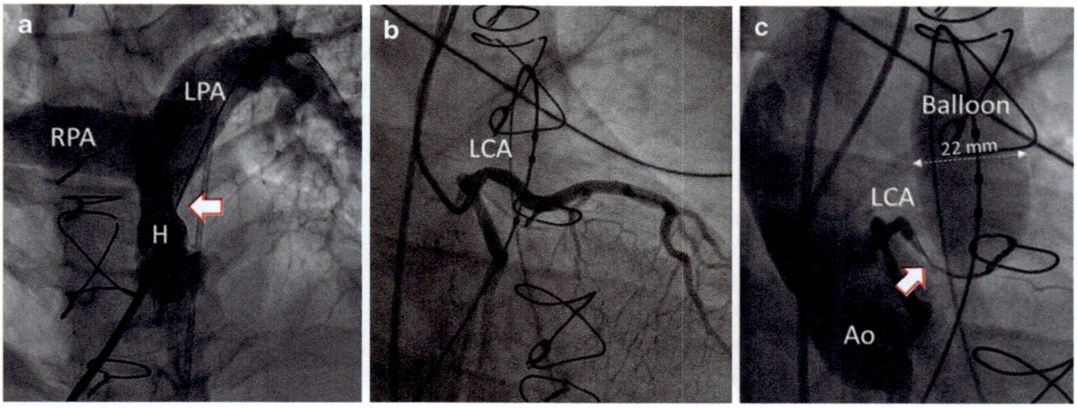

Fig. 16.4 Images from cardiac catheterization: (**a**) angiography of the homograft (H) and pulmonary branches (RPA and LPA) confirmed the severity of the stenosis (*arrow*); (**b**) coronary angiography showed left coronary artery (LCA) patency; (**c**) homograft balloon angioplasty test revealed that the outer BiB balloon, when fully expanded up to 22 mm, markedly reduced the flow in the LCA

studies of the homograft and pulmonary branches, with pressure measurements confirming the severity of the stenosis (Fig. 16.4a), and aortography to evaluate the CAs (Fig. 16.4b). Finally, the RVOT balloon angioplasty test was performed: The outer BiB balloon was slowly and progressively inflated in the homograft, while a coronary angiography was simultaneously performed. A preserved LCA flow was detected up to a balloon diameter of 18 mm; beyond this limit, till the full BiB balloon expansion (i.e., 22 mm), evident signs of LCA compression arose (Fig. 16.4c). The balloon was immediately deflated, stopping the procedure. The patient was transferred to the intensive care unit for monitoring and discharged the day after with no sequelae.

16.6 Discussion

Percutaneous procedures aim at treating cardiovascular diseases through a noninvasive approach. Nevertheless, several life-threatening periprocedural complications can occur during complex interventional cardiac catheterizations [11, 12], thus requiring a dedicated strategy to investigate procedural feasibility and plan the intervention.

The risk of CA compression still represents a critical complication in TPVI, and it is potentially higher in patients with a previous Ross operation due to the combination of extra-anatomic position of the conduit in the RVOT and repositioning of the CAs [6]. Both CMR and CT imaging can assess the conduit morphology and highlight the relative position between the homograft and the surrounding vessels [5, 7]; nonetheless, balloon angioplasty testing along with CA angiography performed during cardiac catheterization still represents the gold standard for the assessment of CA compression, especially in borderline patients [13]. If CA compression is not detected during diagnostic catheterization, TPVI can be completed.

Herein, we exploited a consolidated image-based bioengineering strategy to support interventional cardiologists in the planning of patient-tailored percutaneous procedures [11].

In the reported case, according to CMR and CT preoperative imaging, the LCA was already noted to be in close proximity with the conduit (Fig. 16.1). However, devil lies in the details and, in our case, in the spatial complexity of patient-specific anatomy, which cannot be comprehensively examined from 2D images.

The 3D reconstruction added important insights into the patient-specific anatomy: (1) The homograft, 24 mm in size at the time of implantation, had an oval shape (Fig. 16.2c) with the anteroposterior diameter almost preserved, while the latero-lateral diameter was markedly reduced; and (2) the LCA run posteriorly to the homograft (Fig. 16.2c). Therefore, at the time of TPVI planning there were two possible scenarios: On the one hand, the balloon angioplasty could have rounded the homograft expanding the narrowest latero-lateral diameter preserving its position away from the LCA; on the other hand, the dilation of the homograft could have severely impinged the LCA with the inevitable risk of compression.

The FE model allowed the interventional team to move a step further in the assessment of TPVI feasibility and predicted a potential mechanical interplay between the homograft and LCA, which warranted careful evaluation in vivo. Indeed, the numerical FE inferences proved to be consistent with intraprocedural fluoroscopic evidences: Homograft angioplasty and simultaneous coronary angiography showed that balloon inflation up to 18 mm preserved the flow in the LCA, as confirmed by the FE model; balloon inflation above 18 mm, and up to 22 mm, markedly reduced the flow in the LCA as predicted by the FE model. Conscious of the risks, the interventional team performed the procedure carefully and progressively inflated the balloon monitoring the clinical situation.

Interventional cardiology is a favorable and relevant field for simulation, particularly in patients with congenital heart disease where anatomic derangements are the rule. In the present case, FE analysis was extremely useful to support the interventional team in the evaluation of TPVI feasibility and foresee potential complications during diagnostic cardiac catheterization, providing a reliable prediction of ground truth evidences coming from invasive in vivo tests and measurements.

A more extensive validation has still to be performed, including a consistent number of patients. However, at present, the premises of computer-based predictive simulations have shown to be more than bright, especially for perioperative planning and decision making in those cases considered complex and at high risk of complications. In the near future, preoperative in silico predictions might effectively help clinicians to reduce the need for cardiac catheterization, especially in those patients retained at high or intermediate risk analyzing preoperative imaging.

References

1. Frigiola A, Tsang V, Nordmeyer J, Lurz P, van Doorn C, Taylor AM, Bonhoeffer P, de Leval M. Current approaches to pulmonary regurgitation. Eur J Cardiothorac Surg. 2008;34:576–80. discussion 581–572
2. Hasan BS, Lunze FI, Chen MH, Brown DW, Boudreau MJ, Rhodes J, McElhinney DB. Effects of transcatheter pulmonary valve replacement on the hemodynamic and ventricular response to exercise in patients with obstructed right ventricle-to-pulmonary artery conduits. JACC Cardiovasc Interv. 2014;7:530–42.
3. Lurz P, Nordmeyer J, Muthurangu V, Khambadkone S, Derrick G, Yates R, Sury M, Bonhoeffer P, Taylor AM. Comparison of bare metal stenting and percutaneous pulmonary valve implantation for treatment of right ventricular outflow tract obstruction: use of an x-ray/magnetic resonance hybrid laboratory for acute physiological assessment. Circulation. 2009;119:2995–3001.
4. Aggarwal S, Garekar S, Forbes TJ, Turner DR. Is stent placement effective for palliation of right ventricle to pulmonary artery conduit stenosis? J Am Coll Cardiol. 2007;49:480–4.
5. Malone L, Fonseca B, Fagan T, Gralla J, Wilson N, DiMaria M, Truong U, Browne LP. Preprocedural risk assessment prior to PPVI with CMR and cardiac CT. Pediatr Cardiol. 2017;38:746–53.
6. Morray BH, McElhinney DB, Cheatham JP, Zahn EM, Berman DP, Sullivan PM, Lock JE, Jones TK. Risk of coronary artery compression among patients referred for transcatheter pulmonary valve implantation: a multicenter experience. Circ Cardiovasc Intervent. 2013;6:535–42.
7. Chung R, Taylor AM. Imaging for preintervention planning: transcatheter pulmonary valve therapy. Circ Cardiovasc Imaging. 2014;7(1):182–9. https://doi.org/10.1161/CIRCIMAGING.113.000826.
8. Caimi A, Pasquali M, Sturla F, et al. Prediction of post-stenting biomechanics in coarcted aortas: a pilot finite element study. J Biomech. 2020;105:109796. https://doi.org/10.1016/j.jbiomech.2020.109796.

9. Bosi GM, Capelli C, Cheang MH, et al. Population-specific material properties of the implantation site for transcatheter aortic valve replacement finite element simulations. J Biomech. 2018;71:236–44. https://doi.org/10.1016/j.jbiomech.2018.02.017.
10. Stelzer P. The Ross procedure: state of the art 2011. Semin Thorac Cardiovasc Surg. 2011;23(2):115–23. https://doi.org/10.1053/j.semtcvs.2011.07.003.
11. Caimi A, Sturla F, Pluchinotta FR, et al. Prediction of stenting related adverse events through patient-specific finite element modelling. J Biomech. 2018;79:135–46.
12. Pluchinotta FR, Sturla F, Caimi A, et al. 3-dimensional personalized planning for transcatheter pulmonary valve implantation in a dysfunctional right ventricular outflow tract. Int J Cardiol. 2020;309:33–9.
13. Ansari MM, Cardoso R, Garcia D, et al. Percutaneous pulmonary valve implantation: present status and evolving future. J Am Coll Cardiol. 2015;66(20):2246–55. https://doi.org/10.1016/j.jacc.2015.09.055.

Transcatheter Correction of Sinus Venosus Atrial Septal Defect (SVASD) and Partial Anomalous Pulmonary Venous Drainage with a Covered Stent

17

Eric Rosenthal

Abbreviations

Ao	Aorta
LA	Left atrium
LFV	Left femoral vein
LLPV	Left lower pulmonary vein
LUPV	Left upper pulmonary vein
PAPVD	Partial anomalous pulmonary venous drainage
Pig	Pigtail catheter
RA	Right atrium
RFV	Right femoral vein
RIJV	Right internal jugular vein
RLPV	Right lower pulmonary vein
RMPV	Right middle pulmonary vein
RUPV	Right upper pulmonary vein
SVASD	Sinus venosus ASD
SVC	Superior vena cava

E. Rosenthal (✉)
Evelina London Children's Hospital, London, UK

St Thomas' Hospital, London, UK
e-mail: Eric.Rosenthal@gstt.sthames.nhs.uk

17.1 Case Report

A 54-year-old man with treated hypertension and type II diabetes was seen in a chest pain clinic. A stress echocardiogram was negative for ischemia but revealed a dilated right ventricle. He admitted to mild breathlessness on exertion. A transoesophageal echocardiogram in a non-cardiac centre revealed an intact atrial septum but a positive bubble study. An MRI scan confirmed a dilated right ventricle with a RVEDVI of 153 ml/m2 (compared to LVEDVI of 68 ml/m^2) and a QP:QS of 2:1. The right upper pulmonary vein drained to the SVC-RA junction. A CT scan showed that the RUPV and RMPV drained anomalously, but even with 3D reconstructions it was not absolutely clear that covered stent correction would be possible (Fig. 17.1). In order to characterise the anatomy further, a 3D printed model of his heart was obtained. The model was printed on TangoPlus (Objet–Stratasys, Rehovot, Israel), using polyjet technology on an in-house 3D printer (Objet500 Connex1™, Objet–Stratasys). By placing an uncovered stent in the SVC across the SVASD and into the RA and a vascular sheath from the RUPV behind the stent into the left atrium, it was clear that a covered stent would be suitable for closing the SVASD and diverting the anomalous pulmonary veins to the left atrium (Figs. 17.2 and 17.3).

The patient was keen for transcatheter repair of his defect to avoid the morbidity of cardiac

G. Butera et al. (eds.), *Modelling Congenital Heart Disease*,
https://doi.org/10.1007/978-3-030-88892-3_17

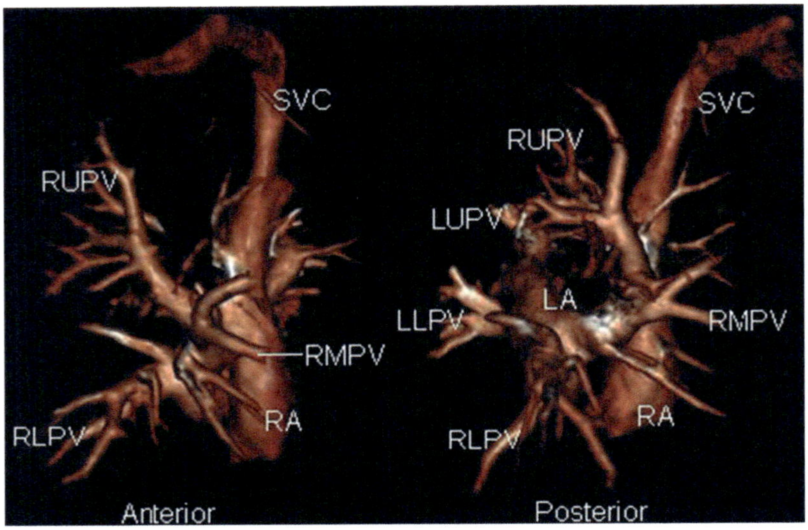

Fig. 17.1 3D reconstruction of CT scan suggests that the RUPV can be diverted to the LA after placement of a covered stent in the SVC but not conclusively

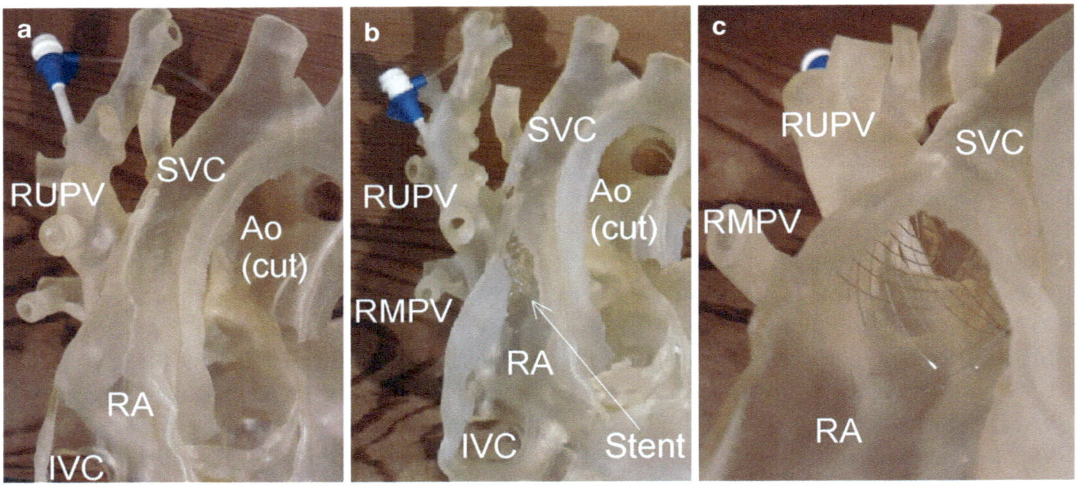

Fig. 17.2 Anterior view of the model shows a vascular sheath in the RUPV, and the stent in the SVC can only be seen dimly through the anterior wall of the SVC (**a**). In (**b**), the SVC is incised to show an uncovered stent at the SVC-RA junction and the vascular sheath can just be made out behind it. In (**c**), the vascular sheath can be clearly seen passing behind the stent lying over the SVASD into the LA

surgery. Cardiac catheterisation proceeding to covered stent implantation was therefore performed. A guide wire circuit was established from the RFV to RIJV to allow stable balloon occlusion of the defect and SVC sizing. From the LFV, a transseptal puncture was performed to place a 5F cut pigtail catheter in the RUPV through an 8F transseptal sheath left in the LA. This allowed pulmonary vein angiography during balloon testing and stent deployment as well as simultaneous pulmonary vein and LA pressure monitoring to confirm the absence of pulmonary vein obstruction by the sizing balloon and then the stent. A 30 mm-by-6 cm Andramed balloon (Andratech) confirmed that stent implantation would be effective without compromising pulmonary blood flow (LA and pulmonary vein pressure rose from a mean of 7 mmHg to 15 mmHg with no gradient between them, while the IVC pressure fell to 4 mm). Transoesophageal

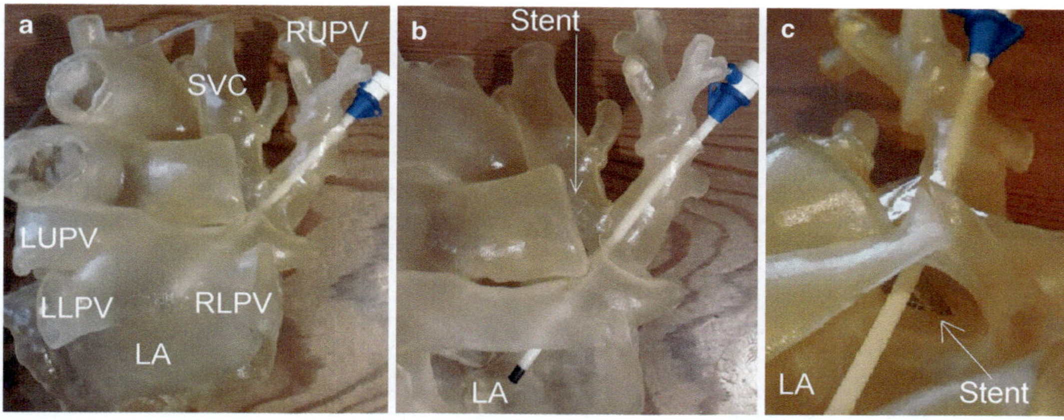

Fig. 17.3 Posterior view of the model shows the vascular sheath in the RUPV and dimly in the LA, while the stent in the SVC is also dimly seen (**a**). In (**b**), a window is cut in the back wall of the LA showing the sheath clearly in the LA. The SVC stent is more easily seen. In (**c**), the incision in the LA is extended into the RLPV and the sheath is seen passing in front of the SVASD (anatomically posterior to) behind, which is the stent in the SVC (anatomically anterior)

echocardiography confirmed defect occlusion and unobstructed pulmonary venous blood flow. A 6 cm custom made covered 10 zig CP stent (Cheatham-Platinum Stent; NUMED Inc., Hopkinton, NY, USA) mounted on a 30 mm-by-6 cm BIB balloon (NUMED Inc.) was placed across the defect through a 20-mm DrySeal sheath (WL Gore) in the RFV. A small residual shunt at the RA end of the stent remained and an additional 5 cm—10 zig CP stent was overlapped with the first stent (Fig. 17.4). The stents were then dilated further with a Coda balloon (Cook). At the end of the procedure, the transoesophageal echocardiogram confirmed a tiny residual defect only. This was no longer apparent on the transthoracic echocardiogram the following day when he was discharged taking clopidogrel for 2 months and aspirin for 6 months.

17.2 Discussion

Repair of a sinus venosus atrial septal defect (SVASD) with partial anomalous pulmonary venous drainage (PAPVD) has traditionally been performed by open heart surgery on cardiopulmonary bypass. Specific complications include sinus node damage with the need for a permanent pacemaker, and pulmonary vein obstruction, SVC obstruction, and residual shunt. In 2014, Garg et al. reported the use of a covered stent in the superior vena cava (SVC) to achieve a repair without surgery [1]. This was a radically different approach, and for many, the concept seemed difficult to understand as the ability to place a device as currently performed for a secundum ASD appeared to be impossible. We therefore explored the feasibility of SVASD and PAPVD repair using a covered stent by producing printed 3D models of the hearts in our patients. The notion that only the SVC "over-rode" the atrial septum was replaced by the concept that both the SVC ***and*** the RUPV over-rode the true atrial septum. This helped to clarify the anatomy of the defect and allowed us to recognise that a SVASD and PAPVD are a result of a deficiency in the wall separating the upper pulmonary veins and SVC at their junction with the right and left atriums allowing a shunt from the left atrium to right atrium.

The ability to place a stent into a model of the heart and demonstrate this to our surgical colleagues helped us to decide on the best approach to repair in each patient. Only after discussing the clinical features (to confirm the need for closure) and anatomical features (confirming suitability of

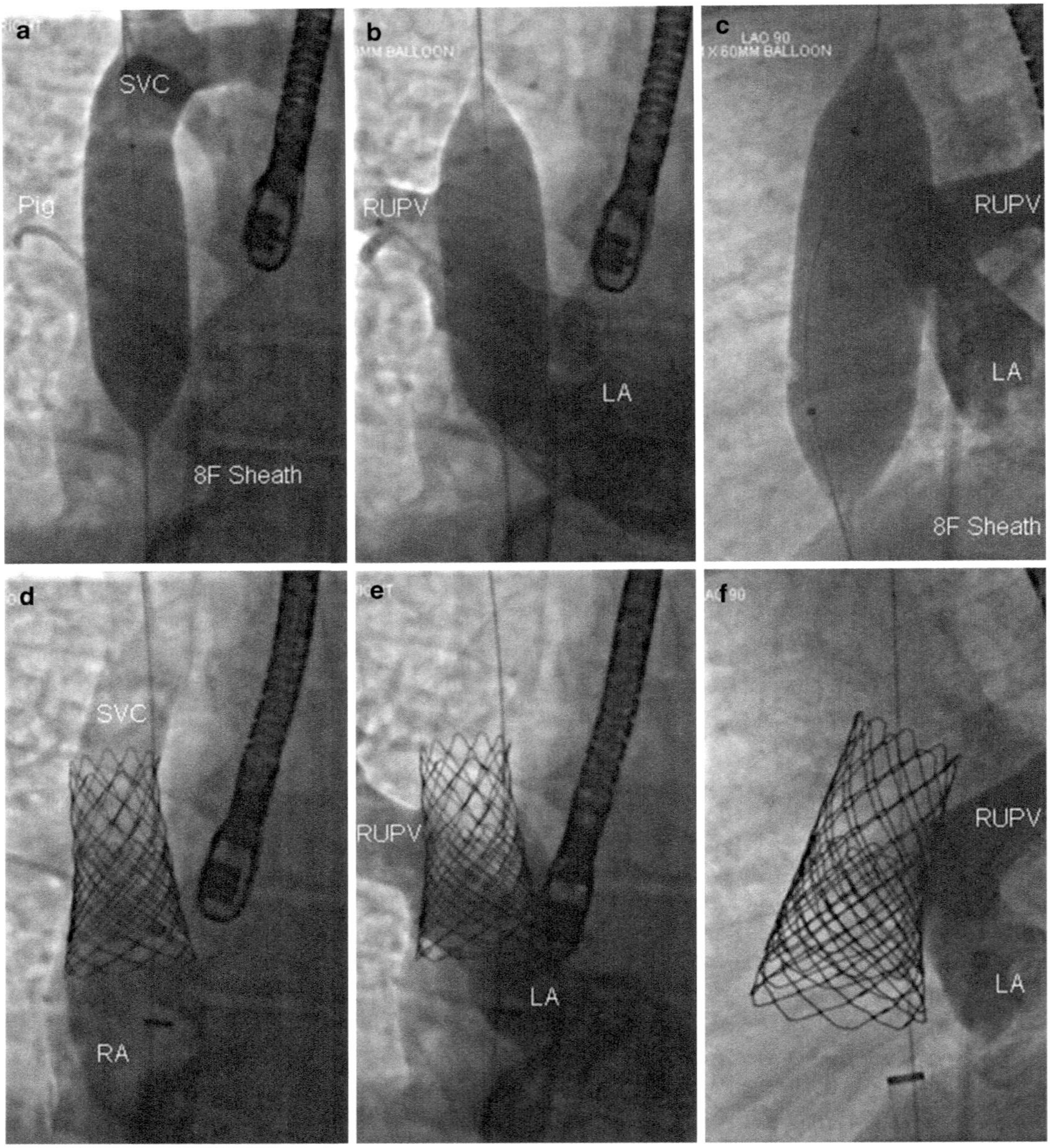

Fig. 17.4 Stent deployment. In (**a**), a 30-mm balloon is inflated across the SVC-RA junction and contrast in the SVC only fills the innominate and azygos veins. The pigtail catheter in the RUPV passes through the transseptal sheath in the LA. Contrast in the RUPV in anterior (**b**) and left lateral (**c**) projections shows unobstructed flow to the LA only. After stent deployment, contrast in the SVC passes to the RA only (**d**). Contrast injection in the RUPV in anterior (**e**) and left lateral (**f**) projections shows unobstructed flow to the LA only

the stent approach) in a multidisciplinary adult congenital heart disease meeting, did we proceed to undertake the procedure [2, 3].

In our unit's experience, stent repair of SVASD and PAPVD is possible in around 3/4 of adults with this condition. A major limitation is when one or more of the anomalous pulmonary vein drains directly into the SVC such that they would either be blocked by the covered stent or left draining freely into the SVC after closure of the sinus venosus defect when placing the stent below the entrance point of the vein. The latter situation may be acceptable if the residual vein is very small (and it is often an additional vein to the 3 main right-sided pulmonary veins) when the surgical approach would also be to leave the vein draining into the SVC. Where the additional vein is larger, then it has been until now a decision taken by the surgeon on the table as to whether it was justified to try diverting this, risking compromise to SVC calibre and or causing pulmonary vein stenosis during the baffling procedure. Since our programme to close these defects has extended, it has become important to decide whether this vein should be incorporated in the repair (surgical or stent) and it is here that the 3D printed model allows both the surgeon and the interventional cardiologist to assess the feasibility of both techniques and which is likely to benefit the patient most.

As illustrated in the case here, inspection of the model confirmed a very high chance of being able to correct the defect with a covered stent when the 3D reconstruction of the CT scan was unclear. What the model cannot do is mimic the effects of biological tissue. Thus, the degree of distension of the SVC is unknown, and in this patient, dilating the stent to 30 mm resulted in a greater degree of stent shortening than anticipated such that a second overlapping stent was required. We are increasingly using 7- and 8-cm stents to avoid this.

References

1. Garg G, Tyagi H, Radha AS. Transcatheter closure of sinus venosus atrial septal defect with anomalous drainage of right upper pulmonary vein into superior vena cava--an innovative technique. Catheter Cardiovasc Interv. 2014;84:473–7.
2. Velasco Forte MN, Byrne N, Valverde I, Gomez C, Hermuzi A, Prachasilchai P, Mainzer G, Pushparajah K, Henningson M, Hussain T, Qureshi S, Rosenthal E. Interventional correction of sinus venosus atrial septal defect and partial anomalous pulmonary venous drainage: procedural planning using 3D printed models. JACC Cardiovasc Imaging. 2018;11:275–8.
3. Riahi M, Velasco Forte MN, Byrne N, Hermuzi A, Jones M, Baruteau AE, Valverde I, Qureshi SA, Rosenthal E. Early experience of transcatheter correction of superior sinus venosus atrial septal defect with partial anomalous pulmonary venous drainage. EuroIntervention. 2018;14:868–76.

18 Modelling the Coronary Anatomy in a Case of Suspected Kawasaki Disease with Giant Coronary Aneurysms

Andrew Shearn, Maria Victoria Ordonez, Massimo Caputo, and Giovanni Biglino

A 58-year-old man with a medical history of well-controlled arterial hypertension and under vasodilator medication presented at the emergency room with atrial fibrillation, which was resolved with rate control medication. During his admission, an echocardiogram identified an incidental anterior echogenic mass between the right atrium (RA) and the right ventricle (RV) without clear origin. The same features were described by transoesophageal echocardiogram.

A cardiac computed tomography (CT) examination was performed a few months later showing a normal origin of the coronary arteries but with two giant aneurysms at the mid- and distal segment of the right coronary artery (RCA) with an anomalous RA termination towards either the RA or superior vein cava (SVC), as shown in Fig. 18.1.

The patient was discussed at the multidisciplinary team meeting in order to evaluate the need for surgical repair vs. clinical surveillance. In this instance, a 3D heart model was requested due to the unclear understanding of the anatomy and particularly to clarify the distal segment of the RCA, before making any final decision.

The CT imaging dataset was imported into commercial software (Mimics v21, Materialise, Leuven, Belgium) and reconstructed following established techniques including thresholding and region growing [1]. Two separate regions of interest ("masks") were created from the imaging of the blood pool, namely one of the aneurysms alone and one of the whole heart subtracting the first mask. The resulting 3D models were then remeshed, smoothed and hollowed in post-processing software (3-Matic, Materialise). Following confirmation that the post-processing had not affected the accuracy of the segmentation by superimposing the models onto the original CT data, a 3D Portable

A. Shearn
Bristol Medical School, Faculty of Health Sciences, University of Bristol, Bristol, UK
e-mail: andrew.shearn@bristol.ac.uk

M. V. Ordonez
University Hospitals Bristol & Weston NHS Foundation Trust, Bristol, UK
e-mail: victoria.ordonez@uhbw.nhs.uk

M. Caputo
Bristol Medical School, Faculty of Health Sciences, University of Bristol, Bristol, UK

University Hospitals Bristol & Weston NHS Foundation Trust, Bristol, UK
e-mail: m.caputo@bristol.ac.uk

G. Biglino (✉)
University of Bristol, Bristol Medical School, Bristol, UK
e-mail: g.biglino@bristol.ac.uk

G. Butera et al. (eds.), *Modelling Congenital Heart Disease*, https://doi.org/10.1007/978-3-030-88892-3_18

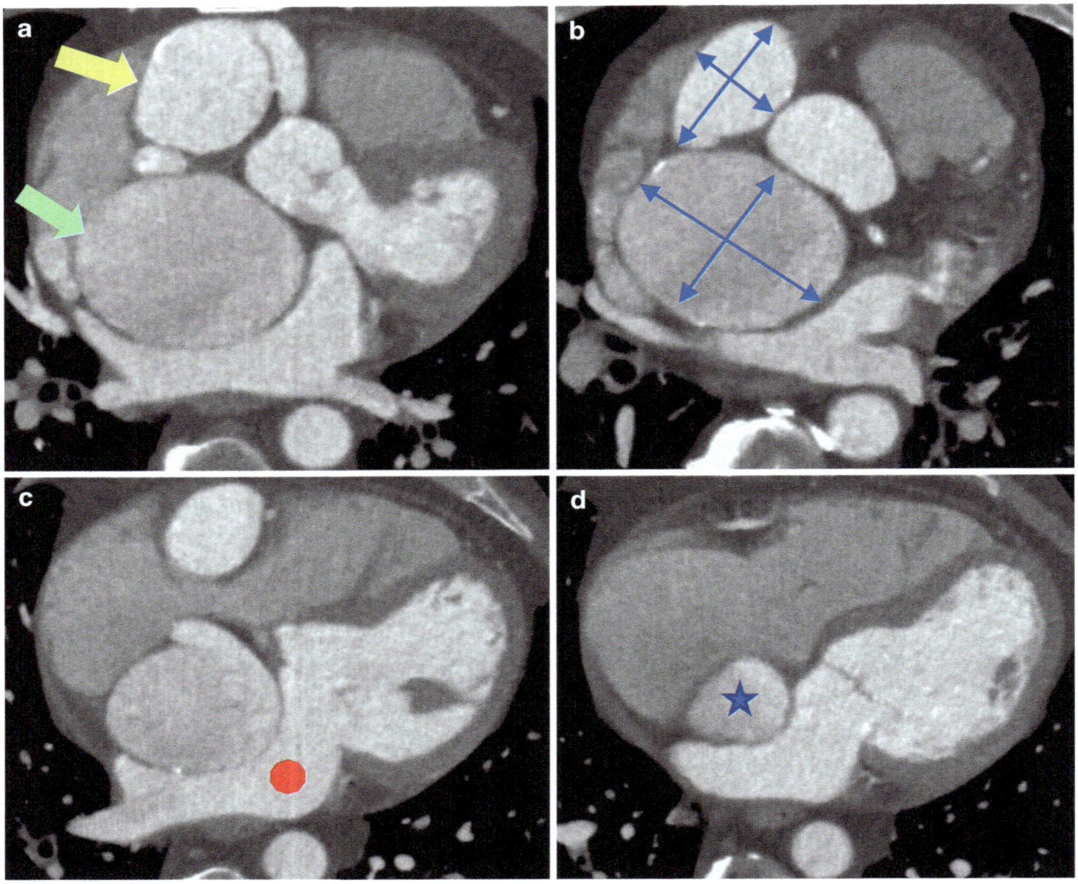

Fig. 18.1 Coronary tomography axial view. (**a**) RCA with two giant aneurysms at the mid- (*yellow arrow*) and distal segment (*green arrow*). (**b**) Mid-aneurysm size of 4.5 cm and distal aneurysm size of 7.5 cm. (**c**) Compression of the left atrium (*red circle*). (**d**) Terminal segment of the RCA into the RA (*blue asterisk*)

Document Format (PDF) file was created and provided to the clinical team. This could then be viewed using Acrobat Reader software (Adobe Inc., San Jose, California, USA). The resulting model is shown in Fig. 18.2.

The format including 3D objects in PDF documents has been used in different contexts and has been reported as helpful in the interpretation of 2D and 3D data, facilitating communication on morphological issues also in other areas of medicine and biology [2]. From the perspective of visualising cardiovascular structures (as in this case the coronary aneurysms), the 3D PDF format allows rendering options such as transparency of the model and switching parts on and off [3], which was found helpful in this case. This means of visualisation has been discussed in the context of other congenital defects, such as tetralogy of Fallot [4]. Whilst in this case the model was not printed for handling, the feasibility of printing case of congenital coronary anomalies and also cases of coronary aneurysm in patients with Kawasaki disease was previously discussed in a case series [5].

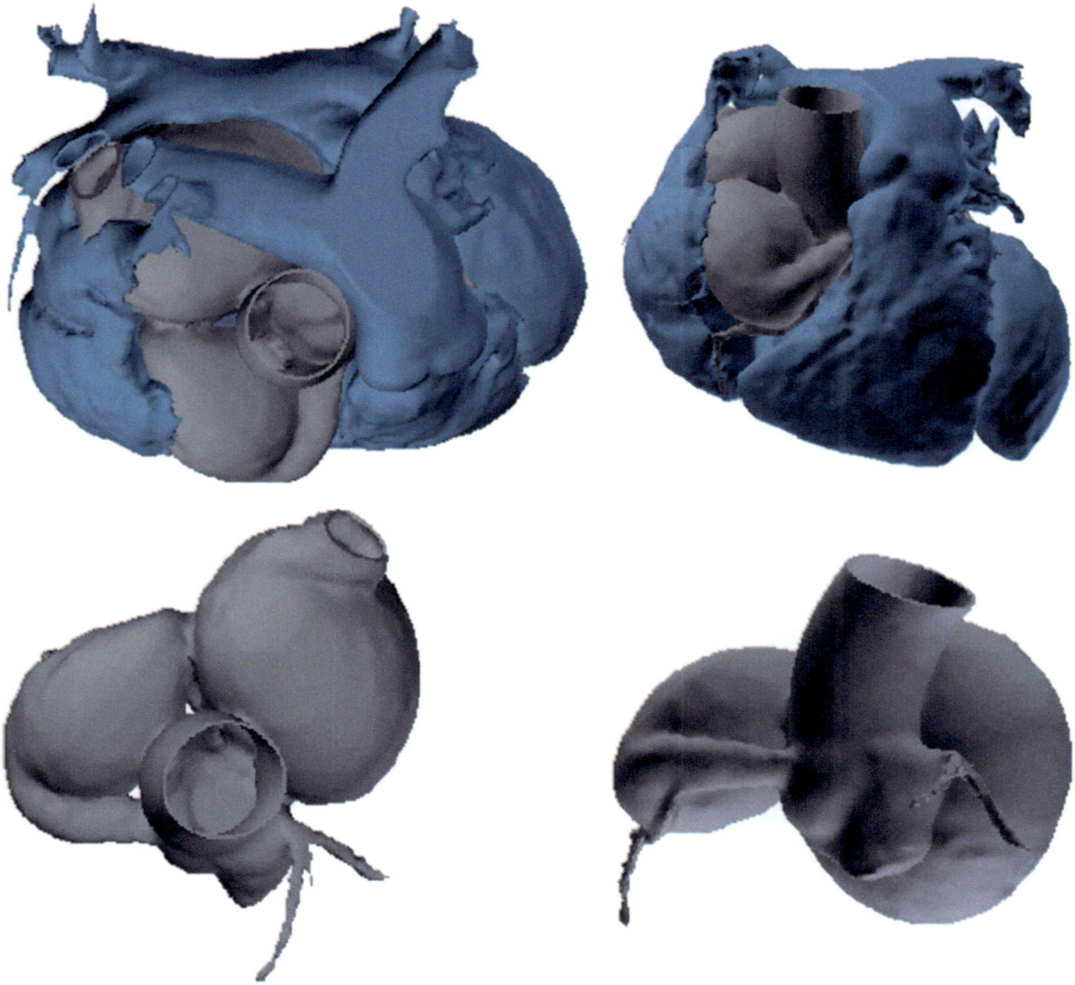

Fig. 18.2 Different views of the 3D reconstructions, highlighting the coronary anatomy including the giant aneurysms (*grey*) with respect to the remained of the blood pool model (*blue*)

References

1. Schievano S, Coats L, Migliavacca F, Norman W, Frigiola A, Deanfield J, Bonhoeffer P, Taylor AM. Variations in right ventricular outflow tract morphology following repair of congenital heart disease: implications for percutaneous pulmonary valve implantation. J Cardiovasc Magn Reson. 2007;9(4):687–95. https://doi.org/10.1080/10976640601187596.
2. de Boer BA, Soufan AT, Hagoort J, Mohun TJ, van den Hoff MJB, Hasman A, Voorbraak FPJM, Moorman AFM, Ruijter JM. The interactive presentation of 3D information obtained from reconstructed datasets and 3D placement of single histological sections with the 3D portable document format. Development. 2011;138(1):159–67.
3. Kiraly L. Three-dimensional modelling and three-dimensional printing in pediatric and congenital cardiac surgery. Transl Pediatr. 2018;7(2):129–38. https://doi.org/10.21037/tp.2018.01.02.
4. Ghosh RM, Mascio CE, Silvestro E, O'Byrne ML, Whitehead KK. A road map for collaterals: use of 3-dimensional techniques in tetralogy of Fallot pulmonary atresia with major aortopulmonary collateral arteries. JTCVS Techniques. 2020;1:82–5.
5. Lee M, Moharem-Elgamal S, Beckingham R, Hamilton M, Manghat N, Milano EG, Bucciarelli-Ducci C, Caputo M, Biglino G. Evaluating 3D-printed models of coronary anomalies: a survey among clinicians and researchers at a university hospital in the UK. BMJ Open. 2019;9(3):e025227. https://doi.org/10.1136/bmjopen-2018-025227.

19 Criss-Cross Heart: Is Biventricular Repair Ever Possible?

Francesca Raimondi

19.1 Introduction

Criss-cross heart is an extremely rare cardiac anomaly accounting for less than 0.1% of congenital heart defects [1]. It consists in the crossing of the long axes of atrioventricular valves.

Criss-cross heart is always associated with other cardiac anomalies such as interventricular septal defect (ISD), abnormal atrioventricular alignments (transposition of great arteries or double outlet right ventricle), hypoplasia of tricuspid valve and right ventricle with pulmonary stenosis [2].

F. Raimondi (✉)
Unité médico-chirurgicale de cardiologie congénitale et pédiatrique, centre de référence des maladies cardiaques congénitales complexes - M3C, Hôpital universitaire Necker-Enfants Malades, Université de Paris, Paris, France

Pediatric Radiology Unit, Hôpital universitaire Necker-Enfants Malades, Université de Paris, Paris, France

Decision and Bayesian Computation, Computation Biology Department, CNRS, URS 3756, Neuroscience Department, CNRS UMR 3571, Institut Pasteur, Paris, France

School of Biomedical Engineering & Imaging Sciences, King's College London, Lambeth Wing St. Thomas' Hospital, London, UK
e-mail: Francesca.raimondi@gmail.com

Only a minority of patients are suitable for a biventricular repair due to the right ventricular hypoplasia, with the majority of patients palliated via a Fontan operation.

We describe here a rare case of a criss-cross heart with SDL, situs solitus, ventricular D-loop, atrioventricular concordance, two balanced ventricles, atrioventricular discordance with L-aorta, pulmonary stenosis and posterior interventricular septal defect.

The 3D models from CT scan data showed clearly the relationship between great vessels and ISD, comforting the possibility to perform biventricular repair within block rotation [3] of the arterial trunk, ISD closure and pulmonary commissurotomy, possibility that was not easy to get so straightly from echocardiographic images.

The patient was successfully operated at 7 months of age, and he is doing very well after 2 years of follow-up.

19.2 CT Scan Acquisition

Cardiac CT imaging was performed using a LightSpeed VCT 64 Slice Scanner (GE Healthcare) after injection of contrast agent (Xenetix® 300 mg/mL, 2 mL/kg). Acquisition of whole heart images was ECG-gated, performed in step-and-shoot (SAS) mode at a single phase (40%) of the cardiac cycle. Heart rate was 100 bpm.

G. Butera et al. (eds.), *Modelling Congenital Heart Disease*,
https://doi.org/10.1007/978-3-030-88892-3_19

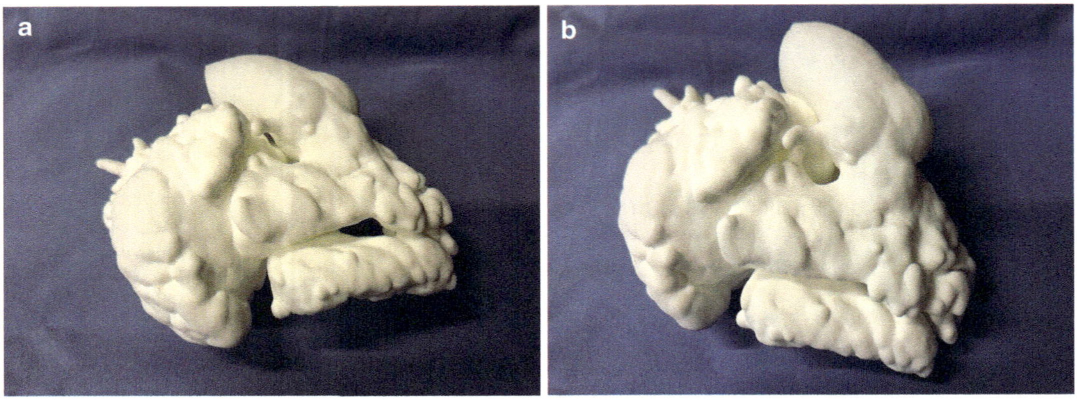

Fig. 19.1 (**a, b**) Photographs of 3D printed model of criss-cross heart in two different views

19.3 3D Rendering Processing

19.3.1 3D Printing

Images were post-processed using Mimics (Materialise) and Meshmixer (Autodesk) to create a 3D reconstruction of the patient blood pool, including atria, ventricles and great vessels. The resulting 3D anatomical model was 3D printed 1:1 scale, in rigid white nylon using selective laser sintering technology (EOS P100) (Fig. 19.1).

19.3.2 3D Pdf

A 3D pdf model was generated from the same subset of data (Fig. 19.2).

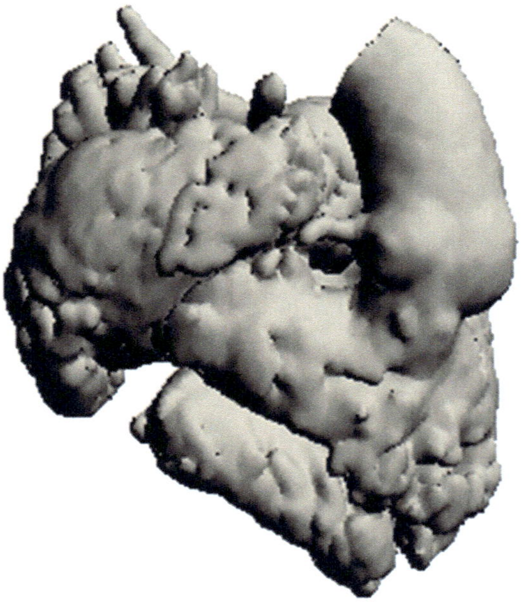

Fig. 19.2 Image from pdf file of the model

19.3.3 Virtual Reality

Virtual reality provides a natural mean to visualise 3D structures, through the integration of motion tracking, stereoscopy and total immersion within the headset. Virtual reality offers the possibility of full volumetric representation while not requiring segmentation of the data. Volume appearance can be designed using Transfer function that leverages image features to define voxels intensity, opacity and colour. In doing so, specific structures can be made visible while leaving all the information stored in the image accessible [4, 5].

In Fig. 19.3, we used the software DIVA [6] to allow visualisation and navigation within the heart of the patient. DIVA provides a dual interface one on the desktop and one in the VR mode. In the desktop, the volumetric representation of the data can be viewed within the 2D screen with an interface allowing modification of visualisation parameters. In the VR interface, image stacks

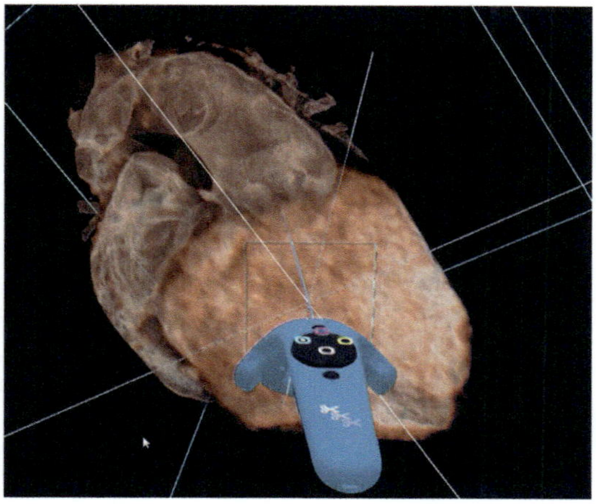

Fig. 19.3 Image from DIVA application on the 3D model derived from CT scan

are represented as "physical objects". The user can grab the volume and navigate inside it via physical manipulation using the VR controller. DIVA renders volumes using a ray-casting approach that can have its sampling resolution adjusted in real time to ensure a sufficient frame rate for the user. A key functionality here is the clipping plane allowing selective removal of a portion of the rendered volume in real time. This function allows "cutting" hearts from all possible angles allowing cleared visualisation of important structures.

19.4 Conclusion

Imaging a criss-cross heart is challenging because of the complex relationship between ventricles and great vessels. Anatomical details are often hardly shown by echocardiography. Often, a Fontan palliation is proposed to overcome complexity of intracardiac arrangement.

Our 3D models showed the complex intracardiac anatomy and the position of the interventricular septal defect in a clear and simple way.

Straight comprehension of intracardiac anatomy may allow to plan the biventricular repair, a rare choice in criss-cross heart, often excluded.

Acknowledgments Jean Baptiste Masson and Mohamed Berhili, Decision and Bayesian Computation, Computational biology and Neuroscience department, Institut Pasteur, Paris.

Silvia Schievano, UCL London.

Nathalie Boddaert, Pediatric Radiology, Hôpital universitaire Necker-Enfants Malades Université de Paris, Paris Cité.

Olivier Raisky, M3C, Hôpital universitaire Necker-Enfants Malades, Université de Paris, Paris Cité.

References

1. Fyler DC. Trends. In: Fyler DC, editor. Nadas' pediatric cardiology. Philadelphia: Hanley and Belfius, Inc.; 1992. p. P273–80.
2. Van Praagh R, Weinberg PM, Van Praagh S. Malposition of the heart. In: Moss AJ, Adams FH, Emmanouilides GE, editors. Heart disease in infants, children and adolescents. 2nd ed. Baltimore: Williams and Wilkins; 1977. p. 395–9.
3. Yamagishi M, Shuntoh K, Matsushita T, et al. Half-turned truncal switch operation for complete transposition of the great arteries with ventricular septal defect and pulmonary stenosis. J Thorac Cardiovasc Surg. 2003;125:966–8.
4. El Beheiry M, et al. Virtual reality: beyond visualization. J Mol Biol. 2019;431(7):1315–21.
5. Blanc T, et al. Genuage: visualizing and analyzing multidimensional point cloud data in virtual reality. Nat Methods. 2020;17:1100–2. bioRxiv
6. El Beheiry M, et al. DIVA: natural navigation inside 3D images using virtual reality. bioRxiv. 2020; https://doi.org/10.1101/2020.04.09.019935.

20 Use of 3D Printing for Congenital Heart Disease

Hannah Tredway, Nikhil Pasumarti, Matthew A. Crystal, Amee M. Shah, and Kanwal M. Farooqi

Three-dimensional (3D) printing has garnered increased interest in the field of pediatric cardiology in the last decade. 3D models play a substantial role in helping to understand intricate intra- and extracardiac anatomy of patients with complex congenital heart disease (CHD). These models can be used for presurgical and precatheterization planning and procedural simulation, with the ultimate goal of reducing operative times, decreasing intraoperative complications, and shortening postoperative hospital stays [1]. We will discuss three cases in which 3D models aided in the understanding and management of patients with CHD.

Supplementary Information The online version of this chapter (https://doi.org/10.1007/978-3-030-88892-3_20) contains supplementary material, which is available to authorized users.

H. Tredway
Department of Pediatrics, New York Presbyterian/Columbia University Irving Medical Center, New York, NY, USA
e-mail: hlt9003@nyp.org

N. Pasumarti · M. A. Crystal · A. M. Shah · K. M. Farooqi (✉)
Division of Pediatric Cardiology, Department of Pediatrics, New York Presbyterian/Columbia University Irving Medical Center, New York, NY, USA
e-mail: np2564@cumc.columbia.edu; mac2376@cumc.columbia.edu; ams79@cumc.columbia.edu; kf2549@cumc.columbia.edu

20.1 Introduction: Case 1

One to three percent of patients with CHD have a complex cardiac malformation known as double outlet right ventricle (DORV), accounting for 3–9 per 100,000 live births. DORV is an abnormality of the ventriculoarterial connection in which both the pulmonary artery and aorta arise either entirely or predominantly from the right ventricle [2]. This can occur as an isolated defect, or can be associated with other intra- or extracardiac anomalies such as atrial or ventricular septal defects (ASD, VSD), atrioventricular (AV) canal, with or without outflow tract obstruction. Patients with DORV undergo varied surgical interventions ranging from single ventricle palliation to biventricular repair that can be staged over a series of operations. The complexity of the intracardiac anatomy, the relationship of the great vessels, presence of associated defects, size and function of the ventricles, AV valves, and ventricular septal defect (VSD) are all taken into consideration when determining the surgical approach [3]. 3D models of DORV provide a clear demonstration of the arrangement of the ventricles, great arteries, and associated defects, which is crucial in preprocedural planning [4, 5].

G. Butera et al. (eds.), *Modelling Congenital Heart Disease*,
https://doi.org/10.1007/978-3-030-88892-3_20

20.1.1 Case History

Our patient is an ex-33-week man with a prenatal diagnosis of DORV with d-malposed great arteries, pulmonic stenosis with a hypoplastic pulmonary valve, confluent mildly hypoplastic branch pulmonary arteries, a large inlet ventricular septal defect, and a large patent ductus arteriosus (PDA). He developed symptoms of pulmonary over-circulation and congestive heart failure and underwent a transcatheter device closure of his PDA at approximately 6 weeks of age. An echo at the time demonstrated unobstructed flow across the aortic valve and mild subpulmonic narrowing caused by conal tissue with an unobstructed pulmonary valve and a good-sized right (RV) and left ventricle (LV). The patient subsequently developed cyanosis in the setting of progressive right ventricular outflow tract (RVOT) obstruction and underwent a bidirectional Glenn procedure at 4 months of age. Given the complexity of baffling the VSD to the great arteries in a small infant of this size, the decision was made to proceed with a bidirectional Glenn procedure at 4 years of age with the ultimate plan for a successful biventricular repair when the child would be larger.

20.1.2 Imaging

Advanced imaging was requested to better delineate the patient's anatomy and to assess candidacy for a biventricular repair given the complexity of the intracardiac anatomy and the small size of the patient. A cardiac computed tomography (CT) angiogram was performed, which demonstrated DORV with d-malposed great vessels, subvalvular muscle bundles obstructing flow below the pulmonary valve, and an inlet type VSD with subaortic extension.

20.1.3 Engineering/3D Printing Tools

A 3D model was requested to better illustrate the patient's anatomy to aid in surgical planning. The Digital Imaging and Communications in Medicine (DICOM) files of the 3D dataset were then imported into Mimics (Materialise, Leuven, Belgium), a commercially available software, for segmentation. The process of segmentation included highlighting the blood pool to specifically delineate the area of interest. A 1-mm layer is applied, and the blood pool is digitally subtracted to allow for visualization of the intracardiac cavity [6]. The program then creates a 3D model, which is imported into and manipulated in 3-Matic (Materialise, Leuven, Belgium). The models are further sectioned along a cut plane for better visualization of intracardiac anatomy, or printed as a whole hollowed model, which can be appropriate for surgical simulation. The digital models are then printed using photopolymeric, thermoplastic, or powder fusion techniques. For higher-end medical grade printers, resolution can be on the order of micrometers, allowing for the complex intracardiac anatomy to be clearly defined in a physical model [7].

20.1.4 Why It Did Help Me?

For this patient, two prints were created: One model was printed to scale to assess for feasibility of the procedure in an infant (Fig. 20.1); and the second was scaled up to better visualize and characterize the anatomy. The models were printed on a uPrint printer (Stratasys, Eden Prairie, Minnesota), using fusion deposition modeling (FDM) technology with soluble supports. These models allowed the surgeons to better identify the relationships of the ventricles, the VSD, and the great arteries from multiple perspectives. Additionally, the different model sizes provided insight into the feasibility of the procedure. Given the appearance of the potential baffle pathways on the model, the patient will likely undergo a biventricular repair at approximately 2 years of age.

20.2 Introduction: Case 2

Heterotaxy results from the failure of the developing embryo to establish left–right lateralization of the thoracoabdominal organs and vessels,

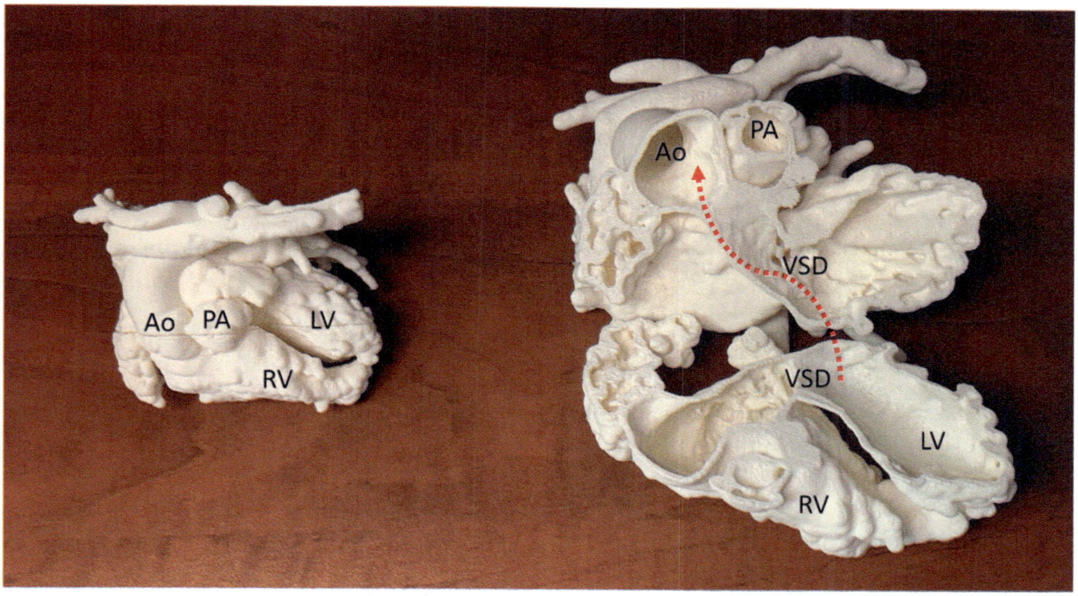

Fig. 20.1 3D printed model of a patient with double outlet right ventricle (DORV). The model on the left was printed true to size of the infant heart, and the model on the right was enlarged to allow assessment of the relevant anatomy more easily. The *red dashed arrow* represents the relatively tortuous pathway from the left ventricle (LV) through the ventricular septal defect (VSD) to the aorta (Ao). Pulmonary artery (PA), right ventricle (RV)

thereby resulting in abnormal symmetry and malposition. This condition also results in complex congenital heart disease and extracardiac defects of midline-associated structures such as the inferior and superior vena cava (IVC, SVC). Heterotaxy has an incidence of one in 6000–20,000 live births and is associated with 1% of all CHD [8].

Cardiovascular defects associated with heterotaxy are extremely variable. The type of cardiac defects is often associated with the type of splenic anomaly, which ranges from asplenia to polysplenia, or a malpositioned spleen. Bilateral SVCs and other systemic venous anomalies are characteristic of asplenia syndrome, which occurs in 46–71% of patients with heterotaxy syndrome. Anomalous connections of the hepatic and pulmonary veins are also often seen [8]. While AV canal defects are the predominant intracardiac defect in this condition, dextrocardia and abnormalities of the ventricular outlet region including DORV, transposition of the great arteries (TGA), subvalvar or valvar stenosis, or valvular atresia can also occur.

20.2.1 Case History

Our patient is a 15-year-old girl with heterotaxy syndrome (asplenia type), transitional atrioventricular canal, and multiple systemic venous anomalies including a left (SVC) to the left atrium (LA) and a left-sided IVC with ipsilateral hepatic veins. She underwent a transitional AV canal repair with complex systemic venous baffling of the malposed IVC and LSVC to the right atrium (RA) at 6 years of age. After an unremarkable postoperative course, she was clinically well and followed with serial echocardiograms. Follow-up echocardiogram demonstrated an unobstructed baffle repair, mildly dilated ventricles, mild-to-moderate RV hypertrophy, and pulmonary hypertension.

20.2.2 Imaging

Eight years after her initial surgery, an echocardiogram demonstrated mildly accelerated flow in the systemic venous baffle pathway and a severely

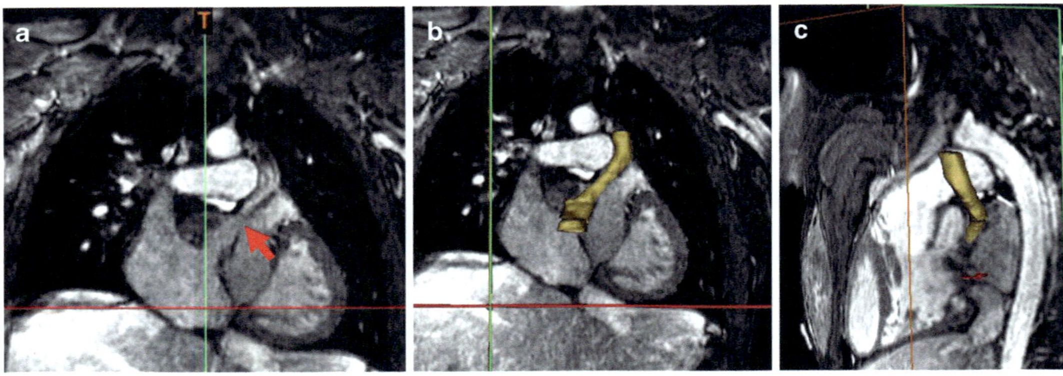

Fig. 20.2 Source images from a noncontrast MRA dataset. (**a**) The red arrow indicates the site of the defect in the baffle from the left superior vena cava to the right atrium. (**b**, **c**) This pathway was segmented alone, represented by the *yellow* 3D digital model overlaid on the MRI images

dilated and hypertrophied RV with mildly depressed systolic function. Cardiac magnetic resonance (MR) imaging was performed to further evaluate biventricular size, function, and intracardiac shunting. It demonstrated evidence of left-to-right shunting, with a Qp:Qs ratio of 2:1, a severely dilated RV (RVEDVi 202 ml/m2) with normal systolic function (EF 52%). Additionally, the bright blood cine images and 3D noncontrast MR angiogram were suggestive of a leak in the baffle from the LSVC to the RA. Given the significant intracardiac shunt resulting from the baffle leak resulting in RV dilation, it was determined that the patient would need occlusion of the baffle leak. After discussion in our surgical conference, she was deemed a candidate for closure of the defect via a transcatheter approach.

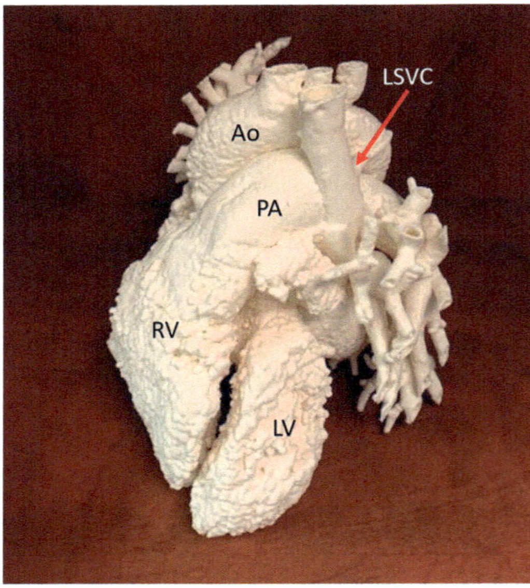

Fig. 20.3 A whole heart model was 3D printed to assess the overall anatomy and the relationship of the LSVC with other cardiac structures

20.2.3 Engineering/3D Printing Tools

The location of the defect, while seemingly amenable to occlusion with a covered stent, also had potential to obstruct flow across the SVC. The use of a flexible covered stent (Gore Viabahn VBX Balloon Expandable Endoprosthesis) was thought to be the best approach given its ability to expand into the somewhat irregular space, which included the baffle pathway and leak. In preparation for this procedure, a 3D digital reconstruction was created from the noncontrast MRA (Fig. 20.2). Using the same segmentation techniques and technology as in the prior case, two 3D models were printed: one of the LSVC to RA baffle pathway and one of the whole heart (Fig. 20.3). Segmentation in this case was particularly challenging as the MRA did not fully opacify the LSVC. As a result, the blood pool within the LSVC required manual editing.

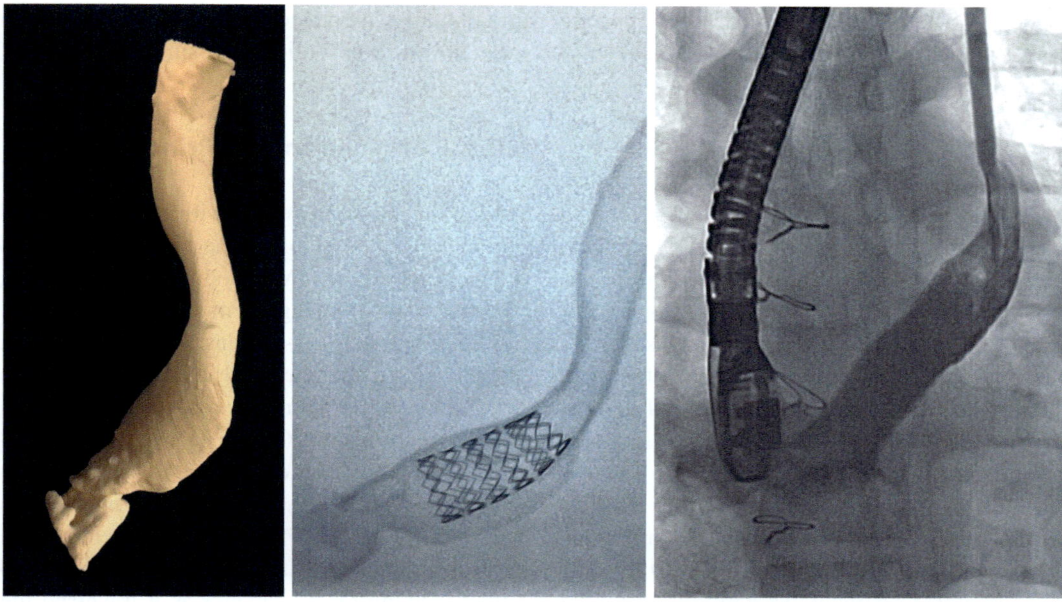

Fig. 20.4 The left superior vena cava (LSVC) was 3D printed alone (**a**) in order to allow preprocedural simulation of deployment of the stent (**b**). The angiogram post-deployment confirmed closure of the baffle leak with an unobstructed pathway from the LSVC to the right atrium (RA)

20.2.4 Why It Did Help Me?

Ultimately, the LSVC model outlined the complex pathway and optimal stent positioning within the vessel. This model was used to practice the covered stent placement and identify the appropriate size stent prior to the procedure (Fig. 20.4). During the catheterization, the stent was successfully positioned within the baffle pathway, covering the area of leak and expanded. A subsequent angiogram confirmed closure of the large leak with a separate tiny residual ASD noted. The patient is now several months postprocedure. She is on coumadin for anticoagulation, has improvement in right ventricular size, and has significantly improved exercise tolerance.

20.3 Introduction: Case 3

Coronary artery fistulas (CAFs) are rare cardiac anomalies that involve an abnormal communication of coronary arteries with cardiac chambers or other vessels [9]. With an incidence of 0.06–0.2% in children, these anomalies may be congenital or acquired, and can cause significant morbidity and mortality among both adult and pediatric patients [10]. CAFs present with widely varied clinical manifestations depending on size of the shunt, occurrence of thromboembolism, and associated myocardial ischemia [11]. In one study, up to 79% of children with CAFs were asymptomatic, 54% presented with a continuous heart murmur, and 8% presented with dyspnea and symptoms of congestive heart failure such as exercise intolerance, dyspnea, angina, and arrhythmia.

Coronary fistulas can result in aneurysmal and tortuous coronary arteries, which are typically diagnosed with color Doppler to delineate the origin, termination, and course of the fistula. Additional imaging such as MR or CT can be employed to further delineate the vascular anatomy [8]. CAFs are prone to aneurysm formation, secondary coronary thrombosis, calcification, and postoperative ventricular aneurysms [11]. Treatment includes percutaneous trans-luminal embolization, surgical ligation, or medical management [10]. Device closure of CAF typically has favorable outcomes when compared to surgical closure.

20.3.1 Case History

Our patient is a 3-year-old girl with a large CAF diagnosed in infancy. She was born at 36 and 5/7 weeks of gestation and was developing normally but noted to have a continuous systolic murmur with radiation to the left back for which she was referred to cardiology. An initial echocardiogram demonstrated a large fistula arising from the left coronary artery measuring 5.5 mm and traveling in a circuitous route into the RA. The fistula appeared unobstructed on echocardiogram at the entry point to the RA with early diastolic flow reversal. Also noted was a mildly dilated RA and LA and moderately dilated LV with normal function. She was followed closely by cardiology and noted to be asymptomatic with no palpitations, dyspnea, tachypnea, syncope, diaphoresis, lethargy, or signs of congestive heart failure.

20.3.2 Imaging

A CT scan was performed in order to further delineate her anatomy in preparation for interventional closure. The CT again demonstrated a severely dilated proximal left main coronary artery with a large coronary artery fistula traveling posterior to the aortic root and slightly inferiorly, narrowing significantly before entering the right atrium. The narrowing into the right atrium made it unlikely that the patient would experience significant heart failure or ischemia secondary to coronary steal. Given the size of her coronary fistula, however, the decision was made to perform a percutaneous device occlusion. Given the complex route of the fistula, a 3D printed model was requested to prepare for interventional closure.

20.3.3 Engineering/3D Printing Tools

The CT was postprocessed with careful segmentation performed to follow the path of the CAF as it emptied into the RA. Blood pool segmentation was used and the digital model was then hollowed, excluding a 1-mm layer left to represent the vessel and chamber walls as described for the previous cases. The model was cropped along a short and coronal axis of the heart, to allow visualization of the CAF course preprocedural planning (Fig. 20.5a, b). A full heart model was printed on Form 3 (Formlabs, Somerville, MA) stereolithography (SLA) printer in a clear resin to allow viewing of the catheter as it traversed the CAF in the model (Fig. 20.6).

20.3.4 Why It Did Help Me?

The 3D printed model allowed our team to consider different approaches to transcatheter closure of this fistula. Practicing the closure on the model demonstrated the feasibility of accessing the fistula from the venous access. If access directly into the fistula from the IVC is not possible, then the approach would require retrograde access from the aorta and then creation of an arteriovenous wire loop to allow venous access into the fistula with a larger sheath. Given our ability to practice performing the procedure prior to the actual catheterization, it will dramatically decrease the procedure time, radiation exposure, and success of the procedure itself. Understanding how best to approach the patient would not have been possible based solely on examining the 2D imaging. The patient is planned to undergo the procedure in the next month.

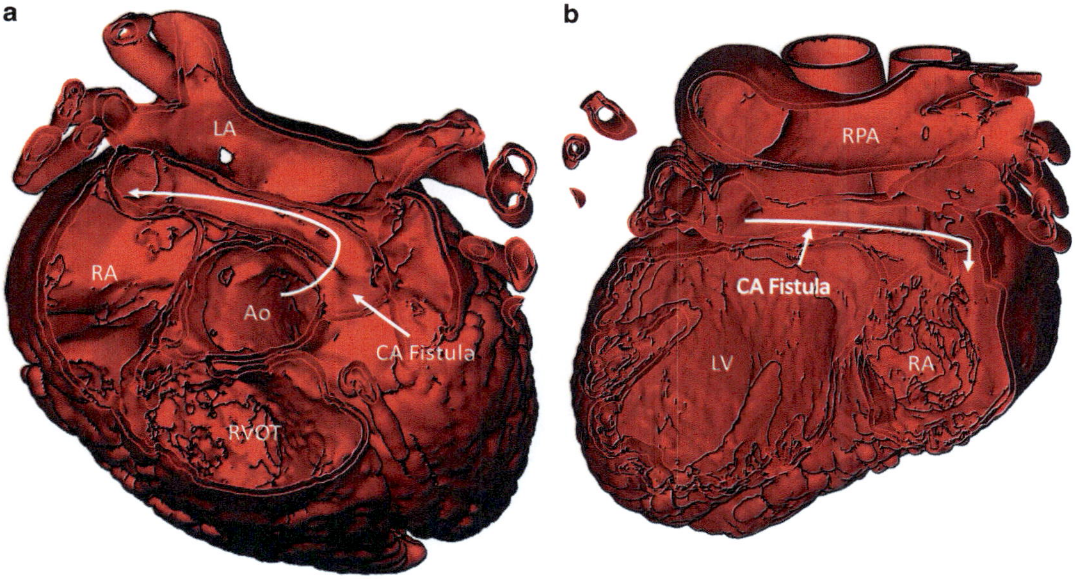

Fig. 20.5 (**a**) The course of the coronary fistula (CAF) is viewed from a short-axis view of the heart. From the leftward aspect of the aortic root, it courses posterior to the aorta (Ao) and rightward to drain into the right atrium (RA). (**b**) Coronal view of the heart as viewed from the posterior aspect reveals the course of the CA fistula, almost parallel to the right pulmonary artery (RPA) from left to right to drain into the right atrium

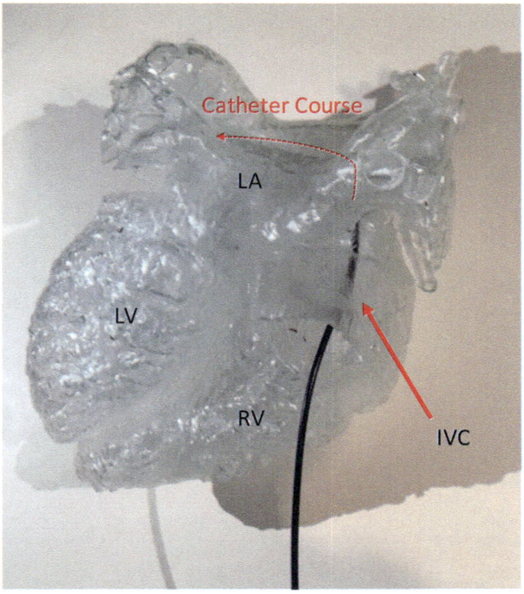

Fig. 20.6 3D printed model of the CA printed in a clear resin (Formlabs, Somerville, MA) allows planning of the transcatheter approach to closure of the fistula as viewed from the posterior aspect. A catheter courses from the inferior vena cava through the fistula (*red dotted line*)

References

1. Bramlet M, et al. Impact of three-dimensional printing on the study and treatment of congenital heart disease. Circ Res. 2017;120(6):904–7.
2. Obler D, et al. Double outlet right ventricle: aetiologies and associations. J Med Genet. 2008;45(8):481–97.
3. Oladunjoye O, et al. Repair of double outlet right ventricle: midterm outcomes. J Thorac Cardiovasc Surg. 2019; https://doi.org/10.1016/j.jtcvs.2019.06.120.
4. Farooqi KM, et al. Use of 3-dimensional printing to demonstrate complex intracardiac relationships in double-outlet right ventricle for surgical planning. Circ Cardiovasc Imaging. 2015;8(5):e003043.
5. Farooqi KM, et al. Application of virtual three-dimensional models for simultaneous visualization of Intracardiac anatomic relationships in double outlet right ventricle. Pediatr Cardiol. 2016;37(1):90–8.
6. Farooqi KM, et al. Blood Pool segmentation results in superior virtual cardiac models than myocardial segmentation for 3D printing. Pediatr Cardiol. 2016;37(6):1028–36.
7. Anwar S, et al. 3D printing is a transformative technology in congenital heart disease. JACC Basic Transl Sci. 2018;3(2):294–312.
8. Bartram U, Wirbelauer J, Speer CP. Heterotaxy syndrome -- asplenia and polysplenia as indicators of visceral malposition and complex congenital heart disease. Biol Neonate. 2005;88(4):278–90.

9. Shimada S, et al. Heart failure and coronary ischemia in a neonate with right coronary artery fistula. Pediatr Int. 2019;61(4):417–8.
10. Said SA, Lam J, van der Werf T. Solitary coronary artery fistulas: a congenital anomaly in children and adults. A contemporary review. Congenit Heart Dis. 2006;1(3):63–76.
11. Wang NK, et al. Coronary arteriovenous fistula in pediatric patients: a 17-year institutional experience. J Formos Med Assoc. 2002;101(3):177–82.

21 Three-Dimensional Printed Model Guiding Transcutaneous Device Closure of a Complex Residual Ventricular Septal Defect: Comparing Apples to Apples

Justin T. Tretter and Puneet Bhatla

21.1 Introduction

Although certainly suggested by multiple case series and case reports, it has been challenging to objectively demonstrate the utility of three-dimensional printed models in various forms of congenital heart disease [1–4]. To justify the time and cost necessary to produce high-quality and anatomically accurate printed models, it must be demonstrated that the printed model adds informative value past the traditional methods of anatomical visualization, the two-dimensional images, and the exquisite detail now available with virtual three-dimensional reconstructions [4–6]. To objectively prove this in a scientific manner in relatively rare lesions utilizing high-cost resources (i.e., three-dimensional printers and printing materials) that are not universally available, or feasible, nor currently reimbursable when used in patient care, this will require a well-designed multicenter study. Awaiting the completion of such an involved effort, our current understanding of how and when three-dimensional printed models may add value in specific forms of congenital heart disease remains experiential and subjective. In this case report, we present a unique case of a complex residual ventricular septal defect (VSD) with initial failed attempt at intervention while relying on two-dimensional and three-dimensional reformatted images, including an ECG-gated cardiac computed tomography (CT) scan. A three-dimensional printed model was subsequently created, guiding a successful second attempt at transcutaneous device closure.

21.2 Case Report

A seven-year-old girl had a history of complex heart disease comprised of multiple VSDs that included multiple large apical and mid-muscular VSDs and a large perimembranous VSD along with a parachute mitral valve with moderate mitral stenosis. At 4 years of age, she underwent surgical patch closure of her perimembranous VSD and periventricular closure of the mid-muscular VSD, which was complicated by perforation of the posterior wall and mitral valvar papillary muscle dysfunction. Two weeks later, she underwent mitral valvar replacement with a 19-mm St. Jude mechanical valve (St. Jude Medical, Minneapolis, MN), as well as surgical closure of an apical muscular VSD with an Amplatzer cribriform septal occluder device (St.

J. T. Tretter (✉)
Heart Institute, Cincinnati Children's Hospital Medical Center, Cincinnati, OH, USA

Department of Pediatrics, University of Cincinnati College of Medicine, Cincinnati, OH, USA
e-mail: justin.tretter@cchmc.org

P. Bhatla
Department of Pediatrics and Radiology, New York Langone Medical Center, New York, NY, USA
e-mail: puneet.bhatla@nyumc.org

G. Butera et al. (eds.), *Modelling Congenital Heart Disease*,
https://doi.org/10.1007/978-3-030-88892-3_21

Jude Medical, Minneapolis, MN) with suturing of the disk to the left ventricular septal surface. Following the procedure, transthoracic echocardiogram demonstrated significant residual left-to-right shunting through the residual muscular VSDs along with moderate tricuspid regurgitation due to impingement of the Amplatzer device on the chordal attachments of the septal leaflet of the tricuspid valve. She continued to have a significant left-to-right shunt, so a year later an 8-mm Amplatzer septal occluder was placed in the residual apical muscular VSD. However, a significant residual shunt remained and another attempt was made thereafter with no success. She then underwent a main pulmonary artery band placement with tricuspid valvar plasty. On follow-up, she continued to have a significant residual VSD with left-to-right shunting, moderate tricuspid valve regurgitation, and right ventricular hypertension, and hence, a decision was made to close the VSDs percutaneously. An ECG-gated cardiac CT angiogram was obtained to better delineate the complex residual defect. It demonstrated clear delineation of the ventricular septum with the previously placed devices in situ and a residual 8- to 10-mm tunnel-like VSD originating between the left ventricular disk of the larger Amplatzer device and the adjacent inferior and posterior walls of the ventricular septum. Another attempt at transcatheter closure of her residual VSD was unsuccessful despite excellent three-dimensional reformatted CT data and a virtual volume rendered model.

Given the multiple failed interventional attempts, a printed three-dimensional model of the interventricular septum was reconstructed from the previously obtained CT data to improve understanding of this complex residual VSD. The existing closure devices were included in the three-dimensional model. This allowed for excellent visualization of the residual VSD and its relationship with the surrounding devices (Fig. 21.1).

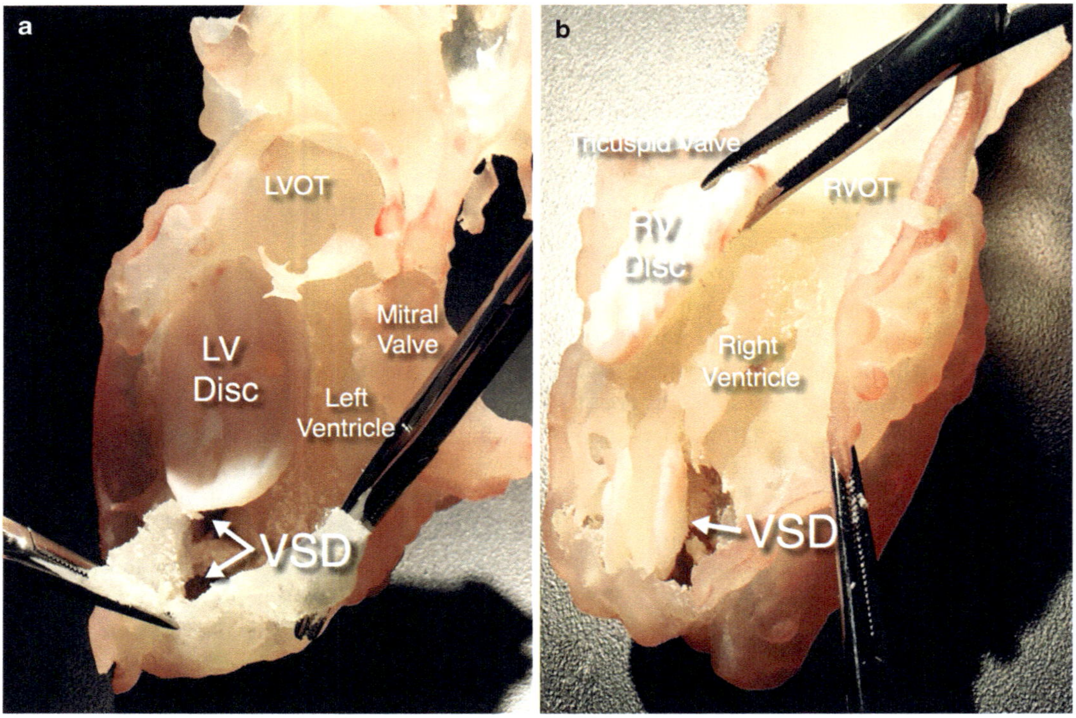

Fig. 21.1 A three-dimensional model of the interventricular septum was printed, including the existing ventricular septal defect closure devices. (**a**) From the left ventricular (LV) aspect of the septum, a complex residual muscular apical ventricular septal defect (VSD) with two entrances is seen inferior to the Amplatzer cribriform device in the apex of the left ventricle. (**b**) From the right ventricular (RV) aspect, the defect is seen to open into the apex of the right ventricle remote from the RV disk of the larger cribriform device, anterior to the smaller 8-mm Amplatzer septal occluder device, which itself is mostly positioned within the RV cavity

This improved understanding of the complex spatial relationships guided the decision to repeat the cardiac catheterization and re-attempt device closure.

Ventricular angiograms obtained during the catheterization were used in association with the 3D model to assess the complex residual VSDs. Although the model was not sterilized and hence not available for direct handling, an assistant held and positioned the model so the interventionalist could alternate visualization between the obtained angiograms and the three-dimensional printed model. The printed three-dimensional model was reviewed extensively during the procedure to improve the understanding of the apparent angiographic and fluoroscopic anatomy. This strategy led to successful placement of four Amplatzer vascular plug type II devices (12 mm × 2, 14 mm × 1, 16 mm × 1) and one 8-mm Amplatzer muscular occluder with exclusion of the apical tunnel formed by the initial cribriform device, the adjacent native septum, and the left ventricular apical free wall. Final angiograms and oximetry demonstrated minimal residual shunting (Fig. 21.2). The pulmonary-to-systemic blood flow ratio decreased from 2:1 to 1.1:1. The echocardiographic mean gradient across the mechanical mitral valve also improved from 16 mmHg to 11 mmHg related to decreased pulmonary blood flow.

21.3 Discussion

Pending more objective confirmation supporting the validity of three-dimensional printing in specific forms of congenital heart disease, we are currently dependent on the increasingly robust experience of many centers around the world [4]. This specific case is very informative in the sense that the interventionalist failed to device close the complex residual VSD when depending solely on the two-dimensional and reformatted three-dimensional imaging data and was subsequently successful with only the additional information provided by a three-dimensional printed model. It was clearly advantageous not only to view the interventricular septal anatomy and surrounding devices, which formed the borders of this complex residual defect, but also the tactile ability to manipulate this model and to view the residual defect in multiple planes. A plastic zip-tie was additionally placed through the defect to better understand the tunneling course through the interventricular septum from left to right ventricle (Fig. 21.3). It was only through the improved resolution afforded by the ECG-gated cardiac CT, which allowed replication of the finer details of the ventricular trabeculations in proximity to the complex residual VSD. Although not tested in any formal manner, there was clear advantage obtained by this printed model over any similar virtual model. The printed model not only allowed for improved pre-procedural planning but also was utilized during the actual case with frequent comparison between the obtained angiograms.

While this is one very specific and unusual case of a complex form of multiple VSDs that had multiple surgeries and interventions, the benefits obtained by the three-dimensional printed model may be generalized to others with complex forms of VSDs, including the so-called swiss cheese VSD. In fact, we have reported on multiple other cases of unoperated complex VSDs where we have found benefit in the additive information provided by three-dimensional printed models [1, 7]. This includes an unusual tunneling VSD in which the three-dimensional printed model actually helped in delaying and hopefully avoiding surgical repair altogether [7]. In this case, the model provided additional insight that the right ventricular exit of the VSD, which was poorly defined by two-dimensional echocardiography due to its unusual tunneling anterior apical course, was in fact restrictive. Unfortunately for now, we must rely on our experiences to help us determine the potential utility of these models weighed against both the time required to create them and their costs. These experiences are often compared to similar, but not identical experiences, adding confounding variables tainting any conclusions. This case report, however, is relatively unique in that we truly are comparing "apples to apples."

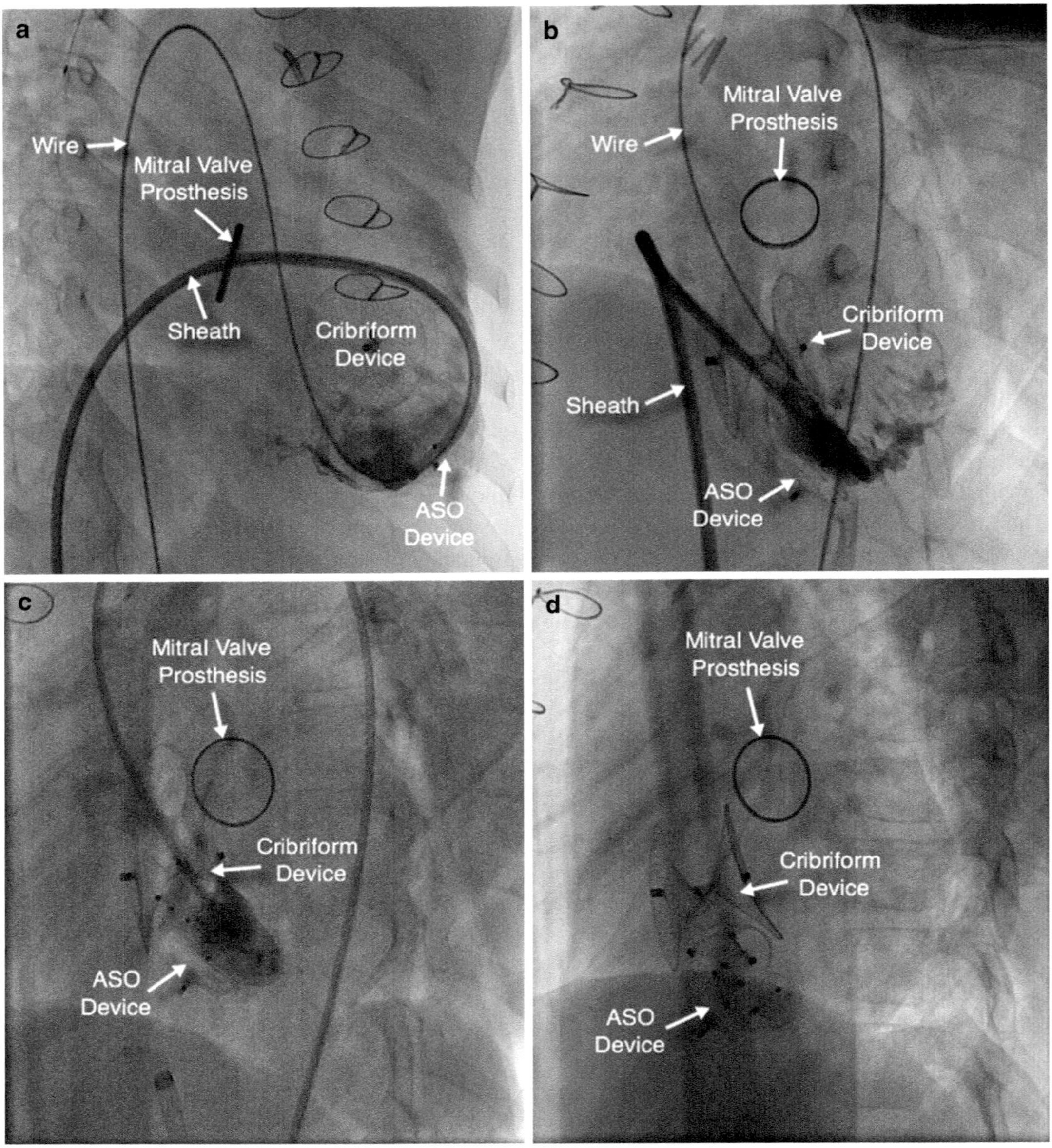

Fig. 21.2 Angiography and fluoroscopic imaging from the cardiac catheterization with device closure of the residual complex ventricular septal defect (VSD). Pre-procedural angiogram in (**a**) right anterior oblique with caudal angulation and (**b**) left anterior oblique with cranial angulation views, with a wire through the VSD forming a femoral vein–femoral artery loop, and the long sheath antegrade across the tricuspid valve into the right ventricle and through the residual VSD into the left ventricular space between the septum and the existing Amplatzer cribriform device. The unusual pocket of space between the septum and cribriform device is demonstrated, along with left to right flow across the complex residual VSD. The more apically located Amplatzer septal occluder (ASO) device is also shown. (**c**) Following placement of 5 additional Amplatzer devices in the complex residual VSD, left anterior oblique angiography with the catheter retrograde from the aorta into the left ventricle demonstrates trivial residual shunting. (**d**) Fluoroscopic imaging in the same view demonstrates the complex arrangement of the newly placed Amplatzer devices occluding the complex residual VSD, between the existing cribriform device and the native septum

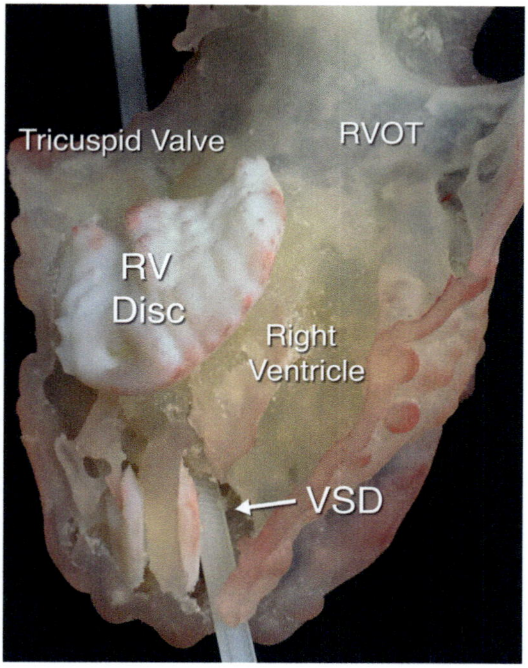

Fig. 21.3 In the same three-dimensional model from Fig. 21.1, a zip-tie has been passed through the complex ventricular septal defect (VSD) and the model is viewed from the right ventricular (RV) aspect of the interventricular septum

References

1. Bhatla P, Tretter JT, Ludomirsky A, Argilla M, Latson LA Jr, Chakravarti S, Barker PC, Yoo SJ, McElhinney DB, Wake N, Mosca RS. Utility and scope of rapid prototyping in patients with complex muscular ventricular septal defects or double-outlet right ventricle: does it Alter Management decisions? Pediatr Cardiol. 2017;38(1):103–14.
2. Ryan J, Plasencia J, Richardson R, Velez D, Nigro JJ, Pophal S, Frakes D. 3D printing for congenital heart disease: a single site's initial three-year experience. 3D Print Med. 2018;4(1):10.
3. Kappanayil M, Koneti NR, Kannan RR, Kottayil BP, Kumar K. Three-dimensional-printed cardiac prototypes aid surgical decision-making and preoperative planning in selected cases of complex congenital heart diseases: early experience and proof of concept in a resource-limited environment. Ann Pediatr Cardiol. 2017;10(2):117–25.
4. Anwar S, Singh GK, Miller J, Sharma M, Manning P, Billadello JJ, Eghtesady P, Woodard PK. 3D Printing is a transformative technology in congenital heart disease. JACC Basic Transl Sci. 2018;3(2):294–312.
5. Mori S, Spicer DE, Anderson RH. Revisiting the anatomy of the living heart. Circ J. 2016;80(1):24–33.
6. Mori S, Tretter JT, Spicer DE, Bolender D, Anderson RH. What is the real cardiac anatomy? Clin Anat. 2019;32(3):288–309.
7. Bhatla P, Mosca RS, Tretter JT. Altering management decisions with gained anatomical insight from a 3D printed model of a complex ventricular septal defect. Cardiol Young. 2017;27(2):377–80.

22 3D "Modeling" and "Printing" in Neonate with Complex Twisted Heart: New Frontier for Clinical Decision and Optimal Surgical Approach

Paolo Ciancarella, Paolo Ciliberti, Luca Borro, and Aurelio Secinaro

Nowadays, modern 3D imaging techniques play a pivotal role in the diagnosis and therapeutic planning of patients with congenital heart disease. Non-invasive representation of patient-specific anatomy can be essential to understand the spatial relationship between heart chambers and vascular structures in complex CHD and to drive clinical/surgical decision-making.

A newborn boy (2.7 kg) with diagnosis of Goldenhar syndrome was affected by complex structural cardio-pulmonary anomalies that included right lung hypoplasia, right-sided heart, twisted superoinferior ventricles with hypoplastic LV, aortic coarctation, and large ventricular septal defect. In order to better define cardiovascular anatomy and hemodynamic consequence related to the complex malformation, the little patient underwent computed tomography angiography (CTA) and cardiovascular magnetic resonance (CMR) (Fig. 22.1).

Computed tomography angiography was acquired with post-iodinate contrast ECG-gated acquisition with a dual-source scanner and high-pitch modality, thus minimizing breath artifacts and reducing radiation exposure. Cardiovascular magnetic resonance was performed under general anesthesia focused on anatomical definition, functional assessment, and hemodynamic evaluation (Fig. 22.2). In particular, CMR anatomy was assessed using a 3D T1-w TSE ECG-gated and respiratory triggered (SPACE). Both CTA and CMR volumetric sequences were acquired in systolic phase avoiding the artifact due to ventricular tachycardia.

CMR results showed morphological and functional parameters considered adequate to support biventricular circulation (left ventricular end-diastolic volume 21 ml/m^2 and aortic forward flow >1 l/min/m^2), with no evidence of endomyocardial fibroelastosis at late gadolinium enhancement imaging.

Data achieved from CTA and CMR were combined using a post-processing software (MIMICS, Materialise) in order to evaluate and isolate cardiac chambers, myocardium, great vessels, and proximal coronary arteries and also to identify

P. Ciancarella (✉) · L. Borro · A. Secinaro
Department of Imaging, Advanced Cardiovascular Imaging Unit, Bambino Gesù Children's Hospital, IRCCS, Rome, Italy
e-mail: paolo.ciancarella@opbg.net; luca.borro@opbg.net; aurelio.secinaro@opbg.net

P. Ciliberti
Department of Pediatric Cardiology and Cardiac Surgery, Bambino Gesù Children's Hospital, IRCCS, Rome, Italy
e-mail: paolo.ciliberti@OPBGMAIL.onmicrosoft.com

G. Butera et al. (eds.), *Modelling Congenital Heart Disease*, https://doi.org/10.1007/978-3-030-88892-3_22

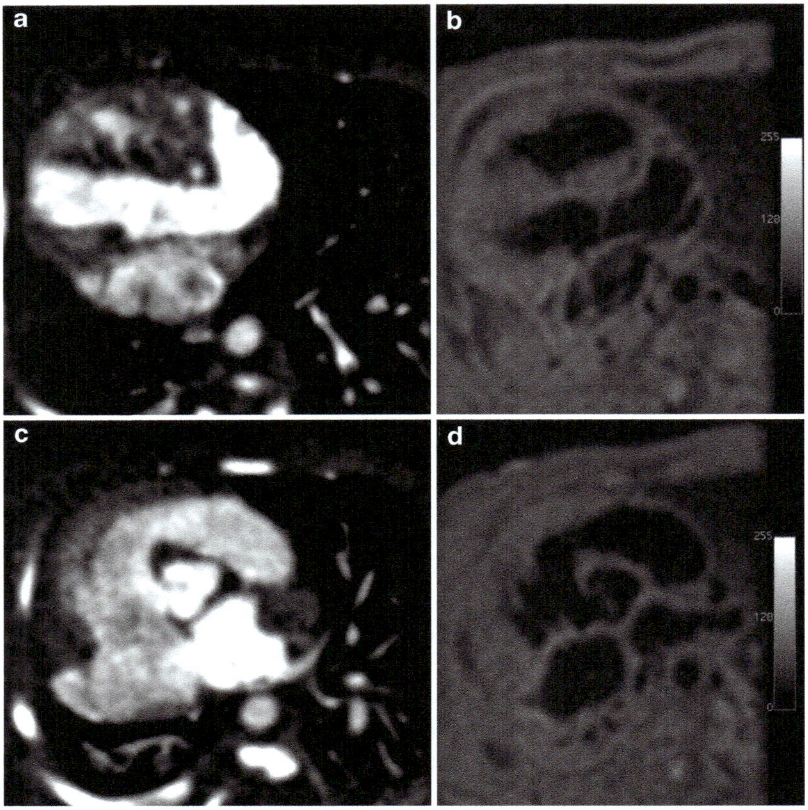

Fig. 22.1 Both CTA (panels **a**–**b**) and CMR (panels **c**–**d**) offer detailed evaluation of cardiac and extracardiac anatomy also in very small pediatric patients with high heart rate

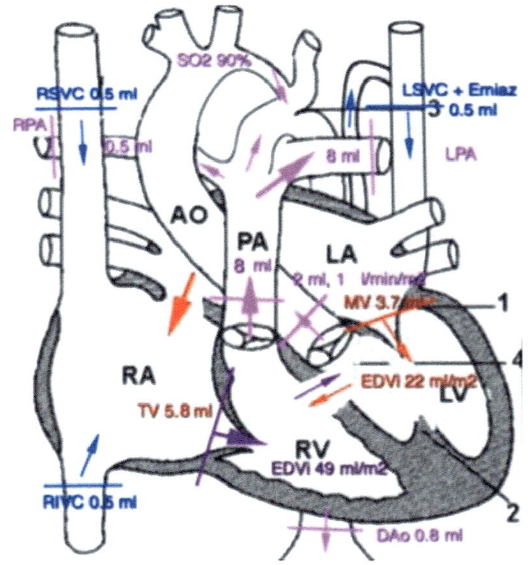

Fig. 22.2 Schematic representation of complex anatomy and hemodynamic. Note the hypoplastic left ventricle (LVEDVi 22 ml/m^2) and presence of intra- and extracardiac shunts with pulmonary overflow

small intracavitary structures like papillary muscles, valvular chords, and accessory muscle bundles (Fig. 22.3).

Stereolithography (STL) file was obtained from 3D data from CTA and CMR. Hollow model was printed out with TuskT (rigid and transparent) material and divided into three pieces in order to optimize intracardiac views and better define relationship between structures. Ventricular septal defect, its margin, and its relationship with surrounding structures such as outflow tracts, RV free wall, and branching right coronary artery course were identified and three-dimensionally displayed (Fig. 22.4). Surgical LV to aorta connectability and VSD closure were considered feasible, and biventricular repair was successfully performed.

In conclusion, modern advanced integrated imaging approach supported by new 3D technologies provides additional insight into complex

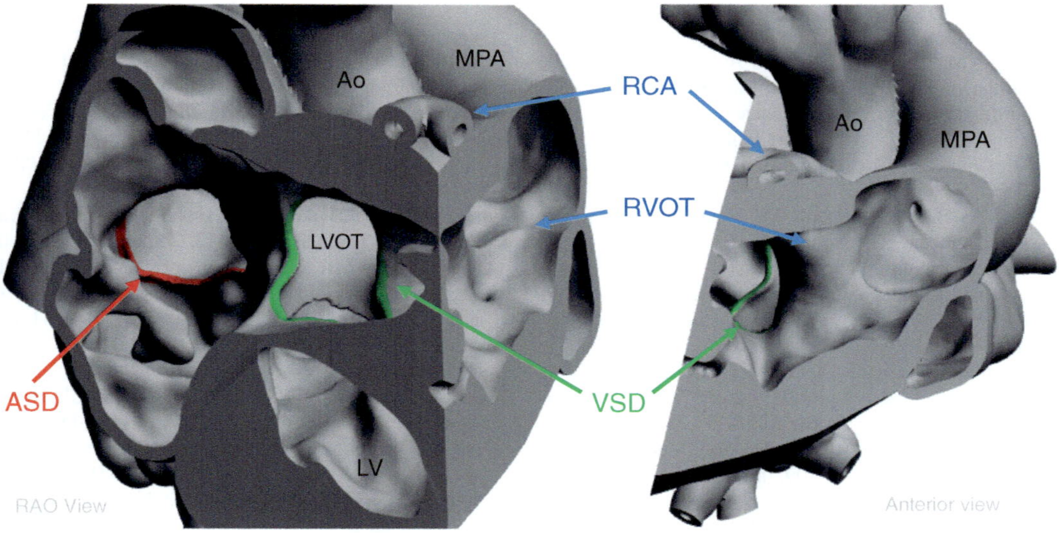

Fig. 22.3 3D patient-specific digital model of the target anatomic structures obtained from CTA and CMR data: cardiac chambers, myocardium, great vessels, and proximal coronary arteries

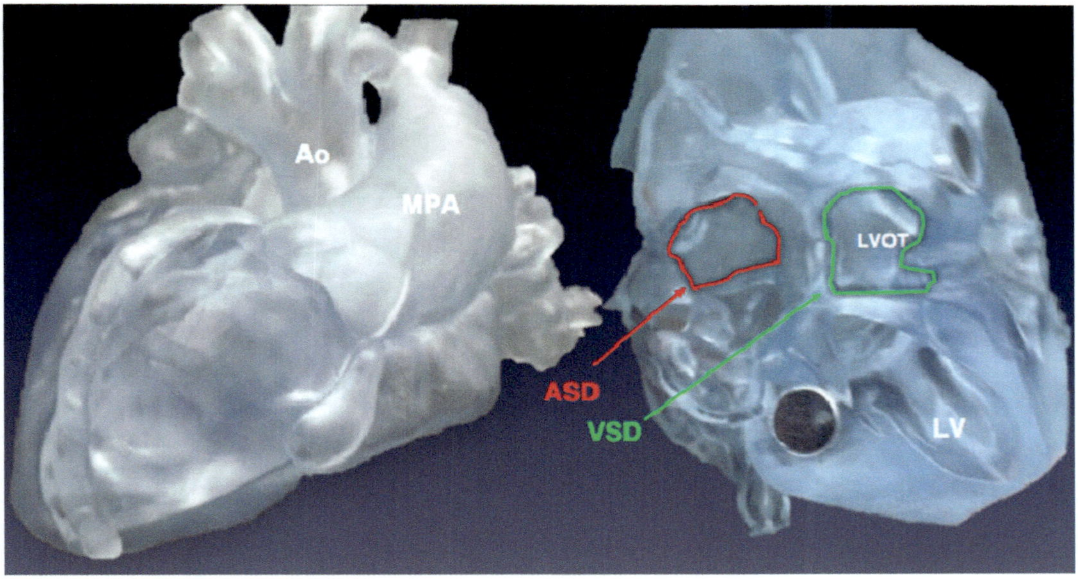

Fig. 22.4 3D printed model made of transparent rigid material and divided into more pieces in order to better delineate spatial relationship between ventricular septal defect, outflow tracts, and other structures

congenital heart disease in neonates, as in this case with crisscross heart. 3D printed models can reduce uncertainties as regards to the patient's specific anatomy and effectively help in the clinical decision-making.

Further Readings

Ciancarella P, Ciliberti P, Santangelo TP, Secchi F, Stagnaro N, Secinaro A. Noninvasive imaging of congenital cardiovascular defects. Radiol Med.

2020;125(11):1167–85. https://doi.org/10.1007/s11547-020-01284-x. Epub 2020 Sep 21

Farooqi KM, Cooper C, Chelliah A, Saeed O, Chai PJ, Jambawalikar SR, Lipson H, Bacha EA, Einstein AJ, Jorde UP. 3D printing and heart failure: the present and the future. JACC Heart Fail. 2019;7(2):132–42. https://doi.org/10.1016/j.jchf.2018.09.011. Epub 2018 Dec 12

Inserra A, Borro L, Spada M, Frediani S, Secinaro A. Advanced 3D "Modeling" and "Printing" for the surgical planning of a successful case of thoraco-omphalopagus conjoined twins separation. Front Physiol. 2020;11:566766. https://doi.org/10.3389/fphys.2020.566766.

Leonardi B, Secinaro A, Calvieri C, Perrone MA, Gimigliano F, Muscogiuri G, Carotti A, Drago F. The role of 3D imaging in the follow-up of patients with repaired tetralogy of Fallot. Eur Rev Med Pharmacol Sci. 2019;23(4):1698–709. https://doi.org/10.26355/eurrev_201902_17132.

Milano EG, Capelli C, Wray J, Biffi B, Layton S, Lee M, Caputo M, Taylor AM, Schievano S, Biglino G. Current and future applications of 3D printing in congenital cardiology and cardiac surgery. Br J Radiol. 2019;92(1094):20180389. https://doi.org/10.1259/bjr.20180389. Epub 2018 Nov 1

Vukicevic M, Mosadegh B, Min JK, Little SH. Cardiac 3D printing and its future directions. JACC Cardiovasc Imaging. 2017;10(2):171–84. https://doi.org/10.1016/j.jcmg.2016.12.001.

Wang DD, Qian Z, Vukicevic M, Engelhardt S, Kheradvar A, Zhang C, Little SH, Verjans J, Comaniciu D, O'Neill WW, Vannan MA. 3D printing, computational modeling, and artificial intelligence for structural heart disease. JACC Cardiovasc Imaging. 2021;14(1):41–60. https://doi.org/10.1016/j.jcmg.2019.12.022. Epub 2020 Aug 26

Yoo SJ, Hussein N, Peel B, Coles J, van Arsdell GS, Honjo O, Haller C, Lam CZ, Seed M, Barron D. 3D modeling and printing in congenital heart surgery: entering the stage of maturation. Front Pediatr. 2021;9:621672. https://doi.org/10.3389/fped.2021.621672.

23 Biventricular Repair of Complex Transposition of Great Arteries Guided by 3D Reconstruction Imaging

Yue-Hin Loke and David N. Schidlow

23.1 Introduction

Three-dimensional (3D) cardiac models, in either printed or digital format, have been used for preoperative planning of complex, variable intracardiac anatomy such as double outlet right ventricle· [1, 2]. Typically, 3D modeling of congenital heart disease requires segmentation of cross-sectional imaging such as cardiac magnetic resonance imaging (cMRI). However, segmentation of atrioventricular valves is still challenging and more successfully performed with 3D echocardiography (3D echo) [3, 4]. For intracardiac anatomy, "hybrid" 3D cardiac models created from both modalities are a potential option [5]. We present a case of using hybrid 3D modeling to guide the preoperative evaluation of D transposition of great arteries (TGAs) and ventricular septal defect (VSD).

23.2 Case Report

23.2.1 Clinical History

The patient was a newborn boy who was diagnosed postnatally after presenting with cyanosis. Two-dimensional transthoracic echocardiography (2D echo) demonstrated TGA with posterior malalignment-type VSD causing subpulmonary stenosis and a hypoplastic pulmonary valve (measuring 6 mm). He was started on prostaglandin E1 and transferred to the cardiac intensive care unit. He subsequently underwent balloon atrial septostomy and eventually a modified right-sided Blalock–Taussig–Thomas shunt at 10 days of age. He had a prolonged hospital course due to medical necrotizing enterocolitis but was eventually discharged at 3 months of life.

On 2D echo, concerns were raised whether an intracardiac surgical repair was feasible, in light of a posteriorly deviated conal septum and elongated potential left ventricle-to-aorta pathway. As a result, the patient underwent advanced cardiac imaging, including 3D echo and cMRI, to clarify the intracardiac anatomy. The image datasets were then used to print a hybrid 3D heart model for preoperative planning.

23.2.2 Clinical Considerations for 3D Modeling

For TGA with posterior malalignment-type VSD and pulmonary valve hypoplasia, the Rastelli operation is the expected surgical approach, consisting of a patch baffle from VSD to the aorta and placement of right ventricle-to-pulmonary artery (RV-PA) conduit [6]. Thus, from a 3D modeling standpoint, display of intracardiac

Y.-H. Loke (✉)
Department of Cardiology, Children's National Hospital, Washington, DC, USA
e-mail: yloke@childrensnational.org

D. N. Schidlow
Department of Cardiology, Boston, MA, USA
e-mail: David.Schidlow@cardio.chboston.org

G. Butera et al. (eds.), *Modelling Congenital Heart Disease*,
https://doi.org/10.1007/978-3-030-88892-3_23

anatomy is critical to guide the complex patch baffle between the VSD to aorta and any potential obstacles to the baffle pathway. For example, the model should show the location and orientation of the tricuspid valve, to ensure adequate separation of tricuspid valve chordae from the baffle pathway. Display of great vessel anatomy is also useful to guide placement of the RV-PA conduit. A hybrid 3D model can ideally provide "the best of both worlds," incorporating the tricuspid valve anatomy by 3D echo and ventricular septal anatomy by cMRI.

23.2.3 Imaging Techniques and Findings

Both 3D echo and cMRI were performed under general anesthesia. With both imaging modalities, careful attention to acquisition parameters should be made to ensure that the blood pool is contrasted and defined clearly.

For 3D echo, probe technique and ultrasound settings should minimize stitch artifact and maximize frame rate. For this patient, images were acquired with an IE33 7X probe (Phillips) placed in subcostal position, angled such that the tricuspid valve was visualized en face. Slight adjustments to time gain compensation provided contrast between the blood pool and tricuspid valve. Elevation and lateral width were optimized to cover the tricuspid valve while maximizing frame rate. A full-volume 3D dataset was acquired using a 2-beat acquisition with an adequate frame rate of 22 Hz. Good temporal resolution is needed to capture the tricuspid valve at its maximal annulus, which changes in size through the cardiac cycle [7]. Based on volume renderings of the acquired dataset, the tricuspid valve appeared remote from the VSD and less likely to interrupt the pathway (Fig. 23.1a).

cMRI was performed on a 1.5 T Aera (Siemens) scanner. Imaging sequences included self-navigated 3D steady-state free precession (3D SSFP) and contrast-enhanced MR angiography (MRA). For this patient, the 3D SSFP provided a detailed view of the ventricular septum in diastole (Fig. 23.1b), while the MRA helped define great vessel orientation (Fig. 23.1c). The cMRI demonstrated a moderate-sized VSD with posterior malalignment of the conal septum. The VSD extended apically into the muscular septum. The conal septum was prominent, impinging on the pulmonic outflow tract and the theoretical pathway of the patch baffle. There was also a right-sided aortic arch and patent right-sided BTT shunt.

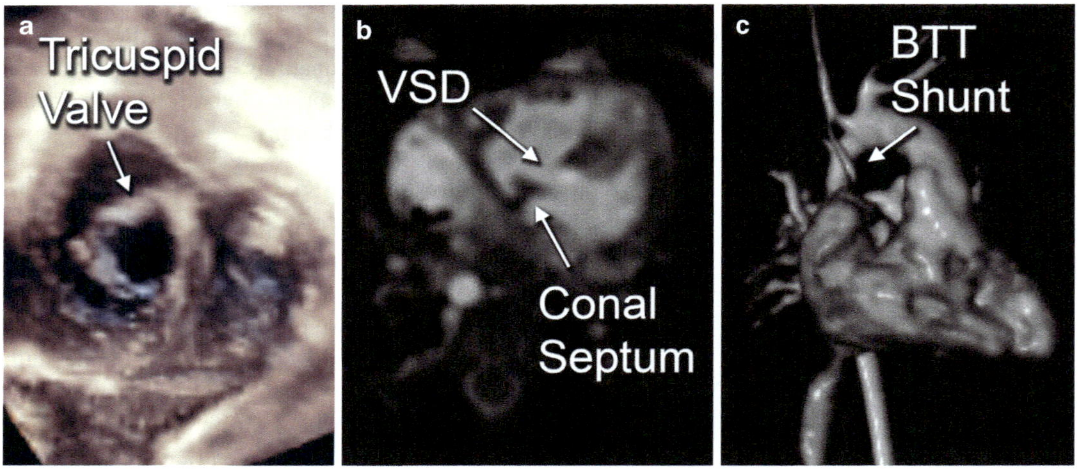

Fig. 23.1 (**a**) *En face* view of tricuspid valve by 3D echocardiography. (**b**) 3D SSFP imaging demonstrating ventricular septal defect (VSD) with posterior alignment of conal septum as shown by 3D SSFP. (**c**) Patent Blalock–Taussig–Thomas (BTT) shunt as shown by MRA

23.2.4 Segmentation of 3D Model

3D segmentation was performed using Mimics (Materialise). Segmentation was performed in a systematic manner, with separate masks representing intracardiac chambers and great vessels, carefully delineating the VSD and conal septum (Fig. 23.2a). A series of Boolean and morphology transformations were then performed, creating a segmentation representing myocardium including ventricular anatomy (Fig. 23.2b). The 3D echo dataset was exported as a Cartesian 3D DICOM and post-processed to filter out ultrasound speckles [3]. The tricuspid valve leaflets were then segmented with Mimics (Fig. 23.2c). To ensure proper alignment of the tricuspid valve from 3D echo with intracardiac anatomy from cMRI, the tricuspid valve annulus was traced and defined; this was also performed on the cMRI dataset (Fig. 23.2d).

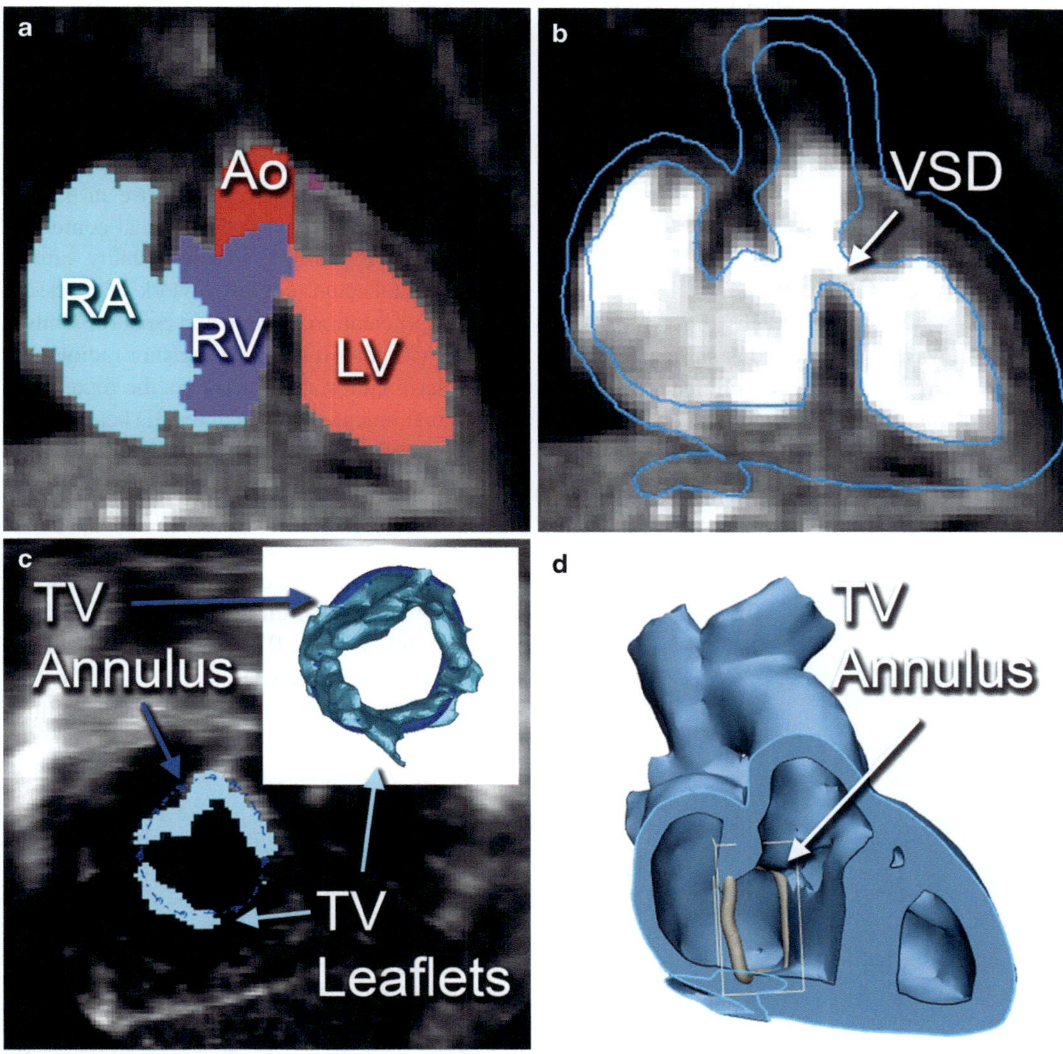

Fig. 23.2 (**a**) 3D Segmentation of cardiac MRI dataset, using separate masks to represent the aorta (Ao), right atrium (RA), right ventricle (RV), and left ventricle (LV). (**b**) A series of Boolean and morphology transformations resulted in a segmentation with ventricular septal defect (VSD) represented. (**c**). Segmentation of tricuspid valve (TV) leaflets and tracing of TV annulus. (**d**) The TV annulus was also traced from the cMRI dataset

23.2.5 3D Editing and Display of 3D Model

3-matic (Materialise) was then used to align the tricuspid valve leaflets with intracardiac anatomy (Fig. 23.3). The resulting model was sectioned, such that both external and internal visual features of the cardiac model were preserved. In this case, cutting the right ventricular-free wall along a coronal oblique plane (similar plane to a theoretical ventriculotomy) allowed visualization of the VSD. Small circular holes were cut into the articulating surface of each section, allowing for placement of magnets that would hold the sections together (Fig. 23.4). The finalized hybrid 3D model components were then printed with a Objet Connex 500 Multi-Material 3D printer (Stratasys).

23.2.6 Clinical Outcome

The patient was first discussed at surgical conference as part of standard clinical practice. Afterward, the 3D model was shared with the surgeon, with one of the cardiologists involved in 3D segmentation providing in-person guidance. The 3D model demonstrated a clear pathway from the VSD to aorta, accessible from a right ventriculotomy approach, with no interference from the tricuspid valve. The decision was made to proceed with a modified Rastelli procedure. At 6 months of age, the patient underwent resection of conal muscle, VSD baffle to aorta, and placement of an RV-PA conduit. He was discharged by postoperative day 10. He has subsequently required replacement of the RV-PA conduit at 2 years of age and has otherwise done well.

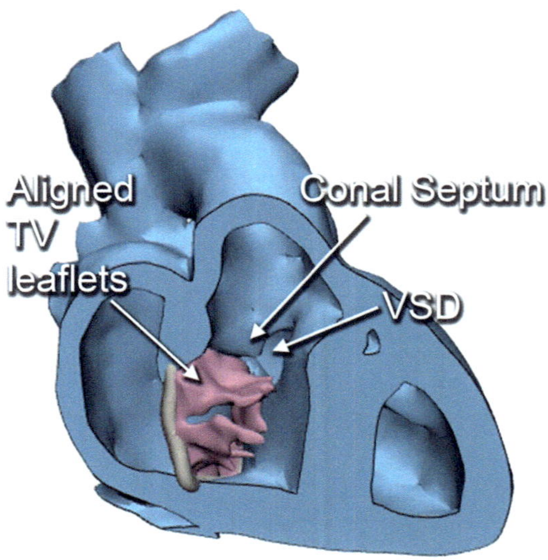

Fig. 23.3 3D Editing of the model, with realignment of tricuspid valve (TV) leaflets using the TV annulus from both cMRI and 3D echo. The model is also sectioned to display the ventricular septal defect (VSD)

23.2.7 Clinical Perspective

For congenital heart disease with complex anatomy, 3D modeling requires expertise in cardiac imaging and familiarity to the clinical context of the anatomy. In addition to good-quality cardiac imaging, intellectual cognitive work is necessary for 3D segmentation, design, and creation. Thus, a provider (whether it be cardiologist or radiologist) should have editorial control over the representation of anatomical elements of the 3D heart model, either by collaboratively reviewing the model or by performing the segmentation themselves.

To effectively use 3D cardiac models as part of preoperative evaluation, the model should be edited as needed to communicate the relevant anatomy for end-user clinician (typically the surgeon). In this case, the RV free wall was sectioned to show the VSD relative to the aorta. The tricuspid valve orientation also required manual adjust-

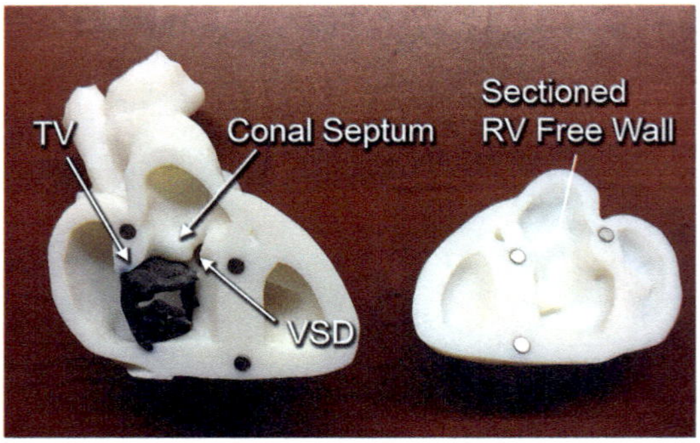

Fig. 23.4 Printed version of the hybrid 3D print model, which was shared with the surgeon

ment, accounting for expected variability between cMRI and 3D echo datasets. Of note, the 3D cardiac model serves as an adjunctive method to highlight specific interpretation of cardiac imaging. To avoid ambiguity, it is critical for the provider with "editorial ownership" to highlight elements of the 3D cardiac model that are accurate, as well as communicate other elements with limited representation during the segmentation process. Ideally, this is performed as an in-person collaboration, promoting a "shared mental model" [8] with other providers in a multidisciplinary environment and facilitating the exchange of complex information within the 3D cardiac model.

23.3 Summary

Hybrid 3D cardiac models can be created from cMRI and 3D echo for the preoperative evaluation for congenital heart disease. For complex intracardiac anatomy, care should be taken through the entire process, including imaging acquisition, segmentation, editing, and presentation. Direct, collaborative clinical input is also helpful to ensure model effectiveness.

References

1. Mottl-Link S, Hübler M, Kühne T, et al. Physical models aiding in complex congenital heart surgery. Ann Thorac Surg. 2008;86:273–7. https://doi.org/10.1016/j.athoracsur.2007.06.001.
2. Farooqi KM, Uppu SC, Nguyen K, et al. Application of virtual three-dimensional models for simultaneous visualization of Intracardiac anatomic relationships in double outlet right ventricle. Pediatr Cardiol. 2016;37:90–8. https://doi.org/10.1007/s00246-015-1244-z.
3. Olivieri LJ, Krieger A, Loke Y-H, et al. Three-dimensional printing of Intracardiac defects from three-dimensional echocardiographic images: feasibility and relative accuracy. J Am Soc Echocardiogr. 2015;28:392–7. https://doi.org/10.1016/j.echo.2014.12.016.
4. Scanlan AB, Nguyen AV, Ilina A, et al. Comparison of 3D echocardiogram-derived 3D printed valve models to molded models for simulated repair of pediatric atrioventricular valves. Pediatr Cardiol. 2017;39(3):538–47. https://doi.org/10.1007/s00246-017-1785-4.
5. Kurup HK, Samuel BP, Vettukattil JJ. Hybrid 3D printing: a game-changer in personalized cardiac medicine? Expert Rev Cardiovasc Ther. 2015;13:1281–4. https://doi.org/10.1586/14779072.2015.1100076.
6. Kreutzer C, De Vive J, Oppido G, et al. Twenty-five-year experience with rastelli repair for transposition of the great arteries. J Thorac Cardiovasc Surg. 2000;120:211–23. https://doi.org/10.1067/mtc.2000.108163.
7. Addetia K, Muraru D, Veronesi F, et al. 3-dimensional echocardiographic analysis of the tricuspid annulus provides new insights into tricuspid valve geometry and dynamics. JACC Cardiovasc Imaging. 2019;12:401–12. https://doi.org/10.1016/j.jcmg.2017.08.022.
8. Lim B-C, Klein KJ. Team mental models and team performance: a field study of the effects of team mental model similarity and accuracy. J Organ Behav. 2006;27:403–18. https://doi.org/10.1002/job.387.

Part IV

Training, Counselling and Miscellanea

Three-Dimensional Printing and Its Applications in Education and Teaching

24

Dimitri Patriki and Andreas A. Giannopoulos

24.1 Introduction

The fast development of medical 3D printing in the last decade has allowed for valuable and affordable applications in numerous fields of medicine and especially cardiology. There is hardly any sub-speciality of cardiology other than congenital heart (CHD) disease, whereby 3D printing modelling has been studied and clinically implemented. Driven also by technical development of imaging modalities, 3D printing technology is being used in diagnosis and management of CHD [1], planning and testing of interventional procedures or implant design for occluder devices in atrial septal defect (ASD), ventricular septal defect (VSD) and even mitral valve prosthetics as well as for surgery planning of arteriovenous malformations [2–6]. Moreover, 3D printing has provided new opportunities in medical research enabling simulation of physiological and pathophysiological processes in 3D printed phantoms [7].

Nevertheless, the potential of 3D printing is not exhausted on clinical applications. Apart from the promising, yet futuristic field of bioprinting, which in theory may allow for generation of implants, prostheses and even organs, anatomical 3D printed models already facilitate anatomical appreciation through enabling a more natural orientation enabling further haptic sensation [4, 8]. The variety of complex 3D relationships within the human body creates the need for educational tools which provide help for medical professionals, students as well as patients and relatives. Especially, the field of CHD can benefit from educational possibilities of 3D printing as new possibilities of diagnosis and treatment emerge, whereby precise understanding of function, anatomy and haemodynamic is essential. Education can refer to patients, family, medical students, medical residents as well as fully trained cardiologists aiming to specialize in new areas.

3D printed models are being used for teaching surgical interventions in double-outlet left ventricle [9] or tetralogy of Fallot [10], as well as cardiac transplantation [11]. Moreover, physical models facilitate the education of patients and family members, teaching anatomy in medical students and training of hospital personnel that handle CHD patients [12–15].

D. Patriki
Department of Nuclear Medicine, Cardiac Imaging, University Hospital Zurich, Zurich, Switzerland
e-mail: dimitri.patriki@usz.ch

A. A. Giannopoulos (✉)
Department of Nuclear Medicine, Cardiac Imaging, University Hospital Zurich, Zurich, Switzerland

Department of Cardiology, University Hospital Zurich, Zurich, Switzerland
e-mail: andreas.giannopoulos@usz.ch

G. Butera et al. (eds.), *Modelling Congenital Heart Disease*,
https://doi.org/10.1007/978-3-030-88892-3_24

24.2 Surgical Preparation and Training

Hardly any medical discipline is as deeply connected to anatomy as it is surgery and in particular cardiovascular surgery. Its practical nature requires hands-on training, which has been traditionally gained by dissection of cadavers in order to avoid complications in patients by unexperienced surgeons. As a result, expertise increases through practical training and through a long-lasting learning curve within the wide field of anatomical variations [16]. The fact that mortality and morbidity are tightly tethered to the surgeons' performance in congenital heart surgery and to the surgical case volumes, emphasizes the importance of a highly trained surgeon [17–19]. Although 3D reconstructions of CT or MRI scan provide additional information, virtual 3D models are just representation of real objects and demand mental 3D reconstruction. On top of that, real-time hands-on training is currently limited on a virtual object. Despite the fact that 3D printers are far away from reproducing 3D models of ideal properties of human tissue, printers using polyjet technology and photopolymer resin materials provide sufficient similarities to human soft tissue that can allow for simulation of surgical procedures [1, 20]. In contrast to rare and expensive cadaveric materials, 3D models can be reproduced and configured according to the needs, allowing repeated procedures on common and rare pathologies or anatomical anomalies.

The remarkable progress and the new opportunities of the ever-evolving CHD surgery require more than ever a deep understanding of anatomy and function of the heart and its pathologies. A universal and unified nomenclature and definition of complex congenital cardiac pathologies are essential and already have been implemented [21]. The use of CHD 3D printed models for training and education has been reported in a number of studies [22–24]. However, participants have pointed out certain drawbacks of the printed models such as sometimes unsatisfactory mechanical properties of the models, especially when using inkjet 3D printing, or inaccuracies of the models.

Baseline imaging resolution and accuracy of the 3D printed model not only are a matter of the printing technique but also depend on the quality of the underlying 3D image data [20]. Nevertheless, it has been shown that hands-on training is helpful and its implementation could be considered in training programmes [22]. Moreover, certain limitations such as the lack of simulation of procedures on a beating heart are being addressed. In another report, a completely artificial replica of the human chest was designed. This was able to simulate the intrathoracic movement including an inflating lung, as well as beating heart with saline rinsed coronaries [23]. As a result, surgeons independent of their skill level were able to train any necessary procedures repeatedly or experiment with new techniques and devices.

Physical 3D printed models have been utilized in the field of surgery for pre-operative planning, simulations and intraoperative navigation [9–11, 25]. Its application as a tool for planning and training has already been introduced in the fields of neurosurgery, cardiovascular surgery, pelvic surgery and maxillofacial surgery [26, 27]. In cases of dysmorphology due to mandibular prognathism, surgeons were able to create 3D models of maxilla and mandible and perform model surgery for planning and prediction of the final outcome. The results demonstrated sufficient accuracy and lead to a decrease in operating time and therefore patient risk [16, 28]. In the treatment of CHD, 3D printing may have similar advantages through potentially reducing surgical and anaesthesia time during cardiovascular surgery and facilitate understanding of complex malformations even for experienced physicians [29]. In a small series of three patients (two ventricular aneurysms and one malignant heart tumour), 3D printed models were used for planning of the surgical procedures. In these cases, 3D models facilitated operative planning and intraoperative orientation, potentially improving surgical outcome [30]. This development may mark the beginning of a new era, where surgeons can rely on more than the usual exposure to graduate clinical experience during the residency [31].

24.3 Student Education

Anatomy is an integral part of medical education. Particularly, students often struggle with complex anatomy and its associated physiological processes, which are traditionally learned with simple 2D plain images and drawings from books and anatomical atlases. Enhanced anatomical comprehension requires time, practice and spatial visualization of the anatomy. Though traditional cadaveric courses exist, they are rather expensive and rare and have only a limited variation of pathologies due to the limited selection of cadavers. Social, cultural and religious factors represent additional boundaries impeding the accessibility to cadaveric material [32]. Considering the above-mentioned and additional financial factors, dissection courses are being reduced if not even abandoned in some countries [33]. Hence, the need for new educational methods arises. Overall, technology-enhanced simulations for education result in improved knowledge, skills and even patient-related outcomes in comparison with traditional educational methods [34].

3D printed models seem to enhance the understanding subjectively and objectively, especially in pathologies with high degree of complexity, such as CHD [24, 35], and have shown advantages even over cadaveric hearts for undergraduate anatomy learning [36]. Additional benefits include durability, cost effectiveness and lack of contagious diseases issues [33]. Moreover, nowadays infrastructure and the World Wide Web allow for fast data exchange from patients from all over the world and the creation of databanks such as the International Medical Image Bank for congenital heart disease (IMIB-CHD) or the National Institutes of Health (NIH) Heart Library NIH 3D Print Heart Library, which store numerous different models and pathologies resulting in an remarkable selection of almost the entire spectrum of CHD. Those advantages affect not only students but also other medical professions such as medical physician trainees, nurses and other healthcare providers [15]. Indeed, further opportunities for 3D printing extend beyond anatomical education. From simulation of haemodynamics of valves to simulations of handling imaging modalities including echocardiography or hands-on training in percutaneous interventions, 3D printing is able to revolutionize our medical education system [35, 37, 38].

24.4 Patient Interaction

Communication is one of the most essential pillars of doctors–patient relationship. Its success may have beneficial effects on patients satisfaction, compliance, knowledge and recovery [39]. Especially, complex health scenarios represent an obstacle for doctors–patient communication as explanation often requires time and in some cases even fundamental medical understanding. Particularly, CHD and its anatomical and pathophysiological challenges demonstrate a fine example. Families of children suffering from CHD have shown to have several knowledge gaps regarding their children's associated heart lesions, medication, possible side effects and complications [40]. A study used 3D models of childrens' hearts with CHD for education and teaching of their parents in order to improve their pathophysiological understanding. One hundred and three patients were divided into two groups with and without 3D model use during the cardiologists' consultation. Although patients who received 3D models showed similar results in questionnaires compared with usual consultation regarding medical knowledge, models were highly appreciated [13]. Several other studies showed similar effects, leading to the conclusion, that 3D models are useful for personalized patient education [41, 42].

24.5 Implementation of 3D Printing in Conferences and Seminars

As a result of the mentioned progression in the field of 3D printing, there is an increasingly high number of organization and companies incorporating 3D printing and accompanying hands-on courses or seminars. Major societies such as

European Society of Cardiology (ESC) and American Heart Association (AHA) support the dissemination of knowledge regarding its applications and provide teaching courses on congresses. Moreover, Radiological Society of North America (RSNA) and its Special Interest Group of 3D printing provide step-by-step guides for the creation of CHD models allowing for advanced appreciation of the 3D printing dedicated freeware and/or commercially available software [43–45]. Furthermore, industry plays a major role in funding innovation of 3D printing techniques and software, while at the same time organizing dedicated seminars. The more 3D printing technology matures and profit increases, the more companies will take part and invest in this promising new technology. Nowadays, multiple companies exist and provide education and teaching courses, in the form of not only hands-on training but also webinars and whitepapers for software and hardware. Their investments in scientific research and education are amongst the pillars of nowadays progress of 3D printing.

24.6 Conclusion

3D printing is a promising new technology for a variety of applications. Though the technology improves and already is in use for certain indications, we are yet a step away from a seamless integration of 3D printing in our everyday clinical routine. This is aggravated by the fact that most processes are not automated yet. Hence, segmentation of didactical valuable 3D models currently has to be done manually or semi-automated, which requires a lot of time and multidisciplinary know-how. Technical infrastructure has to be established, and equipments, especially the proper 3D printers, have to be selected. On top of that, models have to be created from a sufficient modality and areas of interest must be designed and adapted for the current requirements. Though the quality of the printed 3D models is depending on the input source quality, didactic nuances have to be considered. In fact, overly complex or greatly simplified models are counterproductive [33]. As a result, a great amount of expertise in engineering, anatomy and radiology is essential [14, 25, 46].

From a financial point of view, the initially high cost of 3D printing can be counterbalanced by its advantages depending on its use. For surgeons, for example, reduced operating time through better planning understanding and hands-on training with 3D models reduces also possible complications and anaesthesia time, and decreases the likelihood of infections and therefore the use of antibiotics [20]. Nevertheless, 3D printing and associated 3D modelling and segmentation are time-consuming. Automated processes and reduced software and hardware costs may solve this problem in the future, but by now as 3D printing is yet in its infancies and cost–benefit assessment is essential. Collectively, though 3D printing for CHD is complex and still requires a lot of effort, opportunities for advancing education through this disruptive technology grow rapidly and are expected to hold prime role in educational systems in the next decade.

References

1. Yoo SJ, Thabit O, Kim EK, Ide H, Yim D, Dragulescu A, et al. 3D printing in medicine of congenital heart diseases. 3D Print Med. 2015;2(1):3.
2. Armillotta A, Bonhoeffer P, Dubini G, Ferragina S, Migliavacca F, Sala G, et al. Use of rapid prototyping models in the planning of percutaneous pulmonary valved stent implantation. Proc Inst Mech Eng H J Eng Med. 2007;221(4):407–16.
3. Schievano S, Migliavacca F, Coats L, Khambadkone S, Carminati M, Wilson N, et al. Percutaneous pulmonary valve implantation based on rapid prototyping of right ventricular outflow tract and pulmonary trunk from MR data. Radiology. 2007;242(2):490–7.
4. Kim MS, Hansgen AR, Wink O, Quaife RA, Carroll JD. Rapid prototyping: a new tool in understanding and treating structural heart disease. Circulation. 2008;117(18):2388–94.
5. Conti A, Pontoriero A, Iatì G, Marino D, La Torre D, Vinci S, et al. 3D-printing of arteriovenous malformations for radiosurgical treatment: pushing anatomy understanding to real boundaries. Cureus. 2016;8(4):e594.
6. Shah A, Jankharia B, Goel A. Three-dimensional model printing for surgery on arteriovenous malformations. Neurol India. 2017;65(6):1350–4.
7. Canstein C, Cachot P, Faust A, Stalder AF, Bock J, Frydrychowicz A, et al. 3D MR flow analysis in

realistic rapid-prototyping model systems of the thoracic aorta: comparison with in vivo data and computational fluid dynamics in identical vessel geometries. Magn Reson Med. 2008;59(3):535–46.
8. Ventola CL. Medical applications for 3D printing: current and projected uses. Pharm Therapeut. 2014;39(10):704–11.
9. Hadeed K, Dulac Y, Acar P. Three-dimensional printing of a complex CHD to plan surgical repair. Cardiol Young. 2016;26(7):1432–4.
10. Ryan JR, Moe TG, Richardson R, Frakes DH, Nigro JJ, Pophal S. A novel approach to neonatal management of tetralogy of fallot, with pulmonary atresia, and multiple aortopulmonary collaterals. JACC Cardiovasc Imaging. 2015;8(1):103–4.
11. Smith ML, McGuinness J, O'Reilly MK, Nolke L, Murray JG, Jones JFX. The role of 3D printing in preoperative planning for heart transplantation in complex congenitayl heart disease. Irish J Med Sci (1971). 2017;186(3):753–6.
12. Costello JP, Olivieri LJ, Su L, Krieger A, Alfares F, Thabit O, et al. Incorporating three-dimensional printing into a simulation-based congenital heart disease and critical care training curriculum for resident physicians. Congenit Heart Dis. 2015;10(2):185–90.
13. Biglino G, Capelli C, Wray J, Schievano S, Leaver L-K, Khambadkone S, et al. 3D-manufactured patient-specific models of congenital heart defects for communication in clinical practice: feasibility and acceptability. BMJ Open. 2015;5:4.
14. Giannopoulos AA, Mitsouras D, Yoo SJ, Liu PP, Chatzizisis YS, Rybicki FJ. Applications of 3D printing in cardiovascular diseases. Nat Rev Cardiol. 2016;13(12):701–18.
15. Olivieri LJ, Su L, Hynes CF, Krieger A, Alfares FA, Ramakrishnan K, et al. "Just-in-time" simulation training using 3-D printed cardiac models after congenital cardiac surgery. World J Pediatr Congenit Heart Surg. 2016;7(2):164–8.
16. D'Urso PS, Barker TM, Earwaker WJ, Bruce LJ, Atkinson RL, Lanigan MW, et al. Stereolithographic biomodelling in cranio-maxillofacial surgery: a prospective trial. J Cranio-Maxillofac Surg. 1999;27(1):30–7.
17. Nathan M, Karamichalis JM, Liu H, Emani S, Baird C, Pigula F, et al. Surgical technical performance scores are predictors of late mortality and unplanned reinterventions in infants after cardiac surgery. J Thorac Cardiovasc Surg. 2012;144(5):1095–101.e7.
18. Shuhaiber J, Gauvreau K, Thiagarjan R, Bacha E, Mayer J, Del Nido P, et al. Congenital heart surgeon's technical proficiency affects neonatal hospital survival. J Thorac Cardiovasc Surg. 2012;144(5):1119–24.
19. Welke KF, O'Brien SM, Peterson ED, Ungerleider RM, Jacobs ML, Jacobs JP. The complex relationship between pediatric cardiac surgical case volumes and mortality rates in a national clinical database. J Thorac Cardiovasc Surg. 2009;137(5):1133–40.
20. Martelli N, Serrano C, van den Brink H, Pineau J, Prognon P, Borget I, et al. Advantages and disadvantages of 3-dimensional printing in surgery: a systematic review. Surgery. 2016;159(6):1485–500.
21. Giroud JM, Jacobs JP, Spicer D, Backer C, Martin GR, Franklin RCG, et al. Report from the International Society for Nomenclature of Paediatric and congenital heart disease: creation of a visual encyclopedia illustrating the terms and definitions of the international pediatric and congenital cardiac code. World J Pediatr Congenit Heart Surg. 2010;1(3):300–13.
22. Yoo SJ, Spray T, Austin EH 3rd, Yun TJ, van Arsdell GS. Hands-on surgical training of congenital heart surgery using 3-dimensional print models. J Thorac Cardiovasc Surg. 2017;153(6):1530–40.
23. Reuthebuch O, Lang A, Groscurth P, Lachat M, Turina M, Zund G. Advanced training model for beating heart coronary artery surgery: the Zurich heart-trainer. Eur J Cardio-thoracic Surg. 2002;22(2):244–8.
24. White SC, Sedler J, Jones TW, Seckeler M. Utility of three-dimensional models in resident education on simple and complex intracardiac congenital heart defects. Congenit Heart Dis. 3(6):1045–9.
25. Pugliese L, Marconi S, Negrello E, Mauri V, Peri A, Gallo V, et al. The clinical use of 3D printing in surgery. Updat Surg. 2018;70(3):381–8.
26. Rengier F, Mehndiratta A, von Tengg-Kobligk H, Zechmann CM, Unterhinninghofen R, Kauczor HU, et al. 3D printing based on imaging data: review of medical applications. Int J Comput Assist Radiol Surg. 2010;5(4):335–41.
27. Malik HH, Darwood ARJ, Shaunak S, Kulatilake P, El-Hilly AA, Mulki O, et al. Three-dimensional printing in surgery: a review of current surgical applications. J Surg Res. 2015;199(2):512–22.
28. Ayoub AF, Rehab M, O'Neil M, Khambay B, Ju X, Barbenel J, et al. A novel approach for planning orthognathic surgery: the integration of dental casts into three-dimensional printed mandibular models. Int J Oral Maxillofac Surg. 2014;43(4):454–9.
29. Ma XJ, Tao L, Chen X, Li W, Peng ZY, Chen Y, et al. Clinical application of three-dimensional reconstruction and rapid prototyping technology of multislice spiral computed tomography angiography for the repair of ventricular septal defect of tetralogy of Fallot. Genet Molecul Research: GMR. 2015;14(1):1301–9.
30. Jacobs S, Grunert R, Mohr FW, Falk V. 3D-imaging of cardiac structures using 3D heart models for planning in heart surgery: a preliminary study. Interact Cardiovasc Thorac Surg. 2008;7(1):6–9.
31. Halsted WS. The training of the surgeon. J Am Med Assoc. 1904;XLIII(21):1553–4.
32. Estai M, Bunt S. Best teaching practices in anatomy education: a critical review. Ann Anat. 2016;208:151–7.
33. McMenamin PG, Quayle MR, McHenry CR, Adams JW. The production of anatomical teaching resources using three-dimensional (3D) printing technology. Anat Sci Educ. 2014;7(6):479–86.
34. Cook DA, Hatala R, Brydges R, et al. Technology-enhanced simulation for health professions education:

a systematic review and meta-analysis. JAMA. 2011;306(9):978–88.
35. Costello JP, Olivieri LJ, Krieger A, Thabit O, Marshall MB, Yoo S-J, et al. Utilizing three-dimensional printing technology to assess the feasibility of high-Fidelity synthetic ventricular septal defect models for simulation in medical education. World J Pediatr Congenit Heart Surg. 2014;5(3):421–6.
36. Lim KH, Loo ZY, Goldie SJ, Adams JW, McMenamin PG. Use of 3D printed models in medical education: a randomized control trial comparing 3D prints versus cadaveric materials for learning external cardiac anatomy. Anat Sci Educ. 2016;9(3):213–21.
37. Maragiannis D, Jackson MS, Igo SR, Schutt RC, Connell P, Grande-Allen J, et al. Replicating patient-specific severe aortic valve stenosis with functional 3D modeling. Circ Cardiovasc Imaging. 2015;8(10):e003626.
38. Mooney JJ, Sarwani N, Coleman ML, Fotos JS. Evaluation of three-dimensional printed materials for simulation by computed tomography and ultrasound imaging. Simul Healthc. 2017;12(3):182–8.
39. Ong LML, de Haes JCJM, Hoos AM, Lammes FB. Doctor-patient communication: a review of the literature. Soc Sci Med. 1995;40(7):903–18.
40. Cheuk DK, Wong SM, Choi YP, Chau AK, Cheung YF. Parents' understanding of their child's congenital heart disease. Heart. 2004;90(4):435–9.
41. Bernhard J-C, Isotani S, Matsugasumi T, Duddalwar V, Hung AJ, Suer E, et al. Personalized 3D printed model of kidney and tumor anatomy: a useful tool for patient education. World J Urol. 2016;34(3):337–45.
42. Andolfi C, Plana A, Kania P, Banerjee PP, Small S. Usefulness of three-dimensional modeling in surgical planning, resident training, and patient education. J Laparoendosc Adv Surg Tech A. 2017;27(5):512–5.
43. Giannopoulos AA, Chepelev L, Sheikh A, Wang A, Dang W, Akyuz E, et al. 3D printed ventricular septal defect patch: a primer for the 2015 Radiological Society of North America (RSNA) hands-on course in 3D printing. 3D Print Med. 2015;1(1):3.
44. Chepelev L, Hodgdon T, Gupta A, Wang A, Torres C, Krishna S, et al. Medical 3D printing for vascular interventions and surgical oncology: a primer for the 2016 radiological society of North America (RSNA) hands-on course in 3D printing. 3D Print Med. 2015;2(1):5.
45. Chepelev L, Souza C, Althobaity W, Miguel O, Krishna S, Akyuz E, et al. Preoperative planning and tracheal stent design in thoracic surgery: a primer for the 2017 Radiological Society of North America (RSNA) hands-on course in 3D printing. 3D Print Med. 2017;3(1):14.
46. Garcia J, Yang Z, Mongrain R, Leask RL, Lachapelle K. 3D printing materials and their use in medical education: a review of current technology and trends for the future. BMJ Simul Technol Enhanc Learn. 2018;4(1):27–40.

25 Dassault Systèmes' Living Heart Project

Steven Levine, Tom Battisti, Björn Butz, Karl D'Souza, Francisco Sahli Costabal, and Mathias Peirlinck

25.1 Introduction

Cardiovascular disease (CVD) is the leading cause of death in developed countries and exacts a high socioeconomic cost. Computational modeling and simulation (CMS) tools have the potential to decrease these costs and improve current treatment methods by providing advanced diagnostics, root cause analysis, and outcome prediction. Such tools are uniquely capable of accounting for cardiac/vascular tissue mechanics, blood flow, and the interaction between them. CMS tools also can improve diagnosis by augmenting medical imaging, e.g., quantifying shear stresses in an aneurysm, and estimate the impact of treatments on the body, e.g., stresses in the arterial wall after insertion of a stent. Moreover, CMS tools are being used to inform device selection and sizing for a specific patient and to predict the efficacy and possible aftereffects of surgical procedures both short term and long term.

Despite their vast potential, CMS tools are currently underutilized in the study and treatment of CVD. From a clinical perspective, the main reason for slow adoption is the lack of large validation studies necessary to establish reliable models, which have only recently become available due to the engineering challenges of capturing the complexity of the heart including multiple physics (e.g., fluids, solids, electrophysiology) and scales (molecular/tissue/organ).

The use of CMS tools is particularly promising for improving treatments for congenital heart disease (CHD). CHD is the leading cause of birth defect-related deaths. With CHD, the heart and its periphery develop abnormally resulting in highly patient-specific anatomies challenge clinicians due to the uniqueness of each case. Today, a clinician's experience and intuition can often be

S. Levine (✉)
Dassault Systèmes, San Diego, CA, USA
e-mail: Steven.Levine@3ds.com

T. Battisti · K. D'Souza
Dassault Systèmes, Johnston, RI, USA
e-mail: Tom.Battisti@3ds.com; Karl.DSouza@3ds.com

B. Butz
Dassault Systèmes, Karlsruhe, Baden-Württemberg, Germany
e-mail: Bjoern.BUTZ@3ds.com

F. S. Costabal
Department of Mechanical and Metallurgical Engineering, Institute for Biological and Medical Engineering, Pontificia Universidad Católica de Chile, Santiago, Chile

Institute for Biological and Medical Engineering, Pontificia Universidad Católica de Chile, Santiago, Chile

M. Peirlinck
Department of Mechanical Engineering, Stanford University, Stanford, CA, USA

Biofluid, Tissue and Solid Mechanics for Medical Applications Lab (bioMMeda) – Institute of Biomedical Technology, Ghent University, Ghent, Belgium
e-mail: Mathias.Peirlinck@UGent.be

G. Butera et al. (eds.), *Modelling Congenital Heart Disease*,
https://doi.org/10.1007/978-3-030-88892-3_25

the difference between life and death of the patient. CMS can be used in virtual surgery planning (VSP) during which the clinician can at a minimum explore treatment options in advance and at best predetermine the optimal postsurgical anatomy, replacing intuition with science-driven planning.

In this chapter, we will start by introducing the Living Heart Project (LHP). Launched in 2014 and coordinated by Dassault Systèmes (DS), the LHP has the mission to connect biomedical researchers, medical device makers, clinicians, and regulatory bodies to advance the use of CMS for the diagnosis and treatment of CVD, and to translate the technological innovations generated by the LHP community into improved patient care. A cornerstone of the LHP is the development of a CMS "Living Heart Framework" ("Framework") for cardiac modeling that is accessible and shareable by the community. One of the most significant outputs of the LHP has been the creation of the living heart human model (LHHM). The LHHM is a multiphysics computational model of a healthy human built on first principles that simulate the electrophysiology, mechanics, and fluid dynamics of a beating heart. Built on the well-established SIMULIA/Abaqus software suite [2], the LHHM is largely extensible, modular, and customizable, thereby allowing one to modify any of its attributes to represent alternative states (e.g., diseased). Similarly, the framework itself can be used to create entirely new geometric models (e.g., congenital pediatric models and/or animal models). Taken together, the framework and all of the models developed by DS and the community through the LHP constitute what DS refers to as the living heart model (LHM).

We will outline several CVD applications including those involving medical devices and drug interaction for which the LHP community has leveraged the LHM, concluding with one for VSP for a patient suffering from hypoplastic left heart syndrome (HLHS), a single ventricle CHD. At present, such a case study can be used to assist a surgeon to explore treatment options in advance, using realistic models of a patient to understand both the surgical technique and the anatomical interdependencies that govern his interventional choices. Such training can be instrumental to aid optimal surgical choices, and in the future lead to intelligent systems similar to a modern GPS, capable of predicting outcomes more reliably than humanly possible.

25.2 The Living Heart Project

25.2.1 About the Project

Over the past 50 years or more, fundamental research into the physical, chemical, and biological foundations of the human heart has enlightened our understanding of this organ that is so critical to life. The heart's physical and behavioral complexity is its secret for efficiency and resilience, but represents a challenge to those who need to understand the critical cause-and-effect relationships that would lead to new treatment innovations and govern ideal treatment choices in any given patient. In many areas, there is a gap between the observational experience of practicing clinicians and a first principles understanding of the root cause behaviors, representing the rate-limiting step for medical innovations and science-based treatment planning.

The LHP seeks to bridge these gaps by asking those active in the field to collaborate to create the most comprehensive computational models based on a first principles understanding of the phenomena and that replicate the observed in vivo behavior. Such models could serve as a basis for continual refinement through the LHP's clinical and research arms, while being validated through its industrial and regulatory arms. Utilizing the discipline of modern commercial software practices, these models could also be utilized as a clinical training, diagnostic, and treatment guidance system.

Now beginning its fifth year with more than 100 institutions represented, the LHP community has demonstrated that this premise is indeed feasible, delivering a commercial-grade model in the LHHM that is now in daily use by a global community capable of extending the model as required and sharing results for comparison and continual improvement. Under strict version control, the strengths and weaknesses of the LHHM for any given application are openly documented and shared, and thus available to be understood and evaluated by regulatory bodies such as the FDA. The LHHM is freely available to all LHP members worldwide creating an unprecedented opportunity for aggregating and sharing cardiovascular knowledge.

25.2.2 Building the Living Heart Human Model

A realistic model of the human heart must account for both its anatomical and physiological characteristics. The LHHM (Fig. 25.1) is constructed from image data for a living, healthy, adult, 50th percentile, male, and contains well-defined anatomical details of internal (e.g., heart valves) and external (e.g., aortic arch) structures that are necessary to accurately model cardiac hemodynamics and device–heart interactions, as well as microstructural features such as continuously varying muscle fiber orientations that strongly influence the dynamic behavior of the heart. From a functional viewpoint, the heart is essentially a pump wherein blood flow is driven by myocardial contractions that in turn influence cardiac motion. The timing and magnitude of the contractions are controlled by waves of electrical excitation that travel across the heart. To account for these physiological phenomena, we employ multiphysics to simulate its dynamic structural, fluid, and electrical behavior. The structural behavior of cardiac tissue includes both passive and active components. Passive tissue response is modeled with an anisotropic hyperelastic formulation, while that for active tissue response with a time-varying elastance to account for the excitation–contraction coupling. Pathological conditions can be modeled by modifying the material coefficients, changing the loads and boundary conditions, or using alternative user-defined material models to simulate complex phenomena such as tissue growth and remodeling.

A one-dimensional (1D) fluid network is used to model blood flow between the heart chambers and through the external circulation (Fig. 25.2). This network is implicitly coupled with the structural model, and the overall fluid–structure system is calibrated such that clinically relevant metrics (e.g., PV loops) are within their published ranges. Vascular compliances and resistances are modifiable to simulate exercise, hypertension, and other physiological or disease states. While the network model is computationally efficient and adequate for many applications of practical interest, there are situations in which detailed three-dimensional (3D) blood flow modeling is required. For example, when studying artificial heart valves, not only is the structural

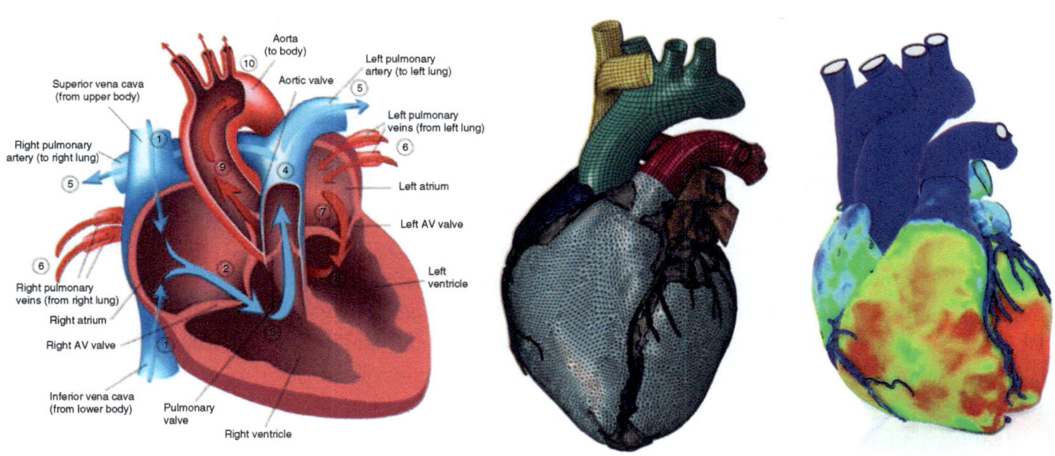

Fig. 25.1 Schematic of human heart (left), FE mesh (center), and property mapping representation of LHHM (right)

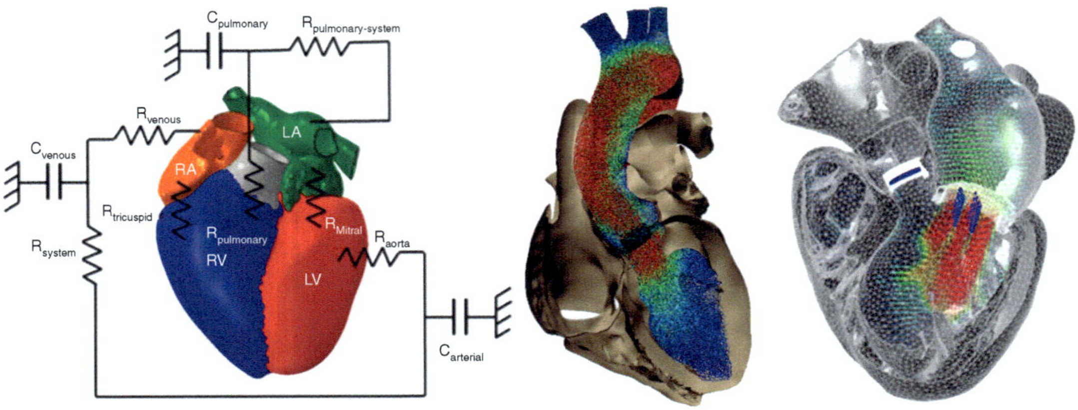

Fig. 25.2 1D blood flow model in the LHHM (left); 3D FSI using SPH (center) and cosimulation with FlowVision (image courtesy of Capvidia) (right) for simulating the hemodynamic characteristics of artificial heart valves

contact with surrounding tissue of concern, but also are the resultant blood flow patterns. If the blood flow pattern is altered from one, which is characteristic of healthy hearts, cardiac remodeling, tissue damage, and other unintended consequences could result, phenomena, which also could be modeled. The LHHM can be augmented to include 3D blood flow representations as necessary. Bidirectional fluid–structure interaction (FSI) with both "traditional" Navier–Stokes solvers and with "meshless" Lattice Boltzmann solvers is possible, and the choice of a flow solver should be based on the specific application and the degree of accuracy required (Fig. 25.2). For intermediate levels of accuracy and run time, smoothed particle hydrodynamics (SPH) can be an attractive alternative to 3D FSI modeling as it can handle complex cardiac phenomena such as flow pinching (as occurs with heart valves) and requires minimal fluid modeling expertise (Fig. 25.2).

Tissue electrical response is characterized by an action potential whose spatiotemporal evolution is governed by both extrinsic (global) and intrinsic (biochemical) variables. In many cases of practical interest, it is common to modify the extrinsic variables (e.g., heart rate, conduction blocks) to study their effects on cardiac function and device efficacy while holding the biochemical variables constant over the duration of the simulation. However, to study some types of arrhythmias, it may be necessary to model the evolution of the biochemical variables explicitly. The LHHM allows the mechanistic modeling of cellular electrophysiological dynamics and uses a multiscale approach wherein cellular dynamics can affect and be affected by organ-level phenomena. In addition to a 3D conduction model for bulk tissue behavior, the LHHM also includes electrically important "1D" anatomical features (e.g., the Purkinje fiber network) to generate physiologically realistic propagation patterns (Fig. 25.3). In summary, the LHHM is an adaptable multiphysics–multiscale tool for cardiac simulation that can facilitate the diagnosis and treatment of CVD.

More details regarding the original and current releases of the LHHM can be found in the following references [2, 3].

25.2.3 LHM Applications

25.2.3.1 Pacemaker Lead Durability

Medical devices for cardiovascular applications have to perform effectively, reliably, and safely in unpredictable conditions to win regulatory approval and meet the exacting standards of today's healthcare industry. Currently, designing a device capable of meeting these rigorous criteria involves high costs, long development times, and a certain degree of luck. Consider, for instance, the design of pacemaker leads. Pacemakers play a key role in cardiac rhythm

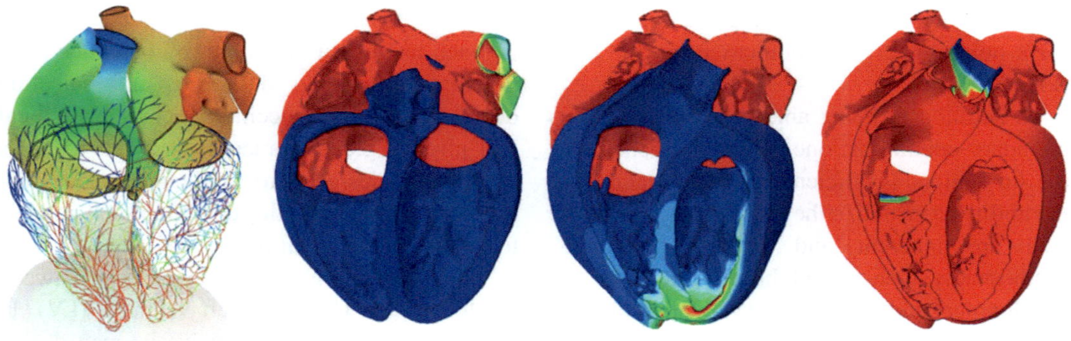

Fig. 25.3 Purkinje fiber network in the LHHM (left); electrical potential at various times over the cardiac cycle (right): maximum (*red*) = 20 mV; minimum (*blue*) = −80 mV

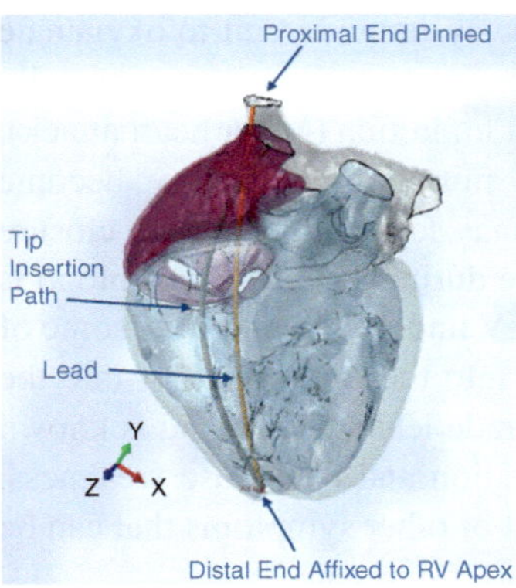

Fixation Site	Apex								Septum							
Redundancy Level	L1				L3				L1				L3			
Lead Stiffness	Low		High		Low		High		Low		High		Low		High	
Ejection Fraction (EF)	20	55	20	55	20	55	20	55	20	55	20	55	20	55	20	55

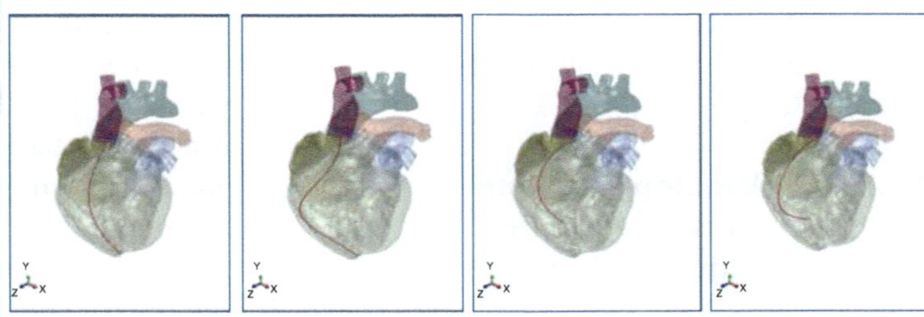

Fig. 25.4 Lead insertion and initial position (left). Parameters studied (right)

management and save thousands of lives. However, the leads for a significant number of these devices undergo in-service failures, thereby causing distress to patients and increasing the overall cost of treatment. Many factors can influence lead reliability, such as lead design, composition, manufacturing process, patient characteristics, and implant configuration. The interdependence of these factors makes lead durability prediction a challenging task.

Lead durability uncertainty reinforces the case for the development of virtual models capable of accurately representing the in vivo environment and studying the various factors that are thought to influence it. The LHM is an ideal tool to help improve pacemaker lead durability. We present a study in which the LHM was used to simulate lead performance under various scenarios and examine in vivo lead motion. Initially, a lead was inserted through the superior vena cava (SVC), right atrium (RA), and right ventricle (RV), and attached to the RV apex using a tip insertion path and appropriate constraints (Fig. 25.4). With the lead in position, a cardiac cycle was simulated to examine the deformation of the lead using an appropriate level of damping to capture the effect of the surrounding blood. The initial data showed that the dynamic response of the lead under cardiac loading conditions was qualitatively similar to what is clinically observed.

With a credible simulation methodology established, a study was designed and executed to explore various factors that may influence lead curvature, which in turn is hypothesized to influence lead durability. The factors that were varied were lead fixation site (apex vs septum), the degree of redundancy (or slack) in the lead when affixed, lead stiffness, and the ejection fraction

(EF) of LV (EF = 20 for a failing heart vs EF = 55 for a normal heart) as summarized in the table (Fig. 25.4).

In addition to being able to simulate in vivo boundary conditions, one of the key benefits of the simulations is efficiency. Lead curvature and the ability to visualize the 3D deformation of the lead are inherent and immediate outputs. Traditionally, one would have to rely on the development and/or usage and validation of complex reconstruction algorithms that operate on 2D image data to produce these desired outputs. Furthermore, based on the simulation results the relative effect of not only each individual factor on lead curvature can be determined, but also any combination of factors, the latter of which is not directly measurable (Fig. 25.5). This study also illustrates the potential to perform simulations for clinical trial design.

More details of this study can be found in the report by Cooke et al. [4].

25.2.3.2 Annuloplasty Ring for Ischemic Mitral Regurgitation

In the previous application, we primarily studied the effects of varying device parameters. In this example, we illustrate modeling a cardiac disease state and treating it with a novel device. The high variability among heart patients means that, even when the underlying disease is the same, a device that works well for some patients may prove ineffective or unsafe for others. Consequently, there is much interest in moving away from one-size-fits-all solutions to designing and selecting devices based on a specific patient's anatomical and physiological characteristics. The LHM is specifically designed to facilitate the development of patient-specific, computational models to assess the treatment of disease.

The mitral valve controls blood flow between the left atrium (LA) and left ventricle (LV). In normal circumstances, the valve is open during ventricular diastole allowing the LV to fill. The valve then closes completely during systole when blood is forced from the LV into the aorta and eventually into the systemic circuit to oxygenate the body. However, in some situations, especially after a myocardial infarction (MI or heart attack), the LV and the mitral annulus may become enlarged, which may lead to incomplete closure of the mitral valve during systole. Since blood is ejected from the LV under high pressure, some of it may flow back into the LA instead of into the aorta. This retrograde leakage of blood is known as mitral regurgitation and can cause weakness, fatigue, and a host of other symptoms that can be life- threatening if left untreated.

Here, the LHM was used to assess the efficacy of a novel medical device being developed to treat ischemic mitral regurgitation. The first step was to induce a MI into the model. Based on real patient post-MI data, the material properties of a region in the LV (including the posterior papil-

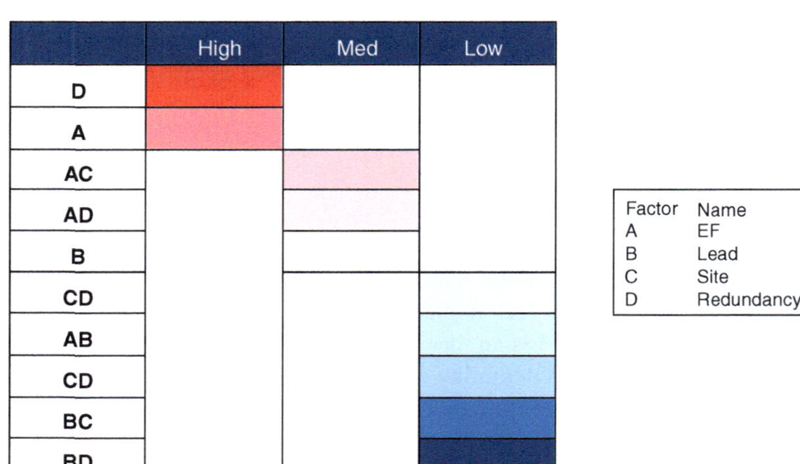

Fig. 25.5 Influence of various factors on lead curvature

lary muscle) were modified to suppress the tissue's active response for the entire cardiac cycle. The diseased model was then run and indeed predicted a lower cardiac ejection fraction in agreement with that observed clinically (Fig. 25.6). Moreover, incomplete closure of the mitral valve during ejection was also clearly visible (Fig. 25.7).

The next step was to simulate the interaction of the device and the infarcted heart. The device under consideration was an annuloplasty ring with a subvalvular component (Fig. 25.8). A subvalvular element was hypothesized to better engage the chordae and hence counterbalance the shifting of the coaptation line in patients with ischemic cardiomyopathy, thereby reestablishing proper leaflet coaptation and valve function. The device was modeled as rigid and affixed to the mitral valve annulus at several distinct points representing suture locations. The heart model predicted improved leaflet coaptation upon implantation of the ring, thus providing conceptual validation of the hypothesis (Fig. 25.9). In addition, we examined the engagement profile of the ring and the chordae and compared the forces generated in the chordae before and after implan-

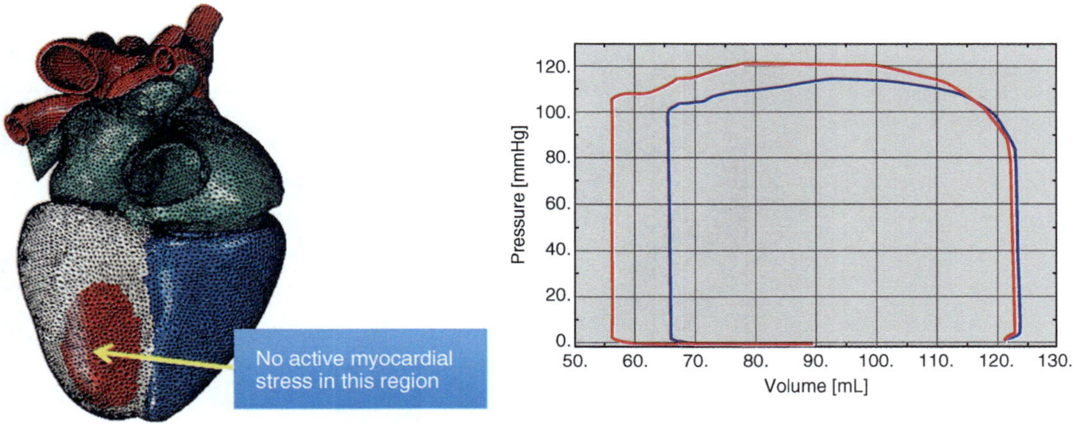

Fig. 25.6 LHM with myocardial infarction in left ventricle (left); modified reduction in ejection fraction from 55% (*red*) to 45% after MI (*blue*) (right). Image reprinted from Ref. [5]

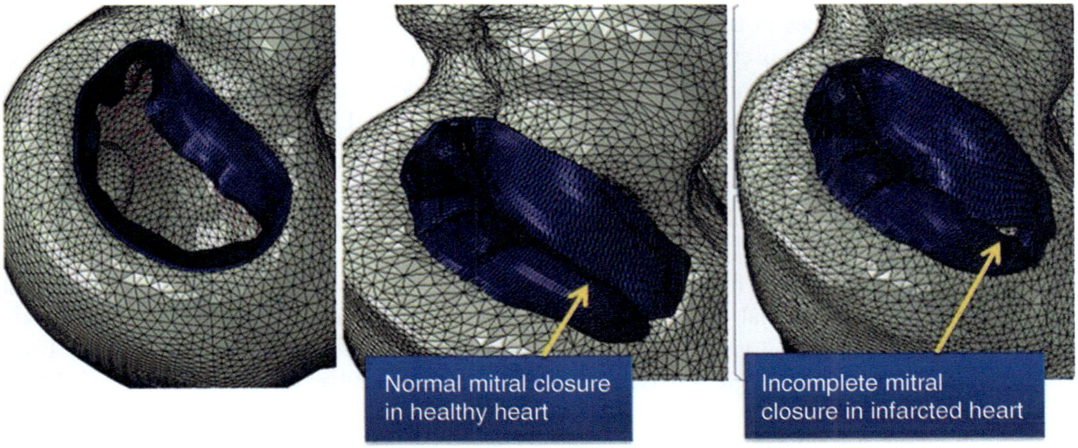

Fig. 25.7 Close-up view of the LHM with LA removed to show the mitral valve open (left), closed in normal heart (middle), and incompletely closed in the infarcted heart (right). Image reprinted from Ref. [5]

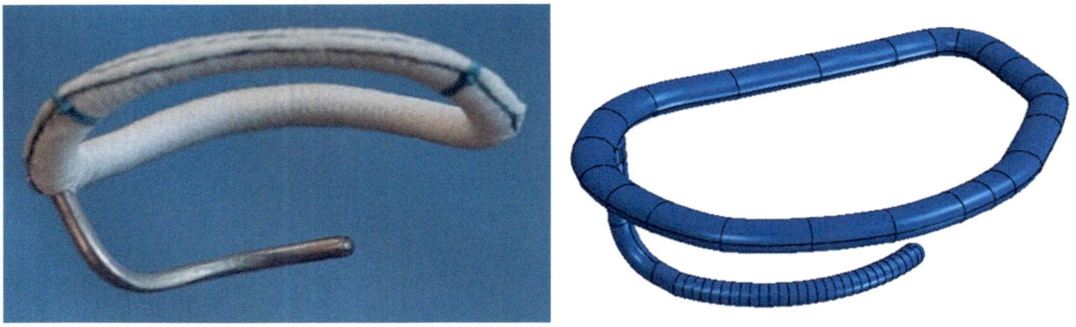

Fig. 25.8 Novel annuloplasty ring with subvalvular element: physical device (left) and finite element model (right). Image reprinted from Ref. [5]

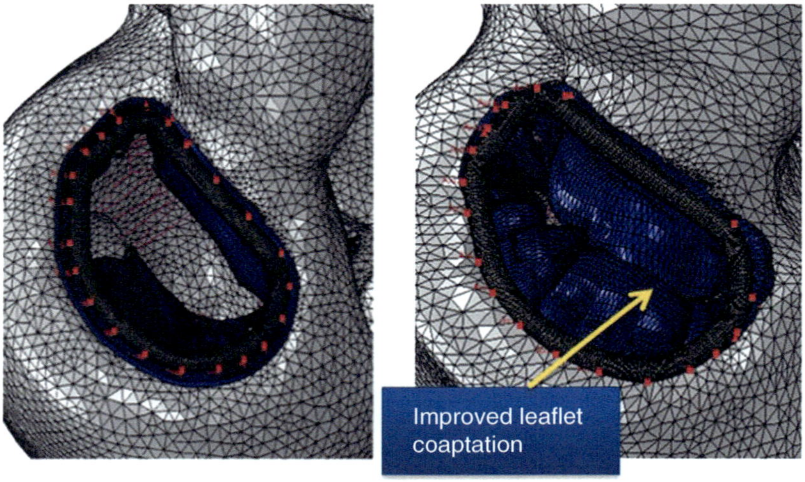

Fig. 25.9 Annuloplasty ring in position in infarcted heart: mitral valve open (left), mitral valve closed (right). Image reprinted from Ref. [5]

tation (Fig. 25.10). We see that the ring was able to generate chordae forces in an infarcted heart that were closer to those observed in a healthy heart, thereby improving confidence in the device.

Using the LHM, device makers can simulate the interaction of a new device with the diseased heart, thereby gaining actionable insight on the influence of the heart on the device and vice versa. Multiple device design modifications can then be made until an optimal design is identified. Moreover, to account for real-world variability, rather than using a single diseased heart model (based on a specific or average patient), multiple diseased models can be generated that collectively represent a target patient population. This methodology can be used to further optimize the device design, or to produce multiple designs each of which is best suited to a subset of the target patient group.

More details of this study can be found in Baillargeon et al. [5].

25.2.4 Drug Safety

Cardiac arrhythmia can be a potentially lethal side effect of a drug's action on the human body whereby the electrical activity of the heart becomes irregular and circulation of blood through the body is compromised. Pharmaceutical companies must assess the risk of arrhythmia posed by a drug for it to be approved. While current early- stage safety assessment protocols are

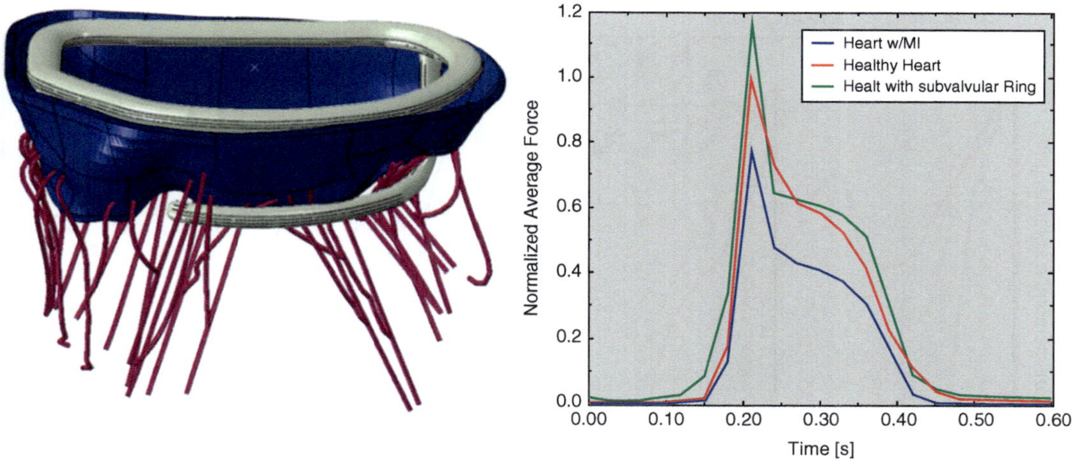

Fig. 25.10 Subvalvular element engaging with mitral chordae (left); restoration of chordae forces with device (right). Image reprinted from Ref. [5]

able to predict the effect of a drug at the cellular level, the response at an organ level may be different. The LHM provides a foundation for virtual safety studies that can offer mechanistic insight into drug action at both the cellular and organ levels.

In this example, cellular response was first studied by creating models of single cardiac cells capable of modeling the activities of individual ion channels (e.g., Na^+, K^+, Ca^{2+}). The O'Hara-Rudy model was used for ventricular cells and the Stewart model for Purkinje cells. The effect of varying concentrations of dofetilide on cardiac action potential dynamics was simulated by selectively blocking the relevant ion channels. The resulting changes in action potentials of endocardial, epicardial, midwall, and Purkinje cells are shown in (Fig. 25.11).

To study response at the organ level, the cellular models were implemented inside of the whole heart model in lieu of the default phenomenological electrical model to investigate how action potential dynamics at the cellular level might affect macroscopic electrical behavior at the whole heart level. The resultant model contains a large number of elements (~10 million) to maintain global spatial accuracy and uses approximately 250 million internal variables to track local ionic current fluctuations. The results show that the overall pattern of electrical propagation can be significantly altered under the administration of dofetilide. Simulations were performed for a range of dosage strengths, and it was observed that as the strength is increased so does the deviation in the electrical pattern compared to that which is normal. The emergence of spiral wave characteristic of Torsade de points was observed to occur at a dofetilide concentration of 5.7× its free plasma concentration. We show the electrical potential on the surface of the model at various times during the cardiac cycle at a baseline (no drug) on the left and at a critical (5.7×) concentration of dofetilide on the right (Fig. 25.12).

While 3D spatiotemporal visualization of cardiac electrical patterns can provide a mechanistic understanding of the cardiotoxic effects of a drug and as such enhance the drug discovery process, the LHM can also be used to generate virtual electrocardiograms (Fig. 25.13) that may be more relevant in clinical settings, for instance to predict potential cardiotoxic effects when multiple drugs act simultaneously. As in the annuloplasty ring example, it is also possible to develop patient or population-specific models using the cellular profiles from genomic data and other sources. Future work in this area could focus on these personalization aspects.

More details of this study can be found in F. Sahli Costabal et al. [6].

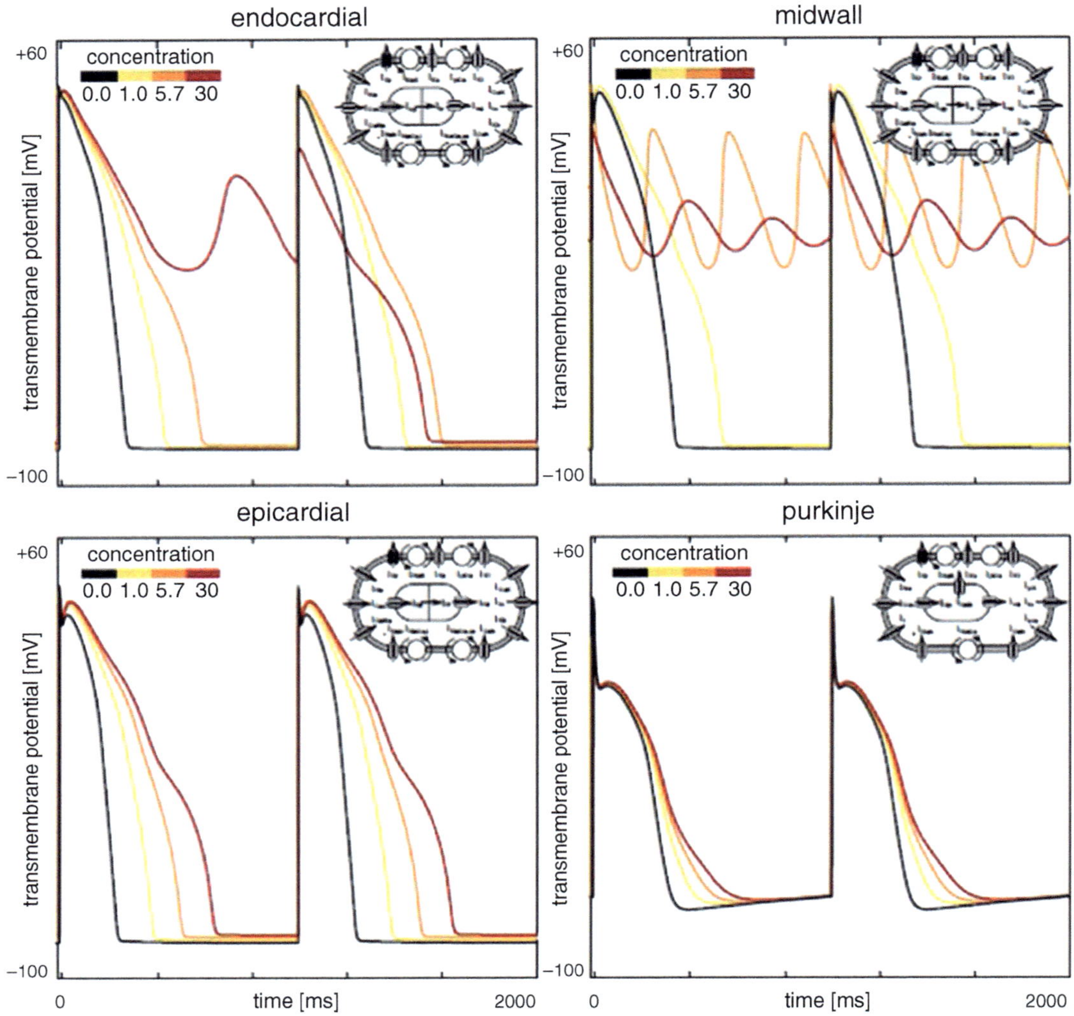

Fig. 25.11 Effect of dofetilide on single-cell action potential of different cell types. Image reprinted from Ref. [6]

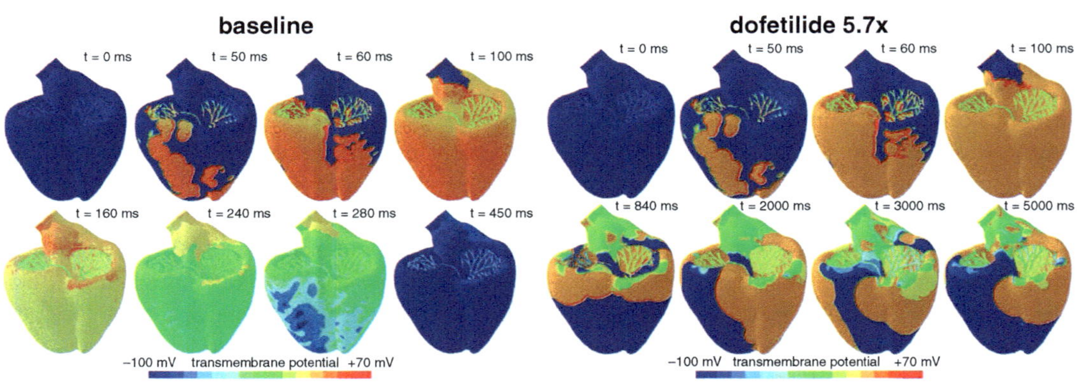

Fig. 25.12 Effect of dofetilide on whole heart electrical propagation, baseline (no drug) left; critical concentration (5.7×) right. Image reprinted from Ref. [6]

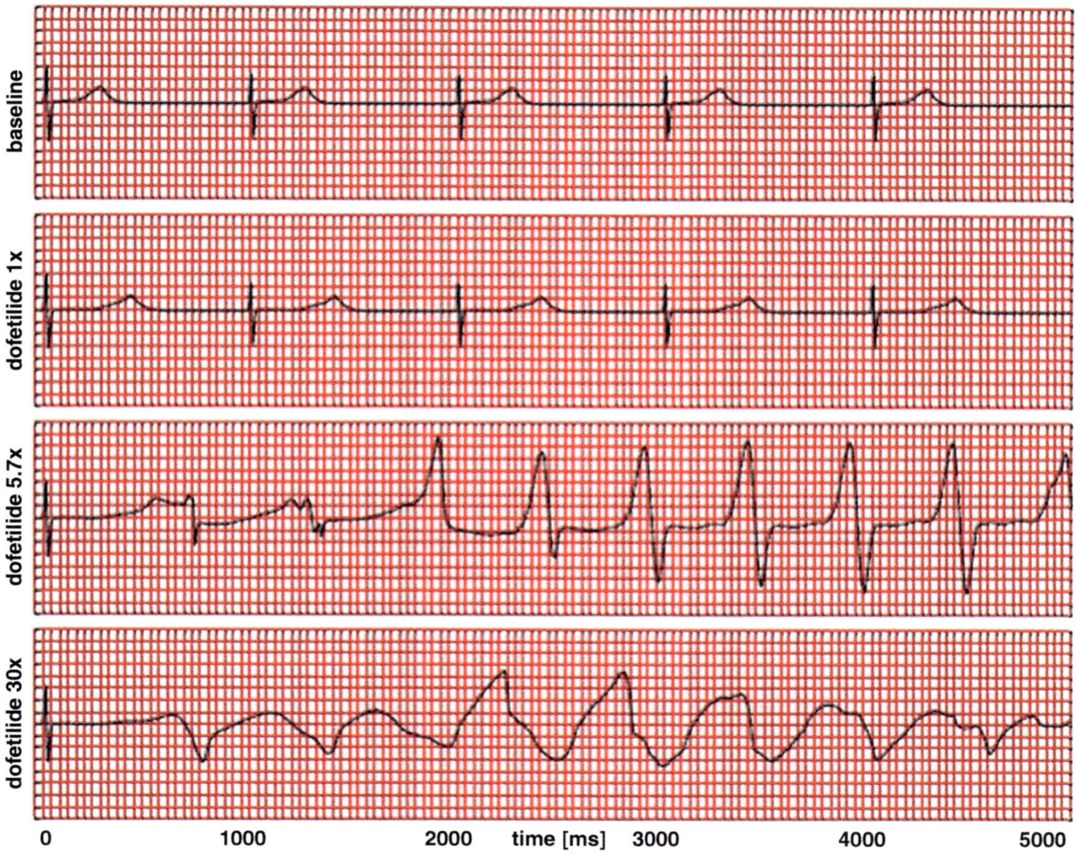

Fig. 25.13 Virtual electrocardiogram recordings for the baseline case and with dofetilide at concentrations of 1×, 5.7×, and 30×. Image reprinted from Ref. [6]

25.2.4.1 Virtual Surgery Planning for the Norwood Procedure

Congenital heart defects are the leading cause of birth defect-related deaths. Neonates with HLHS are born with a non- or underdeveloped left ventricle and aorta. The common treatment approach comprises a series of two to three surgical interventions over a time span of 2–3 years until the full Fontan circulation is established [7]. The healthy right ventricle then works as the sole blood pump for both the systemic and pulmonary circulatory systems [8].

The first stage of surgery (the Norwood procedure) involves a surgical reconstruction of the underdeveloped hypoplastic aorta, which connects the neoaorta to the pulmonary root using a tissue patch. This intervention directs the blood coming from the functioning right ventricle to the systemic circulation, while the pulmonary circulation is driven by the introduction of a systemic-to-pulmonary shunt between the right brachiocephalic trunk (or the right ventricle) and the pulmonary artery. Given that the reconstructed aorta directly influences blood circulation, afterload of the functioning ventricle, and oxygen delivery to the coronary arteries [9], its shape is critical for the child's survival rate and the likelihood of developing adverse effects [7]. Typical complications are low coronary blood flow or kinking of the aortic arch ("gothic arch"), which both lead to unbalanced perfusion of heart and body [10].

Unfortunately, a pre-op estimation of the optimal shape and diameter of the aorta remains a challenging task for the clinician [11] who is faced with multiple degrees of freedom (patch size and shape, patch material, incision points, etc.) in reconstructing the aorta. At the same time, the complex geometrical and structural influence of these choices on the final achieved arch shape and arterial compliance further complicate assessment of postoperative hemodynamic conditions over both the short term and the long term [8]. Furthermore, the uniqueness and wide variation in anatomy and physiology in CHD patients make it hard to learn from previous cases.

Consequently, only basic preoperative planning is done today. Clinicians typically make a first crude assessment of the preoperative state through echocardiography data and develop a rough plan of the target, preferred arch shape. During the operation, they mainly rely on their intuition and experience.

CMS offers the potential to overcome the gaps in today's presurgical planning with so-called VSP [12]. By virtually performing the surgical procedure using computational models prior to the actual surgery on the patient, the postsurgical blood flow and structural behavior of the reconstructed aorta can be simulated and the efficacy of the intervention can be assessed (both short term and long term). This process can be repeated for multiple anatomical variations until the optimal anatomy for the specific patient is determined. Finally, this postsurgical result can be used to guide the surgeon during the procedure in the form of augmented or virtual reality tools or a 3D printed template (Fig. 25.14). The outlined workflow for VSP of the Norwood procedure [13] is shown below (Fig. 25.14).

1. Obtain patient-specific data

   The first step of the workflow is to gather patient-specific data on both the structural and hemodynamic preoperative states. Structurally, the preoperative shape, position, and compliance of the aortic arch with coronary arteries and the pulmonary root need to be captured. As CT and MRI imaging are typically not used for these young infants for ethical reasons (sedation and exposure to radiation), alternative acquisition techniques such as 3D echocardiography should be considered.

   Hemodynamically, the blood flow conditions of the patient need to be measured to derive patient-specific boundary conditions for computational fluid dynamic simulations. Essential quantities are systemic and pulmonary vascular resistance, ejection fraction, and

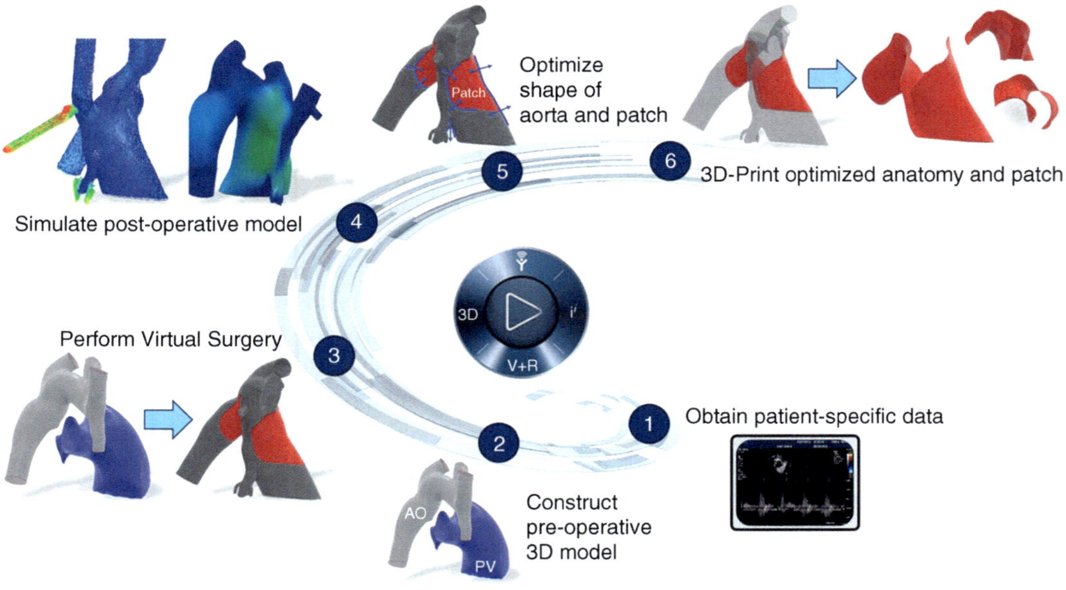

Fig. 25.14 Schematic of the steps of the virtual surgery planning workflow for the Norwood procedure

flow distribution and blood pressure curves among the branches of the aorta [14]. Over time, it may be possible to reasonably estimate these conditions based on validation studies matching models to measured properties.

2. Construct preoperative 3D model.

   Independent of the chosen imaging methodology, the three-dimensional geometry of the patient's vasculature needs to be segmented to construct a 3D model. Given that the focus of this workflow lies on the surgical reconstruction of the aorta, the computational domain is limited to the thoracic aorta and the pulmonary root. For now, the univentricular heart is not considered but can be included in the future, incorporating our experience with modeling realistically beating hearts.

3. **Perform Virtual Surgery.**

   The preoperative 3D model is used as the starting point for the virtual Norwood procedure. This step is performed by a 3D modeling expert under the guidance of an experienced surgeon. Together, they use CAD tools to imitate all relevant steps of the Norwood procedure: Disconnect pulmonary arteries, open the hypoplastic aorta, connect aortic root with pulmonary root, define shape of the reconstructed arch, and connect the shunt to pulmonary arteries.

4. **Simulate postsurgical model.**

   A CFD simulation of the blood flow is set up to study the effect of the postoperative setting on the local and global postoperative hemodynamics. The CFD simulation needs to be coupled to a 0D or 1D lumped parameter network (LPN) of the vasculature in order to obtain physiologically realistic and patient-specific inlet and outlet boundary conditions for the fluid domain and to account for autoregulation [15]. Such LPNs model the impedance and resistance of the vascular tree that is not incorporated in the model, but affect the dynamic pressure response to the pulsatile flow in the aorta. These LPNs are critical given our aim to predict postoperative flow patterns based on presurgical flow and pressure measurements.

   Apart from the dynamically changing hemodynamic boundary conditions at the model's inlets and outlets, the hemodynamic flow patterns are also impacted by the neo-aorta's global and local distensibility, which buffers the pulsatile cardiac output of the heart/single ventricle. Similar to Ref. [16], the importance of accounting for the aorta's elasticity is crucial, especially in light of virtually exploring reconstruction strategies with different potential patch materials, each having their own intrinsic mechanical behavior when subjected to intraluminal pressures (e.g., compliance, thickness, anisotropy). The aortic tissue and patch behavior are accounted for using computational structural and solid dynamics (CSD) simulations computing the deformational hyperelastic response of the reconstructed vessel wall (native and patch tissue) to the luminal pressure fluctuations. The CSD model is then coupled to the CFD/LPN model in a FSI simulation, resulting in a cosimulation framework between three solvers.

5. **Optimize shape of aorta and patch.**

   Now that we have computationally captured the hemodynamic and structural behavior of the reconstructed aorta, we can use parametric optimization methods to vary local diameters, incision points, heights, and the overall patch shape to determine the optimal postsurgical anatomy of the reconstructed arch. Reasonable optimization cost functions take into account target blood flow ratios through each of the branches, and primarily blood flow through the coronary arteries to guarantee sufficient oxygen distribution throughout the body [9, 17]. At the same time, we can optimize the reconstructed aortic arch to ascertain minimum blood residence time or limit wall shear stresses (WSS) to prevent blood clot forming and plaque formation, as regions of low and high WSS are suspected to result in unwanted growth and remodeling in the neoaorta.

6. **3D print optimized anatomy and patch.**

   Using additive manufacturing techniques, the optimized patch geometry can be 3D printed to serve as a stencil for the clinician during surgery. These stencils can guide the clinician and thus limit operation time, which is beneficial for patient outcome [18, 19]. Dependent on the evolutions in tissue engineering and the needed approval of regulatory authorities, the clinician might one day even be able to additively manufacture the tissue patch in a shape that was preoperatively optimized on a patient-specific basis [20].

25.3 Outlook

Over the last decade, the increased use of computational cardiovascular models suitable for the development of new devices and treatments as well as clinical applications has demonstrated that digital simulation can be considered as a part of future medical training programs to augment physical simulation and direct clinical experience. We know from other industries such as aerospace where flight simulators for pilot training were introduced, that once established, such systems can become irreplaceable as tools. Clinicians should benefit from the same on the job, virtual environment providing much of the experiential training necessary for understanding and confidence in a real situation.

The use of CMS tools in the early development and testing of new devices and treatments creates a logical foundation for usage in the clinical environment as the scenarios created to virtually test devices become increasingly realistic. The LHP is providing a unique vehicle to establish this methodology, demonstrating the cumulative effect that a network can yield compared to any one contributor's work.

25.4 Conclusion

In this chapter, we described the pretext and rationale for using a novel project approach to development of the LHM that (1) has the potential to address a vast range of medically relevant applications, (2) is open to be extended, validated, and reviewed, and (3) provides a commercial foundation sufficient for use in a clinical environment. Four applications, ranging from a surrogate virtual patient to inform regulatory standards to medical device and drug testing and eventually VSP of a congenital defect correction, are used to highlight the diversity of the LHM. While each of these individually does not represent a major breakthrough, collectively, they contribute to a body of model experience that allows the LHM limits to be understood and systematically expanded. The LHP community has the opportunity to develop the LHM to become a first-of-a-kind translational framework for the invention of new ways for clinicians to virtually train, diagnose, and devise optimal treatments. However, one critical component to achieving this objective that is still missing is the public availability of a breadth of high-quality physiological data sets that can be used to calibrate and validate models. Our sincere hope is that the growing LHP community will be able to break through this barrier.

References

1. Rubio D, Schoenbaum E, Lee L, Schteingart D, Marantz P, Anderson K. Defining translational research: implications for training. Acad Med. 2010;85(3):470–5.
2. Baillargeon B, Rebelo N, Fox DD, Taylor RL, Kuhl E. The living heart project: a robust and integrative simulator for human heart function. Eur J Mech A/Solids. 2014;48:38–47.
3. SIMULIA Living Heart Human Model R2018X User Guide, Dassault Systèmes; 2017.
4. Effects of Redundancy, Left Ventricular Ejection Fraction, Pacing Site, and Lead Stiffness on Lead Curvature Prepared by: Dan Cooke, Brian Baillargeon, Tom Battisti, and Prabhav Saraswat, Revised 28 April 2017.
5. Baillargeon, B, Costa, I, Leach, JR, Lee, LC, Genet, M, Toutain, A, Wenk, JF, Rausch, MK, Rebelo, N, Acevedo-Bolton, G, Kuhl, E., Navia, JL, Guccione, JM, "Human cardiac function simulator for the optimal Design of a Novel Annuloplasty Ring with a sub-valvular element for correction of ischemic mitral regurgitation," Cardiovasc Eng Technol, 6(2): 105–116. 2015.
6. Costabal FS, Yao J, Kuhl E. Predicting the cardiac toxicity of drugs using a novel multiscale exposure–response simulator. Comput Methods Biomech Biomed Engin. 2018;21(3):232–46.

7. Gewillig M. The Fontan circulation. Heart. 2005;91(6):839–46.
8. Feinstein JA, Benson DW, Dubin AM, Cohen MS, Maxey DM, Mahle WT, et al. Hypoplastic left heart syndrome: current considerations and expectations. J Am Coll Cardiol. 2012;59(Suppl 1):S1–S42.
9. Moghadam ME, Migliavacca F, Vignon-Clementel IE, Hsia TY, Marsden AL, Allianc MCH. Optimization of shunt placement for the norwood surgery using multi-domain modeling. J Biomech Eng-T Asme. 2012;134:5.
10. Hornik CP, He X, Jacobs JP, Li JS, Jaquiss RDB, Jacobs ML, et al. Complications after the Norwood operation: an analysis of the Society of Thoracic Surgeons congenital heart surgery database. Ann Thorac Surg. 2011;92(5):1734–40.
11. Biglino G, Schievano S, Steeden JA, Ntsinjana H, Baker C, Khambadkone S, et al. Reduced ascending aorta distensibility relates to adverse ventricular mechanics in patients with hypoplastic left heart syndrome: noninvasive study using wave intensity analysis. J Thorac Cardiov Sur. 2012;144(6):1307–14.
12. Marsden AL. Simulation based planning of surgical interventions in pediatric cardiology. Phys Fluids. 2013;25(10):101303.
13. Peirlinck M, Butz B, D'Souza K, Baillargeon B, Mertens L, Francois K, et al. Flow optimization in the reconstructed hypoplastic aortic arch. Orlando, FL, USA: Engineering Frontiers in Pediatric and Congenital Heart Disease; 2016.
14. Moura A. The geometrical multiscale modelling of the cardiovascular system: coupling 3D and 1D models. Milano: Politecnico di Milano; 2007.
15. Vignon-Clementel I, Figueroa C, Jansen K, Taylor C. Outflow boundary conditions for three-dimensional finite element modeling of blood flow and pressure in arteries. Comp Methods Appl Mech Eng. 2006;195(29–32):3776–96.
16. Taelman L, Bols J, Degroote J, Muthurangu V, Panzer J, Vierendeels J, et al. Differential impact of local stiffening and narrowing on hemodynamics in repaired aortic coarctation: an FSI study. Med Biol Eng Comput. 2016;54(2):497–510.
17. Rasoul-Arzrumly E, Hung OY, Samady H. Blood flow with boundaries: wall shear stress and plaque progression. J Cardiol Clin Res. 2013;1(2):1009.
18. Vukicevic M, Mosadegh B, Min JK, Little SH. Cardiac 3D printing and its future directions. J Am Coll Cardiol Img. 2017;10(2):171–84.
19. Morrison RJ, Hollister SJ, Niedner MF, Mahani MG, Park AH, Mehta DK, et al. Mitigation of tracheobronchomalacia with 3D-printed personalized medical devices in pediatric patients. Sci Transl Med. 2015;7(285):285ra64-ra64.
20. Xu Y, Hu Y, Liu C, Yao H, Liu B, Mi S. A novel strategy for creating tissue-engineered biomimetic blood vessels using 3D bioprinting technology. Materials. 2018;11(9):1581.

Cardiovascular Simulation as a Decision Support Tool

26

Michael Broomé, Marcus Carlsson, Petter Frieberg, Nina Hakacova, Petru Liuba, and Dirk W. Donker

26.1 Introduction

The great variation of clinical presentations and specifically altered anatomy in individual patients with congenital heart disease necessitate patient-centered individualized approaches, which is much in contrast to the standard of care provided in pediatric and adult cardiology as this is driven by an evidence-based approach. Current practice is based on small clinical studies, case reports, and the experience of individual surgeons and cardiologists relying on their understanding of pathophysiology and interpretation of modern imaging. In order to meet the complexity that this wealth of information might create in every individual case, computer modeling may serve to integrate all detailed patient characteristics. Such a simulation has the potential to deepen understanding of physiology further and enable patient-specific prediction of hemodynamic responses to catheter-based and surgical interventions.

Also, long-term results of interventions are dependent not only on acute hemodynamic changes, but also on secondary remodeling of cardiac chambers and the vascular tree as well as changes in autonomic tone and arrhythmogenicity. It is difficult to build and validate a model covering all these aspects of pediatric cardiology interventions. Clinical model usability is also strongly dependent on model interface and feedback times when testing interventions. The ideal model should be possible to run bedside in real time by the clinician, allowing testing of different treatment options including both surgery, catheter interventions, and pharmacologically induced hemodynamic changes. It is also important to consider the variable features of interventions. Catheter-based closure of a persistent ductus arteriosus (PDA) or persistent foramen ovale (PFO) is different from the staged construction of a Fontan circulation in a single ventricle patient, where the surgically imposed anatomic design of the venous anastomosis will affect the balance of right to left pulmonary blood flow [1]. Also, interventions are

M. Broomé (✉)
ECMO Department, Karolinska University Hospital, Stockholm, Sweden

Anesthesia and Intensive Care, Department of Physiology and Pharmacology, Karolinska Institutet, Stockholm, Sweden
e-mail: michael.broome@ki.se

M. Carlsson · P. Frieberg
Department of Clinical Physiology, Lund University Hospital, Lund, Sweden
e-mail: marcus.carlsson@med.lu.se; petter.frieberg@med.lu.se

N. Hakacova · P. Liuba
Department of Pediatric Cardiology, Lund University Hospital, Lund, Sweden
e-mail: petru.liuba@med.lu.se

D. W. Donker
Intensive Care Center, University Medical Center Utrecht, Utrecht, The Netherlands

Cardiovascular and Respiratory Physiology, TechMed Center, University of Twente, Enschede, The Netherlands
e-mail: d.w.donker@utwente.nl

G. Butera et al. (eds.), *Modelling Congenital Heart Disease*, https://doi.org/10.1007/978-3-030-88892-3_26

sometimes combined with changes in drug therapy adding further complexity to the clinical course of the patient. Another obvious challenge in pediatric medicine is that disease progression and treatments interact with growth of the child.

In the remainder of this section, we present a real-time closed-loop model, Aplysia CardioVascular Lab, which allows convenient interactions via an attractive interface possible to use bedside. The model includes a variety of hemodynamic interactions including selected catheter-based interventions and cardiosurgical procedures as being part of the clinical therapeutic armamentarium of pediatric cardiovascular care. In addition, the model features mechanical support devices such as venovenous and venoarterial extracorporeal membrane oxygenation (ECMO) and various left/right ventricular assist devices (VAD). Automatic scaling algorithms relating pathophysiology to patient size is being developed as well as algorithms for remodeling. The model characteristics are based on a large body of available literature on human physiology and detailed anatomy, including all relevant pathophysiological, physical, and technical principles. The "natural" behavior of the model fits well with clinical experience and pathophysiological understanding of cardiovascular disease, while a formal validation is still far from complete [2].

26.2 Model

A closed-loop real-time 0D model with four cardiac chambers, valves, atrial and ventricular septal interaction, pericardium, 21 system vascular compartments, and 6 pulmonary vascular compartments have been developed as previously described [2–5]. The vascular compartments are shown in Fig. 26.1. The model is based on previous 0D models [6, 7] and further developed in collaboration with experienced clinicians. The software is running on a standard PC with a high-resolution screen and allows real-time modification of model parameters. A selection of current therapeutic interventions and support modalities is also available. Pressures, flows, volumes, and oxygen saturations are updated with a temporal resolution of 4000 Hz. Model parameters are based on published values and physical dimensions of the cardiovascular system as outlined above. Moreover, cardiovascular physiology is modeled to interact with airway mechanics allowing simulation of gas exchange in relation to hemodynamics. The parameters can be scaled to the size of the patient with an intuitive software interface. Beyond pediatric implications, scaling is also possible ranging from a young healthy adult to an octogenarian with inherent processes of aging and disease and back to pediatric, neonatal, and premature patients as based on algorithms that correct for age, weight, and length of the individual patient. In more detail, pathophysiological features such as ventricular contractility, stiffness and relaxation abnormalities, valvular stenosis and regurgitation, atrial and ventricular septal defect shunts, and persistent ductus arteriosus can be added as a single, adjustable feature or in all possible combinations. Next, coronary blood flow and the balance between myocardial oxygen supply and demand can be analyzed in detail. Individually tailored treatment options do not only include adjustable vascular resistance, heart rate, and contractility, but also include surgical interventions as the Blalock–Taussig and Sano shunts, staged repair of single ventricle pathophysiology as well as atrial septal flow regulator devices, as well as left/right and biventricular assist devices, venovenous and venoarterial ECMO with various optional adjunct LV unloading/venting interventions and running in pulsatile/non-pulsatile, and ECG-synchronized/non-ECG-synchronized support modes. An example of the user interface of Aplysia CardioVascular Lab is shown in Fig. 26.2.

The model generates a new steady state after parameter modifications within 5–6 heart beats. It is, however, also possible to activate autoregulatory autonomic and vascular reflexes in order to mimic a more "natural" physiological response. To allow better long-term predictions of treatments and interventions, adaptive remodeling algorithms are included both for disease processes as cardiac chamber hypertrophy and dilatation [8] and for vascular wall thickness and diameters.

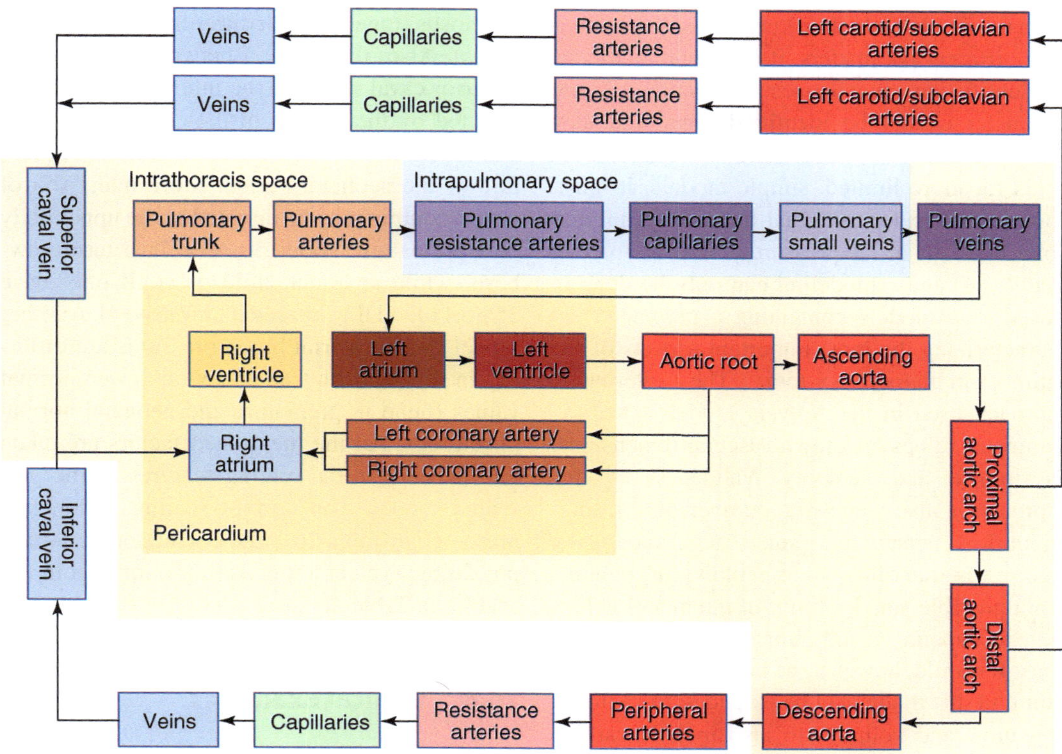

Fig. 26.1 Cardiac and vascular compartments in Aplysia CardioVascular Lab. The cardiac chambers, coronary circulation, and valves are contained in the *dark-yellow* pericardium, which in turn contained the *light-yellow* intrathoracic space. Pulmonary vascular compartments are inside the intrathoracic space and systemic vasculature both inside and outside the chest

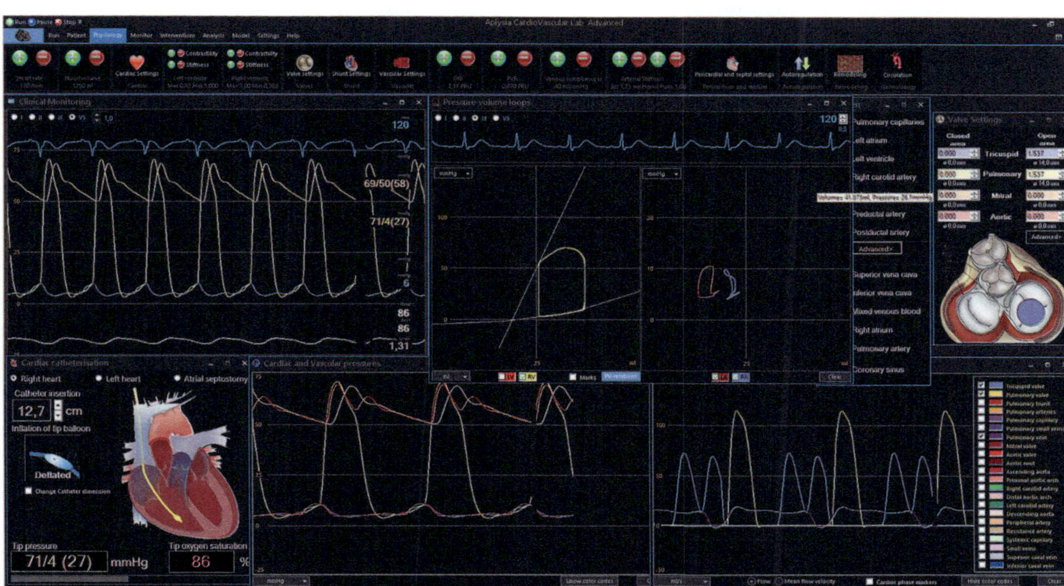

Fig. 26.2 Intuitive user interface of Aplysia CardioVascular Lab showing physiology in single ventricle patient aged 2.5 years with systolic heart failure

In this model, which features more than a hundred different parameters, it is a great challenge to identify the crucial ones. The value of each parameter cannot be identified unambiguously based on clinical measurements in patients as holds for more limited, simple models. It must, however, be understood that meaningful predictions of, e.g., atrial pressures, cardiac loading conditions, and remodeling can only be done in closed-loop models containing a parameter set characterizing the most important aspects of the entire circulatory system. The alternative approach used in the Aplysia model is to base parameter values on known vascular dimensions, physiology, and anatomy. Validation of this approach is also somewhat cumbersome from a scientifical perspective, since, e.g., measuring precise vascular dimensions in individual patients is not possible, and usability of the model will in the end depend on the clinical value of predictions. It should, however, be noticed that the validation in this model can be extended from flows, pressures, and volumes to also include cardiac time intervals, waveforms, and model challenges with changes in blood volume, blood viscosity, and/or vascular resistances in order to further underscore the model's clinical relevance awaiting further validation. Furthermore, when comparing this model with anatomically restricted 3D models or simpler models many "artificial" boundary conditions, such as fixed atrial pressures and rigid vascular walls, can be omitted.

26.3 Scaling

Growth in children interacts with congenital heart diseases in different ways. Growth retardation is often a sign of disease severity. The relation between pulmonary vascular resistance (PVR) and systemic vascular resistance (SVR) changes profoundly over time in the newborn child. The relation PVR/SVR is highest during fetal life and immediately after birth. The rate of decrease in PVR is relevant in the decision making related to staged repair of single ventricle disease and also important in determining shunt flow in atrial septal defects and persistent ductus arteriosus (together with diastolic chamber properties). Also, the relation between flows in the superior caval vein and the inferior caval vein is affected by the growth of the child, where size and blood flow to the head are a dominating feature of the neonatal period. More than 70% of cardiac output may be directed to the upper body in a premature, about 50–60% in a term newborn, while an older child or adult only have 35% of blood flow directed upwards [9]. Another aspect is that arterial blood pressures are different in children and adults, with lower normal values found in premature and neonatal normal physiology. All the mentioned factors are taken into account in the scaling features of the presented model. The current scaling factors are presented in Table 26.1 and a selection of normal physiology values representing output from the scaling in Table 26.2.

26.4 Clinical Examples

26.4.1 Combined Atrial Septal Defect and Patent Ductus Arteriosus in Premature Babies

Premature children often have both an atrial septal defect and a patent ductus arteriosus (PDA). To understand the clinical significance of the PDA in the presence of the atrial shunt can be challenging, which may make it more difficult to take decisions about closure of the PDA. PDA flow is mainly determined by the relation between PVR and SVR, whereas the relative right/left diastolic chamber stiffness is the main determinant of ASD flow. They all have varying degrees of pulmonary hypertension. When aiming to simulate a specific case, all these factors can be adjusted in detail within the model allowing careful evaluation of the upper body and lower body systemic blood flow and oxygenation, respectively. The PDA flow dependency on the ductal diameter and the co-existence of an ASD in a premature baby weighing 1000 grams in gestational week 32 are shown in Fig. 26.3. The ability to simulate a combination of PDA and PFO flows in addition opens possibilities for research stud-

Table 26.1 Selected size scaling factors in cardiovascular model Aplysia CardioVascular Lab

Cardiac parameters	Scaling factor	Comment
Heart rate	$BSA^{-0.3}$	Higher heart rate in small children
Contractility	BSA^{-1}	Higher max elastance in small children
Passive stiffness	BSA^{-1}	Higher min elastance in small children
Myocardial wall volume	BSA	
Valve areas	BSA	
Vascular parameters		
Length	Body length	Small vessels (resistance arteries, capillaries, and capacitance veins) were not scaled
Radius	$BSA^{½}$	
Wall thickness	$BSA^{½}$	
Number of vessels	BSA	Large vessels were not scaled (aorta, caval veins, large arteries)
Young's modulus	$Age^{½}$	Stiffer tissues with age
Typical mean pressure	$BSA^{¼}$	Higher pressures with age
Ratio upper body/lower body vessel numbers	$BSA^{¼}$	Higher proportion of upper body circulation in children
Coronary resistance	$BSA^{0.8}$	
Other		
Blood volume	Body weight	Normal 80 mL/kg
Metabolism	BSA	

Table 26.2 Hemodynamic variables in various ages generated with scaling algorithms in different ages and sizes

	Weight (kg)	Length (cm)	Heart rate (/min)	SAP (mmHg)	LAP (mmHg)	PASP (mmHg)	RAP (mmHg)	CO (L/min)	BV (mL)
Premature w32	2.0	44	152	53	0.3	19	1.0	0.290	160
Newborn	3.6	52	135	70	1.6	22	2.3	0.56	288
1 year	10.5	76	109	97	4.9	26	5.1	1.49	840
4 years	17.0	104	97	95	3.3	22	3.1	2.02	1360
10 years	33.0	141	84	100	3.3	20	2.4	3.10	2640
30 years	85	180	69	117	8.1	27	5.1	5.99	6800

All values are created based on an adult 30-year-old individual, with a length of 180 cm and a weight of 85 kg
Abbreviations: *SAP* systolic arterial pressure, *LAP* left atrial pressure, *PASP* pulmonary arterial systolic pressure, *RAP* right atrial pressure, *CO* cardiac output, *BV* blood volume

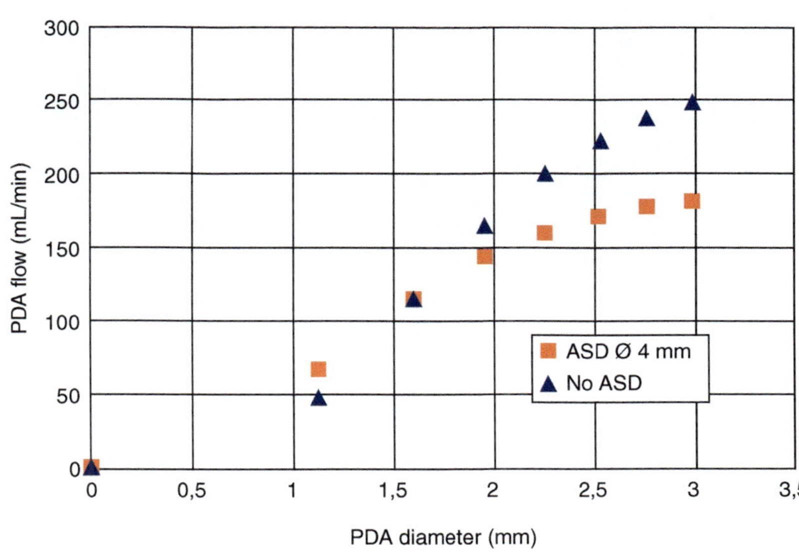

Fig. 26.3 Simulation of combined PDA and ASD in a premature baby 32 GW, weight of 1000 g and length of 32 cm. PDA flow increases with increasing PDA size. PDA flow is lower if an ASD exists

ies that test prediction of the cardiac output with and without atrial or ductal shunts and its use in clinical practice when deciding whether to close the duct.

26.4.2 Tetralogy of Fallot with PDA. Closure of VSD and Dilatation of Pulmonary Valve

Tetralogy of Fallot is one of the most common serious congenital heart malformations consisting of a narrowing of the right ventricular outflow tract (RVOT) and pulmonary valve, a ventricular septal defect in the superior part of the septum, an overriding aorta, and compensatory right ventricular hypertrophy. A full 3D representation of the altered anatomy is beyond the scope of the model, and the overriding aorta and the narrowing of the RVOT cannot be simulated in detail, but assuming a pure valvular stenosis in conjunction with a ventricular septal defect and an open ductal connection in the newborn child suits well to mimic the pathophysiology of the lesion and highlighting the clinical importance of an altered pulmonary vascular resistance (PVR) and consequently the risk of pulmonary over/undercirculation. Figure 26.4 shows the software interface in right heart catheterization of the simulated Fallot patient with a nonrestrictive VSD. A potential use for predictive modeling is to compare effects of different sizes of ductal stents (e.g., 3.0, 3.5, or 4.0 mm) or Blalock–Taussig shunts in patients too small for a definitive surgical correction. Figure 26.5a illustrates the dramatic physiology in Fallot, while Fig. 26.5b, c shows the consequences of partial correction of the syndrome. Finally, in Fig. 26.5d both the VSD, the pulmonary stenosis, and the associated ductus arteriosus are corrected resulting in a normal physiology as evaluated by ventricular pressure–volume loops.

26.4.3 Staged Repair of Single Ventricle Pathophysiology

The disease and the three stages in surgical corrections are presented below in the text. Table 26.3 shows hemodynamic data and Fig. 26.6a–d ventricular pressure–volume loops before and after each stage.

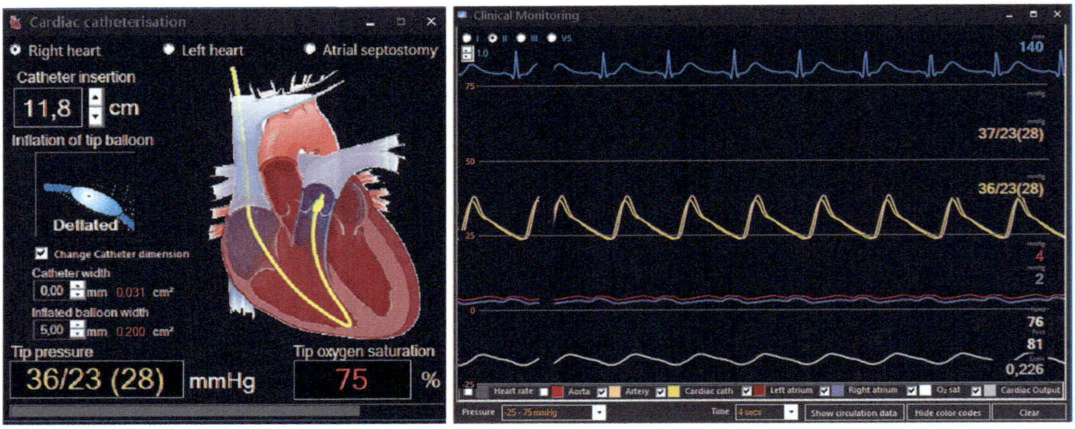

Fig. 26.4 Left panel shows catheterization window and position of right heart catheter in a newborn term child 3.0 kg with a pulmonary stenosis as part of an uncorrected Fallot. The case includes a large unrestrictive VSD and an open duct with moderate pulmonary overcirculation. Right panel shows a clinical monitoring window with ECG, arterial and pulmonary arterial pressures, right and left atrial pressures, and pulse oximetry before treatment. Systemic and pulmonary arterial pressures are almost identical due to a nonrestrictive ductus arteriosus

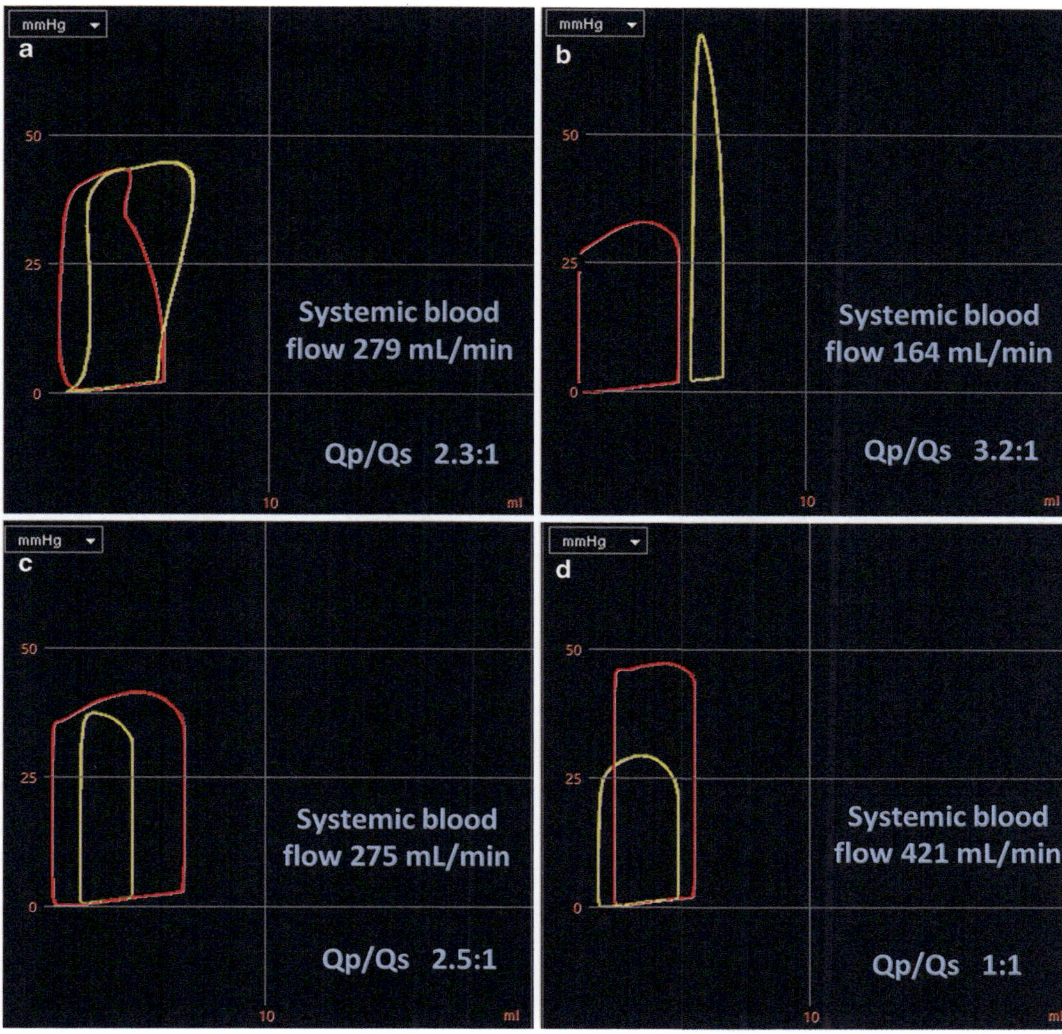

Fig. 26.5 Panel (**a**) shows pressure–volume loops of left (red) and right (yellow) ventricle before correction. Panel (**b**) shows results after only closing the VSD. Panel (**c**) shows results after closure of the VSD and dilatation of the pulmonary valve. Finally, panel (**d**) shows results after finally closing the PDA. The duct is open in simulated scenarios (**a–c**) in this patient. Systemic blood flow and Q_p/Q_s are indicated in *light blue*

Table 26.3 Hemodynamic data from simulation of single ventricle physiology before and after staged repair

	Weight (kg)	Length (cm)	BSA (m²)	Heart rate (/min)	MAP (mmHg)	Cardiac output (L/min)	Blood volume (mL)	RV EF fraction (%)	RAP (mmHg)	SaO_2 (%)
HLHS	3.6	52	0.228	135	41	0.248	342	46	6.7	71
Norwood + BT	3.6	52	0.228	135	49	0.251	342	42	6.6	70
Glenn	8	68	0.389	115	70	0.812	760	29	9.5	69
Fontan	13	88	0.564	103	74	1018	1365	26	10.4	93

Norwood repair done after a few days, Glenn at 6 months, Fontan at 2 years
Abbreviations: *BSA* body surface area, *MAP* mean arterial pressure, *RV EF* right ventricular ejection fraction, *RAP* right atrial pressure, SaO_2 arterial oxygen saturation, *HLHS* hypoplastic left heart syndrome, *BT* Blalock–Taussig shunt

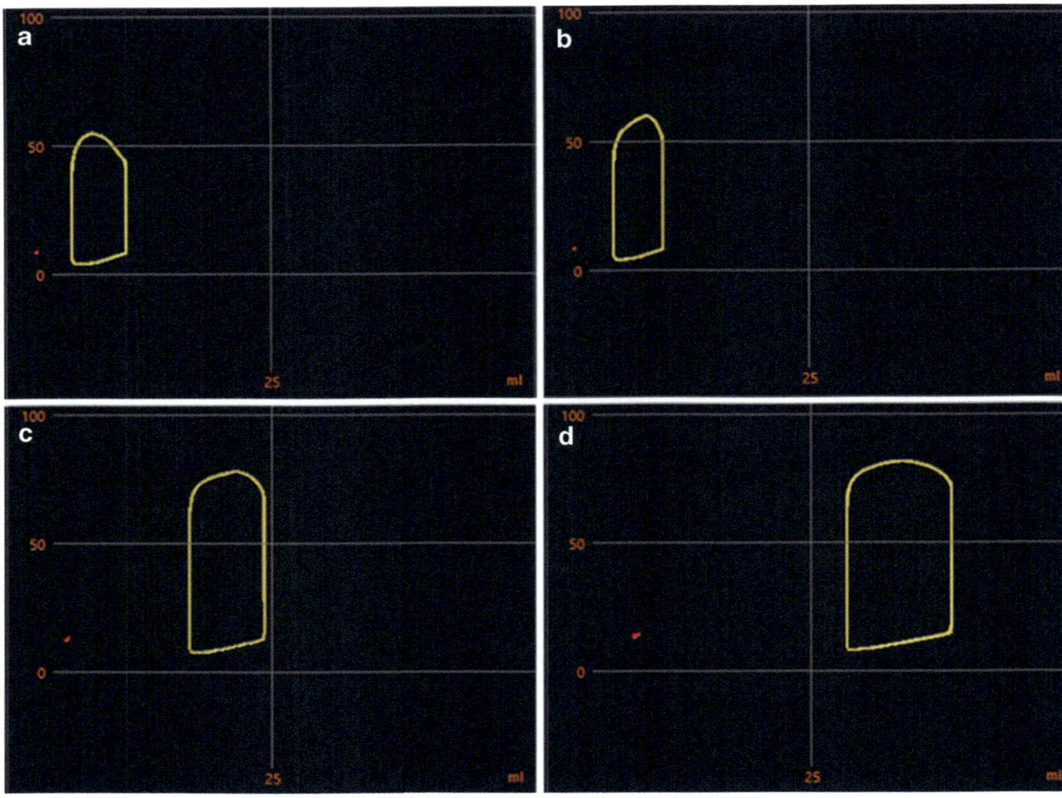

Fig. 26.6 Pressure–volume loops of left (red dot) and right (yellow) ventricle in single ventricle physiology due to hypoplastic left heart syndrome. (**a**) Before correction. (**b**) After Norwood correction with Blalock–Taussig shunt and lower PVR. (**c**) After Glenn correction and further lowering of PVR. (**d**) After Fontan procedure. The child is growing between procedures. Figures (**a**) and (**b**) are both neonatal (3.6 kg/52 cm), whereas figure (**c**) is at 6 months (68 cm/8 kg) and figure (**d**) at 2 years (88 cm/13 kg). A progressive reduction in ejection fraction is seen in this child

26.5 Single Ventricle Pathophysiology

The most common and important clinical presentation is the hypoplastic left heart syndrome (HLHS), where the left ventricle is underdeveloped as an adaptation to lack of flow and/or filling due to defective or absent aortic or mitral valves. The child's ability to stay alive in the immediate postnatal hours relies on the ability of the right ventricle to pump blood not only through the pulmonary circulation but also through the open duct into the systemic circulation. Blood returned to the left atrium will have to pass over to the right atrium through a usually nonrestrictive atrial septal defect. This altered physiology can be created with the model by eliminating left ventricular contractions and creating a left ventricle with a very small and stiff cavity combined with an open duct and a large atrial septal defect.

26.6 Norwood Repair

Norwood repair assures flow from the ventricle into a reconstructed ascending aorta and aortic arch, while the duct is closed. This stage of palliation usually happens within days of birth to avoid the dramatic deterioration expected when the duct closes spontaneously. Reconstruction of the neoaorta is a surgical challenge including the optimization of the anatomical shape of the vessel. In the model, these 3D aspects of physiology cannot be represented, but flows, volumes, pressures, and saturations are still in a realistic range before and after surgery. The pulmonary circulation will be supplied by either a classic Blalock–Taussig (BT) shunt or a Sano shunt from the right ventricle to the pulmonary arteries. These options are available in the model as the last part of the stage I repair (Norwood), when the right ventricle is connected to the aortic root (neoaorta) and blood supply to the pulmonary circulation is urgently needed to allow oxygenation of blood returning from the systemic circulation.

26.7 Glenn Repair

The prerequisites for the Glenn correction, when the superior caval vein is connected to the pulmonary artery and the BT/Sano shunt is closed, are twofold. Firstly, the pulmonary vascular resistance must be sufficiently low to permit blood flow without a pumping chamber. Secondly, the anatomy of the child should provide enough blood flow drained from the upper body to permit oxygenation of the entire body. These prerequisites can be analyzed in the model in all details, and postsurgical Q_p/Q_s and oxygenation can thus be predicted.

26.8 Fontan Repair

With further decline in pulmonary vascular resistance and an increasing proportion of blood flow directed toward the lower part of the body, the next step in the surgical process should be taken. Pulmonary vascular resistance is critical as are systolic and diastolic properties of the ventricle. The clinical course of the patient is sometimes complicated by a regurgitant atrioventricular valve, either due to ventricular dilatation or due to a structurally abnormal valve.

26.8.1 Left Ventricular Unloading during Venoarterial ECMO in Child with Myocarditis

Viral myocarditis in small children may be life-threatening within days. The disease usually affects the left ventricle more severely than the right. Forward failure, pulmonary congestion, and thromboembolic complications are the main clinically encountered problems. Short-term treatment of cardiogenic shock may include VA ECMO or LVAD implantation. Decision making depends on the expected time course of the disease and degree of potential recovery. In our simulated example shown in Fig. 26.7, the lack of unloading during VA ECMO is evident and more outspoken with increasing ECMO flow. The overloaded dilated ventricle imposes a risk of thrombus formation and makes recovery less likely. Many management options can be considered in this scenario, such as interventional/surgical venting and conversion to VAD implantation. In this particular case, an atrial septostomy was tested by simulation showing a successful volume unloading effect, although as a downside smaller stroke volumes and a continued or increased risk of flow stagnation and thrombus formation.

26.9 Vision

The authors believe that simulation will have a considerable impact on clinical decision making in pediatric cardiology and cardiac surgery already within the next few years. A gradual transition from using models in teaching seminars based on generic patients to individual decision support will occur. It is already evident that our current model can improve understanding of optimal cannulation strategies in venovenous [3] and venoarterial ECMO [2, 10–12]. In general, it is not strictly needed to generate a perfectly matching patient-specific simulation in complex clinical situations to draw firm conclusions about the most optimal treatment, and generic simulations may suffice. The most decisive impact of simulation will probably not be in well-defined and well-validated patient groups, but rather when exploring options in more unusual scenarios in complex disease. This prediction may seem like a lack of scientifical ambitions, but simulation is in these situations a rational complement to current beliefs and judgment based on anecdotal experience. Incrementally, models will be validated and further refined. The actual clinical

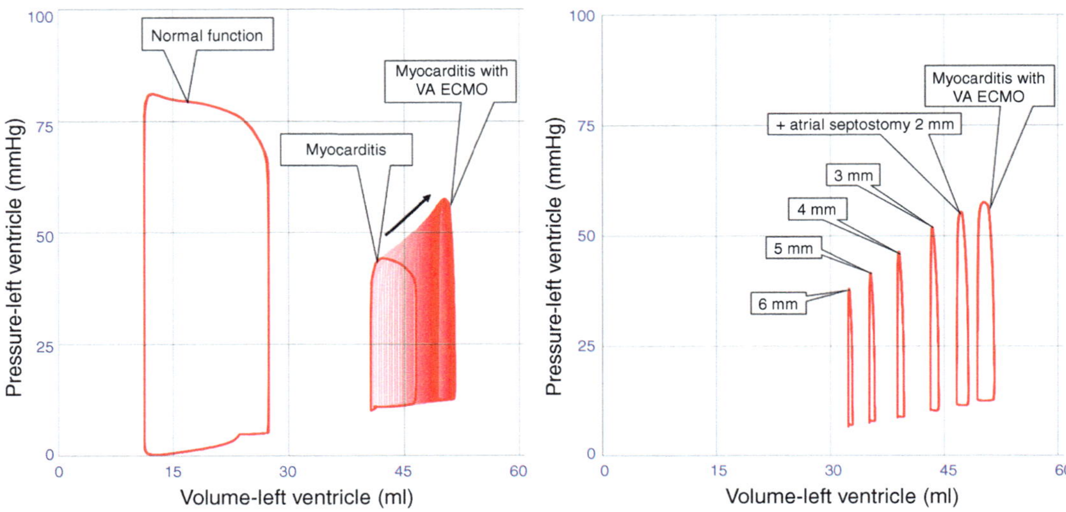

Fig. 26.7 Left ventricular pressure–volume loops in a one-year-old child with severe myocarditis affecting mainly the left ventricle. Left panel shows loop before disease (normal function) and disease with left heart failure before VA ECMO (myocarditis), and *black arrow* indicates increasing VA ECMO flow from 0 to 1.0 L/min with progressive dilatation of the ventricle. Right panel shows the effect of atrial septostomies with various diameters. An atrial septostomy with a diameter of 6 mm unloads the left ventricle well, but the stroke volume is minimal indicating a barely opening aortic valve with risk of flow stagnation and thrombus formation

development, improvement, and implementation will be a result of a stepwise approach to solve well-defined individualized complexities. It is therefore likely that sharing experiences and simulations in case presentations/discussions in scientific conferences and online forums will be important options to spread knowledge and understanding in parallel with traditional scientific publishing.

26.10 Limitations

Fundamentally, it is challenging to validate a complex model over a large range of conditions ranging from new term babies until old age, and with a very broad range of possible disease conditions. This challenge is the main limitation of the model. The large number of parameters cannot be determined in an unequivocally way with available clinical data. Since the majority of parameters are related to anatomy, age and dimensions of the patient best-educated guesses can often be based on generic patients scaled to the size of the patient with help of the software interface. Also, the manifold alternatives for displaying model output makes are possible to evaluate the validity of simulations in relation to previous experience of the doctor. Validation can only take place in selected groups of patients, while the real clinical relevance of simulation will likely be established in complex and rare conditions in individual patients, where simulation results are extrapolated from validated studies. This way of using simulation requires a clinician well acquainted with the model and with sufficient clinical experience and knowledge to allow comparison of model predictions with existing knowledge in a critical fashion. Simulation will never replace experience but provide the clinician additional input in order to further improve understanding and decision making.

26.11 Conclusion

The Aplysia CardioVascular Lab model illustrates the large range of realistic characteristics and clinical possibilities with the use of cardiovascular simulation in congenital heart disease. The closed-loop real-time simulation features render bedside application of the model a realistic tool both to analyze case-based pathophysiology and to explore various treatment options such as mechanical circulatory support as well as valve and shunt interventions. Continued model validation is the greatest challenge we are currently facing in order to implement and reach acceptance of the model in the clinics.

References

1. Frieberg P, Sjoberg P, Revstedt J, Heiberg E, Liuba P, Carlsson M. Simulation of aortopulmonary collateral flow in Fontan patients for use in prediction of interventional outcomes. Clin Physiol Funct Imaging. 2018;38(4):622–9.
2. Broome M, Donker DW. Individualized real-time clinical decision support to monitor cardiac loading during venoarterial ECMO. J Transl Med. 2016;14(1):4.
3. Broman M, Frenckner B, Bjallmark A, Broome M. Recirculation during veno-venous extra-corporeal membrane oxygenation - a simulation study. Int J Artif Organs. 2015;38(1):23–30.
4. Broome M, Frenckner B. Recirculation during veno-venous extra-corporeal membrane oxygenation - a simulation study. Stockholm: EuroELSO; 2013.
5. Broome M, Maksuti E, Bjallmark A, Frenckner B, Janerot-Sjoberg B. Closed-loop real-time simulation model of hemodynamics and oxygen transport in the cardiovascular system. Biomed Eng Online. 2013;12:69.
6. Mynard JP, Davidson MR, Penny DJ, Smolich JJ. A simple versatile model of valve dynamics for use in lumped parameter and one-dimensional cardiovascular models. Int J Numerical Methods Biomed Eng. 2011;28(6–7):626–41.
7. Sun Y, Sjöberg BJ, Ask P, Loyd D, Wranne B. Mathematical model that characterizes transmitral and pulmonary venous flow velocity patterns. Am J Phys. 1995;268(1 Pt 2):H476–H89.
8. Maksuti E, Westerhof B, Ugander M, Donker D, Carlsson M, Broome M. Cardiac remodeling in aortic and mitral valve disease – a simulation study with clinical validation. J Appl Phys. 2019;126:1377–89.
9. Salim MA, DiSessa TG, Arheart KL, Alpert BS. Contribution of superior vena caval flow to total cardiac output in children. A Doppler echocardiographic study. Circulation. 1995;92(7):1860–5.
10. Donker D, Sallisalmi M, Broome M. Right-left ventricular interaction in left-sided heart failure with and without veno-arterial extracorporeal membrane oxygenation support - A simulation study. ASAIO J. 2021;67(3):297–305.
11. Donker DW, Brodie D, Henriques JPS, Broome M. Left ventricular unloading during veno-arterial ECMO - a simulation study. ASAIO J. 2018;65(1):11–20.
12. Lindfors M, Frenckner B, Sartipy U, Bjallmark A, Broome M. Venous cannula positioning in arterial deoxygenation during veno-arterial extracorporeal membrane oxygenation-a simulation study and case report. Artif Organs. 2016;41(1):75–81.

27 Artificial Intelligence in Pediatric Cardiology

Addison Gearhart and Anthony Chang

27.1 Introduction

Historically, medical and surgical management for congenital heart disease (CHD) was comprised of evidence-based interventions determined by the predominant lesion and expected physiology. This "one-size-fits-all" approach was fraught with obvious limitations. The numerous exceptions to management plans for CHD highlighted the need for more sophisticated modeling strategies to accurately capture, understand, and treat patient-specific subtleties in disease progression. This opened the way for artificial intelligence (AI) to offer an innovative solution for complex CHD management. This chapter reviews the history of AI, provides a primer on the aspects of AI and machine learning in pediatric cardiology, discusses current applications of these methods, and explores how pediatric cardiology can incorporate AI in the future.

A. Gearhart
Department of Pediatrics, Boston Children's Hospital, Harvard Medical School, Boston, MA, USA

A. Chang (✉)
Department of Pediatrics, The Sharon Disney Lund Medical Intelligence and Innovation Institute (MI3), Children's Hospital of Orange County, Orange, CA, USA

27.2 Historical Significance of Artificial Intelligence in Medicine

In the 1940s, the British mathematician Alan Turing designed "bombes," electro-mechanical precursors of modern computers, that successfully decoded the German Enigma machine [1]. In his 1950 publication, "Computing Machinery and Intelligence," he described the famous Turing test to determine if a computer could achieve human cognitive capability [2]. The term AI was officially coined at the Dartmouth Conference in 1956 by Stanford computer scientist John McCarthy [3]. AI spread to the medical field in the 1960s with the Dendral experiments, which analyzed mass spectra data and the existing knowledge of chemistry to identify possible chemical structures [4]. In the 1970s, Stanford and Rutgers created SUMEX-AIM Computing Resource to enable researchers to use AI methods to solve issues in medicine internationally [5]. Shortly after, AI suffered a few "AI winters" (first 1974–1980 and then again from 1987 to 1993) when the hype of AI failed to meet expectations. These winters diminished interest and productivity in AI, resulting in skepticism from the medical community [6].

Over the past few decades, advances in data, computing, and algorithmic development resurrected AI from a state of dormancy to an integral technology embedded in almost every aspect of

G. Butera et al. (eds.), *Modelling Congenital Heart Disease*,
https://doi.org/10.1007/978-3-030-88892-3_27

our lives. The highly publicized 1997 IBM Deep Blue's defeat of the Russian world chess champion Boris Kasparov [7] followed in 2011 by IBM Watson's defeat of human contestants in Jeopardy [8]! marked new potential for precision AI in health care.

27.3 Artificial Intelligence Definition

AI implies the use of a computer to algorithmically model human intelligence without human intervention. In medicine, AI enlists computer-based techniques to algorithmically discover complicated patterns in large datasets, and make predictions to answer clinical questions (Fig. 27.1).

27.4 Components of Artificial Intelligence

The end goal of AI in health care is to offer more sophisticated systems to approach disparate sources of information, and offer personalized, intelligence-based diagnostic, prognostic, and therapeutic support to clinicians. The development and strength of algorithms require large pools of data. The main arm of AI is machine learning, further subdivided into deep learning

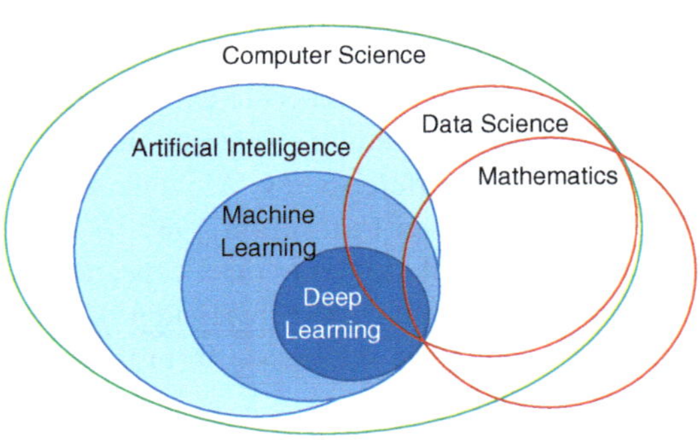

Fig. 27.1 A diagram to orient the reader. AI algorithms are built from big data, and incorporate data mining and data science. The main branch of AI is machine learning and its subtype deep learning. From Chang, AC. Artificial Intelligence in Medicine: Principles and Applications. *Elsevier;* 2020, with permission

(Fig. 27.2). Algorithmic development occurs by a process nicely outlined in Fig. 27.3. The pipeline includes data collection, cleaning and processing the data, feature extraction (selecting imaging data points of interest, medical record information, etc.), task selection of the appropriate computational approach and tools (supervised vs. unsupervised learning), validation (applying the model to unseen data sets outside the training set to test its accuracy), and model deployment (inputting new data previously unseen to the algorithm) to generate automated diagnosis or predictions.

27.5 The Interface of Artificial Intelligence with Big Data

Patients produce 750 quadrillion bytes of data daily, or 30% of the world's **data** production [9]. This surplus of data is slated to worsen with medical information doubling every few months [10]. These data overload, coined "the healthcare data conundrum," is a consequence of an explosion in the volume and complexity of medical, scientific, and patient-generated data [11]. "Big data" refer to infinitely large data sets whose scale, diversity, and complexity require more advanced technology than standard processing and storage methods. The six "V"s of big data include **v**alue (coherent analysis), **v**olume (data size), **v**ariability (data consistency over time), **v**ariety (videos, wearable technology, tweets, etc.), **v**elocity (data accessibility), and **v**eracity (data biases, noise, and uncertainty) [12]. Information collected from mobile phones, wearable technology, social media, environmental and lifestyle-related factors, demographics, "omic" data (genomics, metabolomics, proteomics), and electronic health records (EHRs) all contribute to big data. Physicians cannot cognitively keep up with the current pace of data production, and are inefficiently acting on a small fraction of available information (Fig. 27.4) [13]. AI allows physicians to interpret, organize, and translate data into meaningful, accessible, and clinically actionable information.

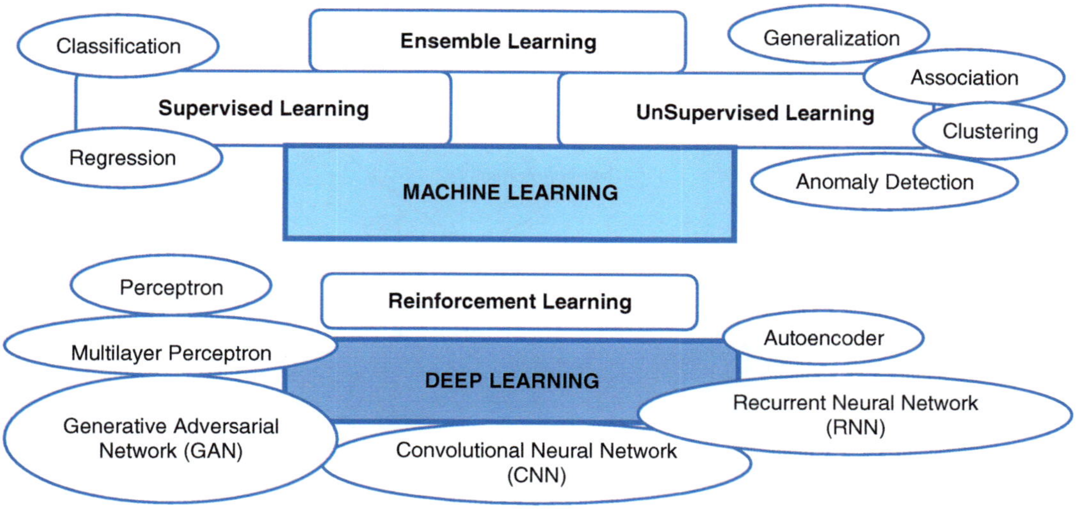

Fig. 27.2 Schematic emphasizing the branches of machine learning and the branches of the main machine learning method, deep learning. Reinforcement learning is considered a hybrid of the two. From Chang, AC. Artificial Intelligence in Medicine: Principles and Applications. *Elsevier;* 2020, with permission

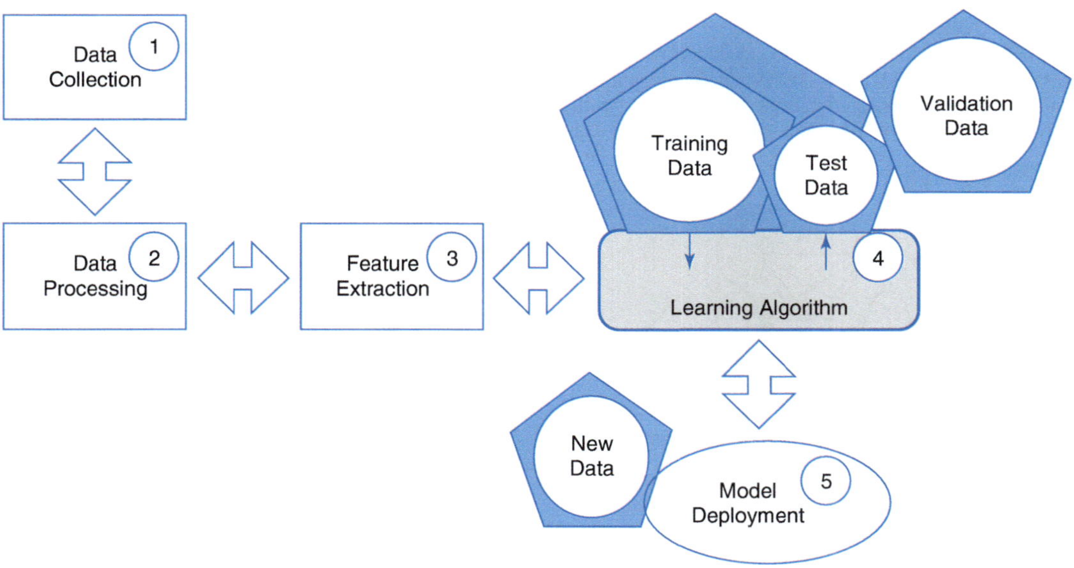

Fig. 27.3 The ML workflow. 1. Data are collected 2. Data are processed. 3. Features are selected 4. Compiled data are used for training data and test data for model development. Algorithms are then validated with separate data from initial cohort data. 5. The model is then exposed to new data. From Chang, AC. Artificial Intelligence in Medicine: Principles and Applications. *Elsevier;* 2020, with permission

27.6 Machine Learning

Machine learning (ML) describes a subfield of AI algorithms capable of improving task performance with experience through learning and adapting to data patterns. ML algorithms can either be trained alone or through a combination of **unsupervised, supervised, and reinforcement learning** techniques (Fig. 27.2). **Deep learning** (**DL**) is the main subtype of ML utilized

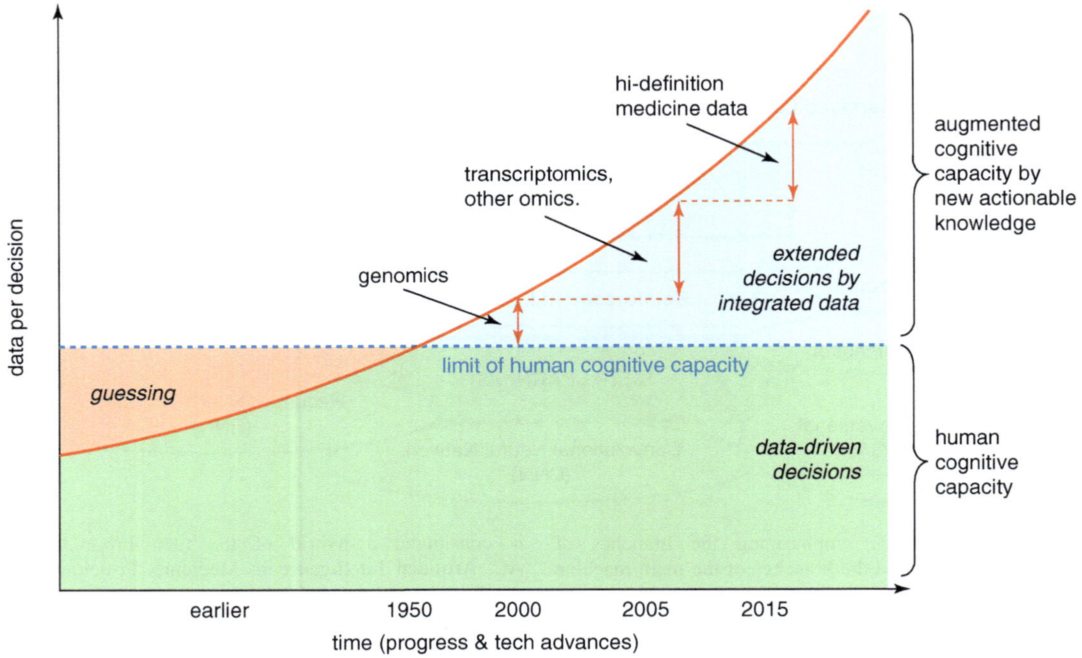

Fig. 27.4 Transformation of the decision-making process. The accelerated rate of data production, volume, and variety now supersedes the limits of human cognitive capacity. From Rossi, RL and Grifantini, RN. Big Data: Challenge and Opportunity for Translational and Industrial Research in Healthcare. *Front. Digit. Humanit*; 2020, with permission

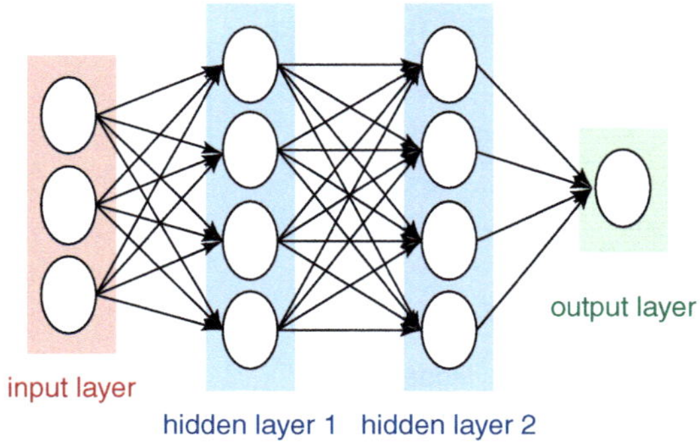

Fig. 27.5 An example of a three-layer neural network with three inputs, two hidden layers of four neurons each, and one output layer. Notice in both cases there are connections (synapses) between neurons across layers, but not within a layer. From Musiol, M. Speeding up Deep Learning. Computational Aspects of Machine Learning. ResearchGate: 2016, with permission

in medicine. **Artificial neural networks (ANNs)** are the main machine learning algorithm currently used (Fig. 27.5). They are inspired by the neuronal and synaptic interconnections in the brain. These networks comprise layers of connections with forward and backward data propagation that perform tasks by considering examples. The two overarching goals of machine learning are automating tasks that might otherwise be performed by humans, and generating clinically important knowledge through harnessing big data. Successful ML techniques used in adult heart failure patients led to improved systems for diagnosis and classification [14, 15], prediction of readmission [16, 17],mortality [17], medication adherence [18, 19], and sudden cardiac arrest [20].

27.7 Unsupervised Learning

Unsupervised learning models capture the natural structure present within unlabeled data points and uncover underlying relationships among these input variables without forcing data to fit an outcome. Classic methodologies include k-means clustering and principal component analysis (Fig. 27.2). Exploratory analysis enables automatic structure identification for messy data sets commonplace in medicine. Future applications include cluster techniques for the surveillance of clinical variables found in the EHR to classify new disease subtypes of CHD.

27.8 Supervised Learning

Supervised learning models train off predictor variables and labeled outcomes. The end goal is the classification of an observation into one more outcomes and to apply the learned outcome to new data to predict future events. Classic methodologies include support vector machines, ANNs, naive Bayesian classifiers, and hidden Markov models. A recent example of supervised machine learning in the adult population by Dawes et al. (2017) includes the integration of clinical and imaging variables (parameters from three-dimensional right ventricular motion on cardiac MRI, exercise capacity, and right heart catheterizations) and the desired outcome (prediction of outcomes in pulmonary hypertension) to discover new regions of RV function on imaging strongly associated with increased survival compared with conventional parameters [21]. Furthermore, Bayesian networks have shown to be powerful prognostic tools in decision making during adult cardiac surgery [22]. Proposed applications in pediatric cardiology similarly include Bayesian techniques to exploit nonlinear relations between several variables to aid in the decision-making processes and compare outcomes for cardiac interventions, such as the Norwood vs. Sano procedure for hypoplastic left heart syndrome.

27.9 Reinforcement Learning

Reinforcement learning attempts to optimize behaviors to achieve maximum rewards with trial and error learning from input data and an outcome. The popularized media example is the Google-trained AlphaGo [23] that beat the human world champion at the Chinese board game Go. In health care, systems can be queried at regular intervals to optimize the decision-making process as clinical changes arise, improving future outcomes. A deep reinforcement learning algorithm utilizing the public MIMIC II intensive care unit (ICU) database learned optimal adult heparin dosing and effectively proposed heparin doses with better expected outcomes than the clinical guidelines through studying sample dosing trials and their associated outcomes [24]. Future applications in pediatric cardiology include determining the optimal dosing of pressors or inotropes to maximize benefits and reduce side effects for postoperative hypotension in the PICU.

27.10 Deep Learning

Deep learning (DL) is the next iteration of ML, performing better than traditional ML by virtue of its ability to learn meaningful abstractions. Major contributions include innovation in image, speech, and text processing. DL attempts to mimic the activity of the layers of neurons in the human neocortex by using ANNs with many stacked processing layers and hidden layers alternating between linear and nonlinear transformations. The two main methodologies of ANNs are **recurrent neural networks** and **convolutional neural networks**.

27.11 Recurrent Neural Networks

Recurrent neural networks (RNNs) house an additional hidden state vector that contains "memory" regarding previously seen data. RNNs cater to tasks heavy in sequential data, such as language recognition and production, and were popularized in voice recognition software. Applications in pediatric cardiology include a RNN applied to ICU data to generate dynamic in-house mortality risk scores with higher accuracy than PRISM 2 or PRISM 3 systems [25]. In adult heart failure, RNNs outperform standard logistical regression for predicting the onset of heart failure using EHR data [26].

27.12 Convolutional Neural Networks

Convolutional neural networks are neural networks with one or more "convolutional layers" followed by one or more fully connected layers or hidden layers of networks (Fig. 27.6). This combination of neural networks has unique applications in discovering and merging local image input features with increasing levels of abstraction for prediction of clinical outcomes. CNNs are at the heart of facial recognition software. A string of electrophysiology studies utilizes DL both for the automated identification of shockable ECG signals [27] and for diagnoses of arrhythmias from ECGs [28]. In addition, a CNN model for an automated tissue classification method for Kawasaki disease achieved a 96% accuracy for classification of coronary arteries on optical coherence tomography (OCT) [29]. Imaging applications of CNNs are robust in pediatric cardiology and therefore will be discussed separately below.

27.13 Imaging with Machine Learning and Deep Learning

Unprecedented amounts of imaging data enable the development of innovative ML algorithms; these algorithms can streamline various stages of the imaging diagnosis workflow and provide automated measurements. Techniques applied to adult cardiac MRIs now automate chamber segmentation [30, 31], view planning for cardiac MRI acquisition [32], and cine MRI analysis [33]. Echocardiogram applications include automatic assessment of left ventricular volume and function [20, 34], classification of coronary artery disease [35], mitral valve regurgitation [36], and assessment of left ventricular ejection fraction and longitudinal strain within 8 s [37]. Automation of accurate, reproducible, and consistent interpretations reduces the risk of human error and decreases time constraints. Faster image acquisition is ideally suited for fussy or critically ill pediatric patients. New automated DL techniques reconstruct images from cardiac MRIs to achieve quality similar to gold standard breath-hold cine techniques for children unable to hold their breath [38].

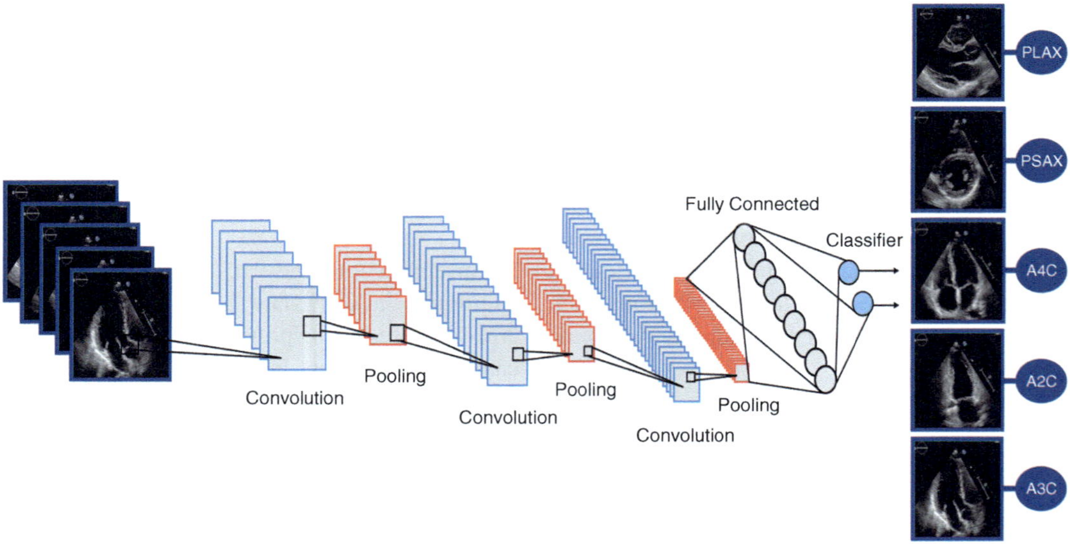

Fig. 27.6 An example of a convolutional neural network model for image classification. Images are acquired and processed, features are extracted, and the CNN model learns to classify these features. Abbreviations: A2C, apical two chamber; A3C, apical three chamber; A4C, apical four chamber; PLAX, parasternal long axis; PSAX, parasternal short axis. From Alsharqi et al. Echo Res Pract. 2018;5(4):R115–R125), with permission

ML also offers diagnostic support with an uncanny ability to recognize subtleties easily missed by the human eye. Diagnostic advancements in echocardiography include algorithms capable of differentiating constrictive from restrictive pericarditis [39] or hypertrophic cardiomyopathy from physiologic cardiomyopathy seen in athletes [40]. The trend in cross-institutional expansion of imaging databases welcomes future opportunities to apply powerful DL methods. Future strategies include DL to detect subtleties to predict the likelihood and time frame for clinical deterioration, allowing more sophisticated treatment strategies. This algorithm could reduce the frequency of surveillance echocardiograms, thereby reducing healthcare expenditure on unnecessary tests.

The other arm of AI imaging and computer vision focuses on generating new clinical insights, through interpreting large reservoirs of imaging data, in the context of clinical data. A recent ML pipeline, using demographics, electrocardiography, and cardiac MRI patient data, successfully uncovered baseline imaging and clinical variables lost to regression analysis for the prediction of deterioration in patients with repaired tetralogy of Fallot [41]. This proposed structure offers clinical decision support to alert the clinician to new important markers of decline that may warrant earlier intervention.

27.14 Virtual Reality

Virtual reality (VR) technology and proficiency-based training methods are advancing the training archetype for surgical and procedural medicine. Such technology translates 2D image sources (e.g., CT, CTA, MRI, MRA, and MRV) into interactive 3D structures, enabling physicians to manipulate unparalleled views of patient-specific organs and tissues (Fig. 27.7). With these new views, physicians can identify and evaluate anatomical and physiologic differences between patients for improved patient selection, presurgical planning, transcatheter structural heart procedural planning, and device placement [42, 43]. Rantner et al. (2013) successfully used VR in a patient with tricuspid atresia by building a patient-specific heart–torso model from MRIs, and conducting simulations on the model to determine optimal ICD placement [44]. At another center in 78 cases of CHD disease, 3D image fusion with live fluoroscopy facilitated improved catheter manipulations and interventions, allowed preselection of ideal projection angles, and reduced fluoroscopic time [45]. Proponents of VR predict these systems will become ubiquitous in-hospital training systems to increase physician proficiency. Benefits of VR training systems include 24/7 user access to procedure rehearsal and objective measurements of procedure competence.

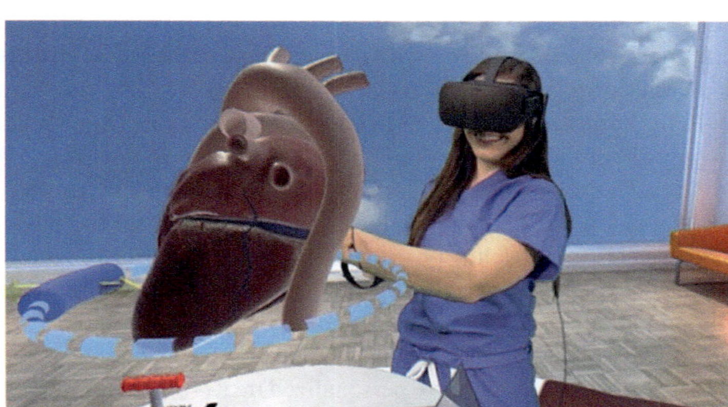

Fig. 27.7 A graphic of the Stanford Virtual Heart to deploy an immersive virtual reality (VR) technology to explain complex congenital heart disease to trainees and families. Graciously provided with permission by Dr. David Axelrod

27.15 Three-Dimensional (3D) Heart

A number of cardiovascular centers own computer-generated, anatomically accurate, three-dimensional models (3D) printing laboratories to facilitate printing of complex anatomy for surgical planning, device sizing, physician training, and patient education [46–50]. The combination of virtual device implantation and 3D printing offers innovative advances for surgical planning and device placement as seen in the repair of a patient with tetralogy of Fallot with pulmonary atresia and pulmonary collaterals [51]. These tools led to successful placement of the SynCardia total artificial heart in at least 32 patients in six countries [49, 52]. Planning interventions with patient-specific materials provides the safest, least invasive, and most efficient operational approach.

27.16 Drug Discovery

AI cloud-based pharmaceutical platforms are designed to process massive amounts of constantly updating medical information (e.g., research papers, patents, clinical trials, patient records) to explore known relationships between biological entities (e.g., genes, symptoms, diseases, proteins, tissues, candidate drugs) to accelerate drug discovery [53]. A Stanford scientist recently announced a DL algorithm, which can predict the properties of potential drug compounds using limited data points, to determine the safest, most effective drug screen for toxicity, side effects, and instability [54]. The acceleration of pharmacogenetic discovery with an abundance of EHR information offers exploration of new therapy targets and dosing.

The CHD population similarly benefits from AI adaptation in the drug discovery pipeline. New avenues for modeling pharmacokinetics and pharmacodynamics in children may mitigate prior limitations of small patient numbers with rare diseases. Opportunities to use genomics to prioritize drug targets promise personalized treatment plans, models of drug–drug interactions, side effect prediction prior to human use, and repurposing existing medications. These may obviate the off-label use of adult drugs for pediatrics. For example, fuzzy logic could be used to smooth out the dosing of vasoconstrictor or vasodilator agents in the CICU.

27.17 Clinical Decision Support Systems

Clinical decision support systems provide clinicians with tools that filter large quantities of clinical knowledge with updated patient-specific information to augment clinical decision making [55, 56]. Support systems such as data mining frameworks are able to discover, extract, and apply relevant evidence from prior patients for future care of similar patients. Currently, the majority of advancements in pediatric cardiology involve clinical scoring systems. These use AI-based neural networks and decision trees to stratify patient-specific risk by adapting to changing clinical variables in real time and offering updated recommendations reflective of the patient's status. The FDA-cleared **Etiometry Platform T3 Data Aggregation and Visualization** [57] collects physiologic data (e.g., laboratory values, monitor streams) from the patient's medical record number (MRN) and assesses the patient's risk of ineffective delivery of oxygen in the ICU. Awareness of changes in clinical status theoretically should prompt physicians to change management strategies sooner, impacting the course of clinical care.

27.18 Precision Medicine

Precision medicine entails a comprehensive approach to disease treatment and prevention encompassing each individual's genetics, anatomy, physiology, and environment to promote a patient-specific treatment strategy (Fig. 27.8). Modern pediatric cardiologists face an escalating volume of patient-specific data, including demographics, surgical and interventional history, laboratory data, remote monitoring, and imaging. This complex field, spanning

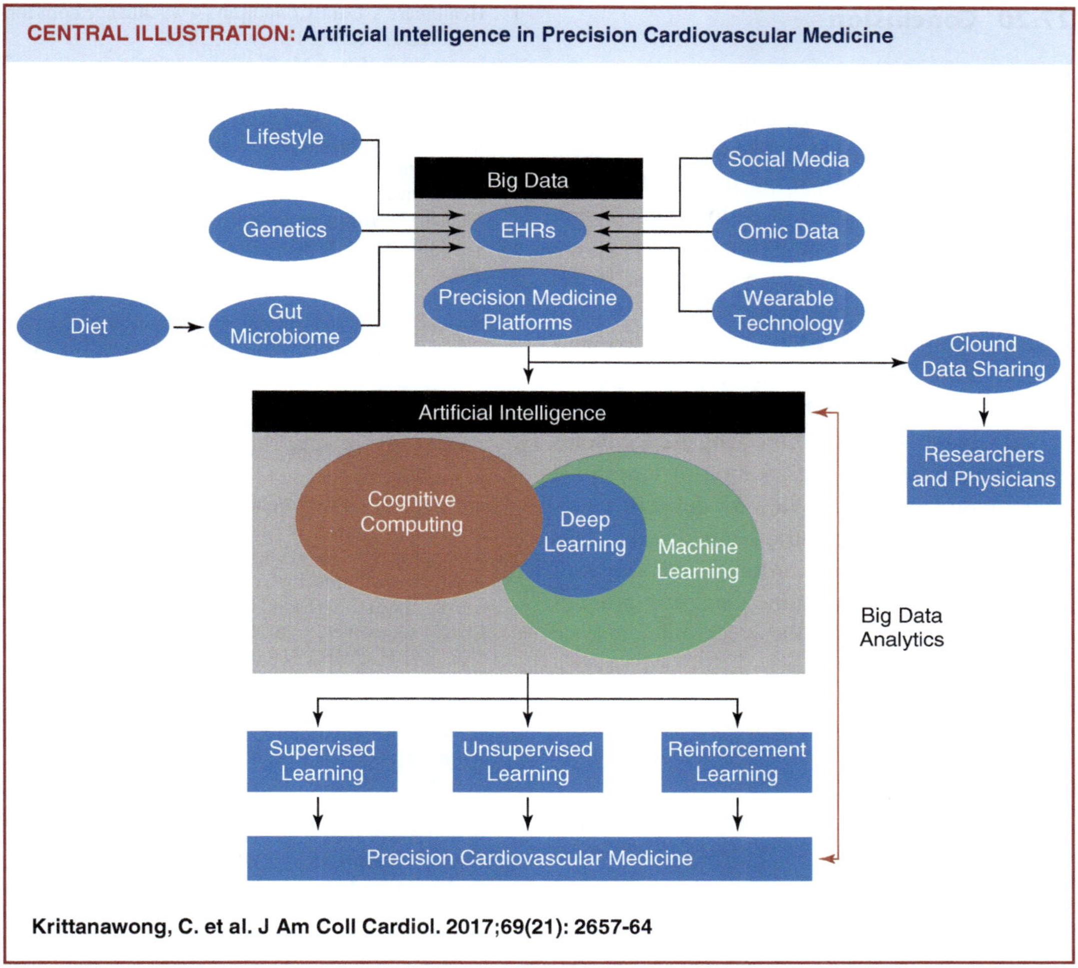

Fig. 27.8 Image showing the interplay between big data, artificial intelligence, and precision cardiovascular medicine. Krittanawong, C. et al. Precision Cardiovascular Medicine. *J AM Coll Cardiolo*: 20167;69(21); 2657–64. Obtained with permission

from fetal to adult medicine, is well suited for multidimensional AI involvement. AI has the opportunity to develop more precise classifications of CHD phenotypes, allowing narrowed predictive analytics for novel interventions and improved treatment plans.

27.19 Limitations

The diversity inherent to medical data poses intrinsic messiness, missing data, and heterogeneity. These obstacles create appropriate skepticism toward the ability for AI to generate meaningful inferences that benefit patient care. Furthermore, the features selected for analysis may bias models by inclusion of irrelevant information. Additional errors occur with enriching data sets to overcome shortage of data quantity. For example, breaking data into smaller pieces of information may provide ML with more usable data points, but the final data set may no longer represent the original. This ongoing debate on the price of pragmatism refers to determining the acceptable compromise in quantity versus quality of data supplied by the EHR.

27.20 Conclusion

The crux of AI and ML in health care is to provide a set of tools to augment the physician's outreach, efficiency, and value. AI and ML provide a library of clinical knowledge derived from the interpretation of volumes of endogenous patient data paired with exogenous sources of big data for the physician to optimize outcomes. It has earned an important role in the transition from a "one-size-fits-all" approach to CHD to patient-specific tailored therapies that acknowledge that subtle variations in CHD lesions require individualized treatment plans. Emerging AI technologies promise to drive the creation of innovative model-based approaches vital for delivering precision medicine to pediatric heart disease. This shift will redefine the cardiologist's scope of practice, while improving the long-term prognosis for patients with these chronic disease processes.

References

1. Copeland BJ, Proudfoot D. Alan Turing's forgotten ideas in computer science. Sci Am. 1999; https://doi.org/10.1038/scientificamerican0499-98.
2. Moor JH. The turing test: the elusive standard of artificial intelligence. Comput Linguist. 2004;30(1950):115–6. https://doi.org/10.1162/089120104773633420.
3. Moor J. The dartmouth college artificial intelligence conference: the next fifty years. AI Mag (American Assoc Artif Intell). 2006;27:4.
4. Lindsay RK. Applications of artificial intelligence for organic chemistry: the DENDRAL project. McGraw-Hill Adv Comput Sci Ser. 1980;xii:194.
5. Nakamura Y. Sumex utility functions. Math Soc Sci. 1996;31(1):39–47. https://doi.org/10.1016/0165-4896(95)00801-2.
6. Lim M. History of AI Winters. Digital Actuaries. https://www.actuaries.digital/2018/09/05/history-of-ai-winters/. Published 2018. Accessed 24 Apr 2019.
7. Hsu FH. IBM's deep blue chess grandmaster chips. IEEE Micro. 1999; https://doi.org/10.1109/40.755469.
8. High R. The era of cognitive systems: an inside look at IBM Watson and how it works. Int Bus Mach Corp. 2012;1(1):1–14. http://www.redbooks.ibm.com/redpapers/pdfs/redp4955.pdf
9. Tech's next big wave: Big data meets biology. Fortune. 2018.
10. Corish B. Medical knowledge doubles every few months; how can clinicians keep up? Elsevier. 2018.
11. Healthcare's Data Conudrum. www.nuance.com/for-healthcare. Accessed May 18, 2018.
12. Ristevski B, Chen M. Big data analytics in medicine and healthcare. J Integr Bioinform. 2018; https://doi.org/10.1515/jib-2017-0030.
13. Rossi RL, Grifantini RM. Big data: challenge and opportunity for translational and industrial research in healthcare. Front Digit Humanit. 2018; https://doi.org/10.3389/fdigh.2018.00013.
14. Tripoliti EE, Papadopoulos TG, Karanasiou GS, Naka KK, Fotiadis DI. Heart failure: diagnosis, severity estimation and prediction of adverse events through machine learning techniques. Comput Struct Biotechnol J. 2017; https://doi.org/10.1016/j.csbj.2016.11.001.
15. Ahmad T, Lund LH, Rao P, et al. Machine learning methods improve prognostication, identify clinically distinct phenotypes, and detect heterogeneity in response to therapy in a large cohort of heart failure patients. J Am Heart Assoc. 2018; https://doi.org/10.1161/JAHA.117.008081.
16. Mortazavi BJ, Downing NS, Bucholz EM, et al. Analysis of machine learning techniques for heart failure readmissions. Circ Cardiovasc Qual Outcomes. 2016; https://doi.org/10.1161/CIRCOUTCOMES.116.003039.
17. Su Z, Brecht T, O'Donovan F, et al. Machine learning enhanced predictions of hospital readmission or death in heart failure. Circulation. 2017;136(Suppl 1):16941.
18. Son YJ, Kim HG, Kim EH, Choi S, Lee SK. Application of support vector machine for prediction of medication adherence in heart failure patients. Healthc Inform Res. 2010; https://doi.org/10.4258/hir.2010.16.4.253.
19. Karanasiou GS, Tripoliti EE, Papadopoulos TG, et al. Predicting adherence of patients with HF through machine learning techniques. Healthc Technol Lett. 2016; https://doi.org/10.1049/htl.2016.0041.
20. Feldman AM, Klein H, Tchou P, et al. Use of a wearable defibrillator in terminating Tachyarrhythmias in patients at High risk for sudden death: results of WEARIT/BIROAD. PACE - Pacing Clin Electrophysiol. 2004;27(1):4–9. https://doi.org/10.1111/j.1540-8159.2004.00378.x.
21. Dawes TJW, de Marvao A, Shi W, et al. Machine learning of three-dimensional right ventricular motion enables outcome prediction in pulmonary hypertension: a cardiac MR imaging study. Radiology. 2017; https://doi.org/10.1148/radiol.2016161315.
22. Verduijn M, Rosseel PMJ, Peek N, de Jonge E, de Mol Bas AJM. Prognostic Bayesian networks: II: an application in the domain of cardiac surgery. J Biomed Inform. 2007;40(6):619–30.
23. Silver D, Schrittwieser J, Simonyan K, et al. Mastering the game of go without human knowledge. Nature. 2017; https://doi.org/10.1038/nature24270.
24. Nemati S, Ghassemi MM, Clifford GD. Optimal medication dosing from suboptimal clinical examples: a deep reinforcement learning approach. In: Proceedings of the Annual International Conference

of the IEEE Engineering in Medicine and Biology Society, EMBS; 2016. doi:https://doi.org/10.1109/EMBC.2016.7591355.
25. Ledbetter D, Aczon M, Ho L, Gunny A, Wetzel R. Recurrent neural networks on electronic medical records to assess PICU mortality risk. Crit Care Med. 2016; https://doi.org/10.1097/01.ccm.0000509170.92653.bd.
26. Rasmy L, Zheng WJ, Xu H, et al. A study of generalizability of recurrent neural network-based predictive models for heart failure onset risk using a large and heterogeneous EHR data set. J Biomed Inform. 2018; https://doi.org/10.1016/j.jbi.2018.06.011.
27. Acharya UR, Fujita H, Oh SL, et al. Automated identification of shockable and non-shockable life-threatening ventricular arrhythmias using convolutional neural network. Futur Gener Comput Syst. 2018; https://doi.org/10.1016/j.future.2017.08.039.
28. Latif S, Usman M, Rana R, Qadir J. Phonocardiographic sensing using deep learning for abnormal heartbeat detection. IEEE Sensors J. 2018; https://doi.org/10.1109/JSEN.2018.2870759.
29. Marsden AL, Feinstein JA. Computational modeling and engineering in pediatric and congenital heart disease. Curr Opin Pediatr. 2015;27(5):587–96. https://doi.org/10.1097/MOP.0000000000000269.
30. Yu L, Cheng JZ, Dou Q, et al. Automatic 3D cardiovascular MR segmentation with densely-connected volumetric convnets. In: Lecture Notes in Computer Science (Including Subseries Lecture Notes in Artificial Intelligence and Lecture Notes in Bioinformatics); 2017. doi:https://doi.org/10.1007/978-3-319-66185-8_33
31. Avendi MR, Kheradvar A, Jafarkhani H. A combined deep-learning and deformable-model approach to fully automatic segmentation of the left ventricle in cardiac MRI. Med Image Anal. 2016; https://doi.org/10.1016/j.media.2016.01.005.
32. Lu X, Jolly MP, Georgescu B, et al. Automatic view planning for cardiac MRI acquisition. In: Lecture Notes in Computer Science (Including Subseries Lecture Notes in Artificial Intelligence and Lecture Notes in Bioinformatics); 2011. doi:https://doi.org/10.1007/978-3-642-23626-6_59
33. Tao Q, Yan W, Wang Y, et al. ORIGINAL RESEARCH • CARDIAC IMAGING deep learning-based method for fully automatic quantification of left ventricle function from cine MR images: a multivendor, multicenter study • content codes. Radiology. 2019; https://doi.org/10.1148/radiol.2018180513.
34. Rahmouni HW, Ky B, Plappert T, et al. Clinical utility of automated assessment of left ventricular ejection fraction using artificial intelligence-assisted border detection. Am Heart J. 2008; https://doi.org/10.1016/j.ahj.2007.11.002.
35. Raghavendra U, Fujita H, Gudigar A, et al. Automated technique for coronary artery disease characterization and classification using DD-DTDWT in ultrasound images. Biomed Signal Process Contr. 2018; https://doi.org/10.1016/j.bspc.2017.09.030.
36. Moghaddasi H, Nourian S. Automatic assessment of mitral regurgitation severity based on extensive textural features on 2D echocardiography videos. Comput Biol Med. 2016; https://doi.org/10.1016/j.compbiomed.2016.03.026.
37. Knackstedt C, Bekkers SCAM, Schummers G, et al. Fully automated versus standard tracking of left ventricular ejection fraction and longitudinal strain the FAST-EFs multicenter study. J Am Coll Cardiol. 2015; https://doi.org/10.1016/j.jacc.2015.07.052.
38. Hauptmann A, Arridge S, Lucka F, Muthurangu V, Steeden JA. Real-time cardiovascular MR with spatio-temporal artifact suppression using deep learning–proof of concept in congenital heart disease. Magn Reson Med. 2019; https://doi.org/10.1002/mrm.27480.
39. Sengupta PP, Huang YM, Bansal M, et al. Cognitive machine-learning algorithm for cardiac imaging; a pilot study for differentiating constrictive pericarditis from restrictive cardiomyopathy. Circ Cardiovasc Imaging. 2016;9:6. https://doi.org/10.1161/CIRCIMAGING.115.004330.
40. Narula S, Shameer K, Salem Omar AM, Dudley JT, Sengupta PP. Machine-learning algorithms to automate morphological and functional assessments in 2D echocardiography. J Am Coll Cardiol. 2016; https://doi.org/10.1016/j.jacc.2016.08.062.
41. Samad MD, Wehner GJ, Arbabshirani MR, et al. Predicting deterioration of ventricular function in patients with repaired tetralogy of Fallot using machine learning. Eur Heart J Cardiovasc Imaging. 2018; https://doi.org/10.1093/ehjci/jey003.
42. Silva JNA, Southworth M, Raptis C, Silva J. Emerging applications of virtual reality in cardiovascular medicine. JACC Basic Transl Sci. 2018;3(3):420–30. https://doi.org/10.1016/j.jacbts.2017.11.009.
43. Dugas CM, Schussler JM. Advanced technology in interventional cardiology: a roadmap for the future of precision coronary interventions. Trends Cardiovasc Med. 2016;26(5):466–73. https://doi.org/10.1016/j.tcm.2016.02.003.
44. Rantner LJ, Vadakkumpadan F, Spevak PJ, Crosson JE, Trayanova NA. Placement of implantable cardioverter-defibrillators in paediatric and congenital heart defect patients: a pipeline for model generation and simulation prediction of optimal configurations. J Physiol. 2013; https://doi.org/10.1113/jphysiol.2013.255109.
45. Glöckler M, Halbfaß J, Koch A, Achenbach S, Dittrich S. Multimodality 3D-roadmap for cardiovascular interventions in congenital heart disease - a single-center, retrospective analysis of 78 cases. Catheter Cardiovasc Interv. 2013;82(3):436–42. https://doi.org/10.1002/ccd.24646.
46. Yoo S-J, Thabit O, Kim EK, et al. 3D printing in medicine of congenital heart diseases. 3D Print Med. 2015;2(1):3. https://doi.org/10.1186/s41205-016-0004-x.

47. Giannopoulos AA, Mitsouras D, Yoo S-J, Liu PP, Chatzizisis YS, Rybicki FJ. Applications of 3D printing in cardiovascular diseases. Nat Rev Cardiol. 2016;13(12):701–18. https://doi.org/10.1038/nrcardio.2016.170.
48. Cantinotti M, Valverde I, Kutty S. Three-dimensional printed models in congenital heart disease. Int J Cardiovasc Imaging. 2017; https://doi.org/10.1007/s10554-016-0981-2.
49. Marsden AL, Feinstein JA. Computational modeling and engineering in pediatric and congenital heart disease. Curr Opin Pediatr. 2015;27(5):587–96. https://doi.org/10.1097/MOP.0000000000000269.
50. Ryan JR. Three dimensional printing and computational visualization for surgical planning and medical education. Arizona: Arizona State University; 2015.
51. Ryan JR, Moe TG, Richardson R, Frakes DH, Nigro JJ, Pophal S. A novel approach to neonatal management of tetralogy of fallot, with pulmonary atresia, and multiple aortopulmonary collaterals. JACC Cardiovasc Imaging. 2015;8(1):103–4. https://doi.org/10.1016/j.jcmg.2014.04.030.
52. Park SS, Sanders DB, Smith BP, et al. Total artificial heart in the pediatric patient with biventricular heart failure. Perfusion. 2014;29(1):82–8. https://doi.org/10.1177/0267659113496580.
53. Kalinin AA, Higgins GA, Reamaroon N, et al. Deep learning in pharmacogenomics: from gene regulation to patient stratification. Pharmacogenomics. 2018; https://doi.org/10.2217/pgs-2018-0008.
54. Kubota T. Stanford researchers create deep learning algorithm that could boost drug development. Stanford News. https://news.stanford.edu/2017/04/03/deep-learning-algorithm-aid-drug-development/. Published 2017. Accessed 4 May 2019.
55. O'Sullivan D, Fraccaro P, Carson E, Weller P. Decision time for clinical decision support systems. Clin Med J R Coll Physicians London. 2014; https://doi.org/10.7861/clinmedicine.14-4-338.
56. Alther M, Reddy CK. Clinical decision support systems. In: Healthcare data analytics; 2015. https://doi.org/10.1201/b18588.
57. Etiometry. Boston Childrens: Research + Innovation. http://www.childrenshospital.org/research-and-innovation/partnerships/etiometry. Published 2017.

Communication in Congenital Heart Disease: A Relevant Application for Engineering Models?

28

Giovanni Biglino, Maria Victoria Ordonez, Andrew Shearn, Sofie Layton, and Jo Wray

28.1 A Delicate Communicative Dynamic

The importance of communication in medicine has been and is extensively discussed in several forums from different viewpoints, with regard to the clarity of language and the impact that communication and interpersonal skills can have on patient care [1]. The interaction between an "expert" and a "non-expert" can often result in lack of shared knowledge, especially when the latter is of a technical nature or being illustrated in a technical way (e.g. an echocardiogram or an MRI dataset presented to a "non-expert"). These instances can often lead to unsuccessful communication [2], where patient–doctor communication is known to play a direct role in patient satisfaction and adherence to treatment [3, 4]. The invisible nature of congenital heart disease and the technical language associated with it present potential barriers to the communicative process, posing a challenge to the specialist to avoid a patronising communicative style [5]. Historically, it was recognised over thirty years ago that "distorted perceptions" of congenital heart lesions can unnecessarily increase anxiety in parents, leading to imposing of inappropriate restrictions, and impacting also on the child's self-perception [6]. More recently, and in the light of the increasing proportion of congenital heart patients who now live into adulthood, it has been increasingly recognised that communication around topics such as employment, health insurance or family planning should be incorporated into a comprehensive process of "transition to adult care" [7].

Research also suggests that it is important to understand "how, why and when" patients participate in largely asymmetric consultations, e.g. situations in which the communication is dominated by one speaker [8]. And a question arises in relation to such asymmetric processes: could tools such as heart models (physical or computational) be used

G. Biglino (✉)
University of Bristol, Bristol Medical School, Bristol, UK
e-mail: g.biglino@bristol.ac.uk

M. V. Ordonez
University Hospitals Bristol and Weston, NHS Foundation Trust, Bristol, UK
e-mail: victoria.ordonez@uhbw.nhs.uk

A. Shearn
Bristol Medical School, University of Bristol, Bristol, UK
e-mail: andrew.shearn@bristol.ac.uk

S. Layton
Royal College of Art, London, UK

J. Wray
Great Ormond Street Hospital for Children, NHS Foundation Trust, London, UK

Institute of Cardiovascular Science, University College London, London, UK
e-mail: jo.wray@gosh.nhs.uk

G. Butera et al. (eds.), *Modelling Congenital Heart Disease*,
https://doi.org/10.1007/978-3-030-88892-3_28

to facilitate these clinical encounters? Before tackling this question, it is important to outline key concepts around transition, transfer and pregnancy in congenital heart patients to better appreciate the nuances and potential of this scenario.

28.2 Transition and Transfer: The Process of Gaining Ownership

With improvements in medical care, 90% of patients with CHD are surviving into adulthood [9]. This has meant that, ultimately, these paediatric patients will need to be transferred to adult services. This has historically presented a problem, as previously paediatric training somewhat neglected the needs of the adolescent [10]. In turn, this also meant that, in some cases, patients with congenital lesions were being cared for by adult specialists with limited training in the long-term care of these patients [10]. In recent years, this has become a recognised problem and efforts have been made to understand and resolve it.

The, now accepted, definition of transition, was put forward in 1993 by the Society for Adolescent Medicine as "the purposeful, planned movement of adolescents and young adults with chronic physical and medical conditions from child-centred to adult-oriented health-care systems" [11]. These adolescents are usually transferred from paediatric care to the adult service around the age of 18, but can be older, depending on the patient [12]. This is often a time of additional upheaval for adolescents, as they are may be sitting major examinations, or leaving home, whether heading up to university or staring in the world of work. There is evidence to show that there is a high risk of these patients being lost to follow-up at this time [13], with subsequent detrimental consequences to their health, such as the need for interventions [14]. This is a problem that is seen in many, if not all, forms of congenital disease; it is not unique to CHD.

The process of preparing adolescents for the process of transition and transfer varies between centres, and there is no consensus on how it should be carried out. However, there is generally a focus on teaching, through improving the understanding of the condition or general wellness issues [15–18]. In order for this to become accepted practice, it is important that proper controlled trials of interventions are carried out. Whilst there has been some effort along these lines, many of these studies include small numbers of patients and have varied outcome measures. Several studies have demonstrated that paying close attention to this process results in an increased understanding by the patient of their condition, and a reduced loss to follow-up [15, 16, 18]. However, when synthesised into systematic reviews, it becomes clear that few conclusions can be drawn from the data. For example, a Cochrane review in 2016 found that only four studies could be included [19]. These four studies were based around different conditions and had different methodologies. Another review was able to include 13 studies, but again struggled to draw conclusions due to the substantial differences in methodologies of the included studies [12]. It is clear that these methodologies need to be standardised before any real conclusions can be drawn on the effectiveness of transition clinics.

Transition and transfer to adult care are clearly important stages in the clinical care of paediatric patients with chronic, long-term conditions such as CHD. There is evidence to suggest that it is important that this process is handled carefully, as otherwise these patients may be lost to life-long follow-up with the potential for detrimental effects on their health. New strategies that will enable this smooth transition are needed and need to be tested in a standardised way, so that their relative effectiveness can be tested robustly. The most effective of these techniques then need to form the basis for new guidelines to allow them to be implemented across a wide range of centres.

28.3 Pregnancy Clinic: A Multidisciplinary Conversation

Given the improvement in surgical outcomes over the last years, more female patients with a spectrum of congenital heart disease (including complex lesions) are now reaching adulthood and therefore childbearing age [20, 21]. This is an important consideration as they need to be understood and followed up according to their clinical complexity.

Women with cardiac disease and their infants are at a greater risk of mortality and morbidity during pregnancy [22]. The European Society of Cardiology and the Royal College of Obstetricians recommend preconception counselling (PCC) for all women with cardiac disease so they are made aware of these risks [23], and PCC plays a fundamental role in their care.

All women with congenital heart disease should seek advice from a congenital cardiologist with expertise in transition to adulthood and pregnancy before contemplating conception [21]. During the pre-pregnancy cardiology consultation, a combination of scoring systems and investigations such as echocardiography, electrocardiogram and stress test is performed in order to stratify the risk of the patient and plan the future pregnancy management [21]. It is important to highlight that contraception and pregnancy discussions should be initiated prior to conception and ideally during teenage years. Clear and thorough communication from a congenital cardiologist regarding risks and outcomes to the patient and family as well as patient understanding are essential at this stage. Moreover, preconception evaluation is important as well as genetic consultation, identifying potential medical and surgical issues before pregnancy in order to improve outcomes [24].

Risk stratification is paramount in all congenital patients as pregnancy physiology carries varying risks even after successful repair [25]. This care has to be carried out by cardiologists and maternal–foetal medicine specialists with expertise in pregnancy and CHD.

Such stratification is based on two elements: (a) maternal risk and (b) foetal risk [26, 27]. Current risks calculators such as CARPREG, ZAHARA and the modified WHO criteria either do not include for instance women with a Fontan circulation, or simply classify them as "high risk". The CARPREG classification includes women with both acquired and congenital heart disease and assigns a point for each predictor (i.e. prior cardiac event, NYHA functional class >II or cyanosis, left heart obstruction, left ventricular ejection fraction <40%). A score of 1 confers 27% risk of maternal cardiovascular complications, and a score > 1 confers a 75% risk [28]. The ZAHARA study also identified predictors of adverse maternal events in patients with congenital heart disease [29]. The modified WHO criteria include specific cardiac lesions in addition to clinical cardiac status, and they apply to women with acquired and congenital heart disease, identifying risk categories ranging from low (group I) to very high risk (group IV) [30, 31].

Individualised and coordinated care is fundamental in this context. A cardiologist with expertise in the congenital field should be included in the pregnancy follow-up plan to improve management and treatment, whilst an obstetrician will provide thorough understanding of the implications of potential cardiac compromise for the pregnant mother and the foetus. A multidisciplinary approach is recommended, and it has been shown to improve mortality and morbidity in the ACHD population. Follow-up should also be carried out by a multidisciplinary team, which might include an anaesthetist, a specialist nurse, a haematologist, geneticists, sonographers, a foetal echocardiographist, a neonatologist and other specialists as appropriate [32].

28.4 A Role for Models in the Realm of Communication?

Given the complex settings briefly outlined above both for transition and for pregnancy in CHD, it is certainly warranted to either design and/or study new tools that can facilitate communication with CHD patients. The potential role and impact of technological innovations in several dimensions of medical communication have been recognised, including delivering information, coordinating care, reinforcing memory and improving adherence, increasing participation in a patient's own care, promoting behavioural changes and improving patients' experience [33].

Whilst engineers may envisage tools such as simulations and models as generating new physiological insights or informing clinical decision-making, tailoring treatment to individual patients and informing risk stratification [34], some of these methods could also be employed in the context of communication. The most emblematic and relevant example is probably that of 3D printing technology. In the light of the ability of a 3D replica not only to display cardiovascular anatomy (such as a "lesion-specific" model) but also to actually present a patient with their own heart —an act imbued with emotional significance —3D printed models have been studied and results suggest a role in facilitating communication. The use of patient-specific models was shown to enhance engagement with patients' parents, improving communication between cardiologists and parents, which could in turn have beneficial repercussions on parents' (and patients') psychological adjustment following treatment [35]. Another study employing 3D patient-specific models with a group of teenagers (aged 16–18) in the transition clinic setting showed that the use of models resulted in significant improvements in patients' confidence in explaining their condition to others, their knowledge of their condition (e.g. correct naming of anatomical terms) and patients' satisfaction with their visit compared to prior appointments [36]. Not only the anatomical descriptions but also the elaboration of the impact on lifestyle was more eloquent after seeing a 3D model, with patients reporting that models helped their understanding. The same study also highlighted a non-negligible (~30%) proportion of patients reporting feeling anxious when viewing their heart model for the first time, which suggest the importance of a psychologist being involved in the team and the need for appropriate model presentation and explanation, providing patients with the opportunity to ask questions and orientate themselves about different features of the models. These studies clearly suggest a potentially positive use of 3D patient-specific models to aid in communicating both with parents of CHD patients and adolescents in the transition clinic, but they lack longitudinal data on long-term use of the model and the long-term impact of repeated viewing. These should be explored in future studies.

As to the pregnancy context, models could potentially also play a role. Heart models could be used, again, to elucidate the anatomy of the patient, perhaps also illustrating complex anatomical arrangements such as a total cavopulmonary connection and its position with respect to other anatomical elements (e.g. placenta), yet resulting in much more complex models than those focusing solely on the heart chambers and the main vessels. The use of 3D printed models here should not be confused with the role these could also play in the foetal clinic, whereby series of CHD foetal heart models have been discussed in the literature [37] but so far derived from imaging data acquired post mortem (microfocus computed tomography data). Modelling of congenital cardiac anomalies from prenatal echocardiography is nevertheless feasible and also compatible with advanced imaging modalities such as 4D flow MRI that can provide insight into and illustrate the haemodynamics of the foetal heart [38]. These have not been thoroughly studied for the purpose of communication, but potentially this represents another valuable application of 3D models to facilitate complex conversations in the clinical setting, as it is also emerging in other medical applications, e.g. renal tumours [39].

The communicative potential of other engineering methodologies such as computational models has been explored less. It could be hypothesised that the ability of advanced visualisation and the opportunity to simulate phenomena such as stent deployment or an ablation could be valuable from a communication perspective, whereby these could be implemented in mobile apps or other easily accessible formats, specifically for CHD patients. Literature in this context is beginning to suggest the potential of tools such as apps for atrial fibrillation developed by the Characterizing Atrial fibrillation by Translating its Causes into Health Modifiers in the Elderly (CATCH ME) Consortium with the European Society of Cardiology (ESC) to increase patient education, improve patient–doctor communication and empower patients to take a more active role in managing their condition [40].

28.5 Advancing the Translation of Models Through Creative Collaborations

According to a theory of social construction of technology, it is important to take into account the perspective of all stakeholders when aiming to successfully translate a technological innovation. When considering congenital heart disease, stakeholders include medical professionals as well as patients and families, and they can be engaged with different approaches in different contexts [41]. Insights on models' features emerged in an artist-led participatory workshop with adolescents with congenital heart disease, where nuanced preferences on models' sizes and materials emerged, with different patients at times having differing views, highlighting the importance of a sensitive approach when considering model use [42]. By collaborating with an artist, this work moved beyond the purely anatomical connotation of the heart models, exploring different scales and casting the models in different materials, e.g. stimulating conversations around the preciousness of the organ when presenting the group with a small bronze heart model that can be held in the palm of one's hand. A creative setting such as this one, involving an interdisciplinary team and framing the process around individual uniqueness rather than an overtly medical lens, can be very valuable for unpacking sensitive themes with CHD patients, thereby taking those conversations to a different level where the meaning of CHD to patients can be explored, rather than purely their understanding. Models can be incorporated in the process, provided there is appropriate collaboration and dialogue amongst team members to ensure that they are presented, described and handled in a safe and sensitive manner [42]. The anatomical representation offered in a 3D model or also a 3D rendering can be successfully incorporated in a creative process as a means to stimulate engagement of adolescents with CHD with concepts of their individual cardiovascular anatomy. This was undertaken in another artist-led workshop supported by a multidisciplinary team, focusing on exploring narratives and perceptions from the adolescents' perspective and resulting in the creation of a first person composite narrative revolving around themes that were central to the patients, including medical references and awareness of the complexity of surgically repaired CHD, elaborating scars and patches into very meaningful and profound metaphors [43]. Collaboration between engineers, medics, artists, psychologists and patients can extend to incorporating models and medical data into original artworks, broadening the conversation to wider audiences and impacting on communication in a larger sense, moving beyond the context of the clinical encounter and stimulating reflections at a societal level, engaging with diverse audiences, both medical and non-medical. The process of artistic representation [44] and the inclusion of overt medical references (such as patients' heart models) sensitively presented after having been filtered through the artist's lens can ultimately lead to what has been defined as "a beautiful, thoughtful, evocative representation of what it means to be a patient, a parent, a doctor, or a scientist involved in treating people with heart disease" [45].

References

1. Dyche L. Interpersonal skill in medicine: the essential partner of verbal communication. J Gen Intern Med. 2007;22:1035–9.
2. Bigi S. The persuasive role of ethos in doctor-patient interactions. Commun Med. 2011;8:67–76.
3. Ha JF, Longnecker N. Doctor-patient communication: a review. Ochsner J. 2010;10:38–43.
4. Martin LR, Williams SL, Haskard KB, Dimatteo MR. The challenge of patient adherence. Ther Clin Risk Manag. 2005;1:189–99.
5. Blackman J, Sahebjalal M. Patient understanding of frequently used cardiology terminology. Br J Cardiol. 2014;21:39.
6. Kaden GG, McCarter RJ, Johnson SF, Ferencz C. Physician-patient communication. Understanding congenital heart disease. Am J Dis Child. 1985;139(10):995–9.
7. Deng LX, Gleason LP, Awh K, Khan AM, Drajpuch D, Fuller S, Goldberg LA, Mascio CE, Partington SL, Tobin L, Kovacs AH, Kim YY. Too little too late? Communication with patients with congenital heart disease about challenges of adult life. Congenit Heart Dis. 2019;14(4):534–40.

8. Bellander T, Karlsson AM. Patient participation and learning in medical consultations about congenital heart defects. PLoS One. 2019;14(7):e0220136.
9. Marelli AJ, Mackie AS, Ionescu-Ittu R, Rahme E, Pilote L. Congenital heart disease in the general population: changing prevalence and age distribution. Circulation. 2007;115:163–72. https://doi.org/10.1161/CIRCULATIONAHA.106.627224.
10. Viner RM. Transition of care from paediatric to adult services: one part of improved health services for adolescents. Arch Dis Child. 2008;93:160–3. https://doi.org/10.1136/adc.2006.103721.
11. Blum RW, Garell D, Hodgman CH, Jorissen TW, Okinow NA, Orr DP, Slap GB. Transition from child-centered to adult health-care systems for adolescents with chronic conditions. A position paper of the Society for Adolescent Medicine. J Adolesc Health. 1993;14:570–6. https://doi.org/10.1016/1054-139X(93)90143-D.
12. Heery E, Sheehan AM, While AE, Coyne I. Experiences and outcomes of transition from pediatric to adult health Care Services for Young People with congenital heart disease: a systematic review. Congenit Heart Dis. 2015;10:413–27. https://doi.org/10.1111/chd.12251.
13. Moceri P, Goossens E, Hascoet S, Checler C, Bonello B, Ferrari E, Acar P, Fraisse A. From adolescents to adults with congenital heart disease: the role of transition. Eur J Pediatr. 2015;174:847–54. https://doi.org/10.1007/s00431-015-2557-x.
14. Reekie J. Flying the nest: are we doing enough to achieve successful transition in young people with congenital heart disease? Heart. 2020;106:87–8. https://doi.org/10.1136/heartjnl-2019-315727.
15. Gaydos SS, Chowdhury SM, Judd RN, Mchugh KE. A transition clinic intervention to improve follow-up rates in adolescents and young adults with congenital heart disease. Cardiol Young. 2020;30(5):633–40. https://doi.org/10.1017/S1047951120000682.
16. Markwart H, Bomba F, Menrath I, Brenk-Franz K, Ernst G, Thyen U, Hildebrandt A, Schmidt S. Assessing empowerment as multidimensional outcome of a patient education program for adolescents with chronic conditions: a latent difference score model. PLoS One. 2020;15:e0230659. https://doi.org/10.1371/journal.pone.0230659.
17. Schmidt S, Herrmann-Garitz C, Bomba F, Thyen U. A multicenter prospective quasi-experimental study on the impact of a transition-oriented generic patient education program on health service participation and quality of life in adolescents and young adults. Patient Educ Couns. 2016;99:421–8. https://doi.org/10.1016/j.pec.2015.10.024.
18. Mackie AS, Rempel GR, Kovacs AH, Kaufman M, Rankin KN, Jelen A, Yaskina M, Sananes R, Oechslin E, Dragieva D, et al. Transition intervention for adolescents with congenital heart disease. J Am Coll Cardiol. 2018;71:1768–77. https://doi.org/10.1016/j.jacc.2018.02.043.
19. Campbell F, Biggs K, Aldiss SK, O'Neill PM, Clowes M, Mcdonagh J, While A, Gibson F. Transition of care for adolescents from paediatric services to adult health services. Cochrane Database Syst Rev. 2016;2016 https://doi.org/10.1002/14651858.CD009794.pub2.
20. Nasir AJ, Hassan A. Adult congenital heart diseases: clinical pattern, management and immediate outcome in Ibn-Al-Bitar Center for Cardiac Surgery, Baghdad 2013. Thi-Qar Medical Journal. 2016;12(2):91–104.
21. Bishop L, Lansbury A, English K. WITHDRAWN: adult congenital heart disease and pregnancy. BJA Education. 2017. doi: https://doi.org/10.1016/j.bjae.2017.10.001
22. Weber RW, et al. Prenatal diagnosis of single ventricle physiology impacts on cardiac morbidity and mortality. Pediatr Cardiol. 2019;40(1):61–70.
23. Regitz-Zagrosek V, et al. 2018 ESC guidelines for the management of cardiovascular diseases during pregnancy. Eur Heart J. 2018;39(34):3165–241.
24. Sayqa A, et al. Pregnancy outcomes in patients with a Fontan circulation and proposal for a risk-scoring system: single centre experience. J Congenit Cardiol. 2017;1:10.
25. Ordonez, Biglino, Curtis
26. Siu SC, et al. Prospective multicenter study of pregnancy outcomes in women with heart disease. Circulation. 2001;104(5):515–21.
27. Cauldwell M, et al. Pre-pregnancy counseling for women with heart disease: a prospective study. Int J Cardiol. 2017;240:374–8.
28. Drenthen W, et al. Predictors of pregnancy complications in women with congenital heart disease. Eur Heart J. 2010;31(17):2124–32.
29. Alvaro GR, et al. Pregnancy in women with a Fontan circulation: a systematic review of the literature. Circ Cardiovasc Qual Outcomes. 2018;11(5):e004575.
30. Cauldwell M, Gatzoulis M, Steer P. Congenital heart disease and pregnancy: a contemporary approach to counselling, pre-pregnancy investigations and the impact of pregnancy on heart function. Obstetric Med. 2017;10(2):53–7.
31. Presbitero P, et al. Pregnancy in cyanotic congenital heart disease. Outcome of mother and fetus. Circulation. 1994;89(6):2673–6.
32. Crooke GA, Le Jemtel TH. Diagnosis and management of adult congenital heart disease. Ann Intern Med. 2018;140(5):408.
33. Boucher JL. Technology and patient-provider interactions: improving quality of care, but is it improving communication and collaboration? Diabetes Spectrum. 2010;23(3):142–4.
34. Biglino G, Capelli C, Bruse J, Bosi GM, Taylor AM, Schievano S. Computational modelling for congenital heart disease: how far are we from clinical translation? Heart. 2017;103(2):98–103.
35. Biglino G, Capelli C, Wray J, Schievano S, Leaver LK, Khambadkone S, Giardini A, Derrick G, Jones A, Taylor AM. 3D-manufactured patient-specific models of congenital heart defects for communication in clin-

ical practice: feasibility and acceptability. BMJ Open. 2015;5(4):e007165.
36. Biglino G, Koniordou D, Gasparini M, Capelli C, Leaver LK, Khambadkone S, Schievano S, Taylor AM, Wray J. Piloting the use of patient-specific cardiac models as a novel tool to facilitate communication during clinical consultations. Pediatr Cardiol. 2017;38(4):813–8.
37. Sandrini C, Lombardi C, Shearn AIU, Ordonez MV, Caputo M, Presti F, Luciani GB, Rossetti L, Biglino G. Three-dimensional printing of fetal models of congenital heart disease derived from microfocus computed tomography: a case series. Front Pediatr. 2020;7:567.
38. Ruedinger KL, Zhou H, Trampe B, Heiser T, Srinivasan S, Iruretagoyena JI, Roldán-Alzate A. Modeling fetal cardiac anomalies from prenatal echocardiography with 3-dimensional printing and 4-dimensional flow magnetic resonance imaging. Circ Cardiovasc Imaging. 2018;11(9):e007705.
39. Bernhard JC, Isotani S, Matsugasumi T, Duddalwar V, Hung AJ, Suer E, Baco E, Satkunasivam R, Djaladat H, Metcalfe C, Hu B, Wong K, Park D, Nguyen M, Hwang D, Bazargani ST, de Castro Abreu AL, Aron M, Ukimura O, Gill IS. Personalized 3D printed model of kidney and tumor anatomy: a useful tool for patient education. World J Urol. 2016;34(3):337–45.
40. Kotecha D, Chua WWL, Fabritz L, Hendriks J, Casadei B, Schotten U, Vardas P, Heidbuchel H, Dean V, Kirchhof P, European Society of Cardiology (ESC) Atrial Fibrillation Guidelines Taskforce, the CATCH ME consortium and the European Heart Rhythm Association (EHRA). European Society of Cardiology smartphone and tablet applications for patients with atrial fibrillation and their health care providers. Europace. 2018;20(2):225–33.
41. Biglino G, Capelli C, Leaver LK, Schievano S, Taylor AM, Wray J. Involving patients, families and medical staff in the evaluation of 3D printing models of congenital heart disease. Commun Med. 2015;12(2–3):157–69.
42. Layton S, Wray J, Leaver LK, Koniordou D, Schievano S, Taylor AM, Biglino G. Exploring the uniqueness of congenital heart disease: an interdisciplinary conversation. J Appl Arts Health. 2016;7(1):77–91.
43. Biglino G, Layton S, Leaver LK, Wray J. Fortune favours the brave: composite first-person narrative of adolescents with congenital heart disease. BMJ Paediatr Open. 2017;1(1):e000186.
44. Biglino G, Layton S, Wray J. Beyond heart symbolism: artistic representation of narratives of congenital heart disease. JAMA. 2017;318(18):1740–2.
45. Coates L. Making peace with our faulty hearts. Lancet. 2018;392(10146):465–6.

29 Three-Dimensional Multimodality Fusion in Minimally Invasive Congenital Heart Interventions

Onno Wink, Alexander Haak, and Sebastian Góreczny

In current clinical practice, a patient has generally been diagnosed and monitored through various means of imaging (e.g., ultrasound, computed tomography (CT), magnetic resonance imaging (MRI), positron emission tomography (PET), and X-ray) before the decision is made to perform an interventional procedure. Previously acquired imaging allows for a better understanding of the anatomy and advanced planning of the procedure, such as the selection of the optimal access site, necessary diagnostic and interventional equipment, and intra-procedural imaging needs [1, 2]. The knowledge gained from pre-catheter imaging may shorten the diagnostic phase of the cardiac catheterization, while optimal angiographic projections, derived from preoperative images, may be utilized during the interventional phase of the procedure. Fusion of three-dimensional (3D) datasets with two-dimensional (2D) fluoroscopy further facilitates the understanding of the anatomy during the treatment in the catheterization laboratory [3–5]. A 3D roadmap projected on fluoroscopy enables a more straightforward cannulation of desired chambers or vessels, positioning of devices, and, in some cases, deployment of stents or occluders without additional angiography [6, 7]. Ultimately, the combination of knowledge gained from pre-catheter imaging and intra-procedural guidance with a 3D roadmap results in a reduction in contrast utilization, radiation exposure, and total procedural time [4, 8, 9].

The major phases (Fig. 29.1) that are involved in order to meaningfully utilize pre-procedural images for the planning and execution of these often complex procedures are as follows:

1. Segmentation
2. Procedure planning
3. Registration
4. Intervention

29.1 Segmentation

During the segmentation phase, the most important structures are extracted from the 3D dataset [10, 11]. The main objective of this process is to build a collection of 3D objects, which can be selectively superimposed on top of the live X-ray.

Supplementary Information The online version of this chapter (https://doi.org/10.1007/978-3-030-88892-3_29) contains supplementary material, which is available to authorized users.

O. Wink · A. Haak
Philips Healthcare, Andover, MA, USA
e-mail: onno.wink@philips.com; alexander.haak@philips.com

S. Góreczny (✉)
Department of Pediatric Cardiology, University Children's Hospital, Faculty of Medicine, Jagiellonian University Medical Collage, Krakow, Poland

Department of Cardiology, Colorado Children's Hospital, Aurora, CO, USA

G. Butera et al. (eds.), *Modelling Congenital Heart Disease*,
https://doi.org/10.1007/978-3-030-88892-3_29

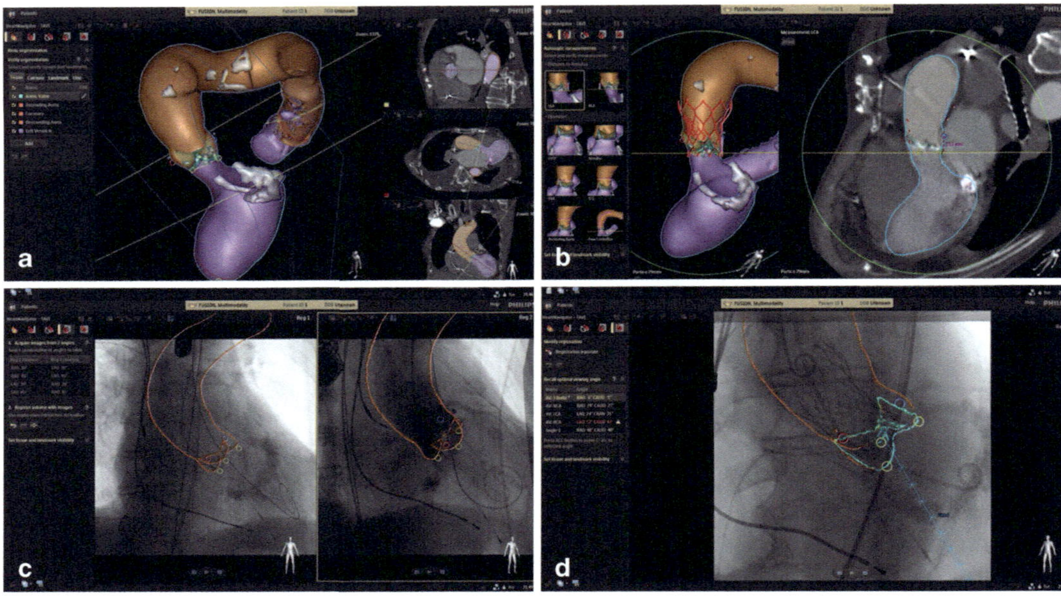

Fig. 29.1 Overview of the major phases involved in multimodality fusion during transcatheter aortic valve replacement: segmentation (**a**), planning (**b**), registration (**c**), and intervention (**d**)

This allows for the visualization of procedure critical structures, which could otherwise be hidden or obscured by bone or ribs if the entire unsegmented 3D images were used. The methods used to perform this segmentation range from simple binary thresholding to atlas-based deformable models. Most often semi-interactive tools are used to keep the user in control, especially if the anatomy is complex or variable as it is the case with congenital diseases. Recently, algorithms have been introduced that utilize statistical models of the anatomy of a population of patients [12]. These algorithms can automatically segment various structures, for instance, heart chambers, but often fall short for patients with anatomies that differ from the trained model. Figure 29.2 gives an example of a fully automatic, atlas-based cardiac segmentation based on a CT dataset and two manual segmentations based on perioperatively acquired CT-like volumes.

29.2 Procedure Planning

During the procedure planning phase, the original 3D dataset (and its associated segmentations) is used to get a better sense of how to navigate through particular anatomy and the type and dimensions of the equipment or implants needed. There are several tools available, ranging from straightforward "in-plane" measurements, to central axis -based curved multi-planar reformats (MPRs) and even simulation of the implanted devices [13]. The previously created segmentation can also be used to print a physical 3D model to get a better sense of the anatomy and allow for the placement of actual devices [14]. In recent years, several services like 3D printing (e.g., Materialise, Stratasys, and Proto Labs), advanced visualization (e.g., EchoPixel, RealView, and Holoxica) and augmented and virtual reality environments have become available to facilitate this planning step [15]. However, basic 3D objects, like points, ellipses, or spheres positioned in the 3D dataset, are often sufficient for guidance in the subsequent registration and interventional steps. Figure 29.3 shows several examples of planning that use a combination of rings, markers, and measurements to suggest the position of a percutaneous pulmonary valve. These also help with the selection of the occluder type and size for treating ventricular septal defect and dimensions of an aortic coarctation.

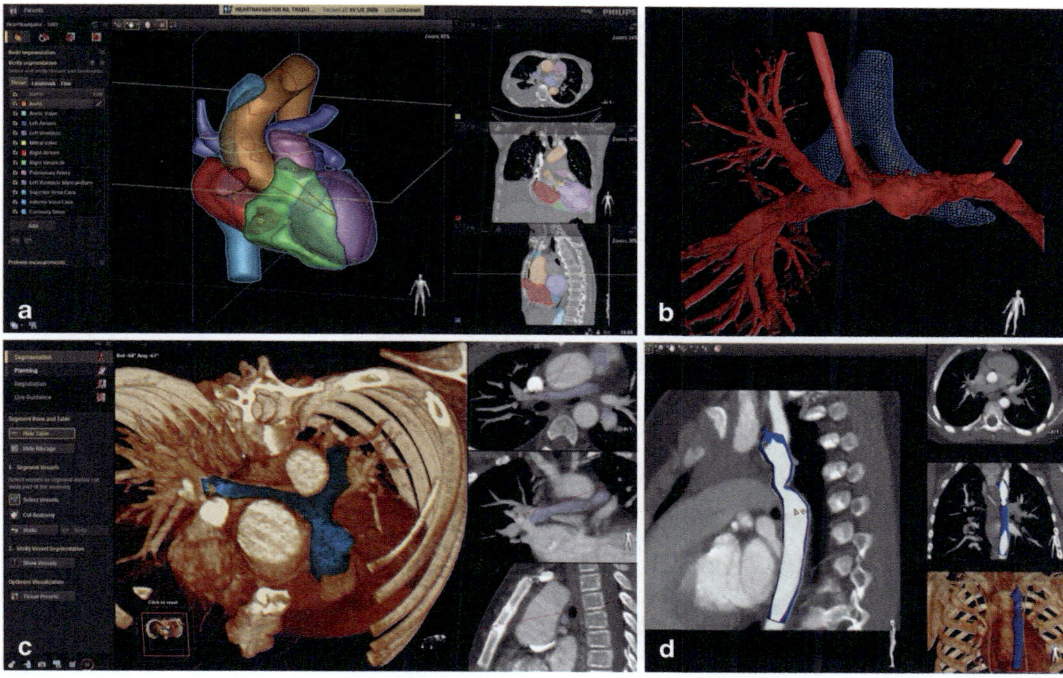

Fig. 29.2 Panel showing various examples of segmentation, including model -based (**a**), semi-automatic (**b**) and manual segmentation on 3D reconstruction (**c**), and multi-planar reformats (MPRs) (**d**)

Next to measurements and device selection, the 3D dataset can be used to determine a sequence of optimal working positions or X-ray projections with a minimum of foreshortening and overlap of adjacent structures at the most important stages of the intervention. For example, a 3D model with a ring marker, highlighting a stenosis or the origin of a side branch, can be moved in various directions to achieve the optimal projection for diagnostic angiography or device placement (Fig. 29.3a) without the need for X-ray. This, in turn, can lead to fewer suboptimal angiographies reducing the overall contrast usage and radiation exposure during the actual intervention.

29.3 Registration

In order to use the pre-procedural images and segmentations as well as the pre-procedural plan for the actual intervention, a registration or alignment to the X-ray system is required. This step is the most critical for successful use of the 3D datasets. Its complexity and associated issues are often underestimated and will be addressed below. In general, there are three common methods that are used to perform the registration:

- 3D to 2D image registration
- 3D to 3D image registration using a linear transformation
- 3D to 3D image registration using a deformable transformation using various levels of automation.

29.3.1 3D to 2D Registration

The most common form of registering the preoperative 3D data set to the X-ray system is to use two 2D X-ray images, acquired at sufficiently different projections (ideally with 90- degree separation). The user usually adjusts the position and orientation of the pre-procedural volume to match both 2D X-ray images best. Figure 29.4

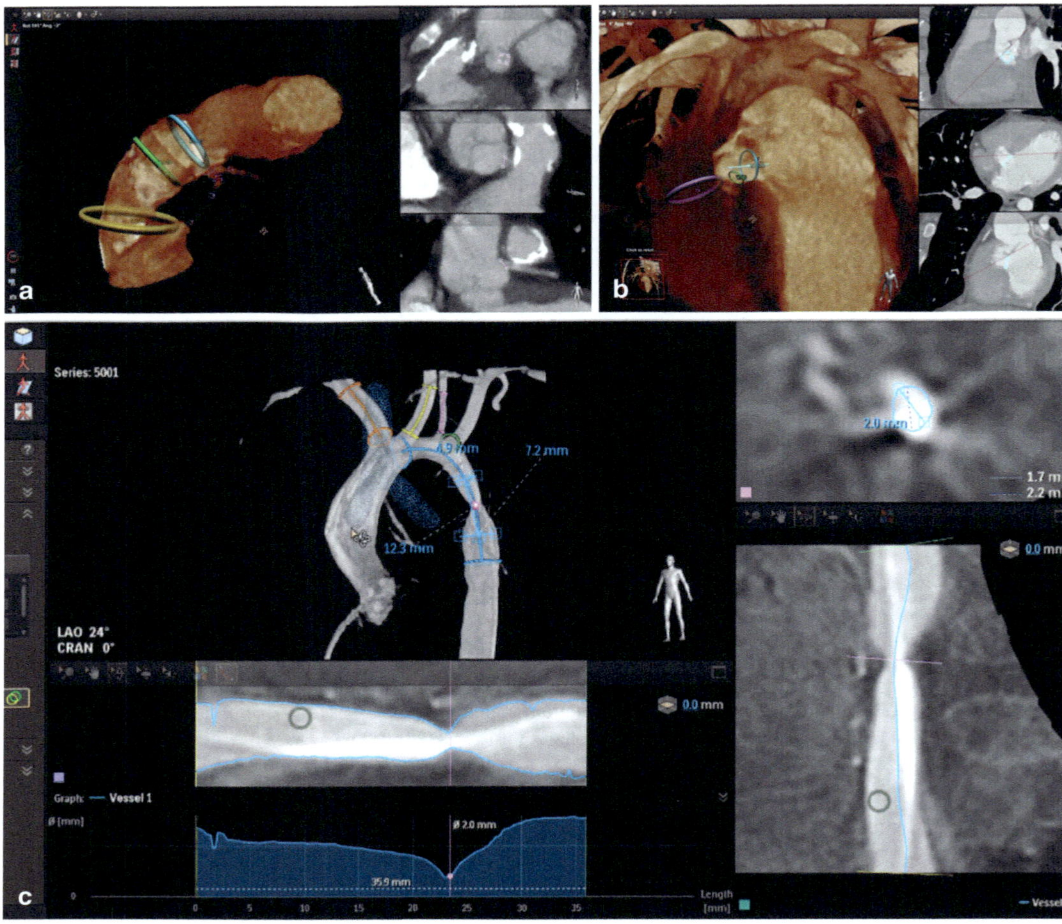

Fig. 29.3 Examples of planning tools utilized for highlighting target lesions and relation to the nearby structures. Ring markers indicate proximal right ventricular outflow tract obstruction (*yellow ring*), the most distal desired stent position (*green ring*), and level (*blue ring*) of adjacent left coronary artery (*magenta rings*) (**a**). Planning of ventricular septal defect (*green ring*) closure by marking of the opening of the defect from the left ventricular surface (*blue ring*) and the tricuspid valve (magenta rings) (**b**). Multilevel measurements of aortic arch and descending aorta prior to dilation of aortic coarctation (**c**)

shows some examples of different anatomical features utilized for registering a CT image to two X-ray projections acquired at two different angles. There are more advanced techniques that use only a single X-ray acquisition to perform this registration. This is achieved by generating a series of virtual X-ray images derived from the 3D volume, which are fit to the actual X-ray using an optimization algorithm using similarity metric. This technique can be used for the entire 3D imaging volume (e.g., Cydar Medical [16]) or a more detailed version of a 3D object, which is visible in the X-ray itself such as a 3DTEE probe (e.g., EchoNavigator [17]).

29.3.2 3D to 3D Image Registration Using a Linear Transformation

Intraoperative 3D imaging capabilities, where the detector rotates quickly around the patient while making a series of X-ray images [18], allow for a direct rigid (linear) registration to the X-ray system. A very basic registration workflow involves

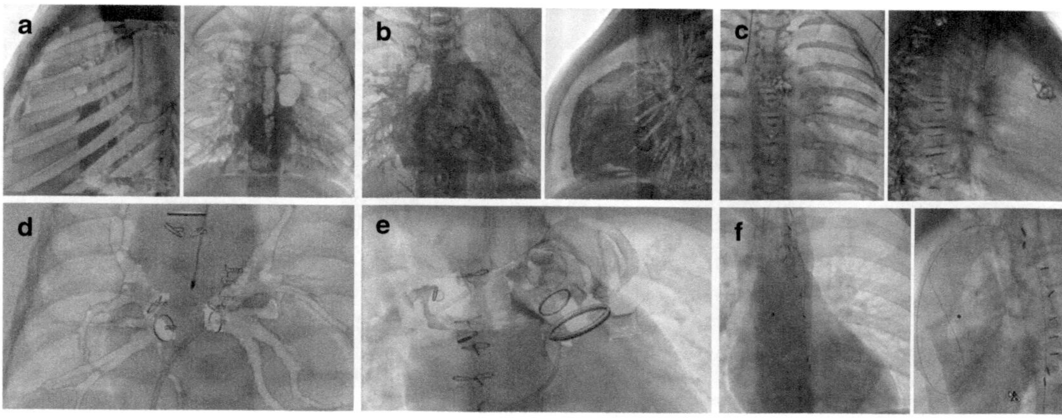

Fig. 29.4 Examples of registration of a preoperative 3D datasets to 2D fluoroscopy taken with 90- degree separation (anterior–posterior and left lateral), using various anatomical landmarks: bony structures (**a**), cardiac and big vessels shadow (**b**), calcifications (**c**), trachea and main bronchi (marked with *blue rings*, **d**), surgical metal clips (*green points*, **e**), and wire in the aorta (**f**)

two steps: the manual rotation and translation of the preoperative dataset and a visual representation of both volumes to allow the user to assess the match. Other methods require at least three user-identified points placed in both volumes on corresponding fiducial markers within the anatomy to define the transformation (Fig. 29.5).

Fully automated 3D-3D techniques require a measure of image alignment quality. Usually, intensity-based methods work reasonably well for registering images of the same modality, like CT to X-ray images, since the same anatomical objects are visible with similar image contrast values in both the source and target images. However, reliance on similar structures, such as bone, is a disadvantage for cardiovascular procedures since the target anatomy is usually not visible with X-ray imaging without using contrast dye. Misregistration of the target anatomy, such as coarctation of the aorta, may be possible despite good alignment of radiopaque structures like ribs, vertebras, or sternal wires.

The registration of MRI and X-ray- based volumes is more difficult as it is hard to find common landmarks due to the intrinsic differences in how the anatomy is represented. There is an entire community of scientists looking into optimizing the registration of multiple datasets from different modalities with increasingly encouraging results [19].

29.3.3 3D to 3D Image Registration Using a Deformable Transformation

As previously discussed, a linear transformation of the preoperative dataset to the target volume is often challenging. One of the confounding factors to this misregistration is the fact that the patient will not be in the exact same posture in the different modalities (e.g., arms up vs. arms down) and laying on differently shaped surfaces (concave vs. flat). Additionally, in the case of cardiovascular imaging, the respiratory and cardiac state of the patient during the acquisition of cardiovascular imaging could have a major impact on the quality of the registration.

In fact, an acceptable linear transformation may not exist and a nonlinear transformation, which allows for a deformation of the volume, may be needed. Unfortunately, these nonlinear transforms are often complex, computationally expensive, and not necessarily suitable for interventional applications. However, there have been recent advances allowing near real -time solutions, which are offered in the public domain [20]. Figure 29.6 shows an example where the preoperative volume is deformed (or warped) to fit with the target volume using both a linear and deformable registration method.

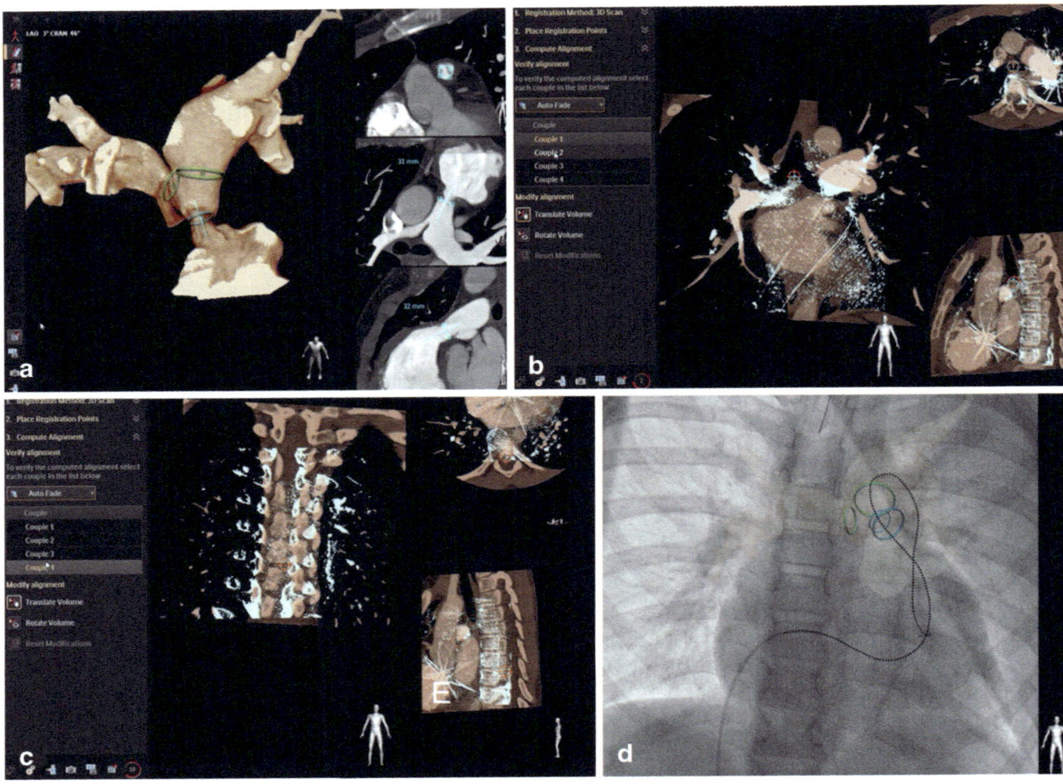

Fig. 29.5 Example of a rigid 3D-3D registration in a patient with discrete stenosis of a right ventricle to pulmonary artery graft. A 3D segmentation of computed tomography scan (**a**) was fused with previous three-dimensional rotational angiography (3DRA) with four pairs of reference points including carina (**b**) and vertebral body (**c**). A soft wire contained in marking rings indicating the narrowest segment of the graft confirmed accurate 3D roadmap alignment (**d**)

29.4 Intervention

After the segmentation, planning, and registration phases, the real work happens in the interventional laboratory. Advanced X-ray systems can track all gantry movements in real time including table movements, source to detector distance changes, viewing angles. As a result, this system information is used to continuously update the rendering of the pre-op dataset to align with the X-ray images. However, several confounding factors may come into play during the actual case that may affect the utility of the preoperative volumes. First, there may be patient movement, even under full anesthesia, mostly due to the respiratory and cardiac motions. Secondly, the anatomy will change when stiff equipment is being introduced (Fig. 29.7a–c). In both situations, the utility of the pre-op dataset becomes less valuable as the registration is invalidated. Nevertheless, in many clinical situations introduction of stiff wires or even long sheaths and/or stents results in only minor misalignment (Fig. 29.7d–e) maintaining utility of the 3D roadmap.

During the case, some form of re-registration of the dataset to update the patient's anatomy is generally available, allowing for small corrections. In actuality, continuous deformable registration would be preferred to deal with these anatomical changes.

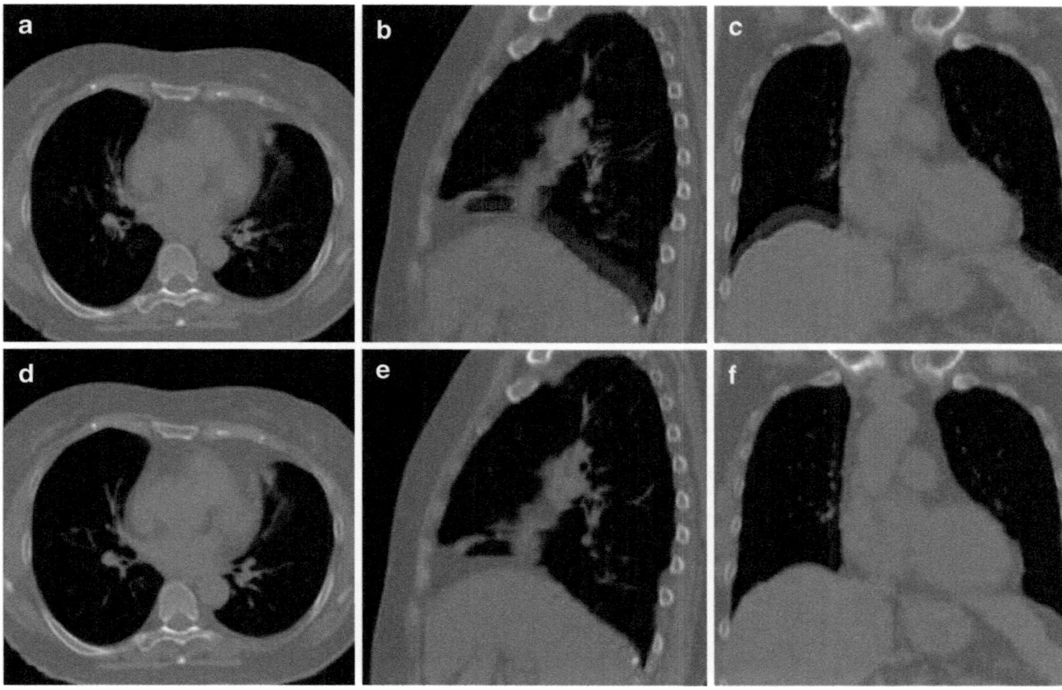

Fig. 29.6 Example of nonlinear 3D deformation. The top row (**a–c**) shows the differences (or registration error in *purple*) with a rigid registration, while the bottom row (**d–f**) shows the result after the nonlinear registration with a significant reduction in the registration error. Image reproduced with permission from de Senneville et al. [21]

29.5 Discussion

Most high-end X-ray systems have the capacity to utilize 3D preoperative datasets. Even in its simplest form, it is helpful to have some processed screen captures of the anatomy and planning available in the room during the case. However, this functionality is typically underutilized. It not only requires additional time and effort to get the preoperative images on the workstation and perform the segmentation on often unique anatomies but is exacerbated because the registration is not fully automated or inaccurate. For example, CT imaging is usually performed during a breath hold and reconstructed for a certain cardiac cycle or cardiac range. However, during the interventional procedure, the patient is either spontaneously breathing or ventilated, and there is cardiac motion. This compound motion will limit the correspondence with the segmentation or pre-procedural data. Additionally, the soft tissue may be deformed by devices and instruments used during the procedure, which can dramatically affect image registration. It is generally not easy to deal with such changes in anatomical state reliably, and there are only a few algorithms that have been used in prototype applications that are trying to address this [23]. Shown in Video 29.1 is an investigational algorithm that compensated for cardiac and breathing motion during aortic valve implantation.

Registration becomes easier once contrast is injected or equipments like wires and catheters are introduced. Unfortunately, this defeats some of the advantages that exist with the use of the preoperative datasets.

In addition, from a practical standpoint, the viewing, segmentation and planning, and actual intervention are generally done at different times and locations (e.g., before the intervention while working from the office). Also, there is no DICOM- like standard yet which allows an easy

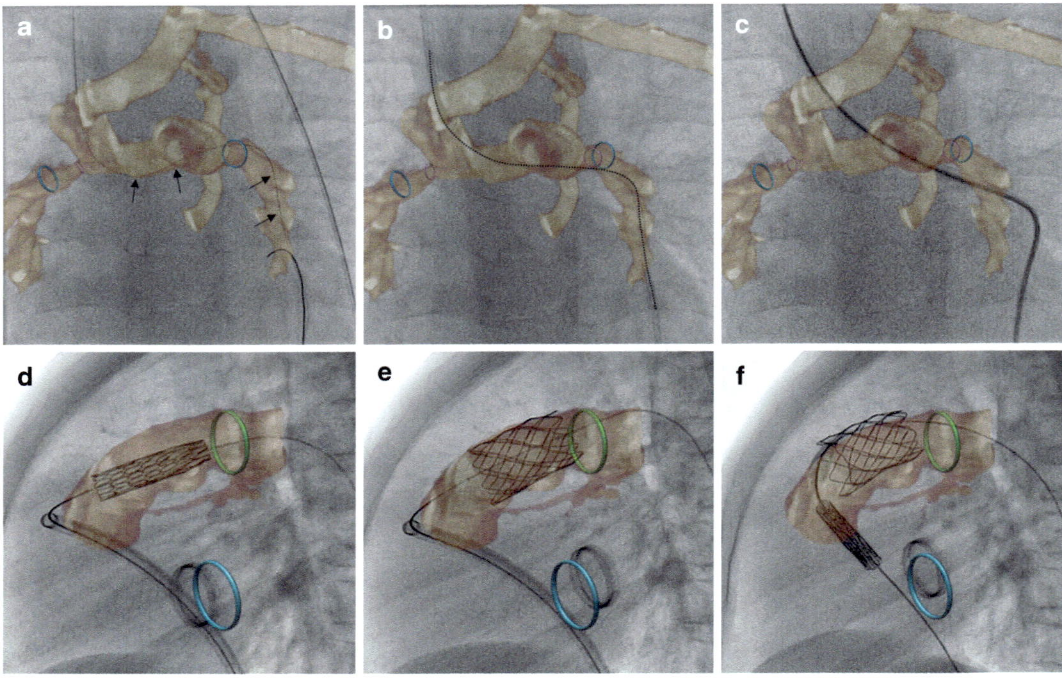

Fig. 29.7 Examples of different degrees of intra-procedural misalignment of the fused three-dimensional (3D) roadmap. A soft wire (*black arrows*) contained in the borders of segmented pulmonary arteries confirmed accurate initial alignment (**a**). Introduction of a diagnostic catheter results in a minor shift of the actual position of pulmonary arteries as compared to the 3D roadmap (**b**), and a stiff wire leads to a significant distortion of the anatomy (**c**). Respiratory and heart motion is reflected by mismatch of the *blue ring* (an artificial mitral valve marked on a pre-procedural CT) and actual position of the mechanic valve (**d**, **e**). Introduction of a long sheath and balloon/stent assembly does not lead to significant misalignment. Even introduction of a percutaneous valve on a 22 Fr delivery system leads to only mild distortion of the anatomy (**f**). With permission from Góreczny et al. [22]

transfer of this information yet, often forcing the interventionalist to re-do some of this work in the laboratory.

Despite some of the challenges, several sites have adapted their workflow to include the use of preoperative data, with documented success. Fusion of pre-catheter imaging performed with different imaging platforms has shown reduction in contrast use, procedure time, and radiation exposure over regular 2D and even 3D rotational angiography (3DRA) for coarctation stenting and percutaneous pulmonary valve implantation [5, 9]. In another study, Goreczny et al. described the application of preregistered 3D datasets for guidance of stent implantation in congenital heart defects [6]. Noninvasive 3D images were effectively reused for positioning and implantation of stents at various locations including the right ventricular outflow tract, pulmonary arteries, and the aorta. All stents were implanted at the desired sites, and in most patients, the stent was positioned and deployed without prior contrast injection.

The combination of several imaging modalities may bring further benefits. A group from Polish Mothers' Memorial Hospital reported application of noninvasive 3D imaging, including CT and ultrasonography, for guidance of percutaneous closure of an arteriovenous fistula [7]. Doppler ultrasound was utilized instead of control angiography for the assessment of the outcome, enabling the elimination of contrast administration and further reduction in radiation exposure during cardiac catheterization. A recent report from Colorado Children's Hospital describes the image fusion of 3D MRI as well as a combination of 3DRA, and intra-cardiac echocardiography (ICE) during percutaneous pulmonary valve implantation. Incorporation of multimodality imaging allowed for the elimination of diagnostic angiography, precise valve implantation, and further contrast and radiation reduction by assessing the outcome with ICE [24].

The future of multimodal interventions is looking bright as new methods and techniques are becoming available. Multimodal interventions allow for fully automatic continuous deformable re-registration, which is capable of handling cardiac and respiratory motion and deformation of structures due to the introduction of stiff equipment like guidewires and catheters. However, even though these tools will eventually be able to handle true congenital anatomy in near real time, and with a minimum amount of user interaction, we feel that multimodality fusion should be part of the clinical routine today. Even with some of the limitations and challenges as described here, it will lead to a better understanding of the anatomy and minimizes the risk to our patients.

Acknowledgments We would like to thank Mr. Adolfo Henriques for editing the text.

References

1. Hascoët S, Warin-Fresse K, Baruteau AE, et al. Cardiac imaging of congenital heart diseases during interventional procedures continues to evolve: pros and cons of the main techniques. Arch Cardiovasc Dis. 2016;109:128–42.
2. Góreczny S, Krings G, Hijazi ZM, Fagan T, Berman D, Kenny D, Morgan GJ. Pros, cons and future perspectives - three questions on three dimensional guidance for cardiac catheterization in congenital heart disease. Postepy Kardiol Interwencyjnej. 2019;15:263–73.
3. Góreczny S, Zahn EM. Giving up knowledge is almost never a good idea: an interview with Dr Evan Zahn. Cardiol Young. 2019;30:1–7. https://doi.org/10.1017/S1047951119002117. [Epub ahead of print]
4. Goreczny S, Dryzek P, Morgan GJ, Lukaszewski M, Moll JA, Moszura T. Novel three-dimensional image fusion software to facilitate guidance of complex cardiac catheterization: 3D image fusion for interventions in CHD. Pediatr Cardiol. 2017;38:1133–42.
5. Ehret N, Alkassar M, Dittrich S, Cesnjevar R, Rüffer A, Uder M, Rompel O, Hammon M, Glöckler M. A new approach of three-dimensional guidance in paediatric cath lab: segmented and tessellated heart models for cardiovascular interventions in CHD. Cardiol Young. 2018;28:661–7.
6. Goreczny S, Moszura T, Lukaszewski M, Podgorski M, Moll JA, Dryzek P. Three-dimensional image fusion of precatheter CT and MRI facilitates stent implantation in congenital heart defects. Rev Esp Cardiol (Engl Ed). 2019;72:512–4.
7. Góreczny S, Podgórski M, Moszura T. Multimodality imaging for no contrast trans-catheter closure of an arteriovenous fistula. Cardiol Young. 2019;29:695–6.
8. Jone PN, Zablah JE, Burkett DA, Schäfer M, Wilson N, Morgan GJ, Ross M. Three-dimensional echocardiographic guidance of right heart catheterization decreases radiation exposure in atrial septal defect closures. J Am Soc Echocardiogr. 2018;31:1044–9.
9. Goreczny S, Moszura T, Dryzek P, Lukaszewski M, Krawczuk A, Moll J, Morgan GJ. Three-dimensional image fusion guidance of percutaneous pulmonary valve implantation to reduce radiation exposure and contrast dose: a comparison with traditional two-dimensional and three-dimensional rotational angiographic guidance. Neth Heart J. 2017;25:91–9.
10. Goreczny S, Dryzek P, Moszura T, Kühne T, Berger F, Schubert S. 3D image fusion for live guidance of stent implantation in aortic coarctation - magnetic resonance imaging and computed tomography image overlay enhances interventional technique. Postepy Kardiol Interwencyjnej. 2017;13:269–72.
11. Góreczny S, Dryżek P, Moszura T, Łukaszewski M, Podgórski M, Nordmeyer S, Kuehne T, Berger F, Schubert S. Magnetic resonance and computed tomography imaging fusion for live guidance of percutaneous pulmonary valve implantation. Postepy Kardiol Interwencyjnej. 2018;14:413–21.
12. Peters J, Ecabert O, Meyer C, Schramm H, Kneser R, Groth A, Weese J. Automatic whole heart segmentation in static magnetic resonance image volumes. Med Image Comput Comput Assist Interv. 2007;10:402–10.
13. de Jaegere P, De Santis G, Rodriguez-Olivares R, Bosmans J, Bruining N, Dezutter T, Rahhab Z, El Faquir N, Collas V, Bosmans B, Verhegghe B, Ren C, Geleinse M, Schultz C, van Mieghem N, De Beule M, Mortier P. Patient-specific computer modeling to predict aortic regurgitation after Transcatheter aortic valve replacement. JACC Cardiovasc Interv. 2016;9:508–12.
14. Kim MS, Hansgen AR, Carroll JD. Use of rapid prototyping in the care of patients with structural heart disease. Trends Cardiovasc Med. 2008;18:210–6.
15. Brun H, Bugge RAB, Suther LKR, Birkeland S, Kumar R, Pelanis E, Elle OJ. Mixed reality holograms for heart surgery planning: first user experience in congenital heart disease. Eur Heart J Cardiovasc Imaging. 2019;20:883–8.

16. Maurel B, Martin-Gonzalez T, Chong D, Irwin A, Guimbretière G, Davis M, Mastracci TM. A prospective observational trial of fusion imaging in infrarenal aneurysms. J Vasc Surg. 2018;68:1706–1713.e1.
17. Ternacle J, Gallet R, Nguyen A, Deux JF, Fiore A, Teiger E, Dubois-Randé JL, Riant E, Lim P. Usefulness of echocardiographic fluoroscopic fusion imaging in adult structural heart disease. Arch Cardiovasc Dis. 2018;111:441–8.
18. Feldkamp LA, Davis LC, Kress J. Practical cone-beam algorithm. J Opt Soc Am. 1984;1:612–9.
19. Viergever MA, Maintz JBA, Klein S, Murphy K, Staring M, Pluim JPW. A survey of medical image registration – under review. Med Image Anal. 2016;33:140–4.
20. Klein S, Staring M, Murphy K, Viergever MA, Pluim JP. Elastix: a toolbox for intensity based medical image registration. IEEE Trans Med Imaging. 2010;29:196–205.
21. Denis de Senneville B, Zachiu C, Ries M, Moonen C. EVolution: an edge based variational method for non-rigid multi-modal image registration. Phys Med Biol. 2016;61:7377–96.
22. Goreczny S, Dryzek P, Moszura T. Novel 3-dimensional image fusion software for live guidance of percutaneous pulmonary valve implantation. Circ Cardiovasc Interv. 2016;9:e003711.
23. Assink N, Izamis M, Nempont O, Verstege M, Sahyoun CP, Haak A, Carroll JD, Messenger JC, Schymik G, Madershahian N, Wahlers TC, Eshuis P. Feasibility of fully automated motion compensated overlay for transcatheter aortic valve implantation. Struct Heart Dis. 2018;4:207–11.
24. Goreczny S, Zablah J, McLennan D, Ross M, Morgan G. Multi-modality imaging for percutaneous pulmonary valve implantation – getting serious about radiation and contrast reduction. Postepy Kardiol Interwencyjnej. 2019;15:110–5.

GPSR Compliance

The European Union's (EU) General Product Safety Regulation (GPSR) is a set of rules that requires consumer products to be safe and our obligations to ensure this.

If you have any concerns about our products, you can contact us on ProductSafety@springernature.com

In case Publisher is established outside the EU, the EU authorized representative is:

Springer Nature Customer Service Center GmbH
Europaplatz 3
69115 Heidelberg, Germany

Batch number: 10732801

Printed by Printforce, the Netherlands